Springer Series in Statistics

Advisors:
P. Bickel, P. Diggle, S. Fienberg,
U. Gather, I. Olkin, S. Zeger

For other titles published in this series, go to
http://www.springer.com/series/692

Paul R. Rosenbaum

Design of Observational Studies

Springer

Paul R. Rosenbaum
Statistics Department
Wharton School
University of Pennsylvania
Philadelphia, PA 19104-6340
USA
rosenbaum@stat.wharton.upenn.edu

ISBN 978-1-4419-1212-1   ISBN 978-1-4419-1213-8 (eBook)
DOI 10.1007/978-1-4419-1213-8
Springer New York Dordrecht Heidelberg London

Library of Congress Control Number: 2009938109

© Springer Science+Business Media, LLC 2010
All rights reserved. This work may not be translated or copied in whole or in part without the written permission of the publisher (Springer Science+Business Media, LLC, 233 Spring Street, New York, NY 10013, USA), except for brief excerpts in connection with reviews or scholarly analysis. Use in connection with any form of information storage and retrieval, electronic adaptation, computer software, or by similar or dissimilar methodology now known or hereafter developed is forbidden.
The use in this publication of trade names, trademarks, service marks, and similar terms, even if they are not identified as such, is not to be taken as an expression of opinion as to whether or not they are subject to proprietary rights.

Printed on acid-free paper

Springer is a part of Springer Science+Business Media (www.springer.com).

For Judy

"Simplicity of form is not necessarily simplicity of experience."
Robert Morris, writing about art.

"Simplicity is not a given. It is an achievement."
William H. Gass, writing about literature.

"Simplicity ... is a very important matter which must be constantly borne in mind."
Sir David Cox, writing about experiments.

Preface

An observational study is an empiric investigation of effects caused by treatments when randomized experimentation is unethical or infeasible. The quality and strength of evidence provided by an observational study is determined largely by its design. Excellent methods of analysis will not salvage a poorly designed study.

The line between design and analysis is easier to draw in practice than it is in theory. In practice, the design of an observational study consists of all activities that precede the examination or use of those outcome measures that will be the basis for the study's conclusions. Unlike experiments, in some observational studies, the outcomes may exist as measurements prior to the design of the study; it is their examination and use, not their existence, that separates design from analysis. Aspects of design include the framing of scientific questions to permit empirical investigation, the choice of a context in which to conduct the investigation, decisions about what data to collect, where and how to collect it, matching to remove bias from measured covariates, strategies and tactics to limit uncertainty caused by covariates not measured, and sample splitting to guide design using individuals who will not be included in the final analysis. In practice, design ends and analysis begins when outcomes are examined for individuals who will be the basis of the study's conclusions. An observational study that begins by examining outcomes is a formless, undisciplined investigation that lacks design.

In theory, design anticipates analysis. Analysis is ever present in design, as any goal is ever present in any organized effort, as a goal is necessary to organize effort. One seeks to ask questions and collect data so that results will be decisive when analyzed. To end well, how should we begin?

Philadelphia, PA *Paul Rosenbaum*
5 August 2009

Acknowledgments

I am in debt to many people: to Jeff Silber, Dylan Small, and Ruth Heller for recent collaborations I describe here in detail; to my teacher, adviser, and coauthor Don Rubin, from whom I learned a great deal; to Ben Hansen, Bo Lu and Robert Greevy for making network optimization algorithms for matching generally accessible inside R; to colleagues, coauthors or former students Katrina Armstrong, Susan Bakewell-Sachs, Lisa Bellini, T. Behringer, Avital Cnaan, Shoshana Daniel, Gabriel Escobar, Orit Even-Shoshan, Joe Gastwirth, Robert Greevy, Sam Gu, Amelia Haviland, Robert Hornik, Abba Krieger, Marshall Joffe, Yunfei Paul Li, Scott Lorch, Bo Lu, Barbara Medoff-Cooper, Lanyu Mi, Andrea Millman, Kewei Ming, Dan Nagin, Dan Polsky, Kate Propert, Tom Randall, Amy Rosen, Richard Ross, Sandy Schwartz, Tom Ten Have, Richard Tremblay, Kevin Volpp, Yanli Wang, Frank Yoon, and Elaine Zanutto, for collaborations I describe more briefly; to Judith McDonald, John Newell, Luke Keele, Dylan Small, and anonymous reviewers for comments on drafts of the book; to Joshua Angrist, David Card, Susan Dynarski, Alan Krueger, and Victor Lavy for making micro-data from their research available in one form or another.

Parts of this book were written while I was on sabbatical. The hospitality of the Department of Statistics at Columbia University and the Department of Economics and the Statistics Program at the National University of Ireland at Galway are gratefully acknowledged. The work was supported in part by the Methodology, Measurement and Statistics Program of the U.S. National Science Foundation.

Of course, my greatest debts are to Judy, Sarah, Hannah, and Aaron.

Contents

Part I Beginnings

1 Dilemmas and Craftsmanship 3
   1.1 Those Confounded Vitamins 3
   1.2 Cochran's Basic Advice 4
       1.2.1 Treatments, covariates, outcomes 5
       1.2.2 How were treatments assigned? 5
       1.2.3 Were treated and control groups comparable? 5
       1.2.4 Eliminating plausible alternatives to treatment effects 6
       1.2.5 Exclusion criteria 6
       1.2.6 Exiting a treatment group after treatment assignment 7
       1.2.7 Study protocol 7
   1.3 Maimonides' Rule 7
   1.4 Seat Belts in Car Crashes 9
   1.5 Money for College 10
   1.6 Nature's 'Natural Experiment' 11
   1.7 What This Book Is About 13
   1.8 Further Reading 18
   References 18

2 Causal Inference in Randomized Experiments 21
   2.1 Two Versions of the National Supported Work Experiment 21
       2.1.1 A version with 185 pairs and a version with 5 pairs 21
       2.1.2 Basic notation 23
   2.2 Treatment Effects in Randomized Experiments 25
       2.2.1 Potential responses under alternative treatments 25
       2.2.2 Covariates and outcomes 26
       2.2.3 Possible treatment assignments and randomization 27
       2.2.4 Interference between units 28
   2.3 Testing the Null Hypothesis of No Treatment Effect 29
       2.3.1 Treated−control differences when the null hypothesis is true 29

        2.3.2   The randomization distribution of the mean difference 31
        2.3.3   The randomization distribution of Wilcoxon's statistic 36
   2.4 Testing Other Hypotheses; Confidence Intervals; Point Estimates... 40
        2.4.1   Testing a constant, additive treatment effect 40
        2.4.2   Confidence intervals for a constant, additive effect 41
        2.4.3   Hodges-Lehmann point estimates of effect 43
        2.4.4   Testing general hypotheses about treatment effects 44
        2.4.5   Multiplicative effects; Tobit effects 46
   2.5 Attributable Effects 49
   2.6 Internal and External Validity 56
   2.7 Summary 57
   2.8 Further Reading 57
   2.9 Appendix: Randomization Distribution of m-statistics 58
   References 61

3 **Two Simple Models for Observational Studies** 65
   3.1 The Population Before Matching 65
   3.2 The Ideal Matching 66
   3.3 A Naïve Model: People Who Look Comparable Are Comparable .. 70
   3.4 Sensitivity Analysis: People Who Look Comparable May Differ ... 76
   3.5 Welding Fumes and DNA Damage 79
   3.6 Bias Due to Incomplete Matching 85
   3.7 Summary 86
   3.8 Further Reading 87
   3.9 Appendix: Exact Computations for Sensitivity Analysis 88
   References 90

4 **Competing Theories Structure Design** 95
   4.1 How Stones Fall 95
   4.2 The Permanent-Debt Hypothesis 98
   4.3 Guns and Misdemeanors 100
   4.4 The Dutch Famine of 1944–1945 100
   4.5 Replicating Effects and Biases 101
   4.6 Reasons for Effects 104
   4.7 The Drive for System 108
   4.8 Further Reading 109
   References 110

5 **Opportunities, Devices, and Instruments** 113
   5.1 Opportunities 113
   5.2 Devices 116
        5.2.1   Disambiguation 116
        5.2.2   Multiple control groups 116
        5.2.3   Coherence among several outcomes 118
        5.2.4   Known effects 121

Contents xv

	5.2.5 Doses of treatment 124
	5.2.6 Differential effects and generic biases 128
5.3	Instruments... 131
5.4	Summary 140
5.5	Further Reading.. 140
	References.. 141

6 Transparency.. 147
References.. 149

Part II Matching

7 A Matched Observational Study 153
  7.1 Is More Chemotherapy More Effective?........................ 153
  7.2 Matching for Observed Covariates 154
  7.3 Outcomes in Matched Pairs 157
  7.4 Summary 159
  7.5 Further Reading.. 161
  References.. 161

8 Basic Tools of Multivariate Matching 163
  8.1 A Small Example 163
  8.2 Propensity Score 165
  8.3 Distance Matrices 168
  8.4 Optimal Pair Matching 172
  8.5 Optimal Matching with Multiple Controls 175
  8.6 Optimal Full Matching 179
  8.7 Efficiency 183
  8.8 Summary 184
  8.9 Further Reading.. 184
  References.. 185

9 Various Practical Issues in Matching 187
  9.1 Checking Covariate Balance 187
  9.2 Almost Exact Matching 190
  9.3 Exact Matching 192
  9.4 Missing Covariate Values 193
  9.5 Further Reading.. 194
  References.. 194

10 Fine Balance... 197
  10.1 What Is Fine Balance? 197
  10.2 Constructing an Exactly Balanced Control Group 198
  10.3 Controlling Imbalance When Exact Balance Is Not Feasible 201
  10.4 Fine Balance and Exact Matching 203
  10.5 Further Reading.. 204

References 204

11 Matching Without Groups 207
- 11.1 Matching Without Groups: Nonbipartite Matching 207
  - 11.1.1 Matching with doses 209
  - 11.1.2 Matching with several groups 210
- 11.2 Some Practical Aspects of Matching Without Groups 211
- 11.3 Matching with Doses and Two Control Groups 213
  - 11.3.1 Does the minimum wage reduce employment? 213
  - 11.3.2 Optimal matching to form two independent comparisons ... 214
  - 11.3.3 Difference in change in employment with 2 control groups . 218
- 11.4 Further Reading 220
- References 220

12 Risk-Set Matching 223
- 12.1 Does Cardiac Transplantation Prolong Life? 223
- 12.2 Risk-Set Matching in a Study of Surgery for Interstitial Cystitis 224
- 12.3 Maturity at Discharge from a Neonatal Intensive Care Unit 228
- 12.4 Joining a Gang at Age 14 231
- 12.5 Some Theory 232
- 12.6 Further Reading 233
- References 234

13 Matching in R 237
- 13.1 R 237
- 13.2 Data 238
- 13.3 Propensity Score 240
- 13.4 Covariates with Missing Values 240
- 13.5 Distance Matrix 242
- 13.6 Constructing the Match 243
- 13.7 Checking Covariate Balance 244
- 13.8 College Outcomes 246
- 13.9 Further Reading 247
- 13.10 Appendix: A Brief Introduction to R 248
- 13.11 Appendix: R Functions for Distance Matrices 250
- References 252

Part III Design Sensitivity

14 The Power of a Sensitivity Analysis and Its Limit 257
- 14.1 The Power of a Test in a Randomized Experiment 257
- 14.2 Power of a Sensitivity Analysis in an Observational Study 265
- 14.3 Design Sensitivity 269
- 14.4 Summary 272
- 14.5 Further Reading 272
- Appendix: Techincal Remarks and Proof of Proposition 14.1 272

References 274

15 Heterogeneity and Causality 275
15.1 J.S. Mill and R.A. Fisher: Reducing Heterogeneity or Introducing Random Assignment 275
15.2 A Larger, More Heterogeneous Study Versus a Smaller, Less Heterogeneous Study 277
15.3 Heterogeneity and the Sensitivity of Point Estimates 281
15.4 Examples of Efforts to Reduce Heterogeneity 282
15.5 Summary 284
15.6 Further Reading 284
References 284

16 Uncommon but Dramatic Responses to Treatment 287
16.1 Large Effects, Now and Then 287
16.2 Two Examples 290
16.3 Properties of a Paired Version of Salsburg's Model 292
16.4 Design Sensitivity for Uncommon but Dramatic Effects 294
16.5 Summary 296
16.6 Further Reading 297
16.7 Appendix: Sketch of the Proof of Proposition 16.1 297
References 298

17 Anticipated Patterns of Response 299
17.1 Using Design Sensitivity to Evaluate Devices 299
17.2 Coherence 299
17.3 Doses 303
17.4 Example: Maimonides' Rule 308
17.5 Further Reading 309
17.6 Appendix: Proof of Proposition 17.1 309
References 310

Part IV Planning Analysis

18 After Matching, Before Analysis 315
18.1 Split Samples and Design Sensitivity 315
18.2 Are Analytic Adjustments Feasible? 317
18.3 Matching and Thick Description 322
18.4 Further Reading 324
References 324

19 Planning the Analysis 327
19.1 Plans Enable 327
19.2 Elaborate Theories 329
19.3 Three Simple Plans with Two Control Groups 330
19.4 Sensitivity Analysis for Two Outcomes and Coherence 339

19.5 Sensitivity Analysis for Tests of Equivalence 341
19.6 Sensitivity Analysis for Equivalence and Difference 343
19.7 Summary 345
19.8 Further Reading 345
19.9 Appendix: Testing Hypotheses in Order 346
References 350

Summary: Key Elements of Design 353

Solutions to Common Problems 355
References 358

Symbols 359

Acronyms 361

Glossary of Statistical Terms 363

Some Books 369
References 369

Suggested Readings for a Course 371
References 371

Index 373

Part I
Beginnings

Chapter 1
Dilemmas and Craftsmanship

Abstract This introductory chapter mentions some of the issues that arise in observational studies and describes a few well designed studies. Section 1.7 outlines the book, describes its structure, and suggests alternative ways to read it.

1.1 Those Confounded Vitamins

On 22 May 2004, the *Lancet* published two articles, one entitled "When are observational studies as credible as randomized trials?" by Jan Vandenbroucke [53], the other entitled "Those confounded vitamins: What can we learn from the differences between observational versus randomized trial evidence?" by Debbie Lawlor, George Smith, Richard Bruckdorfer, Devi Kundu, and Shah Ebrahim [32]. In a randomized experiment or trial, a coin is flipped to decide whether the next person is assigned to treatment or control, whereas in an observational study, treatment assignment is not under experimental control. Despite the optimism of the first title and the pessimism of the second, both articles struck a balance, perhaps with a slight tilt towards pessimism. Vandenbroucke reproduced one of Jim Borgman's political cartoons in which a TV newsman sits below both a banner reading "Today's Random Medical News" and three spinners which have decided that "coffee can cause depression in twins." Dead pan, the newsman says, "According to a report released today...." The cartoon reappeared in a recent report of the Academy of Medical Sciences that discusses observational studies in some detail [43, page 19].

Lawlor et al. begin by noting that a large observational study published in the *Lancet* [30] had found a strong, statistically significant negative association between coronary heart disease mortality and level of vitamin C in blood, having used a model to adjust for other variables such as age, blood pressure, diabetes, and smoking. Adjustments using a model attempt to compare people who are not directly comparable — people of somewhat different ages or smoking habits — removing these differences using a mathematical structure that has elements estimated from the data at hand. Investigators often have great faith in their models, a faith that is expressed in the large tasks they expect their models to successfully perform. Lawlor et al. then note that a large randomized controlled trial published in the *Lancet* [20] compared a placebo pill with a multivitamin pill including vitamin C, finding slightly but not significantly lower death rates under placebo. The randomized trial and the observational study seem to contradict one another. Why is that? There are, of course, many possibilities. There are some important differences between the randomized trial and the observational study; in particular, the treatments are not really identical, and it is not inconceivable that each study correctly answered questions that differ in subtle ways. In particular, Khaw et al. emphasize vitamin C from fruit and vegetable intake rather than from vitamin supplements. Lawlor et al. examine a different possibility, namely that, because of the absence of randomized treatment assignment, people who were not really comparable were compared in the observational study. Their examination of this possibility is indirect, using data from another study, the British Women's Heart and Health Study, in which several variables were measured that were not included in the adjustments performed by Khaw et al. Lawlor et al. find that women with low levels of vitamin C in their blood are more likely to smoke cigarettes, to exercise less than one hour per week, to be obese, and less likely to consume a low fat diet, a high fiber diet, and daily alcohol. Moreover, women with low levels of vitamin C in their blood are more likely to have had a childhood in a "manual social class," with no bathroom or hot water in the house, a shared bedroom, no car access, and to have completed full time education by eighteen years of age. And the list goes on. The concern is that one or more of these differences, or some other difference that was not measured, not the difference in vitamin C, is responsible for the higher coronary mortality among individuals with lower levels of vitamin C in their blood. To a large degree, this problem was avoided in the randomized trial, because there, only the turn of a coin distinguished placebo and multivitamin.

1.2 Cochran's Basic Advice

> The planner of an observational study should always ask himself the question, 'How would the study be conducted if it were possible to do it by controlled experimentation?'
>
> William G. Cochran [9, page 236]
attributing the point to H.F. Dorn.

At the most elementary level, a well designed observational study resembles, as closely as possible, a simple randomized experiment. By definition, the resemblance is incomplete: randomization is not used to assign treatments in an observational study. Nonetheless, elementary mistakes are often introduced and opportunities missed by unnecessary deviations from the experimental template. The current section briefly mentions these most basic ingredients.

1.2 Cochran's Basic Advice

1.2.1 Treatments, covariates, outcomes

Randomized experiment: There is a well-defined treatment, that began at a well-defined time, so there is a clear distinction between covariates measured prior to treatment, and outcomes measured after treatment.

Better observational study: There is a well-defined treatment, that began at a well-defined time, so there is a clear distinction between covariates measured prior to treatment, and outcomes measured after treatment.

Poorer observational study: It is difficult to say when the treatment began, and some variables labeled as covariates may have been measured after the start of treatment, so they might have been affected by the treatment. The distinction between covariates and outcomes is not clear. See [34].

1.2.2 How were treatments assigned?

Randomized experiment: Treatment assignment is determined by a truly random device. At one time, this actually meant coins or dice, but today it typically means random numbers generated by a computer.

Better observational study: Treatment assignment is not random, but circumstances for the study were chosen so that treatment seems haphazard, or at least not obviously related to the outcomes subjects would exhibit under treatment or under control. When investigators are especially proud, having found unusual circumstances in which treatment assignment, though not random, seems unusually haphazard, they may speak of a 'natural experiment.'

Poorer observational study: Little attention is given to the process that made some people into treated subjects and others into controls.

1.2.3 Were treated and control groups comparable?

Randomized experiment: Although a direct assessment of comparability is possible only for covariates that were measured, a randomized trial typically has a table demonstrating that the randomization was reasonably effective in balancing these observed covariates. Randomization provides some basis for anticipating that many covariates that were not measured will tend to be similarly balanced.

Better observational study: Although a direct assessment of comparability is possible only for covariates that were measured, a matched observational study typically has a table demonstrating that the matching was reasonably effective in balancing these observed covariates. Unlike randomization, matching for observed covariates provides absolutely no basis for anticipating that unmeasured covariates are similarly balanced.

Poorer observational study: No direct assessment of comparability is presented.

1.2.4 Eliminating plausible alternatives to treatment effects

Randomized experiment: The most plausible alternatives to an actual treatment effect are identified, and the experimental design includes features to shed light on these alternatives. Typical examples include the use of placebos and other forms of sham or partial treatment, or the blinding of subjects and investigators to the identity of the treatment received by a subject.

Better observational study: The most plausible alternatives to an actual treatment effect are identified, and the design of the observational study includes features to shed light on these alternatives. Because there are many more plausible alternatives to a treatment effect in an observational study than in an experiment, much more effort is devoted to collecting data that would shed light on these alternatives. Typical examples include multiple control groups thought to be affected by different biases, or a sequence of longitudinal baseline pretreatment measurements of the variable that will be the outcome after treatment. When investigators are especially proud of devices included to distinguish treatment effects from plausible alternatives, they may speak of a 'quasi-experiment.'

Poorer observational study: Plausible alternatives to a treatment effect are mentioned in the discussion section of the published report.

1.2.5 Exclusion criteria

Randomized experiment: Subjects are included or excluded from the experiment based on covariates, that is, on variables measured prior to treatment assignment and hence unaffected by treatment. Only after the subject is included is the subject randomly assigned to a treatment group and treated. This ensures that the same exclusion criteria are used in treated and control groups.

Better observational study: Subjects are included or excluded from the experiment based on covariates, that is, on variables measured prior to treatment assignment and hence unaffected by treatment. The same criteria are used in treated and control groups.

Poorer observational study: A person included in the control group might have been excluded if assigned to treatment instead. The criteria for membership in the treated and control groups differ. In one particularly egregious case, to be discussed in §12.1, treatment was not immediately available, and any patient who died before the treatment became available was placed in the control group; then came the exciting news that treated patients lived longer than controls.

1.2.6 Exiting a treatment group after treatment assignment

Randomized experiment: Once assigned to a treatment group, subjects do not exit. A subject who does not comply with the assigned treatment, or switches to another treatment, or is lost to follow-up, remains in the assigned treatment group with these characteristics noted. An analysis that compares the groups as randomly assigned, ignoring deviations between intended and actual treatment, is called an 'intention-to-treat' analysis, and it is one of the central analyses reported in a randomized trial. Randomization inference may partially address noncompliance with assigned treatment by viewing treatment assignment as an instrumental variable for treatment received; see §5.3 and [18].

Better observational study: Once assigned to a treatment group, subjects do not exit. A subject who does not comply with the assigned treatment, or switches to another treatment, or is lost to follow-up, remains in the assigned treatment group with these characteristics noted. Inference may partially address noncompliance by viewing treatment assignment as an instrumental variable for treatment received; see §5.3 and [22].

Poorer observational study: There is no clear distinction between assignment to treatment, acceptance of treatment, receipt of treatment, or switching treatments, so problems that arise in experiments seem to be avoided, when in fact they are simply ignored.

1.2.7 Study protocol

Randomized experiment: Before beginning the actual experiment, a written protocol describes the design, exclusion criteria, primary and secondary outcomes, and proposed analyses.

Better observational study: Before examining outcomes that will form the basis for the study's conclusions, a written protocol describes the design, exclusion criteria, primary and secondary outcomes, and proposed analyses; see Chapter 19.

Poorer observational study: If sufficiently many analyses are performed, something publishable will turn up sooner or later.

1.3 Maimonides' Rule

In 1999, Joshua Angrist and Victor Lavy [3] published an unusual and much admired study of the effects of class size on academic achievement. They wrote [3, pages 533-535]:

[C]ausal effects of class size on pupil achievement have proved very difficult to measure. Even though the level of educational inputs differs substantially both between and within schools, these differences are often associated with factors such as remedial training or students' socioeconomic background ... The great twelfth century Rabbinic scholar, Maimonides, interprets the Talmud's discussion of class size as follows: 'Twenty-five children may be put in charge of one teacher. If the number in the class exceeds twenty-five but is not more than forty, he should have an assistant to help with instruction. If there are more than forty, two teachers must be appointed.' ... The importance of Maimonides' rule for our purposes is that, since 1969, it has been used to determine the division of enrollment cohorts into classes in Israeli public schools.

In most places at most times, class size has been determined by the affluence or poverty of a community, its enthusiasm or skepticism about the value of education, the special needs of students for remedial or advanced instruction, the obscure, transitory, barely intelligible obsessions of bureaucracies, and each of these determinants of class size clouds its actual effect on academic performance. However, if adherence to Maimonides' rule were perfectly rigid, then what would separate a school with a single class of size 40 from the same school with two classes whose average size is 20.5 is the enrollment of a single student.

Maimonides' rule has the largest impact on a school with about 40 students in a grade cohort. With cohorts of size 40, 80, and 120 students, the steps down in average class size required by Maimonides' rule when an additional student enrolls are, respectively, from 40 to 20.5, from 40 to 27, and from 40 to 30.25. For this reason, we will look at schools with fifth grade cohorts in 1991 with between 31 and 50 students, where average class sizes might be cut in half by Maimonides' rule. There were 211 such schools, with 86 of these schools having between 31 and 40 students in fifth grade, and 125 schools having between 41 and 50 students in the fifth grade.

Adherence to Maimonides' rule is not perfectly rigid. In particular, Angrist and Lavy [3, page 538] note that the percentage of disadvantaged students in a school "is used by the Ministry of Education to allocate supplementary hours of instruction and other school resources." Among the 211 schools with between 31 and 50 students in fifth grade, the percentage disadvantaged has a slightly negative Kendall's correlation of -0.10 with average class size, which differs significantly from zero (P-value $= 0.031$), and it has more strongly negative correlations of -0.42 and -0.55, respectively, with performance on verbal and mathematics test scores. For this reason, 86 matched pairs of two schools were formed, matching to minimize to total absolute difference in percentage disadvantaged. Figure 1.1 shows the paired schools, 86 schools with 31 and 40 students in fifth grade, and 86 schools with between 41 and 50 students in the fifth grade. After matching, the upper left panel in Figure 1.1 shows that the percentage of disadvantaged students was balanced; indeed, the average absolute difference within a pair was less than 1%. The upper right panel in Figure 1.1 shows Maimonides' rule at work: with some exceptions, the slightly larger schools had substantially smaller class sizes. The bottom panels of Figure 1.1 show the average mathematics and verbal test performance of these fifth graders, with somewhat higher scores in the schools with between 41 and 50 fifth graders, where class sizes tended to be smaller.

1.4 Seat Belts in Car Crashes

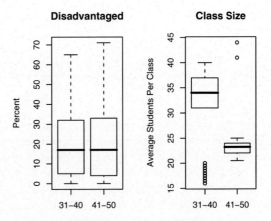

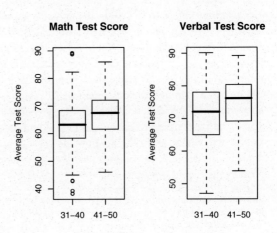

Fig. 1.1 Eighty-six pairs of two Israeli schools, one with between 31 and 40 students in the fifth grade, the other with between 41 and 50 students in the fifth grade, matched for percentage of students in the school classified as disadvantaged. The figure shows that the percentage of disadvantaged students is balanced, that imperfect adherence to Maimonides' rule has yielded substantially different average class sizes, and test scores were higher in the group of schools with predominantly smaller class sizes.

1.4 Seat Belts in Car Crashes

Do safety belts prevent fatalities in automobile accidents? Car crashes vary in severity, depending upon speed, road traction, the reaction time of a driver stepping on the brakes, and physical forces that are rarely, if ever, measured. Wearing safety belts is a precaution. Many people, perhaps most people, who wear safety belts think that a serious accident is possible; this possibility is salient, and small inconveniences seem tolerable if the risk is reduced. In contrast, small inconveniences may seem intolerable if a serious accident is seen as a remote possibility. Does one take a single precaution? Perhaps some people do, but others will take several precautions. If cautious drivers wear seat belts, but also drive at slower speeds, at a greater distance from the car ahead, with greater allowance for road conditions – if

Table 1.1 Crashes in FARS 1975–1983 in which the front seat had two occupants, a driver and a passenger, with one belted, the other unbelted, and one died and one survived.

Driver	Passenger	Not Belted / Belted	Belted / Not Belted
Driver Died	Passenger Survived	189	153
Driver Survived	Passenger Died	111	363

risk-tolerant drivers do not wear seat belts, drive faster and closer, ignore road conditions – then a simple comparison of belted and unbelted drivers may credit seat belts with effects that reflect, in part, the severity of the crash.

Using data from the U.S. Fatal Accident Reporting System (FARS), Leonard Evans [14] looked at crashes in which there were two individuals in the front seat, one belted, the other unbelted, with at least one fatality. In these crashes, several otherwise uncontrolled features are the same for driver and passenger: speed, road traction, distance from the car ahead, reaction time. Admittedly, risk in the passenger seat may differ from risk in the driver seat, but in this comparison there are belted drivers with unbelted passengers and unbelted drivers with belted passengers, so this issue may be examined. Table 1.1 is derived from Evans' [14] more detailed tables. In this table, when the passenger is belted and the driver is not, more often than not, the driver dies; conversely, when the driver is belted and the passenger is not, more often than not, the passenger dies.

Everyone in Table 1.1 is at least sixteen years of age. Nonetheless, the roles of driver and passenger are connected to law and custom, for parents and children, husbands and wives, and others. For this reason, Evans did further analyses, for instance taking account of the ages of driver and passenger, with similar results.

Evans [14, page 239] wrote:

> The crucial information for this study is provided by cars in which the safety belt use of the subject and other occupant differ ... There is a strong tendency for safety belt use or non-use to be the same for different occupants of the same vehicle ... Hence, sample sizes in the really important cells are ... small ...

This study is discussed further in §5.2.6.

1.5 Money for College

To what extent, if any, does financial aid increase college attendance? It would not do to simply compare those who received aid with those who did not. Decisions about the allocation of financial aid are often made person by person, with consideration of financial need and academic promise, together with many other factors. A grant of financial aid is often a response to an application for aid, and the decision to apply or not is likely to reflect an individual's motivation for continued education and competing immediate career prospects.

To estimate the effect of financial aid on college attendance, Susan Dynarski [13] used "a shift in aid policy that affect[ed] some students but not others." Between 1965 and 1982, a program of the U.S. Social Security Administration provided substantial financial aid to attend college for the children of deceased Social Security beneficiaries, but the U.S. Congress voted in 1981 to end the program. Using data from the National Longitudinal Survey of Youth, Dynarski [13] compared college attendance of high school seniors with deceased fathers and high school seniors whose fathers were not deceased, in 1979–1981 when aid was available, and in 1982–1983 after the elimination of the program. Figure 1.2 depicts the comparison. In 1979–1981, while the Social Security Student Benefit Program provided aid to students with deceased fathers, these students were more likely than others to attend college, but in 1982–1983, after the program was eliminated, these students were less likely than others to attend college.

In Figure 1.2, the group that faced a change in incentives exhibited a change in behavior, whereas the group that faced no change in incentives exhibited little change in behavior. In the spirit of §1.2.4, Figure 1.2 studies one treatment using four groups, where only certain patterns of response among the four groups are compatible with a treatment effect; see also [7, 37] and [47, Chapter 5].

Is being the child of a deceased father a random event? Apparently not. It is unrelated to the child's age and gender, but the children of deceased fathers had mothers and fathers with less education and were more likely to be black; however, these differences were about the same in 1979–1981 and 1982–1983, so these differences alone are not good explanations of the shift in college attendance [13, Table 1]. This study is discussed further in Chapter 13.

1.6 Nature's 'Natural Experiment'

In asking whether a particular gene plays a role in causing a particular disease, a key problem is that the frequencies of various forms of a gene (its alleles) vary somewhat from one human community to the next. At the same time, habits, customs, diets, and environments also vary somewhat from one community to the next. In consequence, an association between a particular allele and a particular disease may not be causal: gene and disease may both be associated with some cause, such as diet, that is not genetic. Conveniently, nature has created a natural experiment.

With the exception of sex-linked genes, a person receives two versions of each gene, perhaps identical, one from each parent, and transmits one copy to each child. To a close approximation, in the formation of a fertilized egg, each parent contributes one of two possible alleles, each with probability $\frac{1}{2}$, the contributions of the two parents being independent of each other, and independent for different children of the same parents. (The transmissions of different genes that are neighbors on the same chromosome are not generally independent; see [51, §15.4]. In consequence, a particular gene may be associated with a disease not because it is a cause of the disease, but rather because it is a marker for a neighboring gene that is a cause.)

Several strategies use this observation to create natural experiments that study genetic causes of a specific disease. Individuals with the disease are identified. Richard Spielman, Ralph McGinnis, and Warren Ewens [49] used genetic information on the diseased individual and both parents in their Transmission/Disequilibrium Test (TDT). The test compares the diseased individuals to the known distributions of alleles for the hypothetical children their parents could produce. David Curtis [12], Richard Spielman and Warren Ewens [50], and Michael Boehnke and Carl Langefeld [5] suggested using genetic information on the diseased individual and one or more siblings from the same parents, which Spielman and Ewens called the sib-TDT. If the disease has no genetic cause linked to the gene under study, then the alleles from the diseased individual and her siblings should be exchangeable.

The idea underlying the sib-TDT is illustrated in Table 1.2, using data from Boehnke and Langefeld [5, Table 5], their table being derived from work of Margaret Pericak-Vance and Ann Saunders; see [44]. Table 1.2 gives the frequency of the ε_4 allele of the apolipoprotein E gene in 112 individuals with Alzheimer disease and in an unaffected sibling of the same parents. Table 1.2 counts sibling pairs, not individuals, so the total count in the table is 112 pairs of an af-

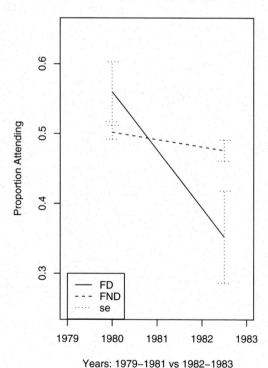

Fig. 1.2 College attendance by age 23 in four groups: before (1979–1981) and after (1982–1983) the end of the Social Security Student Benefit Program for children whose fathers were deceased (FD) or not deceased (FND). Values are proportions with standard errors (se).

1.7 What This Book Is About

Table 1.2 Alzheimer disease and the apolipoprotein E ε_4 allele in 112 sibling pairs, one with Alzheimer disease (affected), the other without (unaffected). The table counts pairs, not individuals. The rows and columns of the table indicate the number (0, 1, or 2) of ApoE alleles for the affected and unaffected sibling.

	# ApoEε_4 Alleles	Unaffected Sib 0	1	2
	0	23	4	0
Affected Sib	1	25	36	2
	2	8	8	6

fected and an unaffected sibling. Each person can receive 0, 1, or 2 copies of the ε_4 allele from parents. For any one pair, write (aff, unaff) for the number of ε_4 alleles possessed by, respectively, the affected and unaffected sibling. In Table 1.2, there are 25 pairs with (aff, unaff) = (1,0). If Alzheimer disease had no genetic link with the apolipoprotein E ε_4 allele, then nature's natural experiment implies that the chance that (aff, unaff) = (2,0), say, is the same as the chance that (aff, unaff) = (0,2), and more generally, the chance that (aff, unaff) = (i, j) equals the chance that (aff, unaff) = (j, i), for $i, j = 0, 1, 2$. In fact, this does not appear to be the case in Table 1.2. For instance, there are eight sibling pairs such that (aff, unaff) = (2,0) and none such that (aff, unaff) = (0,2). Also, there are 25 pairs such that (aff, unaff) = (1,0) and only 4 pairs such that (aff, unaff) = (0,1).

A distribution with the property

$$\Pr\{(\text{aff}, \text{unaff}) = (i, j)\} = \Pr\{(\text{aff}, \text{unaff}) = (j, i)\} \quad \text{for all } i, j$$

is said to be exchangeable. In the absence of a genetic link with disease, nature's natural experiment ensures that the distribution of allele frequencies in affected/unaffected sib pairs is exchangeable. This creates a test, the sib transmission disequilibrium test [50] that is identical to a certain randomization test appropriate in a randomized experiment [31].

1.7 What This Book Is About

Basic structure

Design of Observational Studies has four parts, 'Beginnings,' 'Matching,' 'Design Sensitivity,' and 'Planning Analysis,' plus a brief summary. Part I, 'Beginnings,' is a conceptual introduction to causal inference in observational studies. Chapters 2, 3, and 5 of Part I cover concisely, in about one hundred pages, many of the ideas discussed in my book *Observational Studies* [38], but in a far less technical and less general fashion. Parts II–IV cover material that, for the most part, has not previously appeared in book form. Part II, 'Matching,' concerns the conceptual, practical, and computational aspects of creating a matched comparison that balances many observed covariates. Because matching does not make use of outcome information, it is part of the design of the study, what Cochran [9] called "setting up the comparisons"; that is, setting up the structure of the experimental analog. Even if the matching in Part II is entirely successful, so that after matching, matched treated and control groups are comparable with respect to all observed covariates, the question or objection or challenge will inevitably be raised that subjects who look comparable in observed data may not actually be comparable in terms of covariates that were not measured. Chapters 3 and 5 and Parts III and IV address this central concern. Part III, 'Design Sensitivity,' discusses a quantitative tool for appraising how well competing designs (or data generating processes) resist such challenges. In part, 'Design Sensitivity' will provide a formal appraisal of the design strategies introduced informally in Chapter 5. Part IV discusses those activities that follow matching but precede analysis, notably planning the analysis.

Structure of Part I: Beginnings

Observational studies are built to resemble simple experiments, and Chapter 2 reviews the role that randomization plays in experiments. Chapter 2 also introduces elements and notation shared by experiments and observational studies. Chapter 3 discusses two simple models for observational studies, one claiming that adjustments for observed covariates suffice, the other engaging the possibility that they do not. Chapter 3 introduces the propensity score and sensitivity analysis. Observational studies are built from three basic ingredients: opportunities, devices and instruments. Chapter 5 introduces these ideas in an informal manner, with some of the formalities developed in Part III and others developed in [38, Chapters 4, 6–9].

My impression is that many observational studies dissipate either by the absence of a focused objective or by becoming mired in ornate analyses that may overwhelm an audience but are unlikely to convince anyone. Neither problem is common in randomized experiments, and both problems are avoidable in observational studies. Chapter 4 discusses the first problem, while Chapter 6 discusses the second. In a successful experiment or observational study, competing theories make conflicting predictions; this is the concern of Chapter 4. Transparency means making evidence evident, and Chapter 6 discusses how this is done.

Structure of Part II: Matching

Part II, entitled 'Matching,' is partly conceptual, partly algorithmic, partly data analytic. Chapter 7 is introductory: it presents a matched comparison as it might (and did) appear in a scientific journal. The basic tools of multivariate matching are described and illustrated in Chapter 8, and various common practicalities are discussed in Chapter 9. Later chapters in Part II discuss specific topics in matching, including fine balance, matching with multiple groups or without groups, and risk-set matching. Matching in the computer package R is discussed in Chapter 13.

1.7 What This Book Is About

Structure of Part III: Design Sensitivity

In Chapter 3, it is seen that some observational studies are sensitive to small unobserved biases, whereas other studies are insensitive to quite large unobserved biases. What features of the design of an observational study affect its sensitivity to bias from covariates that were not measured? This is the focus of Part III.

Chapter 14 reviews the concept of power in a randomized experiment, then defines the power of a sensitivity analysis. Design sensitivity is then defined. Design sensitivity is a number that defines the sensitivity of an observational study design to unmeasured biases when the sample size is large. Many factors affect the design sensitivity, including the issues discussed informally in Chapter 5. Chapter 15 revives a very old debate between John Stuart Mill and Sir Ronald Fisher about the relevance to causal inference of the heterogeneity of experimental material. Mill believed it mattered quite a bit; Fisher denied this. Sometimes a treatment has little effect on most people and a dramatic effect on some people. In one sense, the effect is small — on average it is small — but for a few people it is large. Is an effect of this sort highly sensitive to unmeasured biases? Chapter 16 provides the answer. Chapter 17 takes up themes from Chapter 5, specifically coherence and dose-response, and evaluates their contribution to design sensitivity.

Structure of Part IV: Planning Analysis

The sample has been successfully matched — treated and control groups look comparable in terms of measured covariates — and Part IV turns to planning the analysis. Chapter 18 concerns three emprical steps that aid planning the analysis: sample splitting to improve design sensitivity, checking that analytical adjustments are feasible, and thick description of a few matched pairs. After reviewing Fisher's advice — "make your theories elaborate" — Chapter 19 discusses planning the analysis of an observational study.

A less technical introduction to observational studies

The mathematician Paul Halmos wrote two essays, "How to write mathematics" and "How to talk mathematics." In the latter, he suggested that in a good mathematical talk, you don't prove everything, but you do prove something to give the flavor of proofs for the topic under discussion. In the spirit of that remark, *Observational Studies* [38] writes about statistics, where *Design of Observational Studies* talks about statistics. This is done in several ways.

We often develop an understanding by taking something apart and putting it back together. In statistics, this typically means looking at an unrealistically small example in which it is possible to see the details of what goes on. For this reason, I discuss several unrealistically small examples in parallel with real examples of practical size. For instance, Chapter 2 discusses two versions of a paired randomized experiment, one with five pairs, the other with 185 pairs. The five pairs are a random sample from the 185 pairs. With 5 pairs, there are $2^5 = 32$ possible treatment assignments, and it is possible to see what is happening. With 185 pairs, there are $2^{185} = 4.9 \times 10^{55}$ possible treatment assignments, and it is not possible to see what is happening, although nothing new is happening beyond what you saw happening with five pairs. The larger experiment is just larger, somewhat awkward to inspect, but no different conceptually. In a similar way, Chapter 7 discusses the construction of 344 pairs matched for many covariates, while Chapter 8 discusses the construction of 21 pairs matched for three covariates. With 21 pairs, you can see what is happening, whereas with 344 pairs you cannot see as much, but nothing new is happening.

Chapter 2 discusses a number of very old, very central concepts in statistics. These include: the role of randomization in experiments, the nature of randomization tests, obtaining confidence intervals by inverting hypothesis tests, building an estimator using an estimating equation, and so on. This material is so old and central to the subject that an article in a statistical journal might reduce the entire chapter to a paragraph, and that would be fine for someone who had been around this track a few times. My goal in Chapter 2 is not concise expression. My goal in Chapter 2 is to take you around this track a few times.

To prove something, not everything, I develop statistical theory only for the case of matched pairs with continuous responses. The case of matched pairs is the simplest nontrivial case. All of the important concepts appear in the case of matched pairs, but most of the technical issues are easy. Randomization distributions for matched pairs are little more than a series of independent coin flips. Everybody can do coin flips. In a randomized experiment, the coin flips are fair, but in a sensitivity analysis, the coin flips may be biased. The matching methods in Part II are not restricted to pair matching — matching with multiple controls, matching with variable controls, full matching, risk set matching are all there — however, if you want to work through the derivations of the associated statistical analyses for continuous, discrete, censored, and multivariate responses, you will need to turn to [38] or the references discussed in 'Further Reading.'

Focusing the theoretical presentation on matched pairs permits discussion of key concepts with the minimum of mathematics. Unlike statistical analysis, research design yields decisions rather than calculations — decisions to ask certain questions, in certain settings, collecting certain data, adding certain design elements, attending to certain patterns — and for such decisions, the concepts are more important than details of general computations. What is being left out by focusing on matched pairs? For one thing, sensitivity analyses in other cases are easy to do but require more mathematical machinery to justify. Some of this machinery is aesthetically pleasing, for instance exact results using Holley's inequality [1, 6, 25] in I. R. Savage's [45] finite distributive lattice of rank orders or samples; see [35, 36] and [38, §4]. Some of this machinery uses large sample approximations or asymptotics that work easily and well even in small samples, but discussion of these approximations means a step up in the level of technical detail; see [16] and [38, §4]. To get a feeling for the difference between the paired case and other cases, see [39], where

1.7 What This Book Is About

both cases are discussed in parallel, one right after the other. If you need to do a sensitivity analysis for a situation other than matched pairs, see [38, §4] or other references in Further Reading. Another item that is left out of the current discussion is a formal notation and model for multiple groups, as opposed to a treated and a control group; see [38, §8] for such a notation and model. Such a model adds more subscripts and symbols with few additional concepts. The absence of such a notation and model has a small effect on the discussion of multiple control groups (§5.2.2), differential effects (§5.2.6), the presentation of the matching session in R for a difference-in-differences example in Chapter 13, and the planned analysis with two control groups (§19.3); specifically, these topics are described informally with reference to the literature for formal results.

Dependence among chapters

Design of Observational Studies is highly modular, so it is not necessary to read chapters in order. Part II may be read before or after Part I or not at all; Part II is not needed for Parts III and IV. Part III may be read before or after Part IV.

In Part I, Chapter 5 depends on Chapter 3, which in turn depends on Chapter 2. The beginning of Chapter 2, up through §2.4.3, is needed for Chapters 3 and 5, but §2.5 is not essential except for Chapter 16, and the remainder of Chapter 2 is not used later in the book. Chapters 4 and 6 may be read at any time or not at all.

In Part II, most chapters depend strongly on Chapter 8 but only weakly on each other. Read the introductory Chapter 7 and Chapter 8; then, read what you like in Part II.

The situation is similar in Part III. All of the chapters of Part III depend upon Chapter 14, which in turn depends on Chapter 5. The remaining chapters of Part III may be read in any order or not at all.

The two chapters in Part IV may be read out of sequence, and both depend upon Chapter 5.

At the back of the book, there is a list of symbols and a glossary of statistical terms. In the index, a **bold page number** locates the definition of a technical term or symbol.

Some books (e.g., [38]) contain practice problems for you to solve. My sense is that the investigator planning an observational study has problems enough, so instead of further problems, at the back of the book there is a list of solutions.

As a scholar at a research university, I fall victim to periodic compulsions to make remarks that are largely unintelligible and totally unnecessary. These remarks are found in appendices and footnotes. Under no circumstances read them. If you read a footnote, you will suffer a fate worse than Lot's wife.[1]

---

[1] She turned into a pillar of salt when she looked where her husband instructed her not to look. Opposed to the story of Lot and his wife is the remark of Immanuel Kant: "Sapere aude" (Dare to know) [27, page 17]. You are, by the way, off to a bad start with these footnotes.

1.8 Further Reading

One might reasonably say that the distinction between randomized experiments and observational studies was introduced by Sir Ronald Fisher's [15] invention of randomized experimentation. Fisher's book [15] of 1935 is of continuing interest. As noted in §1.2, William Cochran [9] argued that observational studies should be understood in relation to experiments; see also the important paper in this spirit by Donald Rubin [41]. A modern discussion of quasi-experiments is given by William Shadish, Thomas Cook, and Donald Campbell [47], and Campbell's [8] collected papers are of continuing interest. See also [17, 33, 54]. Natural experiments in medicine are discussed by Jan Vandenbroucke [53] and the report edited by Michael Rutter [43], and in economics by Joshua Angrist and Alan Kruger [2], Timothy Besley and Anne Case [4], Daniel Hamermesh [19], Bruce Meyer [33], and Mark Rosenzweig and Kenneth Wolpin [40]. Natural experiments are prominent also in recent developments in genetic epidemiology [5, 12, 31, 49, 50]. The papers by Jerry Cornfield and colleagues [11], Austin Bradford Hill [23], and Mervyn Susser [52] remain highly influential in epidemiology and are of continuing interest. Miguel Hernán and colleagues [22] illustrate the practical importance of adhering to the experimental template in designing an observational study. For a general discussion of observational studies, see [38].

References

1. Anderson, I.: Combinatories of Finite Sets. New York: Oxford University Press (1987)
2. Angrist, J.D., Krueger, A.B.: Empirical strategies in labor economics. In: Ashenfelter, O., Card, D. (eds.) Handbook of Labor Economics, Volume 3, pp. 1277–1366. New York: Elsevier (1999)
3. Angrist, J.D., Lavy, V.: Using Maimonides' rule to estimate the effect of class size on scholastic achievement. Q J Econ **114**, 533–575 (1999)
4. Besley, T., Case, A.: Unnatural experiments? Estimating the incidence of endogenous policies. Econ J **110**, 672–694 (2000)
5. Boehnke, M., Langefeld, C.D.: Genetic association mapping based on discordant sib pairs: The discordant alleles test. Am J Hum Genet **62**, 950–961 (1998)
6. Bollobás, B.: Combinatorics. New York: Cambridge University Press (1986)
7. Campbell, D.T.: Factors relevant to the validity of experiments in social settings. Psychol Bull **54**, 297–312 (1957)
8. Campbell, D.T.: Methodology and Epistemology for Social Science: Selected Papers. Chicago: University of Chicago Press (1988)
9. Cochran, W.G.: The planning of observational studies of human populations (with Discussion). J Roy Statist Soc A **128**, 234–265 (1965)
10. Cook, T.D., Shadish, W.R.: Social experiments: Some developments over the past fifteen years. Annu Rev Psychol **45**, 545–580 (1994)
11. Cornfield, J., Haenszel, W., Hammond, E., Lilienfeld, A., Shimkin, M., Wynder, E.: Smoking and lung cancer: Recent evidence and a discussion of some questions. J Natl Cancer Inst **22**, 173–203 (1959)
12. Curtis, D.: Use of siblings as controls in case-control association studies. Ann Hum Genet **61**, 319–333 (1997)

13. Dynarski, S.M.: Does aid matter? Measuring the effect of student aid on college attendance and completion. Am Econ Rev **93**, 279–288 (2003)
14. Evans, L.: The effectiveness of safety belts in preventing fatalities. Accid Anal Prev **18**, 229–241 (1986)
15. Fisher, R.A.: Design of Experiments. Edinburgh: Oliver and Boyd (1935)
16. Gastwirth, J.L., Krieger, A.M., Rosenbaum, P.R.: Asymptotic separability in sensitivity analysis. J Roy Statist Soc B **62**, 545–555 (2000)
17. Greenstone, M., Gayer, T.: Quasi-experimental and experimental approaches to environmental economics. J Environ Econ Manag **57**, 21–44 (2009)
18. Greevy, R., Silber, J.H., Cnaan, A., Rosenbaum, P.R.: Randomization inference with imperfect compliance in the ACE-inhibitor after anthracycline randomized trial. J Am Statist Assoc **99**, 7–15 (2004)
19. Hamermesh, D.S.: The craft of labormetrics. Indust Labor Relat Rev **53**, 363–380 (2000)
20. Heart Protection Study Collaborative Group.: MRC/BHF Heart Protection Study of antioxidant vitamin supplementation in 20,536 high-risk individuals: A randomised placebo-controlled trial. Lancet **360**, 23–33 (2002)
21. Heckman, J.J.: Micro data, heterogeneity, and the evaluation of public policy: Nobel lecture. J Polit Econ **109**, 673–748 (2001)
22. Hernán, M.A., Alonso, A., Logan, R., Grodstein, F., Michels, K.B., Willett, W.C., Manson, J.E., Robins, J.M.: Observational studies analyzed like randomized experiments: an application to postmenopausal hormone therapy and coronary heart disease (with Discussion). Epidemiology **19**, 766–793 (2008)
23. Hill, A.B.: The environment and disease: Association or causation? Proc Roy Soc Med **58**, 295–300 (1965)
24. Holland, P.W.: Statistics and causal inference. J Am Statist Assoc **81**, 945–960 (1986)
25. Holley, R.: Remarks on the FKG inequalities. Comm Math Phys **36**, 227–231 (1974)
26. Imbens, G.W., Wooldridge, J.M.: Recent developments in the econometrics of program evaluation. J Econ Lit **47**, 5–86 (2009)
27. Kant, I.: What is enlightenment? In: I. Kant, Toward Perpetual Peace and Other Writings. New Haven, CT: Yale University Press (1785, 2006)
28. Katan, M.B.: Apolipoprotein E isoforms, serum cholesterol, and cancer. Lancet **1**, 507–508 (1986) Reprinted: Int J Epidemiol **33**, 9 (2004)
29. Katan, M.B.: Commentary: Mendelian randomization, 18 years on. Int J Epidemiol **33**, 10–11 (2004)
30. Khaw, K.T., Bingham, S., Welch, A., Luben, R., Wareham, N., Oakes, S., Day, N.: Relation between plasma ascorbic acid and mortality in men and women in EPIC-Norfolk prospective study. Lancet **357**, 657–663 (2001)
31. Laird, N.M., Blacker, D., Wilcox, M.: The sib transmission/disequilibrium test is a Mantel-Haenszel test. Am J Hum Genet **63**, 1915 (1998)
32. Lawlor, D.A., Smith, G.D., Bruckdorfer, K.R., Kundo, D., Ebrahim, S.: Those confounded vitamins: What can we learn from the differences between observational versus randomized trial evidence? Lancet **363**, 1724–1727 (2004)
33. Meyer, B.D.: Natural and quasi-experiments in economics. J Business Econ Statist **13**, 151–161 (1995)
34. Rosenbaum, P.R.: The consequences of adjustment for a concomitant variable that has been affected by the treatment. J Roy Statist Soc A **147**, 656–666 (1984)
35. Rosenbaum, P.R.: On permutation tests for hidden biases in observational studies: An application of Holley's inequality to the Savage lattice. Ann Statist **17**, 643–653 (1989)
36. Rosenbaum, P.R.: Quantiles in nonrandom samples and observational studies. J Am Statist Assoc **90**, 1424–1431 (1995)
37. Rosenbaum, P.R.: Stability in the absence of treatment. J Am Statist Assoc **96**, 210–219 (2001)
38. Rosenbaum, P.R.: Observational Studies (2nd ed.). New York: Springer (2002)
39. Rosenbaum, P.R.: Sensitivity analysis for m-estimates, tests, and confidence intervals in matched observational studies. Biometrics **63**, 456–464 (2007)

40. Rosenzweig, M.R., Wolpin, K.I.: Natural 'natural experiments' in economics. J Econ Lit **38**, 827–874 (2000)
41. Rubin, D.B.: Estimating causal effects of treatments in randomized and nonrandomized studies. J Educ Psychol **66**, 688–701 (1974)
42. Rubin, D.B.: The design versus the analysis of observational studies for causal effects: Parallels with the design of randomized trials. Statist Med **26**, 20–36 (2007)
43. Rutter, M.: Identifying the Environmental Causes of Disease: How Do We Decide What to Believe and When to Take Action? London: Academy of Medical Sciences (2007)
44. Saunders, A.M., Strittmatter, W.J., Schmechel, D., et al.: Association of apolipoprotein E allele epsilon 4 with late-onset familial and sporadic Alzheimer's disease. Neurology **43**, 1467–1472 (1993)
45. Savage, I.R.: Contributions to the theory of rank order statistics: Applications of lattice theory. Rev Int Statist Inst **32**, 52–63 (1964)
46. Sekhon, J.S.: Opiates for the matches: Matching methods for causal inference. Annu Rev Pol Sci **12**, 487-508 (2009)
47. Shadish, W. R., Cook, T. D., Campbell, D.T.: Experimental and Quasi-Experimental Designs for Generalized Causal Inference. Boston: Houghton-Mifflin (2002)
48. Shadish, W. R., Cook, T. D.: The renaissance of field experimentation in evaluating interventions. Annu Rev Psychol **60**, 607–629 (2009)
49. Spielman, R.S., McGinnis, R.E., Ewens, W.J.: Transmission test for linkage disequilibrium. Am J Hum Genet **52**, 506–516 (1993)
50. Spielman, R.S., Ewens, W.J.: A sibship test for linkage in the presence of association: The sib transmission/disequilibrium test. Am J Hum Genet **62**, 450–458 (1998)
51. Strachan, T., Read, A.P.: Human Molecular Genetics. New York: Garland (2004)
52. Susser, M.: Epidemiology, Health and Society: Selected Papers. New York: Oxford University Press (1987)
53. Vandenbroucke, J.P.: When are observational studies as credible as randomized trials? Lancet **363**, 1728–1731 (2004)
54. West, S.G., Duan, N., Pequegnat, W., Gaist, P., Des Jarlais, D.C., Holtgrave, D., Szapocznik, J., Fishbein, M., Rapkin, B., Clatts, M., Mullen, P.D.: Alternatives to the randomized controlled trial. Am J Public Health **98**, 1359–1366 (2008)

Chapter 2
Causal Inference in Randomized Experiments

Abstract An observational study is an empiric investigation of treatment effects when random assignment to treatment or control is not feasible. Because observational studies are structured to resemble simple randomized experiments, an understanding of the role randomization plays in experiments is important as background. As a prelude to the discussion of observational studies in later chapters, the current chapter contains a brief review of the logic of causal inference in a randomized experiment. Only one simple case is discussed in detail, namely a randomized paired experiment in which subjects are paired before randomization and one subject in each pair is picked at random to receive treatment, the other receiving control. Although a foundation for later chapters, much of the material in this chapter is quite old, dating from Sir Ronald Fisher's work in the 1920s and 1930s, and it is likely to be familiar from other contexts, such as a course in the design of experiments.

2.1 Two Versions of the National Supported Work Experiment

2.1.1 A version with 185 pairs and a version with 5 pairs

Though discussed daily in the newspaper, unemployment is a curious phenomenon. Defined abstractly, it implies that someone entered an active labor market intending to sell their labor but was unable to find a buyer, often for long periods of time. Of course, the abstract definition leaves out most of what is happening.

Robert LaLonde [30] reviews several decades of

> public sector sponsored employment and training programs ... [intended to] enhance participants' productive skills and, in turn, increase their future earnings and tax payments and reduce their dependence on social welfare benefits. [...The] primary recipients of public sector sponsored training [have been] economically disadvantaged or dislocated workers. [30, page 149]

The National Supported Work Demonstration (NSW) included a randomized experiment evaluating the effects of one such program [7, 24]. Kenneth Couch writes:

> The NSW provided work experiences primarily in service occupations for females and construction for males. The jobs were designed to be consistent with a concept known as graduated stress. Stress within the working environment increased gradually during the training period until it simulated the workplace norms of the private sector. At that point, not more than 18 month after entry, individuals who received services provided by the NSW had to attempt a transition to unsubsidized employment. Screening criteria for the NSW limited participation to individuals severely handicapped in the labor market. ... After screening by a local social service agency and referral into the program, each participant was randomly assigned to a treatment (experimental) or control group. Treatments received the services offered by the NSW. Controls continued enrollment in other available social programs [7, pages 381–382].

The current chapter will use pieces of the NSW experiment to illustrate the logic of randomized experimentation. A reader interested not in randomized experimentation but rather in the full NSW program and its effects might begin with Couch's [7] study of the program's long term effects. The NSW became important in thinking about methodology for observational studies in economics because of a series of studies [12, 19, 29, 57], beginning with LaLonde's 1986 study [29], in which the randomized control group was set aside and various analytical methods were applied to nonrandomized controls from survey data.

Because the goal is to discuss the logic of randomized experiments in a manner that prepares for the discussion of observational studies in later chapters, several adjustments and simplifications were made. First, the data set is what Rajeev Dehejia and Sadek Wahba [12, Table 1] call the "RE74 subset," which in turn is a subset of the data used by LaLonde [29]. The subset consists of males who were randomized to treatment or control after December 1975 and who had left the program before January 1978, with annual earnings recorded in 1974, 1975 and 1978. Because of these requirements, earnings in 1974 and 1975 are pretreatment covariates unaffected by the NSW program, and earnings in 1978 is an outcome that may be affected by treatment. Furthermore, to emphasize the parallel with matched observational studies, the randomized treated and control groups were matched: specifically, all 185 treated men were matched to 185 untreated controls using eight covariates, forming 185 matched pairs.[1]

For the 185 matched pairs, Table 2.1 shows the distribution of the eight covariates. Before treatment, the treated and control groups looked fairly similar.

---

[1] As mentioned in Chapter 1, to keep the theoretical technicalities to a minimum, statistical theory is developed only for the case of matched pairs, so I have slightly reshaped the NSW experiment into a matched pair study. Randomization inference for unmatched randomized experiments is similar and is discussed in [44, Chapter 2]. The matching used precisely the methods described later in §8.4, using a penalty function to implement calipers on the propensity score, with the rank-based Mahalanobis distance used within calipers. Multivariate matching before randomization can improve the efficiency of a randomized experiment; see [17]. For illustration only in this expository chapter, the matching is done after random assignment, which entails discarding some of the randomized controls, something I would not do if the goal were to perform the most efficient analysis of the NSW experiment.

2.1 Two Versions of the National Supported Work Experiment

Table 2.1 Pretreatment covariates for 185 matched pairs from the NSW randomized experiment. For age and years of education, means, median (50%) and quartiles (25% and 75%) are given. For earnings, the mean of all earnings (including zero earnings) and the percentage of zero earnings are given. For binary variables, the percentage is given.

Covariate	Group	Mean	25%	50%	75%
Age	Treated	25.82	20	25	29
	Control	25.70	20	25	29
Years of	Treated	10.35	9	11	12
Education	Control	10.19	9	10	11
Covariate	Group	Mean		Percent $0	
Earnings in $	Treated	2096		71%	
in 1974	Control	2009		75%	
Earnings in $	Treated	1532		60%	
in 1975	Control	1485		64%	
Covariate	Group	Percent			
Black	Treated	84%			
	Control	85%			
Hispanic	Treated	6%			
	Control	5%			
Married	Treated	19%			
	Control	20%			
No High School	Treated	71%			
Degree	Control	77%			

Figure 2.1 displays the earnings in 1978 after treatment for the 185 matched pairs of men. In the boxplot of 185 matched pair differences, the dotted lines are at 0 and ±5000. It appears that treated men earned somewhat more.

In general, in this book, I will use small examples to illustrate the details of what goes on, and examples of practical size to illustrate analysis and interpretation. In that spirit, Table 2.2 is a random sample of five of the 185 pairs; it will be used to illustrate the details of randomization inference. The sample happens to consist of pairs 15, 37, 46, 151, and 181 of the 185 pairs.

2.1.2 Basic notation

Table 2.2 exemplifies notation that will be used throughout the book. The index of the pair is i, $i = 1, 2, \ldots, 5 = I$, and the index of the person in the pair is j, $j = 1, 2$. In Table 2.2 and throughout the book, Z indicates treatment, $Z = 1$ for treated, $Z = 0$ for control, x is an observed covariate — there are eight covariates, x_{ijk}, $k = 1, \ldots, 8 = K$, in Table 2.2 — and R indicates a response, in this case, earnings after the end of treatment in 1978. The first pair in Table 2.2 consists of two unmarried, young black men with seven or eight years of education and no earnings in 1974 and 1975, before the start of treatment; after treatment, in 1978, the treated man earned more. The treated-minus-control matched pair difference in 1978 earnings

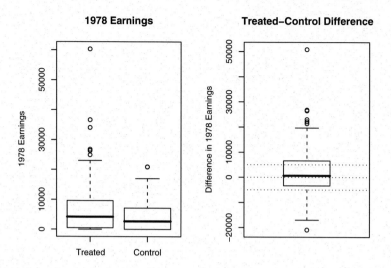

Fig. 2.1 Earnings in 1978, after treatment, for 185 pairs of men in the NSW randomized experiment. The dotted lines are at −$5000, $0, and $5000.

Table 2.2 Five pairs sampled at random from the 185 pairs in the NSW randomized experiment. The variables are: id = pair number among the 185 pairs; pairs 1 to 5; person 1 or 2 in a pair; treat=1 if treated, 0 if control; age in years; edu=education in years; black=1 if black, 0 otherwise; hisp=1 if hispanic, 0 otherwise; married=1 if married, 0 otherwise; no degree=1 if no high school degree, 0 otherwise; and re74, re75, and re78 are earnings in dollars in 1974, 1975, and 1978. Also, in pair i, Y_i is the treated-minus-control matched pair difference in 1978 earnings, $Y_i = (Z_{i1} - Z_{i2})(R_{i1} - R_{i2})$.

id	i	j	treat Z_{ij}	age x_{ij1}	edu x_{ij2}	black x_{ij3}	hisp x_{ij4}	married x_{ij5}	nodegree x_{ij6}	re74 x_{ij7}	re75 x_{ij8}	re78 R_{ij}	Y_i
15	1	1	1	17	7	1	0	0	1	0	0	3024	1456
15	1	2	0	18	8	1	0	0	1	0	0	1568	
37	2	1	1	25	5	1	0	0	1	0	0	6182	3988
37	2	2	0	24	7	1	0	0	1	0	0	2194	
46	3	1	1	25	11	1	0	1	1	0	0	0	−45
46	3	2	0	25	11	1	0	1	1	0	0	45	
151	4	1	1	28	10	1	0	0	1	0	2837	3197	−2147
151	4	2	0	22	10	1	0	0	1	0	2175	5344	
181	5	1	1	33	12	1	0	1	0	20280	10941	15953	3173
181	5	2	0	28	12	1	0	1	0	10585	5551	12780	

is $Y_i = (Z_{i1} - Z_{i2})(R_{i1} - R_{i2})$, so $Y_1 = (1-0)(3024 - 1568) = \$1456$ for the first pair of men.

It is convenient to have a symbol that represents the K covariates together, and vector notation does this, so $\mathbf{x}_{ij} = (x_{ij1}, \ldots, x_{ijK})^T$ contains the K covariate values for the jth person or unit in the ith pair. If you are not familiar with vector notation, it will not be a problem here. Vectors and vector notation have many uses, but for

2.2 Treatment Effects in Randomized Experiments

the most part, in the current book they are used simply to give concise names to arrays of data. For instance, in Table 2.2, $\mathbf{x}_{11} = (17,7,1,0,0,1,0,0)^T$ contains the covariate values for the first man in the first pair.

In the same way, $\mathbf{Z}$ indicates the treatment assignments for all $2I$ subjects in the I matched pairs, $\mathbf{Z} = (Z_{11}, Z_{12}, Z_{21}, \ldots, Z_{I2})^T$. For the $I = 5$ pairs in Table 2.2, $\mathbf{Z} = (1,0,1,0,1,0,1,0,1,0)^T$. The notation is slightly redundant, because $Z_{i2} = 1 - Z_{i1}$, so that, now and then, when compact expression is needed, only the Z_{i1}'s are mentioned. Also, $\mathbf{R} = (R_{11}, R_{12}, \ldots, R_{I2})^T$ and $\mathbf{Y} = (Y_1, Y_2, \ldots, Y_I)^T$.

It is possible to match for observed covariates, but not for a covariate that was not observed. There is an important sense in which a failure to match or control for an unobserved covariate presents no special problem in a randomized experiment but can present substantial problems in an observational study. Clarification of this distinction is one goal of the current chapter. For this purpose, it is convenient to give a name, u_{ij}, to the unobserved covariate. In the current chapter, which is focused on inference in randomized experiments, u_{ij} could be any unmeasured covariate (or any vector containing several unmeasured covariates). Success in finding and keeping a job may depend on aspects of personality, intelligence, family and personal connections, and physical appearance; then, in this chapter in connection with the NSW experiment, u_{ij} could be a vector of measurements of these attributes, a vector that was not measured by the investigators. Also, $\mathbf{u} = (u_{11}, u_{12}, \ldots, u_{I2})^T$.

To aid in reading the table, subject $j = 1$ is always the treated subject and subject $j = 2$ is always the control subject; however, strictly speaking, one should not do this. Strictly speaking, ij is the 'name' of a unique person; that person had his name, ij, before treatments were randomly assigned. In a paired randomized experiment, Z_{i1} is determined by I independent flips of a fair coin, so $Z_{11} = 1$ happens with probability $\frac{1}{2}$ and $Z_{11} = 0$ happens with probability $\frac{1}{2}$; then $Z_{i2} = 1 - Z_{i1}$. That is, in Table 2.2, Z_{i1} is 1 and Z_{i2} is 0 for every i, but this would happen by chance in a paired randomized experiment with $I = 5$ pairs with probability $\left(\frac{1}{2}\right)^5 = \frac{1}{32} = 0.03125$. Strictly speaking, some of the Z_{i1}'s should be 1 and some of the Z_{i2}'s should be 1. Quantities in statistical computations, such as the treated-minus-control difference in responses, $Y_i = (Z_{i1} - Z_{i2})(R_{i1} - R_{i2})$, are unaffected by the ordering of the table. So there is a small inconsistency between the logic of random assignment and the form of a readable table. Having mentioned this once, I will adhere to the logic of random assignment in theoretical discussions, present tables in readable form, and ignore the small inconsistency.

2.2 Treatment Effects in Randomized Experiments

2.2.1 Potential responses under alternative treatments

In Table 2.2, the first man, $(i,j) = (1,1)$, was randomly assigned to treatment, $Z_{11} = 1$, and had earnings of $R_{11} = \$3024$ in 1978, but things might have been different. Had the coin fallen differently, the first man, $(i,j) = (1,1)$, might have been randomly assigned to control, $Z_{11} = 0$, and in this case his earnings in 1978 might have been different. We will never know what this first man's earnings would have been had the coin fallen differently, had he been assigned to control. Perhaps the treatment was completely ineffective, and perhaps this first man would have held the same job, with the same earnings, namely $3024, had he been assigned to control. Or perhaps the treatment raised his earnings, and his earnings under control would have been lower. We will never know about this one man, but with 370 men in 185 pairs, half randomly assigned to treatment, the others to control, we can say something about what would have happened to the 370 men under treatment and under control.

Not only the first man but each man (i,j) has two potential responses, a level of earnings he would exhibit in 1978 if assigned to treatment, r_{Tij}, and a level of earnings he would exhibit in 1978 if assigned to control, r_{Cij}. We see one of these. Specifically, if (i,j) were assigned to treatment, $Z_{ij} = 1$, we would see r_{Tij}, but if (i,j) were assigned to control, $Z_{ij} = 0$, we would see r_{Cij}. The response, R_{ij}, we actually observe from (i,j) — that is, the 1978 earnings actually recorded in Table 2.2 — equal r_{Tij} if $Z_{ij} = 1$ or r_{Cij} if $Z_{ij} = 0$; that is, in a formula, $R_{ij} = Z_{ij} r_{Tij} + (1 - Z_{ij}) r_{Cij}$. Also, $\mathbf{r}_T = (r_{T11}, r_{T12}, \ldots, r_{TI2})^T$ and $\mathbf{r}_C = (r_{C11}, r_{C12}, \ldots, r_{CI2})^T$.

To say that the treatment has no effect on this response from (i,j) is to say $r_{Cij} = r_{Tij}$. To say that the treatment caused (i,j)'s earnings to increase by $1000 is to say $r_{Tij} = r_{Cij} + 1000$ or $r_{Tij} - r_{Cij} = 1000$. We will never be in a position to confidently assert either of these things about a single man. It is, however, a very different thing to assert that all 370 men were unaffected by the treatment — to assert that $r_{Cij} = r_{Tij}$ for $i = 1, 2, \ldots 185$, $j = 1, 2$; in a randomized experiment, we may be in a position to confidently deny that. The hypothesis that the treatment had no effect on anyone, namely $H_0 : r_{Cij} = r_{Tij}$ for $i = 1, 2, \ldots 185$, $j = 1, 2$, is known as Ronald Fisher's [13] sharp null hypothesis of no effect. This hypothesis may be written compactly as $H_0 : \mathbf{r}_T = \mathbf{r}_C$.

The notation that expresses treatment effects as comparisons of potential responses under alternative treatments was introduced into the design of experiments by Jerzy Neyman [35] in 1923 and was used to solve various problems in randomized experiments; e.g., [62, 64, 9, 40] and [28, § 8.3]. Donald Rubin [50] first advocated use of this notation in observational studies.

2.2.2 Covariates and outcomes

In §1.2, the distinction between covariates and outcomes was emphasized. A covariate, such as $\mathbf{x}_{ij}$ or u_{ij}, is a pretreatment quantity, so there is only one version of a covariate. A response or outcome has two potential values, (r_{Tij}, r_{Cij}), one of which is observed, namely R_{ij}, depending upon the treatment assignment Z_{ij}, that is, $R_{ij} = Z_{ij} r_{Tij} + (1 - Z_{ij}) r_{Cij}$.

2.2 Treatment Effects in Randomized Experiments

Notice that $(r_{Tij}, r_{Cij}, \mathbf{x}_{ij}, u_{ij})$ do not change when treatments are assigned at random, that is, when Z_{ij} is determined by a coin flip or a random number, but in general the observed response, R_{ij}, does change. It is convenient to have a symbol that represents the quantities that are not changed when treatments are randomized. Let $\mathscr{F}$ denote the array of quantities $\{(r_{Tij}, r_{Cij}, \mathbf{x}_{ij}, u_{ij}), i = 1, 2, \ldots, I, j = 1, 2\}$ that do not change when Z_{ij} is determined. (These quantities are fixed in Fisher's theory of randomization inference; hence, the symbol is $\mathscr{F}$.)

2.2.3 Possible treatment assignments and randomization

The observed treatment assignment in Table 2.2 is $\mathbf{Z} = (1,0,1,0,1,0,1,0,1,0)^T$, but random assignment of treatments within pairs might have picked a different assignment. Write $\mathscr{Z}$ for the set of the 2^I possible values of $\mathbf{Z}$; that is, $\mathbf{z} \in \mathscr{Z}$ if and only if $\mathbf{z} = (z_{11}, z_{12}, \ldots, z_{I2})^T$ with $z_{ij} = 0$ or $z_{ij} = 1$ and $z_{i1} + z_{i2} = 1$ for each i,j. For Table 2.2, there are $2^I = 2^5 = 32$ possible values $\mathbf{z} \in \mathscr{Z}$. Generally, if A is a finite set, then $|A|$ is the number of elements of A, so that, in particular, $|\mathscr{Z}| = 2^I$. Table 2.3 lists, in abbreviated form, the 32 possible treatment assignments for Table 2.2; the abbreviation consists in listing z_{i1} but not z_{i2} because $z_{i2} = 1 - z_{i1}$. The observed treatment assignment, $\mathbf{Z} = (1,0,1,0,1,0,1,0,1,0)^T$ in Table 2.2, corresponds with the first row of Table 2.3. In the second row of Table 2.3, the treatment assignment for the fifth pair in Table 2.2 has been reversed, so the second man, not the first man, in that pair is assigned to treatment.

For the $I = 185$ pairs in Table 2.1, the set $\mathscr{Z}$ of possible treatment assignments contains $2^I = 2^{185} \doteq 4.9 \times 10^{55}$ possible treatment assignments, $\mathbf{z} \in \mathscr{Z}$. It would be inconvenient to list them.

What does it mean to randomly assign treatments? At an intuitive level, one assignment $\mathbf{z} \in \mathscr{Z}$ is picked at random, each having probability 2^{-I}, or $2^{-5} = \frac{1}{32} = 0.03125$ for Table 2.3. For instance, one might flip a fair coin independently five times to determine $\mathbf{Z}$ in Table 2.3. The intuition is that randomization is making a statement about how $\mathbf{Z}$ alone was determined, that is, about the marginal distribution of $\mathbf{Z}$. This intuition is not quite correct; in an important way, randomization means more than this. Specifically, in a randomized experiment, the information in $(r_{Tij}, r_{Cij}, \mathbf{x}_{ij}, u_{ij})$ (or in $\mathscr{F}$) is of no use in predicting Z_{ij}. That is, the coin is fair not just in coming up heads half the time, independently in different pairs, but more importantly the coin knows nothing about the individual and is impartial in its treatment assignments. The design of the paired randomized experiment forces $\mathbf{Z}$ to fall in $\mathscr{Z}$ — that is, it forces the event $\mathbf{Z} \in \mathscr{Z}$ to occur — which is denoted concisely by saying the event $\mathscr{Z}$ occurs. Randomization in the paired randomized experiment means:

$$\Pr(\mathbf{Z} = \mathbf{z} \mid \mathscr{F}, \mathscr{Z}) = \frac{1}{|\mathscr{Z}|} = \frac{1}{2^I} \text{ for each } \mathbf{z} \in \mathscr{Z}. \tag{2.1}$$

Table 2.3 The set $\mathscr{Z}$ of $32 = 2^5$ possible treatment assignments for the small version of the NSW experiment with $I = 5$ pairs. Only Z_{i1} is listed because $Z_{i2} = 1 - Z_{i1}$.

Label	Pair 1 Z_{11}	Pair 2 Z_{21}	Pair 3 Z_{31}	Pair 4 Z_{41}	Pair 5 Z_{51}
1	1	1	1	1	1
2	1	1	1	1	0
3	1	1	1	0	1
4	1	1	1	0	0
5	1	1	0	1	1
6	1	1	0	1	0
7	1	1	0	0	1
8	1	1	0	0	0
9	1	0	1	1	1
10	1	0	1	1	0
11	1	0	1	0	1
12	1	0	1	0	0
13	1	0	0	1	1
14	1	0	0	1	0
15	1	0	0	0	1
16	1	0	0	0	0
17	0	1	1	1	1
18	0	1	1	1	0
19	0	1	1	0	1
20	0	1	1	0	0
21	0	1	0	1	1
22	0	1	0	1	0
23	0	1	0	0	1
24	0	1	0	0	0
25	0	0	1	1	1
26	0	0	1	1	0
27	0	0	1	0	1
28	0	0	1	0	0
29	0	0	0	1	1
30	0	0	0	1	0
31	0	0	0	0	1
32	0	0	0	0	0

In words, even if you knew the information in $\mathscr{F}$ — that is, the information in $\left(r_{Tij}, r_{Cij}, \mathbf{x}_{ij}, u_{ij} \right)$ for every person, $i = 1, 2, \ldots, I$, $j = 1, 2$ — and even if you knew the structure of the paired randomized experiment, namely $\mathbf{Z} \in \mathscr{Z}$, you would be unable to use that information to predict treatment assignments because, given all of that information, the 2^I possible assignments $\mathbf{z} \in \mathscr{Z}$ have equal probabilities, 2^{-I}.

2.2.4 Interference between units

The notation for treatment effects appears innocuous, but it actually entails a fairly strong assumption, called "no interference between units" by David Cox [8, §2.4],

2.3 Testing the Null Hypothesis of No Treatment Effect

namely "that the observation on one unit should be unaffected by particular assignment of treatments to other units." There would be interference between units in Table 2.2 if switching man $(i, j) = (2, 1)$ — i.e., the first man in the second pair — from treatment to control would affect the earnings of the man we discussed before, $(i, j) = (1, 1)$, that is, the first man in the first pair. Interference is probably not widespread in the NSW experiment, but it cannot be entirely ruled out. For instance, a potential employer might hire two subsidized workers under NSW, but retain for unsubsidized permanent employment only the better of the two employees. In this case, switching the treatment for one man might cause another man to gain or lose a job. The two-outcomes notation makes no provision for interference between units; each unit or man has only two potential responses depending exclusively on his own treatment assignment. If there were interference between units, the response of each man would depend upon all of the treatment assignments, $(Z_{11}, Z_{12}, \ldots, Z_{I2})$. In Table 2.2, there are two ways to assign treatments in each of five pairs, so there are $32 = 2^5$ possible treatment assignments. If there were interference between units, each man would have not two potential responses but 32 potential responses, depending upon all of the treatment assignments, $(Z_{11}, Z_{12}, \ldots, Z_{52})$.

Interference between units is likely to be very limited in the NSW experiment because most men don't interfere with each other. The situation is different when the same person is studied repeatedly under different treatments in a randomized order. For instance, this is common in many experiments in cognitive neuroscience, in which a person's brain is watched using functional magnetic resonance imaging while she performs different cognitive tasks [36]. In this case, an experimental unit is one time for one person. In experiments of this kind, interference between different times for the same person is likely.

In randomized experiments with interference between units, inference is possible but requires additional care [26, 47, 58]. There is considerable simplification if only units in the same pair interfere with one another [46, §6].

2.3 Testing the Null Hypothesis of No Treatment Effect

2.3.1 Treated−control differences when the null hypothesis is true

What would happen if the null hypothesis of no effect were true?

As discussed in §2.2.1, the null hypothesis of no treatment effect says each person would exhibit the same response whether assigned to treatment, $Z_{ij} = 1$, or to control, $Z_{ij} = 0$; that is, the hypothesis asserts $H_0 : r_{Cij} = r_{Tij}$ for all i, j, or concisely $H_0 : \mathbf{r}_T = \mathbf{r}_C$. In Table 2.2, if $H_0 : \mathbf{r}_T = \mathbf{r}_C$ were true, then the first man in the first pair would have earned \$3024 in 1978 whether he was assigned to treatment or to control. Similarly, if $H_0 : \mathbf{r}_T = \mathbf{r}_C$ were true, the second man in the first pair would have earned \$1568 whether he was assigned to treatment or control. This null hypothesis is not implausible; it might be true. Many treatments you or I receive during the course of a year do not change our earnings during the course of the year: we work at the same job, receive the same pay. Then again, the NSW might have had a beneficial effect, raising the earnings for treated men; in this case, $H_0: \mathbf{r}_T = \mathbf{r}_C$ would be false. A basic question is: To what extent do the data from the randomized experiment provide evidence against the null hypothesis of no treatment effect?

If the null hypothesis of no treatment effect, $H_0: \mathbf{r}_T = \mathbf{r}_C$, were true, then depending upon the fall of the coin for the first pair in Table 2.2, the treated-minus-control difference in 1978 earnings in the first pair, namely Y_1, would be either $3024 - 1568 = 1456$ or $1568 - 3024 = -1456$. In other words, if the null hypothesis of no treatment effect were true, then the treatment does not do anything to the responses, and it affects the treated-minus-control difference in responses only by labeling one man as treated, the other as control. Recall that the observed response for man j in pair i is $R_{ij} = Z_{ij} r_{Tij} + (1 - Z_{ij}) r_{Cij}$ and the treated-minus-control difference in responses in pair i is $Y_i = (Z_{i1} - Z_{i2})(R_{i1} - R_{i2})$. If the treatment had no effect, so that $r_{Tij} = r_{Cij}$ for all i, j, then $Y_i = (Z_{i1} - Z_{i2})(R_{i1} - R_{i2})$ is $Y_i = (Z_{i1} - Z_{i2})(r_{Ci1} - r_{Ci2})$ which equals $r_{Ci1} - r_{Ci2}$ if $(Z_{i1}, Z_{i2}) = (1,0)$ or $-(r_{Ci1} - r_{Ci2})$ if $(Z_{i1}, Z_{i2}) = (0,1)$. If the treatment has no effect, randomization just randomizes the labels 'treated' and 'controls'; however, it does not change anybody's earnings, and it simply changes the sign of the treated-minus-control difference in earnings.

What experimental results might occur if the null hypothesis of no effect were true?

Under the null hypothesis of no treatment effect, $H_0: \mathbf{r}_T = \mathbf{r}_C$, Table 2.4 displays the possible treatment assignments Z_{i1} and possible treated-minus-control differences in responses Y_i for the five pairs in Table 2.2. Each row of Table 2.4 is one possible outcome of the experiment when the null hypothesis, H_0, is true. The random numbers actually gave us the first row of Table 2.4, but had the coins fallen differently, another row would have occurred. If H_0 were true, each row of Table 2.4 has probability $\frac{1}{2^5} = \frac{1}{32}$ of turning out to be the results of the experiment: flip the five coins and you get one row of Table 2.4 as your experiment. We know this without assumptions. We know that each of the 32 treatment assignments in Table 2.3 has probability $\frac{1}{2^5} = \frac{1}{32}$ because we randomized: $\mathbf{Z}$ was determined by the independent flips of five fair coins. We know 'if H_0 were true, each row of Table 2.4 has probability $\frac{1}{2^5} = \frac{1}{32}$,' because if H_0 were true, then different $\mathbf{Z}$'s simply change the signs of the Y_i's as in Table 2.4. Nothing is assumed.

2.3 Testing the Null Hypothesis of No Treatment Effect

Table 2.4 The set $\mathscr{Z}$ of $32 = 2^5$ possible treatment assignments and the corresponding possible treated-minus-control differences in responses $Y_i = (Z_{i1} - Z_{i2})(R_{i1} - R_{i2})$ under the null hypothesis of no treatment effect, $H_0 : \mathbf{r}_T = \mathbf{r}_C$, for the small version of the NSW experiment with $I = 5$ pairs. Only Z_{i1} is listed because $Z_{i2} = 1 - Z_{i1}$. Under the null hypothesis, different treatment assignments relabel men as treated or control, but do not change their earnings, so the sign of Y_i changes accordingly.

Label	Pair 1 Z_{11}	Pair 2 Z_{21}	Pair 3 Z_{31}	Pair 4 Z_{41}	Pair 5 Z_{51}	Pair 1 Y_1	Pair 2 Y_2	Pair 3 Y_3	Pair 4 Y_4	Pair 5 Y_5
1	1	1	1	1	1	1456	3988	−45	−2147	3173
2	1	1	1	1	0	1456	3988	−45	−2147	−3173
3	1	1	1	0	1	1456	3988	−45	2147	3173
4	1	1	1	0	0	1456	3988	−45	2147	−3173
5	1	1	0	1	1	1456	3988	45	−2147	3173
6	1	1	0	1	0	1456	3988	45	−2147	−3173
7	1	1	0	0	1	1456	3988	45	2147	3173
8	1	1	0	0	0	1456	3988	45	2147	−3173
9	1	0	1	1	1	1456	−3988	−45	−2147	3173
10	1	0	1	1	0	1456	−3988	−45	−2147	−3173
11	1	0	1	0	1	1456	−3988	−45	2147	3173
12	1	0	1	0	0	1456	−3988	−45	2147	−3173
13	1	0	0	1	1	1456	−3988	45	−2147	3173
14	1	0	0	1	0	1456	−3988	45	−2147	−3173
15	1	0	0	0	1	1456	−3988	45	2147	3173
16	1	0	0	0	0	1456	−3988	45	2147	−3173
17	0	1	1	1	1	−1456	3988	−45	−2147	3173
18	0	1	1	1	0	−1456	3988	−45	−2147	−3173
19	0	1	1	0	1	−1456	3988	−45	2147	3173
20	0	1	1	0	0	−1456	3988	−45	2147	−3173
21	0	1	0	1	1	−1456	3988	45	−2147	3173
22	0	1	0	1	0	−1456	3988	45	−2147	−3173
23	0	1	0	0	1	−1456	3988	45	2147	3173
24	0	1	0	0	0	−1456	3988	45	2147	−3173
25	0	0	1	1	1	−1456	−3988	−45	−2147	3173
26	0	0	1	1	0	−1456	−3988	−45	−2147	−3173
27	0	0	1	0	1	−1456	−3988	−45	2147	3173
28	0	0	1	0	0	−1456	−3988	−45	2147	−3173
29	0	0	0	1	1	−1456	−3988	45	−2147	3173
30	0	0	0	1	0	−1456	−3988	45	−2147	−3173
31	0	0	0	0	1	−1456	−3988	45	2147	3173
32	0	0	0	0	0	−1456	−3988	45	2147	−3173

2.3.2 The randomization distribution of the mean difference

Randomization test of no effect using the mean as the test statistic

In general, the null distribution of a statistic is its distribution if the null hypothesis were true. A P-value or significance level is computed with reference to a null distribution. This section considers the null distribution of the sample mean difference; it was considered by Fisher [13, Chapter 3] in the book in which he introduced randomized experimentation.

Table 2.5 The possible treated-minus-control differences in responses $Y_i = (Z_{i1} - Z_{i2})(R_{i1} - R_{i2})$ and their mean, $\overline{Y}$, under the null hypothesis of no treatment effect, $H_0 : \mathbf{r}_T = \mathbf{r}_C$, for the small version of the NSW experiment with $I = 5$ pairs.

Label	Pair 1 Y_1	Pair 2 Y_2	Pair 3 Y_3	Pair 4 Y_4	Pair 5 Y_5	Mean $\overline{Y}$
1	1456	3988	−45	−2147	3173	1285.0
2	1456	3988	−45	−2147	−3173	15.8
3	1456	3988	−45	2147	3173	2143.8
4	1456	3988	−45	2147	−3173	874.6
5	1456	3988	45	−2147	3173	1303.0
6	1456	3988	45	−2147	−3173	33.8
7	1456	3988	45	2147	3173	2161.8
8	1456	3988	45	2147	−3173	892.6
9	1456	−3988	−45	−2147	3173	−310.2
10	1456	−3988	−45	−2147	−3173	−1579.4
11	1456	−3988	−45	2147	3173	548.6
12	1456	−3988	−45	2147	−3173	−720.6
13	1456	−3988	45	−2147	3173	−292.2
14	1456	−3988	45	−2147	−3173	−1561.4
15	1456	−3988	45	2147	3173	566.6
16	1456	−3988	45	2147	−3173	−702.6
17	−1456	3988	−45	−2147	3173	702.6
18	−1456	3988	−45	−2147	−3173	−566.6
19	−1456	3988	−45	2147	3173	1561.4
20	−1456	3988	−45	2147	−3173	292.2
21	−1456	3988	45	−2147	3173	720.6
22	−1456	3988	45	−2147	−3173	−548.6
23	−1456	3988	45	2147	3173	1579.4
24	−1456	3988	45	2147	−3173	310.2
25	−1456	−3988	−45	−2147	3173	−892.6
26	−1456	−3988	−45	−2147	−3173	−2161.8
27	−1456	−3988	−45	2147	3173	−33.8
28	−1456	−3988	−45	2147	−3173	−1303.0
29	−1456	−3988	45	−2147	3173	−874.6
30	−1456	−3988	45	−2147	−3173	−2143.8
31	−1456	−3988	45	2147	3173	−15.8
32	−1456	−3988	45	2147	−3173	−1285.0

Perhaps the most familiar test statistic is the mean of the I treated-minus-control differences in response, $\overline{Y} = \frac{1}{I} \sum_{i=1}^{I} Y_i$. The null distribution of $\overline{Y}$ is the distribution of $\overline{Y}$ when the null hypothesis of no treatment effect, $H_0 : \mathbf{r}_T = \mathbf{r}_C$, is true. Using Table 2.4, we may compute $\overline{Y}$ from each of the 32 possible experimental outcomes; see Table 2.5. For instance, in the experiment as it was actually performed, $\overline{Y} = 1285$ in the first row of Table 2.5: on average, the five treated men earned \$1,285 more than their matched controls. If the null hypothesis, H_0, were true, and the treatment assignment in the $i = 5$th pair were reversed, as in the second row of Table 2.5, then in that different randomized experiment, the mean would have been $\overline{Y} = 15.8$, or \$15.80. Similar considerations apply to all 32 possible randomized experiments.

2.3 Testing the Null Hypothesis of No Treatment Effect

In fact, Tables 2.5 and 2.6 give the null randomization distribution of the mean difference $\overline{Y}$. That is, the tables give the distribution of $\overline{Y}$ derived from the random assignment of treatments and the null hypothesis, H_0, that the treatment is without effect. Table 2.5 lists the results in the order of Table 2.3, while Table 2.6 sorts Table 2.5 into increasing order of the possible values of $\overline{Y}$. It is the same distribution, but Table 2.6 is easier to read. In both tables, each possible outcome or row has probability $\frac{1}{2^5} = \frac{1}{32} = 0.03125$. The tail probability, $\Pr\left(\overline{Y} \geq y \mid \mathscr{F}, \mathscr{Z}\right)$, used in computing significance levels or P-values, is also given in Table 2.6. The one-sided P-value for testing no treatment effect in Table 2.2 is $\Pr\left(\overline{Y} \geq 1285 \mid \mathscr{F}, \mathscr{Z}\right) = 0.1875 = \frac{6}{32}$ because 6 of the 32 random assignments produce means $\overline{Y}$ of 1285 or more when the null hypothesis of no effect is true. The two-sided P-value is twice this one-sided P-value,[2] or $2 \times \Pr\left(\overline{Y} \geq 1285 \mid \mathscr{F}, \mathscr{Z}\right) = 0.375 = \frac{12}{32}$, which equals $\Pr\left(\overline{Y} \leq -1285 \mid \mathscr{F}, \mathscr{Z}\right) + \Pr\left(\overline{Y} \geq 1285 \mid \mathscr{F}, \mathscr{Z}\right)$ for the distribution in Table 2.6 because the distribution is symmetric about zero.

The reasoned basis for inference in experiments

Fisher [13, Chapter 2] spoke of randomization as the "reasoned basis" for causal inference in experiments. By this, he meant that the distribution in Table 2.6 for the mean difference $\overline{Y}$ and similar randomization distributions for other test statistics provided a valid test of the hypothesis of no effect caused by the treatment. Moreover, these tests required no assumptions whatsoever. Fisher placed great emphasis on this, so it is worthwhile to consider what he meant and why he regarded the issue as important.

The distribution in Table 2.6 was derived from two considerations. The first consideration is that the experimenter used coins or random numbers to assign treatments, picking one of the 32 treatment assignments in Table 2.3, each with probability $1/32$. Provided the experimenter is not dishonest in describing the experiment, there is no basis for doubting this first consideration. The first consideration is not an assumption; it is a fact describing how the experiment was conducted. The second consideration is the null hypothesis of no treatment effect. Table 2.6 is the null distribution of the mean difference, $\overline{Y}$; that is, Table 2.6 is the distribution $\overline{Y}$ would have if the null hypothesis of no treatment effect *were* true. The null hypothesis may well be false — perhaps the experiment was undertaken in the hope of showing it is false — but that is beside the point because Table 2.6 is the distribution $\overline{Y}$ *would*

---

[2] In general, if you want a two-sided P-value, compute both one-sided P-values, double the smaller one, and take the minimum of this value and 1. This approach views the two-sided P-value as a correction for testing twice [10]. Both the sample mean in the current section and Wilcoxon's signed rank statistic in the next section have randomization distributions under the null hypothesis that are symmetric, and with a symmetric null distribution there is little ambiguity about the meaning of a two-sided P-value. When the null distribution is not symmetric, different definitions of a two-sided P-value can give slightly different answers. As discussed in [10], the view of a two-sided P-value as a correction for testing twice is one sensible approach in all cases. For related results, see [55].

Table 2.6 The randomization distribution of the sample mean $\overline{Y}$ of $I = 5$ matched pair differences in 1978 earnings under the null hypothesis of no treatment effect in the NSW experiment. Because the observed value of $\overline{Y}$ is \$1285, the one-sided P-value is 0.1875 and the two-sided P-value is twice that, or 0.375. With $I = 5$ pairs, the difference would be significant at the conventional 0.05 level in a one-sided test only if all five differences were positive, in which case $\overline{Y} = 2161.8$ and the one-sided P-value is 0.03125. Notice the that null distribution of $\overline{Y}$ is symmetric about zero.

y	$\Pr\left(\overline{Y} = y \mid \mathscr{F}, \mathscr{Z}\right)$	$\Pr\left(\overline{Y} \geq y \mid \mathscr{F}, \mathscr{Z}\right)$
2161.8	0.03125	0.03125
2143.8	0.03125	0.06250
1579.4	0.03125	0.09375
1561.4	0.03125	0.12500
1303.0	0.03125	0.15625
1285.0	0.03125	0.18750
892.6	0.03125	0.21875
874.6	0.03125	0.25000
720.6	0.03125	0.28125
702.6	0.03125	0.31250
566.6	0.03125	0.34375
548.6	0.03125	0.37500
310.2	0.03125	0.40625
292.2	0.03125	0.43750
33.8	0.03125	0.46875
15.8	0.03125	0.50000
−15.8	0.03125	0.53125
−33.8	0.03125	0.56250
−292.2	0.03125	0.59375
−310.2	0.03125	0.62500
−548.6	0.03125	0.65625
−566.6	0.03125	0.68750
−702.6	0.03125	0.71875
−720.6	0.03125	0.75000
−874.6	0.03125	0.78125
−892.6	0.03125	0.81250
−1285.0	0.03125	0.84375
−1303.0	0.03125	0.87500
−1561.4	0.03125	0.90625
−1579.4	0.03125	0.93750
−2143.8	0.03125	0.96875
−2161.8	0.03125	1.00000

have if the null hypothesis of no treatment effect *were* true. One tests a null hypothesis using $\overline{Y}$ by contrasting the actual behavior of $\overline{Y}$ with the behavior $\overline{Y}$ would have if the null hypothesis were true. A small P-value indicates that the behavior $\overline{Y}$ actually exhibited would have been remarkably improbable if the null hypothesis were true. In this sense too, the second consideration, the null hypothesis, is not an assumption. One does not assume or believe the null hypothesis to be true when testing it; rather, certain logical consequences of the null hypothesis are worked out for the purpose of evaluating its plausibility. In brief, nothing was assumed to be

2.3 Testing the Null Hypothesis of No Treatment Effect

true in working out the distribution in Table 2.6, that is, in testing the hypothesis that the treatment caused no effect.

You sometimes hear it said that "You cannot prove causality with statistics." One of my professors, Fred Mosteller, would often say: "You can only prove causality with statistics." When he said this, he was referring to randomization as the reasoned basis for causal inference in experiments.

Some history: Random assignment and Normal errors

Concerning Table 2.6, one might reasonably ask how this null randomization distribution of the mean treated-minus-control difference, $\overline{Y}$, compares with the t-distribution that might be used if the Y_i were assumed to be Normally distributed with constant variance. The null distribution in Table 2.6 made no assumption that the Y_i were Normal with constant variance; rather, it acknowledged that the experimenter did indeed randomize treatment assignments. Questions of this sort received close attention in early work on randomization inference; see, for instance, [62]. This attention focused on the moments of the distributions and their limiting behavior as the number, I, of pairs increases. Generally, the two approaches to inference gave similar inferences for large I when the (r_{Cij}, r_{Tij}) were sufficiently well-behaved, but could diverge in other situations. In early work [13, 62], this was often understood to mean that randomization formed a justification for inferences based on Normal models without requiring the assumptions of those models to be true. However, it was not long before attention focused on the divergences between randomization inference and Normal theory, so the concern ceased to be with justifying Normal theory when applied to well-behaved data from a randomized experiment; instead, concern shifted to inferences that would be valid and efficient whether or not the data are well-behaved.[3] This leads to the randomization distribution of statistics other than the mean difference, $\overline{Y}$, because the mean was found to be one of the least efficient statistics for responses that are not so well-behaved [1].

Two such statistics will be discussed: Frank Wilcoxon's signed rank statistic [63] in §2.3.3 and the randomization distribution of the statistics used in Peter Huber's m-estimation [25, 33] in §2.9. The close connection between rank statistics and m-estimates is developed in detail in [27]. Sections 2.3.3 and 2.9 play different roles in this book, and it is not necessary to read §2.9. The signed rank statistic will be discussed at many points in the book. In contrast, m-estimation is mentioned here to emphasize that the basic issues discussed here and later are not tied to any one statistic and may be developed for large classes of statistics in an entirely parallel way.

---

[3] I write 'Normal' distribution rather than 'normal' distribution because Normal is the name, not the description, of a distribution, in the same way that 'Sitting Bull' is the name of a man, not a description of that man. The exponential distribution is exponential, the logistic distribution is logistic, but the Normal distribution is not normal. Under normal circumstances, data are not Normally distributed.

2.3.3 The randomization distribution of Wilcoxon's statistic

Computing the randomization distribution under the null hypothesis of no effect

Among robust alternatives to the mean treated-minus-control difference, $\overline{Y}$, by far the most popular in practice is Wilcoxon's signed rank statistic, T; see [63] or [32, §3.2]. As with $\overline{Y}$, Wilcoxon's statistic uses the treated-minus-control differences in responses, $Y_i = (Z_{i1} - Z_{i2})(R_{i1} - R_{i2})$, in the I pairs. First, the absolute differences $|Y_i|$ are calculated; then these absolute differences are assigned ranks q_i from smallest to largest, from 1 to I; then Wilcoxon's signed rank statistic, T, is the sum of those ranks q_i for which Y_i is positive. With $Y_1 = 1456$, $Y_2 = 3988$, $Y_3 = -45$, $Y_4 = -2147$, $Y_5 = 3173$ in Table 2.2, the absolute differences are $|Y_1| = 1456$, $|Y_2| = 3988$, $|Y_3| = 45$, $|Y_4| = 2147$, $|Y_5| = 3173$, with ranks $q_1 = 2$, $q_2 = 5$, $q_3 = 1$, $q_4 = 3$, and $q_5 = 4$, and Y_1, Y_2, Y_5 are positive, so $T = q_1 + q_2 + q_5 = 2 + 5 + 4 = 11$.

In general, Wilcoxon's signed rank statistic is $T = \sum_{i=1}^{I} \text{sgn}(Y_i) \cdot q_i$, where $\text{sgn}(a) = 1$ if $a > 0$, $\text{sgn}(a) = 0$ if $a \leq 0$, and q_i is the rank of $|Y_i|$.

As in §2.3.2 and Table 2.5, if the null hypothesis of no treatment effect, $H_0 : \mathbf{r}_T = \mathbf{r}_C$, were true then $Y_i = (Z_{i1} - Z_{i2})(r_{Ci1} - r_{Ci2})$, and randomization simply changes the signs of the Y_i, as in Table 2.4. For each row of Table 2.4, Wilcoxon's signed rank statistic, T, may be computed, and this is done in Table 2.7. Under the null hypothesis of no treatment effect, $H_0 : \mathbf{r}_T = \mathbf{r}_C$, each row of Table 2.7 has a probability of $\frac{1}{2^5} = \frac{1}{32}$ of becoming the observed experimental results in the five pairs, and 7 rows produce values of T that are 11 or more, so the one-sided P-value is $\Pr(T \geq 11| \mathscr{F}, \mathscr{Z}) = \frac{7}{32} = 0.21875$. Table 2.8 reorganizes Table 2.7, sorting by the possible values of the test statistic and removing repetitions of values that occur several times. In parallel with §2.3.2, the two-sided P-value is twice this one-sided P-value[4], or $2 \times \Pr(T \geq 11| \mathscr{F}, \mathscr{Z}) = \frac{14}{32} = 0.4375$ which equals $\Pr(T \leq 4| \mathscr{F}, \mathscr{Z}) + \Pr(T \geq 11| \mathscr{F}, \mathscr{Z})$ for the distribution in Table 2.7 because the distribution is symmetric about its expectation $I(I+1)/4 = 5(5+1)/4 = 7.5$.

The null distribution of Wilcoxon's T is, in many ways, a very simple distribution. In Table 2.7, as the signs change, the absolute values $|Y_i|$ stay the same, so the ranks of the absolute values also stay the same; see the bottom of Table 2.7. That is, if $H_0 : \mathbf{r}_T = \mathbf{r}_C$, were true, then the $|Y_i| = |r_{Ci1} - r_{Ci2}|$ and q_i are

---

[4] In practice, statistical software is used to do the calculations. In the statistical package R, the command wilcox.test(.) is used. Consider, first, the $I = 5$ matched pair differences in 1978 earnings in Table 2.7, which are contained in the vector dif5.

```
> dif5
[1] 1456 3988  -45 -2147 3173
> wilcox.test(dif5)
 Wilcoxon signed rank test
data: dif5
V = 11, p-value = 0.4375
```

In the case of $I = 5$, wilcox.test performs calculations that exactly reproduce the value $T = 11$ and the two-sided P-value of 0.4375. By adjusting the call to wilcox.test, either of the two one-sided P-values may be obtained.

2.3 Testing the Null Hypothesis of No Treatment Effect

Table 2.7 The possible treated-minus-control differences in responses $Y_i = (Z_{i1} - Z_{i2})(R_{i1} - R_{i2})$ and Wilcoxon's signed rank statistic, T, under the null hypothesis of no treatment effect, $H_0: \mathbf{r}_T = \mathbf{r}_C$, for the small version of the NSW experiment with $I = 5$ pairs.

Label	Pair 1 Y_1	Pair 2 Y_2	Pair 3 Y_3	Pair 4 Y_4	Pair 5 Y_5	Wilcoxon's T
1	1456	3988	−45	−2147	3173	11
2	1456	3988	−45	−2147	−3173	7
3	1456	3988	−45	2147	3173	14
4	1456	3988	−45	2147	−3173	10
5	1456	3988	45	−2147	3173	12
6	1456	3988	45	−2147	−3173	8
7	1456	3988	45	2147	3173	15
8	1456	3988	45	2147	−3173	11
9	1456	−3988	−45	−2147	3173	6
10	1456	−3988	−45	−2147	−3173	2
11	1456	−3988	−45	2147	3173	9
12	1456	−3988	−45	2147	−3173	5
13	1456	−3988	45	−2147	3173	7
14	1456	−3988	45	−2147	−3173	3
15	1456	−3988	45	2147	3173	10
16	1456	−3988	45	2147	−3173	6
17	−1456	3988	−45	−2147	3173	9
18	−1456	3988	−45	−2147	−3173	5
19	−1456	3988	−45	2147	3173	12
20	−1456	3988	−45	2147	−3173	8
21	−1456	3988	45	−2147	3173	10
22	−1456	3988	45	−2147	−3173	6
23	−1456	3988	45	2147	3173	13
24	−1456	3988	45	2147	−3173	9
25	−1456	−3988	−45	−2147	3173	4
26	−1456	−3988	−45	−2147	−3173	0
27	−1456	−3988	−45	2147	3173	7
28	−1456	−3988	−45	2147	−3173	3
29	−1456	−3988	45	−2147	3173	5
30	−1456	−3988	45	−2147	−3173	1
31	−1456	−3988	45	2147	3173	8
32	−1456	−3988	45	2147	−3173	4
All 1–32	\multicolumn{5}{c} Absolute Differences		
	1456	3988	45	2147	3173	
All 1–32	\multicolumn{5}{c} Ranks of Absolute Differences		
	2	5	1	3	4	

fixed (i.e., determined by $\mathscr{F}$) as the treatment assignment $\mathbf{Z}$ varies, so Wilcoxon's T is changing only as the I signs change, and the I signs are independent. In a formula, under the null hypothesis of no effect, the signed rank statistic is $T = \sum_{i=1}^{I} \text{sgn}\{(Z_{i1} - Z_{i2})(r_{Ci1} - r_{Ci2})\} \cdot q_i$ where $\text{sgn}(a) = 1$ if $a > 0$, $\text{sgn}(a) = 0$ if $a \leq 0$, and q_i is the rank of $|r_{Ci1} - r_{Ci2}|$. The expectation $E(T|\mathscr{F}, \mathscr{Z})$ and variance $\text{var}(T|\mathscr{F}, \mathscr{Z})$ of the null distribution in Table 2.8 have simple formulas. Write $s_i = 1$ if $|Y_i| > 0$ and $s_i = 0$ if $|Y_i| = 0$. If $H_0: \mathbf{r}_T = \mathbf{r}_C$ were true in a randomized experiment, then $E(T|\mathscr{F}, \mathscr{Z}) = (1/2) \sum_{i=1}^{I} s_i q_i$ and $\text{var}(T|\mathscr{F}, \mathscr{Z}) =$

Table 2.8 The randomization distribution of Wilcoxon's signed rank statistic, T, for $I = 5$ matched pair differences in 1978 earnings under the null hypothesis of no treatment effect in the NSW experiment. Count is the number of treatment assignments that yield $T = t$; for instance, there are two ways to obtain $T = 12$. Because the observed value of T is 11, the one-sided P-value is 0.21875 and the two-sided P-value is twice that, or 0.4375. With $I = 5$ pairs, the difference would be significant at the conventional 0.05 level in a one-sided test only if all five differences were positive, in which case $T = 15$ and the one-sided P-value is 0.03125. Notice the that null distribution of T is symmetric about its expectation, $I(I+1)/4 = 5(5+1)/4 = 7.5$.

t	Count	$\Pr(T=t \mid \mathscr{F}, \mathscr{Z})$	$\Pr(T \geq t \mid \mathscr{F}, \mathscr{Z})$
15	1	0.03125	0.03125
14	1	0.03125	0.06250
13	1	0.03125	0.09375
12	2	0.06250	0.15625
11	2	0.06250	0.21875
10	3	0.09375	0.31250
9	3	0.09375	0.40625
8	3	0.09375	0.50000
7	3	0.09375	0.59375
6	3	0.09375	0.68750
5	3	0.09375	0.78125
4	2	0.06250	0.84375
3	2	0.06250	0.90625
2	1	0.03125	0.93750
1	1	0.03125	0.96875
0	1	0.03125	1.00000

$(1/4)\sum_{i=1}^{I}(s_i q_i)^2$. When there are no ties, these formulas simplify because the ranks are then $1, 2, \ldots, I$. If there are no ties, in the sense that $|Y_i| > 0$ for all i and all of the $|Y_i|$ are distinct, then the formulas simplify to $E(T \mid \mathscr{F}, \mathscr{Z}) = I(I+1)/4$ and $\mathrm{var}(T \mid \mathscr{F}, \mathscr{Z}) = I(I+1)(2I+1)/24$; see [32, §3.2] Moreover, under very mild conditions, as $I \to \infty$, the null distribution of $\{T - E(T \mid \mathscr{F}, \mathscr{Z})\}/\sqrt{\mathrm{var}(T \mid \mathscr{F}, \mathscr{Z})}$ converges in distribution to the standard Normal distribution [32].

Testing no effect with all $I = 185$ matched pairs

If Wilcoxon's test is applied to all $I = 185$ matched pairs in Figure 2.1, then the two-sided P-value for testing no effect is 0.009934. When all $I = 185$ pairs are considered, the difference in 1978 earnings in Figure 2.1 appears to be an effect caused by the intervention. The difference in 1978 earnings is too large and systematic to be due to chance, the flip of the coin that assigned one man to treatment, the next to control.[5]

---

[5] Performing the calculations in R is straightforward. The $I = 185$ paired differences in 1978 earnings are contained in the vector dif.
```
> length(dif)
[1] 185
> dif[1:6]
[1] 3889.711 -4733.929 18939.192 7506.146 -6862.342 -8841.886
```

2.3 Testing the Null Hypothesis of No Treatment Effect

The procedure used in Table 2.5 for the mean, $\overline{Y}$, and in Table 2.7 for Wilcoxon's statistic, T, are entirely parallel and work in a similar way for any test statistic.[6]

The null distribution of Wilcoxon's T in Table 2.8 has a curious property not shared by the null distribution of the mean difference, $\overline{Y}$, in Table 2.6. The null distribution of $\overline{Y}$ in Table 2.6 depends on the numeric values of Y_i, so this distribution could not have been written down before the experiment was conducted. You need to have the Y_i's to construct Table 2.6. In contrast, the distribution of T in Table 2.8 does not depend upon the differences in earnings, the Y_i's, provided the $|Y_i|$ are nonzero and different (or untied) so that they can be ranked 1, 2, 3, 4, 5 and each Y_i is either positive or negative. Provided the $I = 5$ earnings differences are untied, the null distribution in Table 2.8 could be written down before the experiment is conducted, and indeed the distribution in Table 2.8 appears at the back of many textbooks.

Ties and distribution-free statistics

What happens to Wilcoxon's T when some $|Y_i|$ are tied? If several $|Y_i|$ are equal, it does not seem reasonable to assign them different ranks, so instead they are assigned the average of the ranks they would have had if they had differed ever so slightly. If $|Y_4| = |Y_5|$ and they would have had ranks 3 and 4 had they not been tied, then they are both assigned rank 3.5. Under the null hypothesis of no effect, if $|Y_i| = 0$ then $Y_i = 0$ no matter how treatments are assigned, and pair i contributes 0 to Wilcoxon's T for every treatment assignment. With ties, the null distribution of T is determined as has been done twice before in Tables 2.5 and 2.7; however, the null distribution now depends not only on the sample size I but also on the specific pattern of ties, so the distribution is not known in advance of the experiment.

The phrase 'distribution-free statistic' is used in a variety of related but slightly distinct senses. In one of these senses, both the sample mean, $\overline{Y}$, and Wilcoxon's signed rank statistic, T, always have randomization distributions, as computed in Tables 2.5 and 2.7, and if there are no ties among the $|Y_i|$, then T is distribution-free, in the sense that its null distribution does not depend upon the values of the Y_i.

```
> wilcox.test(dif)
  Wilcoxon signed rank test with continuity correction
data: dif
V = 9025, p-value = 0.009934
alternative hypothesis: true location is not equal to 0
```

Now, with $I = 185$ pairs, the statistic is $T = 9025$ and the two-sided P-value is 0.009934. In computing this P-value, R has taken appropriate account of ties of both kinds and has made use of a Normal approximation to the null distribution of T. It has also employed a 'continuity correction' intended to improve the accuracy of approximate P-values obtained from the Normal distribution.

[6] This statement is true conceptually. That is, the statement would be true if we had an infinitely fast computer. As a practical matter, producing a table like Table 2.5 for large I is quite tedious, even for a computer, so randomization distributions are obtained either with clever algorithms [37] or with the aid of large sample approximations [32]. It is not necessary to focus on these technical details in a chapter whose focus is on concepts. In understanding concepts, we can pretend for the moment that we have an infinitely fast computer.

2.4 Testing Other Hypotheses; Confidence Intervals; Point Estimates

2.4.1 Testing a constant, additive treatment effect

Section 2.3 tested Fisher's [13] sharp null hypothesis of no treatment effect, which asserts that each subject would exhibit the same response whether assigned to treatment or control, that is, $H_0 : r_{Tij} = r_{Cij}$ for $i = 1,\ldots,I$, $j = 1,2$, or equivalently $H_0 : \mathbf{r}_T = \mathbf{r}_C$. This test for no effect quickly provides the basis for testing any hypothesis that specifies an effect, and from this to confidence intervals.

The simplest case is considered first, namely a constant, additive treatment effect, τ, in which $r_{Tij} = r_{Cij} + \tau$ for $i = 1,\ldots,I$, $j = 1,2$, or equivalently $\mathbf{r}_T = \mathbf{r}_C + \mathbf{1}\tau$, where $\mathbf{1} = (1,1,\ldots,1)^T$. In this case, the observed response, $R_{ij} = Z_{ij}r_{Tij} + (1-Z_{ij})r_{Cij}$ from the jth subject in pair i becomes $R_{ij} = Z_{ij}(r_{Cij} + \tau) + (1-Z_{ij})r_{Cij} = r_{Cij} + \tau Z_{ij}$, that is, the observed response, R_{ij}, equals the response to control, r_{Cij}, plus the treatment effect τ if ij is assigned to treatment, $Z_{ij} = 1$. Of course, R_{ij} and Z_{ij} are observed, but r_{Cij} and τ are not.

The hypothesis of a constant, additive effect is not strictly plausible for any $\tau \neq 0$ in the NSW study in §2.1. For instance, in Table 2.2, the $j = $ 1st man in the $i = $ 3rd pair was treated but earned $0.00 in 1978. Because earnings are positive, the only positive constant effect τ that can yield $0.00 for a treated man is $\tau = 0$. Indeed, in Figure 2.1, a quarter of the controls had $0.00 of earnings in 1978. Moreover, even if negative earnings were possible, the boxplots in Figure 2.1 for all 185 pairs of men strongly suggest the treatment effect is not an additive constant. In a very large randomized experiment with an additive, constant effect τ, the two boxplots of responses, R_{ij}, on the left in Figure 2.1 would look the same, except that the treated boxplot would be shifted up by τ, and the boxplot of pair differences, Y_i, on the right in Figure 2.1 would be symmetric about τ. In fact, neither of these two conditions appears to be true in Figure 2.1. After illustrating the most commonly used methods for an additive effect, three alternative approaches will be considered for the NSW study: a multiplicative effect, a truncated additive effect similar to that found in a Tobit model, and 'attributable effects' of various kinds. As will be seen, although the hypothesized effects are different, the method of inference is much the same.

Suppose that we hypothesized that the additive effect τ is some particular number, τ_0. For instance, in §2.1, a rather extreme hypothesis says that the NSW program increases earnings by $5000, or $H_0 : \tau = 5000$, or $H_0 : \tau = \tau_0$ with $\tau_0 = 5000$. It will soon be seen that this hypothesis is quite implausible. If this hypothesis were true, the observed response would be $R_{ij} = r_{Cij} + \tau_0 Z_{ij} = r_{Cij} + 5000 Z_{ij}$, so r_{Cij} could be computed from the observed (R_{ij}, Z_{ij}) together with the supposedly true hypothesis, $H_0 : \tau = 5000$, as $r_{Cij} = R_{ij} - \tau_0 Z_{ij} = R_{ij} - 5000 Z_{ij}$. Moreover, the matched treated-minus-control difference in observed earnings is $Y_i = (Z_{i1} - Z_{i2})(R_{i1} - R_{i2})$, so if the hypothesis $H_0 : \tau = \tau_0$ were true,

$$Y_i = (Z_{i1} - Z_{i2})(R_{i1} - R_{i2})$$

2.4 Testing Other Hypotheses; Confidence Intervals; Point Estimates

Table 2.9 Five matched pair differences from the NSW experiment adjusted for the hypothesis $H_0 : \tau = 5000$.

i	1	2	3	4	5
Y_i	1456	3988	−45	−2147	3173
$\tau_0 = 5000$	5000	5000	5000	5000	5000
$Y_i - \tau_0$	−3544	−1012	−5045	−7147	−1827

$$= (Z_{i1} - Z_{i2}) \{ (r_{Ci1} + \tau_0 Z_{i1}) - (r_{Ci2} + \tau_0 Z_{i2}) \}$$
$$= \tau_0 + (Z_{i1} - Z_{i2})(r_{Ci1} - r_{Ci2}),$$

and $Y_i - \tau_0 = (Z_{i1} - Z_{i2})(r_{Ci1} - r_{Ci2})$. In other words, if $H_0 : \tau = \tau_0$ were true with $\tau_0 = 5000$, then the observed difference in earnings, Y_i, in pair i, less $\tau_0 = 5000$, would equal $\pm (r_{Ci1} - r_{Ci2})$ depending upon the random treatment assignment $Z_{i1} - Z_{i2}$ in pair i. In particular, if $H_0 : \tau = 5000$ were true, then the sign of $Y_i - 5000$ would be determined by a coin flip, and we would expect about half of the $Y_i - 5000$ to be positive, half to be negative. If $H_0 : \tau = 5000$ were true for the five pairs in Table 2.2, then, for pair $i = 1$, the adjusted difference is $r_{C11} - r_{C12} = Y_1 - 5000 = 1456 - 5000 = -3544$. The five adjusted differences are listed in Table 2.9. Even with just the five pairs in Table 2.9, the hypothesis $H_0 : \tau = 5000$ does not look plausible: all five adjusted differences are negative, whereas if the hypothesis were true, the signs would be ± 1 independently with probability $\frac{1}{2}$. Only one of the 32 treatment assignments $\mathbf{z} \in \mathscr{Z}$ would give negative signs to all five adjusted responses, and that one assignment has probability $1/32 = .03125$ in a randomized experiment. If $H_0 : \tau = 5000$ were true, to produce Table 2.9 would require quite a bit of bad luck when $\mathbf{Z}$ is picked at random from $\mathscr{Z}$.

Table 2.10 displays all 32 treatment assignments, $\mathbf{z} \in \mathscr{Z}$, together with all possible adjusted responses, $Y_i - \tau_0 = (Z_{i1} - Z_{i2})(r_{Ci1} - r_{Ci2})$, when $H_0 : \tau = 5000$ is true, together with Wilcoxon's signed rank statistic T computed from these adjusted responses. In the first row, for the actual treatment assignment, $T = 0$ because all five differences are negative. From Table 2.10, $\Pr(T \leq 0 | \mathscr{F}, \mathscr{Z}) = 1/32 = .03125$ if $H_0 : \tau = 5000$ is true, and the two-sided P-value is $\Pr(T \leq 0 | \mathscr{F}, \mathscr{Z}) + \Pr(T \geq 15 | \mathscr{F}, \mathscr{Z}) = 2/32 = .0625$.

2.4.2 Confidence intervals for a constant, additive effect

If the treatment has an additive effect, $r_{Tij} = r_{Cij} + \tau$ for $i = 1, \ldots, I$, $j = 1, 2$, then a 95% confidence set for the additive treatment effect, τ, is formed by testing each hypothesis $H_0 : \tau = \tau_0$ and retaining for the confidence set the values of τ_0 not rejected at the 5% level. In general, a $1 - \alpha$ confidence set is the set of hypothesized values of a parameter not rejected by a level α test [31, §3.5]. If the parameter is a number, such as τ, then a $1 - \alpha$ confidence interval is the shortest interval containing the $1 - \alpha$ confidence set, and this interval may be a half-line or the entire line. For

Table 2.10 Using Wilcoxon's signed rank statistic T to test the null hypothesis that the training program increased wages by an additive constant of \$5000, that is, $H_0 : r_{Tij} = r_{Cij} + 5000$ for all i, j.

Label	Pair 1 Z_{11}	Pair 2 Z_{21}	Pair 3 Z_{31}	Pair 4 Z_{41}	Pair 5 Z_{51}	Pair 1 $Y_1 - \tau_0$	Pair 2 $Y_2 - \tau_0$	Pair 3 $Y_3 - \tau_0$	Pair 4 $Y_4 - \tau_0$	Pair 5 $Y_5 - \tau_0$	T
1	1	1	1	1	1	−3544	−1012	−5045	−7147	−1827	0
2	1	1	1	1	0	−3544	−1012	−5045	−7147	1827	2
3	1	1	1	0	1	−3544	−1012	−5045	7147	−1827	5
4	1	1	1	0	0	−3544	−1012	−5045	7147	1827	7
5	1	1	0	1	1	−3544	−1012	5045	−7147	−1827	4
6	1	1	0	1	0	−3544	−1012	5045	−7147	1827	6
7	1	1	0	0	1	−3544	−1012	5045	7147	−1827	9
8	1	1	0	0	0	−3544	−1012	5045	7147	1827	11
9	1	0	1	1	1	−3544	1012	−5045	−7147	−1827	1
10	1	0	1	1	0	−3544	1012	−5045	−7147	1827	3
11	1	0	1	0	1	−3544	1012	−5045	7147	−1827	6
12	1	0	1	0	0	−3544	1012	−5045	7147	1827	8
13	1	0	0	1	1	−3544	1012	5045	−7147	−1827	5
14	1	0	0	1	0	−3544	1012	5045	−7147	1827	7
15	1	0	0	0	1	−3544	1012	5045	7147	−1827	10
16	1	0	0	0	0	−3544	1012	5045	7147	1827	12
17	0	1	1	1	1	3544	−1012	−5045	−7147	−1827	3
18	0	1	1	1	0	3544	−1012	−5045	−7147	1827	5
19	0	1	1	0	1	3544	−1012	−5045	7147	−1827	8
20	0	1	1	0	0	3544	−1012	−5045	7147	1827	10
21	0	1	0	1	1	3544	−1012	5045	−7147	−1827	7
22	0	1	0	1	0	3544	−1012	5045	−7147	1827	9
23	0	1	0	0	1	3544	−1012	5045	7147	−1827	12
24	0	1	0	0	0	3544	−1012	5045	7147	1827	14
25	0	0	1	1	1	3544	1012	−5045	−7147	−1827	4
26	0	0	1	1	0	3544	1012	−5045	−7147	1827	6
27	0	0	1	0	1	3544	1012	−5045	7147	−1827	9
28	0	0	1	0	0	3544	1012	−5045	7147	1827	11
29	0	0	0	1	1	3544	1012	5045	−7147	−1827	8
30	0	0	0	1	0	3544	1012	5045	−7147	1827	10
31	0	0	0	0	1	3544	1012	5045	7147	−1827	13
32	0	0	0	0	0	3544	1012	5045	7147	1827	15

many simple statistics, including Wilcoxon's signed rank statistic, the confidence set and the confidence interval for τ are the same [32, Chapter 4].

The five pairs in Table 2.10 cannot reject at the 5% level any two-sided hypothesis using Wilcoxon's statistic, T, because the smallest two-sided significance level is $\Pr(T \leq 0| \mathscr{F}, \mathscr{Z}) + \Pr(T \geq 15| \mathscr{F}, \mathscr{Z}) = 2/32 = .0625$. However, a $1 - .0625 = 93.75\%$ confidence interval for τ may be constructed. The endpoints of the confidence interval are determined in Table 2.11, where Wilcoxon's signed rank statistic, T, rejects hypotheses below $H_0 : \tau = -2147$ and above $H_0 : \tau = 3988$ in a two-sided, 0.0625 level test, and the 93.75% confidence interval for τ is $[-2147, 3988]$.

2.4 Testing Other Hypotheses; Confidence Intervals; Point Estimates

Table 2.11 The endpoints of the 93.75% confidence interval for an additive treatment effect τ in five pairs from the NSW experiment. Wilcoxon's signed rank statistic T takes steps at -2147 and 3988, which define the endpoints of the two-sided confidence interval, $[-2147, 3988]$.

τ_0	-2147.0001	-2147	3987.9999	3988
T	15	14	1	0
P-value	.0625	0.1250	0.1250	.0625

With the full sample of $I = 185$ pairs, the 95% confidence interval for τ is $[391, 2893]$ or between \$391 and \$2893, a long interval but one that excludes zero treatment effect.[7]

2.4.3 Hodges-Lehmann point estimates of effect

In §2.4.1, the hypothesis of an additive treatment effect, $H_0 : r_{Tij} = r_{Cij} + \tau_0$ for $i = 1, \ldots, I$, $j = 1, 2$, was tested by applying Wilcoxon's signed rank statistic to the adjusted pair differences, $Y_i - \tau_0$, which equal $(Z_{i1} - Z_{i2})(r_{Ci1} - r_{Ci2})$ if H_0 is true. In §2.4.2, a $1 - \alpha$ confidence interval for an additive effect was formed by testing every value τ_0 and retaining for the interval all values not rejected by the Wilcoxon test at level α. The same logic was used by Joseph Hodges and Erich Lehmann [21, 32] to create a point estimate. They asked: What value τ_0 when subtracted from Y_i makes Wilcoxon's statistic T computed from $Y_i - \tau_0$ behave the way we expect Wilcoxon's statistic to behave with a true hypothesis? As noted in §2.3.3, when Wilcoxon's signed rank statistic T is computed from $(Z_{i1} - Z_{i2})(r_{Ci1} - r_{Ci2})$, it has expectation $E(T | \mathscr{F}, \mathscr{Z}) = I(I+1)/4$ provided $|r_{Ci1} - r_{Ci2}| > 0$ for all i. Intuitively, one finds a value, $\widehat{\tau}$, such that Wilcoxon's statistic T equals $I(I+1)/4$ when T is computed from $Y_i - \widehat{\tau}$. The intuition does not quite work, because the

---

[7] Continuing the discussion from Note 5, calculating the confidence interval in R is straightforward: the call to `wilcox.test(.)` includes the option `conf.int=T`.
```
>wilcox.test(dif,conf.int=T)
Wilcoxon signed rank test with continuity correction
data: dif
V = 9025, p-value = 0.009934
alternative hypothesis: true location is not equal to 0
95 percent confidence interval:
 391.359 2893.225
sample estimates:
(pseudo)median
      1639.383
```
In addition to the confidence interval, R supplies the Hodges-Lehmann point estimate 1639.383 of τ, as discussed in §2.4.3. In the R output, the Hodges-Lehmann estimate is labeled a (pseudo) median.

In calculating this confidence interval, R makes use of various computational short-cuts. Because these computational shortcuts do not change the confidence interval, because they make the subject seem more complex than it is, and because they do not generalize to other situations, I do not discuss such shortcuts in this book.

Wilcoxon statistic is discrete, taking on a finite number of values, so it may never equal $I(I+1)/4$ or it may equal $I(I+1)/4$ for an interval of values of τ_0. So Hodges and Lehmann added two very small patches. If the signed rank statistic never equals $I(I+1)/4$, then there is a unique value of τ_0 where T passes $I(I+1)/4$, and that unique value is the Hodges-Lehmann estimate, $\hat{\tau}$. If the signed rank statistic equals $I(I+1)/4$ for an interval of values of τ_0, then the midpoint of that interval is the Hodges-Lehmann estimate, $\hat{\tau}$. In Table 2.2, $I = 5$, so $I(I+1)/4 = 5(5+1)/4 = 7.5$. Wilcoxon's T is $T = 8$ if computed from $Y_i - 1455.9999$, but it is $T = 7$ if computed from $Y_i - 1456.0001$, so $\hat{\tau} = \$1456$. For all $I = 185$ pairs, the point estimate[8] of the increase in earnings is $\hat{\tau} = \$1639$.

Although $\hat{\tau}$ is one of the two well-known estimates that are called 'Hodges-Lehmann (HL) estimates,' Hodges and Lehmann [21] had actually proposed a general method for constructing estimates from tests of no treatment effect. The test statistic need not be the Wilcoxon statistic, and the parameter need not be an additive treatment effect. One equates a test statistic to its null expectation and solves the resulting equation for the point estimate. The method is applicable with other statistics. In a mildly dull way, if you take the sample mean of $Y_i - \tau_0$ as a test statistic, as in §2.3.2, then the mean of $Y_i - \tau_0$ has expectation zero when H_0 is true, so the estimating equation is $(1/I)\sum(Y_i - \hat{\tau}) = 0$ with solution $\hat{\tau} = \overline{Y}$; that is, the Hodges-Lehmann estimate derived from the mean is the mean of the Y_i. Almost, but not quite, the same thing happens with the randomization distribution of m-tests in §2.9. In §2.4.5 there are Hodges-Lehmann estimates of a multiplicative effect [2] and a Tobit effect.

2.4.4 Testing general hypotheses about treatment effects

Any hypothesis that specifies the $2I$ treatment effects, $r_{Tij} - r_{Cij}$, may be tested using the same reasoning as in §2.4.1. Let τ_{0ij} be $2I$ specific numbers, and let $\theta_0 = (\tau_{011}, \tau_{012}, \ldots, \tau_{0I2})^T$ collect them into a vector of dimension $2I$. Consider the hypothesis $H_0 : r_{Tij} = r_{Cij} + \tau_{0ij}$ for $i = 1, \ldots, I$ and $j = 1, 2$, or equivalently $H_0 : \mathbf{r}_T = \mathbf{r}_C + \theta_0$. Any specific hypothesis about effects may be expressed in this way.

For instance, a basic hypothesis of this form entails interaction between the treatment and a covariate, or so-called 'effect-modification.' In Table 2.1, one might entertain a hypothesis in which there are two values of τ_{0ij} depending upon whether the jth man in the ith pair has or does not have a high school degree. The hypothesis might assert that $\tau_{0ij} = \$1000$ if ij has a high school degree and $\tau_{0ij} = \$2000$ if ij does not have such a degree, so the program has a larger benefit for men without a degree.

The logic of §2.4.1 barely changes. If $H_0 : \mathbf{r}_T = \mathbf{r}_C + \theta_0$ were true, the observed response $R_{ij} = Z_{ij} r_{Tij} + (1 - Z_{ij}) r_{Cij}$ from the jth person in pair i would be $R_{ij} =$

---

[8] To compute the Hodges-Lehmann estimate in R, see Note 7.

2.4 Testing Other Hypotheses; Confidence Intervals; Point Estimates

$r_{Cij} + Z_{ij}\tau_{0ij}$, so that the observable data, (R_{ij}, Z_{ij}), together with the hypothesis permit the computation of $r_{Cij} = R_{ij} - Z_{ij}\tau_{0ij}$, in parallel with §2.4.1. Indeed, using the fact that $Z_{i2} = 1 - Z_{i1}$, if $H_0 : \mathbf{r}_T = \mathbf{r}_C + \theta_0$ were true, the treated-minus-control matched-pair difference in responses is

$$Y_i = (Z_{i1} - Z_{i2})(R_{i1} - R_{i2})$$
$$= (Z_{i1} - Z_{i2})\{(r_{Ci1} + \tau_{0i1}Z_{i1}) - (r_{Ci2} + \tau_{0i2}Z_{i2})\}$$
$$= (Z_{i1} - Z_{i2})(r_{Ci1} - r_{Ci2}) + (\tau_{0i1}Z_{i1} + \tau_{0i2}Z_{i2})$$

so the adjusted difference

$$Y_i - (\tau_{0i1}Z_{i1} + \tau_{0i2}Z_{i2}) = (Z_{i1} - Z_{i2})(r_{Ci1} - r_{Ci2}) \tag{2.2}$$

is $\pm(r_{Ci1} - r_{Ci2})$ depending upon the random assignment of treatments $Z_{i1} - Z_{i2}$. The rest of the argument in §2.4.1 is as before, with a statistic, such as Wilcoxon's signed rank statistic, T, computed from the the adjusted differences (2.2), and compared to its randomization distribution formed from all 2^I assignments of signs to $(Z_{i1} - Z_{i2})(r_{Ci1} - r_{Ci2})$.

In abstract principle, a 95% confidence set for $\theta = (r_{T11} - r_{C11}, \ldots, r_{TI2} - r_{CI2})^T$ could be constructed by testing every hypothesis $H_0 : \mathbf{r}_T = \mathbf{r}_C + \theta_0$ and retaining for the confidence set the values of θ_0 that are not rejected by some 5% level test. In practice, such a confidence set would not be intelligible, for it would be a subset of $2I$-dimensional space. In the NSW data in §2.1, the confidence set would be a subset of $2I = 2 \times 185 = 370$ dimensional space and would be beyond human comprehension. In brief, it is straightforward to make valid statistical inferences that are so complex, so faithful to the minute detail of reality, that they are unintelligible and of no practical use whatsoever. The 1-dimensional confidence interval for the single (scalar) constant effect τ in §2.4.2 provides insight into the $2I$-dimensional confidence set for θ because the constant model $\mathbf{r}_T = \mathbf{r}_C + \mathbf{1}\tau$ defines a line in $2I$-dimensional space as τ varies over the real numbers. If all the tests were performed using the same test statistic, say Wilcoxon's signed rank statistic T, then τ_0 is excluded from the 1-dimensional confidence interval for an additive effect if and only if $\theta_0 = (\tau_0, \tau_0, \ldots, \tau_0)^T$ is excluded from the $2I$-dimensional confidence set for θ; after all, it is the same hypothesis tested by the same test. In this sense, a 1-dimensional model for the $2I$ dimensional effect, such as the constant effect model, may be understood as an attempt to glean insight into the $2I$ dimensional confidence set for θ while recognizing that any 1-dimensional model, indeed any intelligible model, is to some degree an oversimplification. Understanding of θ is often aided by contrasting several intelligible models, rather than discarding them. Two other 1-dimensional effects are considered in §2.4.5. Arguably, the joint consideration of three 1-dimensional models for the $2I$-dimensional parameter θ provides more humanly accessible insight into θ than would a $2I$-dimensional confidence set.

An alternative approach to understanding the $2I$-dimensional confidence set for θ uses 'attributable effects.' In typical practice, an attributable effect provides a 1-

dimensional summary statement about the $2I$-dimensional θ, but this summary does not determine θ uniquely. Attributable effects are discussed in §2.5.

2.4.5 Multiplicative effects; Tobit effects

A constant multiplicative effect β

As noted in §2.4.1, the hypothesis of a positive, constant additive effect, $H_0 : r_{Tij} = r_{Cij} + \tau$ for $i = 1,\ldots,I$, $j = 1,2$, has the disagreeable property that it implies that a man with zero earnings under treatment would have negative earnings under control. In this section, two other 1-dimensional families of hypotheses for the $2I$-dimensional effect $\theta = (r_{T11} - r_{C11},\ldots,r_{TI2} - r_{CI2})^T$ are considered; they do not produce negative earnings.

The first hypothesis asserts that the effect is a constant multiplier rather than a constant addition, $H_0 : r_{Tij} = \beta_0 r_{Cij}$ for $i = 1,\ldots,I$, $j = 1,2$ with $\beta_0 \geq 0$. This hypothesis avoids the implication of negative earnings under one of the treatments. The multiplicative effect hypothesis is a special case of the hypothesis in §2.4.4, namely the hypothesis $H_0 : \mathbf{r}_T = \mathbf{r}_C + \theta_0$ with $\tau_{0ij} = r_{Tij} - r_{Cij} = \beta_0 r_{Cij} - r_{Cij} = (\beta_0 - 1) r_{Cij}$.

For an additive effect, the treatment effect is removed by subtraction, but for a multiplicative effect, the effect is removed by division [2]. If $H_0 : r_{Tij} = \beta_0 r_{Cij}$ were true for all ij, then $R_{ij}/\beta_0^{Z_{ij}} = r_{Cij}$; that is, if $Z_{ij} = 0$ then $R_{ij} = r_{Cij}$ and $\beta_0^{Z_{ij}} = 1$, while if $Z_{ij} = 1$, then $R_{ij} = r_{Tij}$, $\beta_0^{Z_{ij}} = \beta_0$ and $R_{ij}/\beta_0^{Z_{ij}} = r_{Tij}/\beta_0 = r_{Cij}$. The hypothesis $H_0 : r_{Tij} = \beta_0 r_{Cij}$ is tested by applying Wilcoxon's signed rank statistic to the I adjusted, treated-minus-control matched pair differences, $(Z_{i1} - Z_{i2})\left(R_{i1}/\beta_0^{Z_{i1}} - R_{i2}/\beta_0^{Z_{i2}}\right)$, which equal $(Z_{i1} - Z_{i2})(r_{Ci1} - r_{Ci2})$ if the hypothesis is true. The set of values of β_0 not rejected in a 5%-level test is a 95% confidence set for a constant multiplicative effect β.

In the NSW experiment with $I = 185$ pairs, the two-sided, 95% confidence interval for β is $[1.08, 1.95]$, or between an 8% increase and a 95% increase. The Hodges-Lehmann point estimate is $\widehat{\beta} = 1.45$.

Although a multiplicative effect avoids the implication of negative earnings, it has peculiarities of its own. If the effect is multiplicative, then a man who has $r_{Cij} = 0$ earnings under control also has $r_{Tij} = \beta r_{Cij} = 0$ earnings under treatment. Many controls did have zero earnings in 1978; see Figure 2.1. Among its several goals, the NSW program intended to bring into employment men who would otherwise be out of work, and a multiplicative effect says this cannot happen. This leads to the consideration of a 'Tobit effect.'

2.4 Testing Other Hypotheses; Confidence Intervals; Point Estimates 47

Tobit effects

Two men with the same positive earnings, r_{Cij}, under control in 1978 have something in common: the labor market attached the same dollar value to their labor. Two men with zero earnings under control in 1978 have less in common: the labor market declined to attach a positive value to the labor they might supply. One might entertain the thought that two unemployed men are unequally unemployed: perhaps the market attaches negative but unequal values to the labor they might supply. Expressed in operational terms, the subsidy that an employer would require before employing the two unemployed men might be different because one man could supply labor of greater net value to the employer than the other. In a simple economic model, workers are paid the values of the marginal products of their labor, and a worker is unemployed if the marginal product of his labor is negative. (We have all known people like that.) 'Tobit effects,' named for James Tobin [60], attempt to express this idea.

Suppose the NSW program raises the marginal value of every worker's labor by the same constant, τ_0, but the worker is employed, with positive earnings, only if the marginal value of the labor is positive. In this case, if the jth man in the ith pair would have earnings of r_{Tij} in the NSW program, and if $r_{Tij} \geq \tau_0$, then $r_{Cij} = r_{Tij} - \tau_0$, but if $r_{Tij} < \tau_0$ then $r_{Cij} = 0$; that is, in general, $r_{Cij} = \max(r_{Tij} - \tau_0, 0)$. This hypothesis is another special case of the general hypothesis in §2.4.4.

If the hypothesis, $H_0 : r_{Cij} = \max(r_{Tij} - \tau_0, 0)$ for $i = 1, \ldots, I$, $j = 1, 2$, were true, then $\max(R_{ij} - \tau_0 Z_{ij}, 0) = r_{Cij}$. The hypothesis $r_{Cij} = \max(r_{Tij} - \tau_0, 0)$ may be tested by applying Wilcoxon's signed rank test to the I adjusted treated-minus-control differences, $(Z_{i1} - Z_{i2}) \{\max(R_{ij} - \tau_0 Z_{ij}, 0) - \max(R_{ij} - \tau_0 Z_{ij}, 0)\}$, which equal $(Z_{i1} - Z_{i2})(r_{Ci1} - r_{Ci2})$ if the hypothesis is true. The test is the same as before; see §2.4.4.

For the Tobit effect $r_{Cij} = \max(r_{Tij} - \tau, 0)$, the 95% confidence interval for τ is [\$458, \$3955], and the Hodges-Lehmann point estimate is $\widehat{\tau} = \$2114$.

Comparing the three effects: additive, multiplicative, tobit

Figure 2.2 compares an additive effect, $r_{Cij} = r_{Tij} - \tau$, a multiplicative effect, $r_{Cij} = r_{Tij}/\beta$, and a Tobit effect, $r_{Cij} = \max(r_{Tij} - \tau, 0)$. In each of the three panels of Figure 2.2, the 1978 earnings of the 185 treated men ($Z_{ij} = 1$) and the 185 control men ($Z_{ij} = 0$) are displayed after removing the estimated treatment effect from the earnings of the treated men. In each case, the Hodges-Lehmann estimate is used to estimate the treatment effect. For an additive effect, $R_{ij} - \widehat{\tau}Z_{ij}$ is plotted; for a multiplicative effect, $R_{ij}/\widehat{\beta}^{Z_{ij}}$ is plotted; for a Tobit effect, $\max(R_{ij} - \widehat{\tau}Z_{ij}, 0)$ is plotted. If the number I of pairs were large, then the pair of treated-control boxplots would look the same with an effect of the correct form. The boxplots for an additive effect suggest the effect is not, in fact, additive, in part because the boxplots look quite different, and in part because negative earnings are produced. The boxplots for the multiplicative effect look better, although the lower quartile for the treated

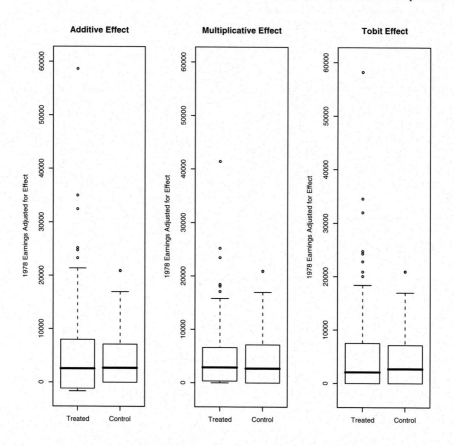

Fig. 2.2 Earnings in 1978 after adjustment to remove three possible treatment effects, an additive effect of $1639, a multiplicative effect of 1.45, and a Tobit effect of $2114. If the correct effect were removed, the two boxplots would look similar, differing only by chance.

group is positive while for the control group it is zero. The boxplots for the Tobit effect are better than for the additive effect, and the lower quartile in both groups is now zero. Although the Tobit effect is, arguably, the best of the three, it misses the long right tail in the treated group: a handful of treated men earned much more than is accounted for by the Tobit effect. This was clearly visible in Figure 2.1 as well. More will be said about the tail in §2.5.

2.5 Attributable Effects

Why use attributable effects?

As seen in §2.4.4, it is straightforward to test a hypothesis that specifies all $2I$ treatment effects, $r_{Tij} - r_{Cij}$, say the hypothesis $H_0 : r_{Tij} = r_{Cij} + \tau_{0ij}$ for $i = 1,\ldots,I$ and $j = 1, 2$, or equivalently the hypothesis $H_0 : \mathbf{r}_T = \mathbf{r}_C + \boldsymbol{\theta}_0$ with $\boldsymbol{\theta}_0 = (\tau_{011}, \tau_{012}, \ldots, \tau_{0I2})^T$. We were barred from using such a test to produce a confidence set for $\boldsymbol{\theta} = (r_{T11} - r_{C11}, \ldots, r_{TI2} - r_{CI2})^T$ in part by our inability to make practical sense of a confidence set that is a subset of $2I$-dimensional space. In §2.4.5, some insight into $\boldsymbol{\theta}$ was gleaned by contrasting three 1-dimensional effects, that is, three paths or curves through the $2I$-dimensional space of possible effects. By contrast, 'attributable effects' make summary statements about $\boldsymbol{\theta}$ without restricting the form of $\boldsymbol{\theta}$ in any way. Attributable effects are a general concept [42, 43] but, in the current section, only the case of matched pairs is considered using Wilcoxon's signed rank statistic [46] and an extension of it by W. Robert Stephenson [59, 49]. The discussion is simplified by assuming that there are no ties of any kind and discussing ties separately at the end.

Aligned responses: Shifting attention from pairs to individuals

If no simplifying form is imposed upon $\boldsymbol{\theta} = (r_{T11} - r_{C11}, \ldots, r_{TI2} - r_{CI2})$, then it is easiest to understand a treatment effect on one man, $r_{Tij} - r_{Cij}$, rather than on the matched pair difference between two men. In simple cases, such as an additive effect, the difference of two matched men has a simple relationship with the separate responses of the two individual men (§2.4.1), but in general this is not so. At the same time, the men were carefully matched for important characteristics, such as education, age and pretreatment earnings, and it seems wise to retain the comparison of two similar men. These two considerations lead to 'aligned responses,' introduced by Hodges and Lehmann [20] in an extension of Wilcoxon's signed rank statistic. With matched pairs, the aligned response for the jth man in the ith pair is the difference between his observed response, R_{ij}, and the average response in his pair, $\overline{R}_i = (R_{i1} + R_{i2})/2$, so the aligned response is $R_{ij} - \overline{R}_i$. Talk about aligned responses is, in a sense, talk about one man while taking account of the pairing. Of course, the treated-minus-control pair difference of aligned responses $R_{ij} - \overline{R}_i$ equals the treated-minus-control difference in the responses R_{ij} themselves because $\overline{R}_i$ appears twice and cancels. If Wilcoxon's signed rank statistic were computed from the aligned responses, it would equal Wilcoxon's signed rank statistic computed from the responses themselves. So far, the switch to aligned responses is a switch in the way we talk rather than a switch in what we calculate.[9]

---

[9] At this point, we part company with Hodges and Lehmann [20]. They were interested in creating a generalization of Wilcoxon's signed rank statistic so that the matched sets were no longer pairs and might, for example, include several controls. Their statistic, the aligned rank statistic, ranks

If the null hypothesis of no treatment effect, $H_0 : r_{Tij} = r_{Cij}$ for $i = 1,\ldots,I$ and $j = 1,2$, were true, then the pair of aligned responses in pair i would be

$$\left(r_{Ci1} - \frac{r_{Ci1} + r_{Ci2}}{2}, r_{Ci2} - \frac{r_{Ci1} + r_{Ci2}}{2} \right) \tag{2.3}$$

no matter how treatments were assigned. In contrast, if the treatment has an effect, then the pair of aligned responses in pair i would be

$$\left(r_{Ti1} - \frac{r_{Ti1} + r_{Ci2}}{2}, r_{Ci2} - \frac{r_{Ti1} + r_{Ci2}}{2} \right) \tag{2.4}$$

if the first man were the treated man, $Z_{i1} = 1$, $Z_{i2} = 0$, and the aligned responses would be

$$\left(r_{Ci1} - \frac{r_{Ci1} + r_{Ti2}}{2}, r_{Ti2} - \frac{r_{Ci1} + r_{Ti2}}{2} \right) \tag{2.5}$$

if the second man were the treated man, $Z_{i1} = 0$, $Z_{i2} = 1$, so in either case the pair of aligned responses is

$$\left(R_{i1} - \overline{R}_i, R_{i2} - \overline{R}_i \right) = Z_{i1} \left(r_{Ti1} - \frac{r_{Ti1} + r_{Ci2}}{2}, r_{Ci2} - \frac{r_{Ti1} + r_{Ci2}}{2} \right) \tag{2.6}$$

$$+ Z_{i2} \left(r_{Ci1} - \frac{r_{Ci1} + r_{Ti2}}{2}, r_{Ti2} - \frac{r_{Ci1} + r_{Ti2}}{2} \right). \tag{2.7}$$

For instance, it might happen that $r_{Ti1} - (r_{Ti1} + r_{Ci2})/2 = \$1000$ in (2.4) and $r_{Ti2} - (r_{Ci1} + r_{Ti2})/2 = \$5000$ in (2.5), so the effect in pair i is positive and substantial but far from constant; rather it depends upon which man is treated. Can Wilcoxon's statistic (and Stephenson's generalization) help to draw inferences about effects that vary in size from person to person?

Thinking about heterogeneous effects by focusing on small groups of pairs

To think about effects that are not constant, that are heterogeneous, it is natural to look at several pairs at once. If the effect varies, variations in effect have to part of the story. Let $\mathscr{I} \subseteq \{1,\ldots,I\}$ be a subset of m pairs of the I pairs. Wilcoxon's statistic looks at pairs $m = 2$ at a time, so $\mathscr{I} = \{i,i'\}$ for $i \neq i'$, $1 \leq i < i' \leq I$.

---

the aligned responses (not the absolute differences) and sums the ranks of treated subjects. For matched pairs, they would rank the $R_{ij} - \overline{R}_i$ from 1 to $2I$ and sum the I ranks for the I treated subjects. Hodges and Lehmann [20] show that, for matched pairs, Wilcoxon's signed rank statistic and the aligned rank statistic produce virtually the same inferences. Can you see why? In pairs, how are $R_{i1} - \overline{R}_i$ and $R_{i2} - \overline{R}_i$ related? Think about the pair with largest $R_{ij} - \overline{R}_i$, the pair that contains the aligned response that gets the largest aligned rank. How is that pair related to the pair that has the largest $|R_{i1} - R_{i2}|$, the one that has the largest rank in Wilcoxon's statistic?

2.5 Attributable Effects

Stephenson [59] suggested using other values of m, $1 \leq m \leq I$. For the moment, focus on one subset $\mathscr{I}$, perhaps $\mathscr{I} = \{1,2\}$; later, the statistic will average over all subsets.

Among the m pairs in $\mathscr{I}$, there are $2m$ men, or $2m = 4$ men for Wilcoxon's statistic. Which one man among these $2m$ men seems to have the most exceptionally positive earnings, in the sense that his observed aligned response, $R_{ij} - \overline{R}_i$, is the largest, the most positive, of the $2m$ aligned responses? That one man seems to be an interesting man, deserving special attention, because he entered or reentered the workforce with high earnings in 1978 relative to someone else of similar age, education and so forth. This man is the most interesting of $2m$ men. Now that we have focused on one interesting man, it is time to ask: was he a treated man? We ought to tally one success for the NSW treatment if this man — the man with the highest aligned earnings among the $2m$ men in $\mathscr{I}$ — is a treated man. So write $H_{\mathscr{I}} = 1$ if indeed he is a treated man, and $H_{\mathscr{I}} = 0$ if he is a control.

Even if the treatment had no effect, so that the aligned responses were always (2.3), the random assignment of treatments would grant a success, $H_{\mathscr{I}} = 1$, to the treatment with probability $\frac{1}{2}$. One of the two values in the pair (2.3) is positive, and even if the treatment is without effect, randomization picks the positive entry for treatment half the time. That is true for each pair, so it is true for the pair $i \in \mathscr{I}$ with the largest $r_{Cij} - (r_{Ci1} + r_{Ci2})/2$ in (2.3). Evidently, some work needs to be done to distinguish apparent successes that are merely due to luck of the random assignment and successes that were actually caused by the treatment.

To distinguish successes caused by treatment from successes that happen by luck, a new quantity, $H_{C\mathscr{I}}$, is introduced. Imagine, for a brief moment, that we could observe the response to control, r_{Cij}, for all $2I$ subjects, so that, in particular, we could calculate the pair of aligned control responses in (2.3). In point of fact, we observe only half of the r_{Cij} because half of the men receive treatment, so we observe (2.4) or (2.5) not (2.3). If we could calculate (2.3), we could calculate a quantity $H_{C\mathscr{I}}$ which is similar to $H_{\mathscr{I}}$, but is computed from the aligned control responses in (2.3). Specifically, with (2.3) in hand, we could find the man with the largest, most positive of the $2m$ aligned responses to control (2.3) for the m pairs $i \in \mathscr{I}$ and score $H_{C\mathscr{I}} = 1$ if that man was a treated man and $H_{C\mathscr{I}} = 0$ if that man was a control. Like $H_{\mathscr{I}}$, the quantity $H_{C\mathscr{I}}$ depends on the random assignment of treatments, $\mathbf{Z}$, so $H_{\mathscr{I}}$ and $H_{C\mathscr{I}}$ are random variables; however, $H_{\mathscr{I}}$ is observed but $H_{C\mathscr{I}}$ is not.

If $H_{\mathscr{I}} - H_{C\mathscr{I}} = 1 - 0 = 1$, then in the set $\mathscr{I}$ an effect of the treatment caused a success to be tallied that would not have been tallied if all $2m$ men had received the control. If $H_{\mathscr{I}} - H_{C\mathscr{I}} = 0 - 1 = -1$, then in the set $\mathscr{I}$ an effect of the treatment prevented a success from being tallied that would have been tallied if all $2m$ men had received the control. If $H_{\mathscr{I}} - H_{C\mathscr{I}} = 0$, then in the set $\mathscr{I}$ the effects of the treatment did not change whether a success was tallied. One might say, with some imprecision, that $H_{\mathscr{I}}$ gives credit for successes that occur by chance, but $H_{\mathscr{I}} - H_{C\mathscr{I}}$ does not. We cannot calculate quantities like $H_{\mathscr{I}} - H_{C\mathscr{I}}$ because we see $H_{\mathscr{I}}$ but not $H_{C\mathscr{I}}$; however, if we could calculate $H_{\mathscr{I}} - H_{C\mathscr{I}}$, we would know whether a success or failure was attributable to effects caused by the treatment. That we never see $H_{C\mathscr{I}}$ turns out to be less of a problem than one might suppose.

It seems arbitrary to focus on just the $2m$ men in $\mathscr{I}$ when there are data on $2I$ men. For Wilcoxon's statistic, with $m = 2$, this would mean focusing on $2m = 4$ men when there are $2I = 370$ men to consider. To drive out the arbitrary element of what otherwise seems to be a promising idea, it is natural to consider every subset $\mathscr{I}$ of m distinct pairs, to compute $H_\mathscr{I}$ for each one, and to tally up the total number of successes for the treatment in all of these comparisons. So let $\mathscr{K}$ be the collection of all subsets $\mathscr{I}$ of m distinct pairs. For Wilcoxon's statistic, with $m = 2$, the collection is $\mathscr{K}$ contains $\{1,2\}, \{1,3\}, \{2,3\}, \ldots, \{I-1,I\}$. In general, $\mathscr{K}$ contains $|\mathscr{K}| = I!/\{m!(I-m)!\}$ subsets, or 17,020 subsets of two pairs for the Wilcoxon statistic in the NSW experiment. The statistic $\widetilde{T} = \sum_{\mathscr{I} \in \mathscr{K}} H_\mathscr{I}$ tallies up the successes for the treatment in all possible comparisons of m distinct pairs.[10] In a parallel, the unobservable quantity $\widetilde{T}_C = \sum_{\mathscr{I} \in \mathscr{K}} H_{C\mathscr{I}}$ tallies the success for the treatment that would have occurred had all $2I$ men received the control. Finally, $A = \widetilde{T} - \widetilde{T}_C = \sum_{\mathscr{I} \in \mathscr{K}} H_\mathscr{I} - H_{C\mathscr{I}}$ tallies the number of successes attributable to effects of the treatment, that is, the net increase in successes due to effects of the treatment. If the treatment has no effect, then $H_\mathscr{I} = H_{C\mathscr{I}}$ for every $\mathscr{I} \in \mathscr{K}$ and $A = 0$. In general, A is an integer between $-|\mathscr{K}|$ and $|\mathscr{K}|$. It is often convenient to think of the proportions, $\widetilde{T}/|\mathscr{K}|$, $\widetilde{T}_C/|\mathscr{K}|$, and $A/|\mathscr{K}|$ rather than the counts.

Computations; null distribution

Computing $\widetilde{T}$ sounds as if it would be an enormous amount of work: even for $m = 2$, there are 17,020 terms $H_\mathscr{I}$ for the $I = 185$ pairs in the NSW experiment. Actually, it is easy. Stephenson [59] wrote $\widetilde{T}$ in the form of a signed rank statistic, similar to the form for Wilcoxon's T in §2.3.3, specifically $\widetilde{T} = \sum_{i=1}^{I} \text{sgn}\{(Z_{i1} - Z_{i2})(r_{Ci1} - r_{Ci2})\} \cdot \widetilde{q}_i$, where $\text{sgn}(a) = 1$ if $a > 0$, $\text{sgn}(a) = 0$ if $a \leq 0$, and $\widetilde{q}_i$ is a new quantity that will be defined in just a moment. With pairs, it is always true that $R_{i1} - \overline{R}_i = -R_{i2} + \overline{R}_i$, so $|R_{i1} - \overline{R}_i| = |R_{i2} - \overline{R}_i|$. Rank the pairs from 1 to I based on the value of $|R_{i1} - \overline{R}_i|$; therefore, pair i gets rank q_i where this rank q_i is the same as the rank q_i in §2.3.3 of $|Y_i|$ that appears in Wilcoxon's signed rank statistic. The pairs with ranks $q_i = 1$, $q_i = 2$, ..., $q_i = m-1$, never contain the largest observed aligned response $R_{ij} - \overline{R}_i$ in a subset of m pairs, so set $\widetilde{q}_i = 0$ for these pairs. The pair with rank $q_i = m$ contains the largest observed aligned response $R_{ij} - \overline{R}_i$ in exactly one subset of m pairs, so set $\widetilde{q}_i = 1$ for this pair. The pair with rank $q_i = m+1$ contains the largest observed aligned response $R_{ij} - \overline{R}_i$ in exactly m subsets $\mathscr{I}$ of m pairs, so set $\widetilde{q}_i = m$ for this pair. In general, the pair with rank q_i contains the largest aligned response in exactly $\widetilde{q}_i$ pairs, where

---

[10] A statistic built in this way is known as a U-statistic, a type of statistic invented by Wassily Hoeffding [22], and for $m = 2$, the statistic $\widetilde{T}$ is the U-statistic that most closely resembles Wilcoxon's signed rank statistic; see [32, Chapter 4] or [39, §3.5]. Wilcoxon's statistic T does not quite equal $\widetilde{T}$, but the behavior of the two statistics is virtually identical when I is large. If Wilcoxon had ranked the I pairs from 0 to $I-1$, he would have produced $\widetilde{T}$ rather than T [39, §3.5]. If the collection $\mathscr{K}$ were expanded to include the I self-pairs, $(1,1), (2,2), \ldots, (I,I)$, then $\widetilde{T}$ would equal T [32, Chapter 4].

2.5 Attributable Effects

$$\widetilde{q}_i = \binom{q_i - 1}{m - 1} \text{ for } q_i \geq m, \ \widetilde{q}_i = 0 \text{ for } q_i < m. \tag{2.8}$$

With $\widetilde{q}_i$ so defined,

$$\widetilde{T} = \sum_{\mathscr{J} \in \mathscr{K}} H_{\mathscr{J}}$$

$$= \sum_{i=1}^{I} \text{sgn}\{(Z_{i1} - Z_{i2})(r_{Ci1} - r_{Ci2})\} \cdot \widetilde{q}_i.$$

With $m = 2$ for Wilcoxon's statistic,

$$\binom{q_i - 1}{m - 1} = \binom{q_i - 1}{1} = q_i - 1, \tag{2.9}$$

so T and $\widetilde{T}$ are very similar, except the ranks in $\widetilde{T}$ are $0, 1, \ldots, I-1$ rather than $1, 2, \ldots, I$; see Note 10 and [39, §3.5]. With $m = 5$, the ranks $\widetilde{q}_i$ are $0, 0, 0, 0, 1, 5, 15, 35, 70, 126, \ldots$.

If the null hypothesis of no treatment effect, $H_0 : r_{Tij} = r_{Cij}$ for $i = 1, \ldots, I$ and $j = 1, 2$, were true, then the exact null distribution of $\widetilde{T}$ could be determined by direct enumeration, as was done for Wilcoxon's T in Tables 2.7 and 2.8. If H_0 is true, then $\widetilde{T}$ has the distribution of the sum of I independent random variables taking the values 0 and $\widetilde{q}_i$ each with probability $\frac{1}{2}$, for $i = 1, \ldots, I$. Also, the central limit theorem would provide a Normal approximation to the null distribution of $\widetilde{T}$ based on its null expectation and variance.

This null distribution of $\widetilde{T}$ shares with Wilcoxon's statistic the curious property noted in §2.3.3, namely that, in the absence of ties, the null distribution may be written down before the experiment is conducted, without seeing any of the responses. This turns out to be convenient. In particular, even though we cannot observe the $2I$ potential responses to control in (2.3), we do know that, by definition, they satisfy the null hypothesis of no treatment effect. In light of this, even though we cannot observe any of the $H_{C\mathscr{J}}$, we do know the distribution of

$$\widetilde{T}_C = \sum_{\mathscr{J} \in \mathscr{K}} H_{C\mathscr{J}}. \tag{2.10}$$

The distribution of the unobservable $\widetilde{T}_C$ is the known null distribution of $\widetilde{T}$. That is, even when the treatment has an effect, in the absence of ties, the distribution of $\widetilde{T}_C$ is available by direct enumeration with the ranks, $\widetilde{q}_i$; moreover,

$$E\left(\widetilde{T}_C \middle| \mathscr{F}, \mathscr{Z}\right) = (1/2) \sum_{i=1}^{I} \widetilde{q}_i, \tag{2.11}$$

$$\text{var}\left(\widetilde{T}_C \middle| \mathscr{F}, \mathscr{Z}\right) = (1/4) \sum_{i=1}^{I} \widetilde{q}_i^2 \tag{2.12}$$

and the distribution of $\left\{\tilde{T}_C - E\left(\tilde{T}_C \mid \mathscr{F}, \mathscr{Z}\right)\right\} / \sqrt{\text{var}\left(\tilde{T}_C \mid \mathscr{F}, \mathscr{Z}\right)}$ converges in distribution to the standard Normal distribution as $I \to \infty$ with m fixed. Using this known distribution of $\tilde{T}_C$, find the smallest value t_α such that $\Pr\left(\tilde{T}_C \leq t_\alpha \mid \mathscr{F}, \mathscr{Z}\right) \geq 1 - \alpha$. For large I, the critical value t_α is approximately

$$E\left(\tilde{T}_C \mid \mathscr{F}, \mathscr{Z}\right) + \Phi^{-1}(1-\alpha)\sqrt{\text{var}\left(\tilde{T}_C \mid \mathscr{F}, \mathscr{Z}\right)} \qquad (2.13)$$

where $\Phi^{-1}(\cdot)$ is the standard Normal quantile function, or

$$E\left(\tilde{T}_C \mid \mathscr{F}, \mathscr{Z}\right) + 1.65\sqrt{\text{var}\left(\tilde{T}_C \mid \mathscr{F}, \mathscr{Z}\right)} \qquad (2.14)$$

for $\alpha = 0.05$.

Our interest is in the number $A = \tilde{T} - \tilde{T}_C$ of successes attributable to effects of the treatment, where $\tilde{T}$ can be calculated from the data and $\tilde{T}_C$ is not observed. It is easy to see that $\Pr\left(A \geq \tilde{T} - t_\alpha \mid \mathscr{F}, \mathscr{Z}\right) \geq 1 - \alpha$, where $\tilde{T} - t_\alpha$ is something that can be calculated, so $A \geq \tilde{T} - t_\alpha$ may be asserted with $1 - \alpha$ confidence.[11] In terms of proportions of successes rather than counts, the confidence statement is $A/|\mathscr{K}| \geq \left(\tilde{T} - t_\alpha\right)/|\mathscr{K}|$.

Ties

Ties are a minor inconvenience. Among the $I = 185$ pairs in the NSW experiment, there are 14 ties among the $|Y_i|$, all with $|Y_i| = 0$. Briefly, ties make the procedure conservative, so $A \geq \tilde{T} - t_\alpha$ occurs with probability at least $1 - \alpha$ for large I. Specifics follow. Although the distribution of $\tilde{T}_C$ is known in the absence of ties, it is not known if ties occur, because the pattern of ties in the unobservable (2.3) affects the distribution. For the observable $\tilde{T}$, it is natural to use average ranks, averaging the $\tilde{q}_i$ for pairs with tied $|Y_i|$, and setting $\text{sgn}(a) = 1$ for $a > 0$, $\text{sgn}(a) = \frac{1}{2}$ for $a = 0$, $\text{sgn}(a) = 0$ for $a < 0$. Speaking informally, a tie for the highest $R_{i1} - \bar{R}_i$ for $i \in \mathscr{I}$ means one success is shared as a fractional success equally among the tied values. The same thing cannot be done for $\tilde{T}_C$ because the ties in (2.3) are not observed. Suppose that the unobserved $\tilde{T}_C$ allowed for ties in the manner proposed for the observed $\tilde{T}$, but the distribution used to determine t_α ignored ties, using instead the ranks $\tilde{q}_i$ in (2.8) for $q_i = 1, 2, \ldots, I$; then the expectation is correct, $E\left(\tilde{T}_C \mid \mathscr{F}, \mathscr{Z}\right) = (1/2)\sum_{i=1}^{I} \tilde{q}_i$ or $E\left(\tilde{T}_C/|\mathscr{K}| \mid \mathscr{F}, \mathscr{Z}\right) = \frac{1}{2}$,

---

[11] To see this, observe that: $\Pr\left(A \geq \tilde{T} - t_\alpha \mid \mathscr{F}, \mathscr{Z}\right) = \Pr\left(\tilde{T} - \tilde{T}_C \geq \tilde{T} - t_\alpha \mid \mathscr{F}, \mathscr{Z}\right) = \Pr\left(\tilde{T}_C \leq t_\alpha \mid \mathscr{F}, \mathscr{Z}\right) \geq 1 - \alpha$ by the definition of t_α. The assertion $A \geq \tilde{T} - t_\alpha$ is a confidence interval for an unobservable random variable. Confidence sets for random variables are less familiar than confidence sets for parameters, but they are hardly new; see, for instance, Weiss [61].

2.5 Attributable Effects

the variance is too large, $\text{var}\left(\widetilde{T}_C\middle|\mathscr{F},\mathscr{Z}\right) \leq (1/4)\sum_{i=1}^{I}\widetilde{q}_i^2$, the approximate $t_\alpha = E\left(\widetilde{T}_C\middle|\mathscr{F},\mathscr{Z}\right) + \Phi^{-1}(1-\alpha)\sqrt{\text{var}\left(\widetilde{T}_C\middle|\mathscr{F},\mathscr{Z}\right)}$ is too large for large I and $\alpha < \frac{1}{2}$, so $\Pr\left(\widetilde{T}_C \leq t_\alpha \middle| \mathscr{F},\mathscr{Z}\right) \geq 1-\alpha$, and $A \geq \widetilde{T} - t_\alpha$ may be asserted with at least $1-\alpha$ confidence. For $m=2$ in the NSW experiment, the ratio of the tied to untied variances for $\widetilde{T}$ is 0.9996 for $m=2$, and is closer to 1 for larger m.

Results for the NSW experiment: Big effects, now and then

Using the Wilcoxon signed rank statistic means looking at pairs $m=2$ at a time. For the $I = 185$ pairs, there are 17020 subsets of two pairs. In 60.8% of these 17020 subsets, the man with the highest aligned earnings $R_{i1} - \overline{R}_i$ was a treated man, whereas 50% is expected by chance, so the point estimate is a 10.8% increase attributable to effects of the treatment, but we are 95% confident of only a 3.8% increase.

More interesting is what happens in groups of $m = 20$ pairs or $2m = 40$ men. Here, there are 3.1×10^{26} comparisons $\mathscr{I}$ in $\mathscr{K}$. In 85.7% of these comparisons, the man with the highest aligned earnings $R_{i1} - \overline{R}_i$ was a treated man, whereas 50% were expected by chance, so the point estimate is a 35.7% increase attributable to effects of the treatment, but we are 95% confident of only a 15.8% increase. When you look at 40 matched men and you pick the man with the most impressive increase in aligned earnings, that man is typically a treated man: it happened 85.7% of the time. When there are big gains in earnings, mostly they occur in the treated group. As Figure 2.1 suggests, the treatment has a big effect on a subset of men, with perhaps little effect on many others. Indeed, the possibility of detecting large but rare effects is the motivation for considering $m > 2$.

Uncommon but dramatic responses to treatment

If the effect of a treatment is not constant, then it is larger for some people than for others. Perhaps the NSW training program changed the lives of some people and had little or no effect on many others. David Salsburg [52] argued that common statistical procedures were poorly equipped to detect uncommon but large treatment effects, but he argued that such effects were often important; see also [53]. William Conover and David Salsburg [5] proposed a type of mixture model in which a small fraction of the population responds strongly to treatment, so $r_{Tij} - r_{Cij}$ is large, while the rest of the population is unaffected by treatment, so $r_{Tij} - r_{Cij} = 0$. Under this model, they derived the locally most powerful rank test. The ranks they obtained are difficult to interpret, so they do not lend themselves to use with attributable effects. It turns out, however, that for large I, after rescaling, their ranks are virtually the same as the ranks (2.8) that Stephenson [59] proposed on the basis of other considerations. As has been seen, Stephenson's ranks have a straightforward interpretation in terms of attributable effects. This subject is developed in detail in Chapter 16 and [49]. In short, the methods described in this section are well adapted to the study of uncommon but dramatic responses to treatment of the type that seem to have occurred in the NSW experiment.

2.6 Internal and External Validity

In Fisher's theory of randomized experiments, the inference is from the responses the $2I$ subjects did exhibit under the treatments they did receive, and it is to the $2I$ unobserved responses these same individuals would have exhibited had they received the alternative treatment. In brief, the inference concerns the unobservable causal effects $r_{Tij} - r_{Cij}$ for the $2I$ individuals in the experiment. See, for instance, Welch [62].

A distinction is often made between 'internal' and 'external' validity [54]. A randomized experiment is said to have a high level of 'internal validity' in the sense that randomization provides a strong or 'reasoned' basis for inference about the effects of the treatment, $r_{Tij} - r_{Cij}$, on the $2I$ individuals in the experiment. Causal inference is challenging even if one were only interested in these $2I$ individuals, and randomization meets that specific challenge. Often, one is interested not just in these $2I$ people, but in people generally, people on other continents, people who will be born in the future. 'External' validity refers to the effects of the treatment on people not included in the experiment.

For instance, suppose that four large hospitals, say in Europe and the United States, cooperated in a randomized trial of two forms of surgery, with sixty day mortality as the outcome. Properly conducted, the trial would provide a sound basis for inference about the effects of the two forms of surgery on the $2I$ patients in this one clinical trial. One might plausibly think that similar effects would be found at other large hospitals in Europe and the United States, Canada, Japan and other countries, but nothing in a statistical analysis based on four hospitals provides a warrant for this plausible extrapolation. The extrapolation sounds plausible because of what we know about surgeons and hospitals in these countries, not because data from four cooperating hospitals provides a sound basis for extrapolation. Would similar effects be found at the small ambulatory surgical centers that are increasingly common in the United States and elsewhere? Will the same effects be found at large hospitals ten years from now as other technologies improve? Statistical analysis of a clinical trial at four large hospitals provides no warrant for these extrapolations, although surgeons may be able to offer an informed guess. These questions refer to the degree of external validity of the clinical trial.

The common view [54], which I share, is that internal validity comes first. If you do not know the effects of the treatment on the individuals in your study, you are not well-positioned to infer the effects on individuals you did not study who live in circumstances you did not study. Randomization addresses internal validity. In actual practice, external validity is often addressed by comparing the results of several internally valid studies conducted in different circumstances at different times. These issues also arise in observational studies, and so will be discussed at various points later on.

2.7 Summary

The goal of causal inference in the NSW experiment is not to say that the earnings of treated and control men differed systematically, but rather that the treatment was the cause of the difference, so that if the treatment had been withheld from the treated men, there would have been no systematic difference between the two groups. Many statistical procedures can recognize a systematic difference, providing the sample size is sufficiently large. Randomization plays a key role in experiments that study the effects actually caused by a treatment.

The current chapter has done several things. It has introduced many elements that randomized experiments and observational studies have in common, such as treatment effects, treatment assignments and pretreatment covariates unaffected by treatment. The chapter has discussed the large role played by random treatment assignment in experiments in justifying, or warranting, or being the 'reasoned basis' for inference about effects caused by treatment. Without random assignment, in observational studies, this justification or warrant or reasoned basis is absent. Therefore, the chapter has framed the problem faced in an observational study, where the goal is again to draw inferences about the effects caused by treatment, but treatments were not assigned at random [4].

2.8 Further Reading

Randomized experimentation is due to Sir Ronald Fisher [13], whose book of 1935 is of continuing interest. David Cox and Nancy Reid [11] present a modern discussion of the theory of randomized experiments. For randomized clinical trials in medicine, see [16, 38], and for experiments in the social sciences, see [3, 6]. Among books that discuss permutation and randomization inferences, Erich Lehmann's [32] book is especially careful in distinguishing properties that follow from random assignment and properties that involve sampling a population. For causal effects as comparisons of potential outcomes under alternative treatments, see the papers by Jerzy Neyman [35], B.L. Welch [62] and Donald Rubin [50]. Donald Rubin's paper [50] is the first parallel, formal discussion of causal effects in randomized and observational studies. For three important illustrations of the enhanced clarity provided by this approach, see [14, 15, 51]. Fisher [13], Welch [62], Kempthorne [28], Wilk [64], Cox [9], Robinson [40], and many others viewed sampling models as approximations to randomization inferences. Randomization inference is, of course, not limited to matched pairs [32, 44] and it may be used in conjunction with general forms of covariance adjustment [45, 56].

In labor economics, the National Supported Work experiment became a small laboratory for methods for observational studies when Robert LaLonde [29] set aside the randomized control group and replaced it with a nonrandomized control group drawn from a survey; see [12, 18, 19, 29, 57] for discussion.

2.9 Appendix: Randomization Distribution of m-statistics

Giving observations controlled weight using a ψ function

Sections 2.3.2 and 2.3.3 were similar in structure, even though the sample mean $\overline{Y}$ and Wilcoxon's signed rank statistic are not similar as statistics. It was suggested in §2.3.3 that randomization inference is very general and virtually any statistic could be handled in a parallel manner. This is illustrated using another statistic, an m-test, that is, a test associated with Peter Huber's [25] m-estimates. The discussion here derives from a paper by J. S. Maritz [33]; see also [34, 48]. It should be emphasized that the statistics used in m-tests differ in small but consequential ways from the statistics used in m-estimation, and the description that follows is for m-tests.[12]

As in Huber's m-estimation of a location parameter, Maritz's test statistic involves a function $\psi(\cdot)$ that gives weights to scaled observations. Rather than 'reject outliers' from a statistical analysis, a suitable ψ-function gradually reduces the weight of observations as they become increasingly extreme. Because we want an observation that is high by 3 units to pull up as strongly as an observation that is low by 3 units pulls down, the function $\psi(\cdot)$ is required to be an odd function, meaning $\psi(-y) = -\psi(y)$ for every $y \geq 0$, so in particular $\psi(-3) = -\psi(3)$ and $\psi(0) = 0$. For testing and confidence intervals, it is wise to insist that $\psi(\cdot)$ be monotone increasing, so $y < y'$ implies $\psi(y) \leq \psi(y')$. One such function, similar to a trimmed mean, was proposed by Huber; it has $\psi(y) = \max\{-1, \min(y, 1)\}$, so that $\psi(y) = -1$ for $y < -1$, $\psi(y) = y$ for $-1 \leq y \leq 1$, and $\psi(y) = 1$ for $y > 1$. In effect, the mean $\overline{Y}$ has a $\psi(\cdot)$ function of $\psi(y) = y$.

Consider using the treated-minus-control differences in paired responses, $Y_i = (Z_{i1} - Z_{i2})(R_{i1} - R_{i2})$, in testing the hypothesis of an additive treatment effect, $H_0 : r_{Tij} = \tau_0 + r_{Cij}$, for all ij, where τ_0 is a specified number. When $\tau_0 = 0$, this is

---

[12] The difference involves the scaling factor. In estimation, the scaling factor might be the median absolute deviation from the sample median. In testing, the scaling factor might be the median absolute deviation from the hypothesized median, which for testing no effect is zero. In testing a hypothesis, one can presume the hypothesis is true for the purpose of testing it. By using the hypothesized median rather than the sample median, the scaling factor becomes fixed as the treatment assignments vary — as illustrated later — with the consequence that Maritz's statistic has a null randomization distribution that is the sum of independent bounded random variables, and this greatly simplifies the creation of a large sample approximation to the null randomization distribution. This small difference has various consequences that should be considered before designing a new m-test; see [33, 48] for discussion.

2.9 Appendix: Randomization Distribution of m-statistics

Fisher's sharp null hypothesis of no treatment effect, and this is the case presented in detail in Table 2.12. As seen in §2.4, if H_0 were true, then the treated-minus-control difference in responses in pair i is $Y_i = \tau_0 + (Z_{i1} - Z_{i2})(r_{Ci1} - r_{Ci2})$. In particular, if H_0 were true, $Y_i - \tau_0$ would be $\pm(r_{Ci1} - r_{Ci2})$ and $|Y_i - \tau_0| = |r_{Ci1} - r_{Ci2}|$ would be fixed[13] not varying with the randomly assigned treatment Z_{ij}. This is visible for $\tau_0 = 0$ at the bottom of Table 2.12: assuming $H_0 : r_{Tij} = \tau_0 + r_{Cij}$ with $\tau_0 = 0$ for the purpose of testing it, the $|Y_i - \tau_0|$ are the same for all $|\mathscr{Z}| = 32$ treatment assignments in $\mathscr{Z}$.

Scaling

The observations are scaled before the $\psi(\cdot)$ function is calculated. The scaling factor s_{τ_0} used by Maritz's [33] statistic is a specific quantile of the I absolute differences, $|Y_i - \tau_0|$, most commonly the median, $s_{\tau_0} = \text{median}\,|Y_i - \tau_0|$. When used with Huber's $\psi(y) = \max\{-1, \min(y, 1)\}$, scaling by $s_{\tau_0} = \text{median}\,|Y_i - \tau_0|$ trims half the observations and is analogous to a midmean. If a larger quantile of the $|Y_i - \tau_0|$ is used for scaling, say the 75th or the 90th quantile rather than the median, then there is less trimming. If the maximum $|Y_i - \tau_0|$ is used for scaling with Huber's $\psi(y) = \max\{-1, \min(y, 1)\}$, then Maritz's statistic becomes the sample mean and it reproduces the randomization inference in §2.3.2.

Because s_{τ_0} is calculated from $|Y_i - \tau_0|$, if $H_0 : r_{Tij} = \tau_0 + r_{Cij}$, were true, then s_{τ_0} would have the same value for each of the 32 treatment assignments, or $s_{\tau_0} = 2147$ in Table 2.12. In the example in Table 2.12, the calculations continue with Huber's function $\psi(y) = \max\{-1, \min(y, 1)\}$, specifically the scaled absolute responses, $|Y_i - \tau_0|/s_{\tau_0}$, and the scored, scaled absolute responses, $\psi(|Y_i - \tau_0|/s_{\tau_0})$. Notice that $\psi(|Y_i - \tau_0|/s_{\tau_0})$ takes the same value for all 32 treatment assignments.

Randomization test of a hypothesized additive effect, $H_0 : r_{Tij} = r_{Cij} + \tau_0$

For testing the hypothesis of an additive treatment effect, $H_0 : r_{Tij} = \tau_0 + r_{Cij}$, Maritz's [33] m-test statistic is $T^* = \sum_{i=1}^{I} \psi\{(Y_i - \tau_0)/s_{\tau_0}\}$. Recall that $\psi(\cdot)$ is an odd function, $\psi(-y) = -\psi(y)$ for $y \geq 0$, so $\psi(y) = \text{sign}(y) \cdot \psi(|y|)$ where $\text{sign}(y) = 1$ if $y > 0$, $\text{sign}(y) = 0$ if $y = 0$, $\text{sign}(y) = -1$ if $y < 0$. If $H_0 : r_{Tij} = \tau_0 + r_{Cij}$ were true, then $Y_i - \tau_0 = (Z_{i1} - Z_{i2})(r_{Ci1} - r_{Ci2})$, so $T^* = \sum_{i=1}^{I} \text{sign}(Y_i - \tau_0) \cdot \psi(|Y_i - \tau_0|/s_{\tau_0})$ or equivalently

$$T^* = \sum_{i=1}^{I} \text{sign}\{(Z_{i1} - Z_{i2})(r_{Ci1} - r_{Ci2})\} \cdot \psi(|r_{Ci1} - r_{Ci2}|/s_{\tau_0}). \qquad (2.15)$$

---

[13] That is, $|Y_i - \tau_0| = |r_{Ci1} - r_{Ci2}|$ is determined by $\mathscr{F}$.

Table 2.12 The possible treated-minus-control differences in adjusted responses $Y_i - \tau_0 = (Z_{i1} - Z_{i2})(r_{Ci1} - r_{Ci2})$ and Huber-Maritz m-statistic, T^*, under the null hypothesis of an additive treatment effect, $H_0 : r_{Tij} = r_{Cij} + \tau_0$, for $\tau_0 = 0$ for the small version of the NSW experiment with $I = 5$ pairs.

Label	Pair 1 $Y_1 - \tau_0$	Pair 2 $Y_2 - \tau_0$	Pair 3 $Y_3 - \tau_0$	Pair 4 $Y_4 - \tau_0$	Pair 5 $Y_5 - \tau_0$	Huber-Maritz T^*
1	1456	3988	−45	−2147	3173	1.66
2	1456	3988	−45	−2147	−3173	−0.34
3	1456	3988	−45	2147	3173	3.66
4	1456	3988	−45	2147	−3173	1.66
5	1456	3988	45	−2147	3173	1.70
6	1456	3988	45	−2147	−3173	−0.30
7	1456	3988	45	2147	3173	3.70
8	1456	3988	45	2147	−3173	1.70
9	1456	−3988	−45	−2147	3173	−0.34
10	1456	−3988	−45	−2147	−3173	−2.34
11	1456	−3988	−45	2147	3173	1.66
12	1456	−3988	−45	2147	−3173	−0.34
13	1456	−3988	45	−2147	3173	−0.30
14	1456	−3988	45	−2147	−3173	−2.30
15	1456	−3988	45	2147	3173	1.70
16	1456	−3988	45	2147	−3173	−0.30
17	−1456	3988	−45	−2147	3173	0.30
18	−1456	3988	−45	−2147	−3173	−1.70
19	−1456	3988	−45	2147	3173	2.30
20	−1456	3988	−45	2147	−3173	0.30
21	−1456	3988	45	−2147	3173	0.34
22	−1456	3988	45	−2147	−3173	−1.66
23	−1456	3988	45	2147	3173	2.34
24	−1456	3988	45	2147	−3173	0.34
25	−1456	−3988	−45	−2147	3173	−1.70
26	−1456	−3988	−45	−2147	−3173	−3.70
27	−1456	−3988	−45	2147	3173	0.30
28	−1456	−3988	−45	2147	−3173	−1.70
29	−1456	−3988	45	−2147	3173	−1.66
30	−1456	−3988	45	−2147	−3173	−3.66
31	−1456	−3988	45	2147	3173	0.34
32	−1456	−3988	45	2147	−3173	−1.66
All 1–32	Absolute Differences $|Y_i - \tau_0|$					
	1456	3988	45	2147	3173	
All 1–32	Scaled Absolute Differences $|Y_i - \tau_0|/s_{\tau_0}$					
	0.68	1.86	0.02	1.00	1.48	
All 1–32	Scored, Scaled Absolute Differences $\psi(|Y_i - \tau_0|/s_{\tau_0})$					
	0.68	1.00	0.02	1.00	1.00	

In words, if $H_0 : r_{Tij} = \tau_0 + r_{Cij}$ were true, T^* would be the sum of I independent quantities that take the fixed values $\psi\left(|r_{Ci1} - r_{Ci2}|/s_{\tau_0}\right)$ and $-\psi\left(|r_{Ci1} - r_{Ci2}|/s_{\tau_0}\right)$ each with probability $\frac{1}{2}$.

For instance, the observed value of T^* is 1.66 in the first row of Table 2.12; that is, $T^* = 1.66 = 0.68 + 1.00 + (-0.02) + (-1.00) + 1.00$. Ten of the 32 equally probable treatment assignments $\mathbf{z} \in \mathscr{Z}$ yield values of T^* that are at least 1.66, so the one-sided P-value for testing $H_0 : r_{Tij} = \tau_0 + r_{Cij}$ with $\tau_0 = 0$ is $\frac{10}{32} = 0.3125 = \Pr(T^* \geq 1.66 | \mathscr{F}, \mathscr{Z})$. The two-sided P-value is $\frac{20}{32} = 0.625$ which equals both $2 \cdot \Pr(T^* \geq 1.66 | \mathscr{F}, \mathscr{Z})$ and $\Pr(T^* \leq -1.66 | \mathscr{F}, \mathscr{Z}) + \Pr(T^* \geq 1.66 | \mathscr{F}, \mathscr{Z})$ because the null distribution of T^* given $\mathscr{F}, \mathscr{Z}$ is symmetric about zero.

Table 2.12 calculates the exact null distribution of T^*, but a large sample approximation using the central limit theorem is easy to obtain as $I \to \infty$. The null distribution of T^* is the distribution of the sum of I independent terms taking the fixed values $\pm \psi\left(|r_{Ci1} - r_{Ci2}|/s_{\tau_0}\right)$ each with probability $\frac{1}{2}$. Therefore, in a randomized experiment under the null hypothesis, $E(T^* | \mathscr{F}, \mathscr{Z}) = 0$ and $\text{var}(T^* | \mathscr{F}, \mathscr{Z}) = \sum_{i=1}^{I} \left\{ \psi\left(|r_{Ci1} - r_{Ci2}|/s_{\tau_0}\right) \right\}^2$. With a bounded $\psi(\cdot)$ function, such as Huber's $\psi(y) = \max\{-1, \min(y, 1)\}$, as $I \to \infty$, the null distribution of $T^*/\sqrt{\text{var}(T^* | \mathscr{F}, \mathscr{Z})}$ converges in distribution to the standard Normal distribution provided $\Pr(s_{\tau_0} > 0) \to 1$; see [33].

The discussion in this section has focused on matched pairs, but essentially the same reasoning may be applied when matching with multiple controls [48].

m-tests in the NSW experiment

As seen in §2.5, in the NSW experiment, much of the treatment effect is in the tail, so less trimming is better. In testing the hypothesis of no effect, $H_0 : \tau_0 = 0$, scaling by $s_{\tau_0} = \text{median} |Y_i - \tau_0|$ trims half the observations and yields a two-sided P-value of 0.040, while scaling by the 90th percentile of $|Y_i - \tau_0|$ yields a two-sided P-value of 0.0052. Using the 90th percentile, the 95% confidence interval for an additive effect is \$475 to \$2788.

References

1. Andrews, D.F., Bickel, P.J., Hampel, F.R., Huber, P.J., Rogers, W.H., Tukey, J.W.: Robust Estimates of Location. Princeton, NJ: Princeton University Press (1972)
2. Bennett, D.M.: Confidence limits for a ratio using Wilcoxon's signed rank test. Biometrics **21**, 231–234 (1965)
3. Boruch, R.: Randomized Experiments for Planning and Evaluation. Thousand Oaks, CA: Sage (1997)
4. Cochran, W.G.: The planning of observational studies of human populations (with Discussion). J Roy Statist Soc A **128**, 234–265 (1965)
5. Conover, W.J., Salsburg, D.S.: Locally most powerful tests for detecting treatment effects when only a subset of patients can be expected to 'respond' to treatment. Biometrics **44**, 189–196 (1988)

6. Cook, T.D., Shadish, W.R.: Social experiments: Some developments over the past fifteen years. Annu Rev Psychol **45**, 545–580 (1994)
7. Couch, K.A.: New evidence on the long-term effects of employment training programs. J Labor Econ **10**, 380–388. (1992)
8. Cox, D.R.: Planning of Experiments. New York: Wiley (1958)
9. Cox, D.R.: The interpretation of the effects of non-additivity in the Latin square. Biometrika **45**, 69–73. (1958)
10. Cox, D.R.: The role of significance tests. Scand J Statist **4**, 49–62 (1977)
11. Cox, D.R., Reid, N.: The Theory of the Design of Experiments. New York: Chapman and Hall/CRC (2000)
12. Dehejia, R.H., Wahba, W.: Causal effects in nonexperimental studies: Reevaluating the evaluation of training programs. J Am Statist Assoc **94**, 1053–1062 (1999)
13. Fisher, R.A.: Design of Experiments. Edinburgh: Oliver and Boyd (1935)
14. Frangakis, C.E., Rubin, D.B.: Principal stratification in causal inference. Biometrics **58**, 21–29 (2002)
15. Freedman, D.A.: Randomization does not justify logistic regression. Statist Sci **23**, 237–249 (2008)
16. Friedman, L.M., DeMets, D.L., Furberg, C.D.: Fundamentals of Clinical Trials. New York: Springer (1998)
17. Greevy, R., Lu, B., Silber, J.H., Rosenbaum, P.R.: Optimal matching before randomization. Biostatistics **5**, 263–275 (2004)
18. Hämäläinen, K., Uusitalo, R., Vuori, J.: Varying biases in matching estimates: Evidence from two randomized job search training experiments. Labour Econ **15**, 604–618 (2008)
19. Heckman, J.J., Hotz, V.J.: Choosing among alternative nonexperimental methods for estimating the impact of social programs: The case of manpower training (with Discussion). J Am Statist Assoc **84**, 862–874 (1989)
20. Hodges, J.L., Lehmann, E.L.: Rank methods for combination of independent experiments in analysis of variance. Ann Math Statist **33**, 482–497 (1962)
21. Hodges, J.L., Lehmann, E.L.: Estimates of location based on ranks. Ann Math Statist **34**, 598–611 (1963)
22. Hoeffding, W.: A class of statistics with asymptotically normal distributions. Ann Math Statist **19**, 293–325 (1948)
23. Holland, P.W.: Statistics and causal inference. J Am Statist Assoc **81**, 945–960 (1986)
24. Hollister, R., Kemper, P., Maynard, R., eds.: The National Supported Work Demonstration. Madison: University of Wisconsin Press (1984)
25. Huber, P.J.: Robust estimation of a location parameter. Ann Math Statist **35**, 73–101 (1964)
26. Hudgens, M.G., Halloran, M.E.: Toward causal inference with interference. J Am Statist Assoc **482**, 832–842 (2008)
27. Jurečková, J., Sen, P.K.: Robust Statistical Procedures. New York: Wiley (1996)
28. Kempthorne, O.: Design and Analysis of Experiments. New York: Wiley (1952)
29. LaLonde, R.J.: Evaluating the econometric evaluations of training programs with experimental data. Am Econ Rev **76**, 604–620 (1986)
30. LaLonde, R.J.: The promise of public sector-sponsored training programs. J Econ Perspec **9**, 149–168 (1995)
31. Lehmann, E.L.: Testing Statistical Hypotheses. New York: Wiley (1959)
32. Lehmann, E.L.: Nonparametrics. San Francisco: Holden-Day (1975)
33. Maritz, J.S.: A note on exact robust confidence intervals for location. Biometrika **66**, 163–166 (1979)
34. Maritz, J.S.: Distribution-Free Statistical Methods. London: Chapman and Hall (1995)
35. Neyman, J.: On the application of probability theory to agricultural experiments: Essay on principles, Section 9. In Polish, but reprinted in English with Discussion by T. Speed and D. B. Rubin in Statist Sci **5**, 463–480 (1923, reprinted 1990)
36. Nichols, T.E., Holmes, A.P.: Nonparametric permutation tests for functional neuroimaging. Human Brain Mapping **15**, 1–25.

37. Pagano, M., Tritchler, D.: On obtaining permutation distributions in polynomial time. J Am Statist Assoc **78**, 435–440 (1983)
38. Piantadosi, S.: Clinical Trials. New York: Wiley (2005)
39. Pratt, J.W., Gibbons, J.D.: Concepts of Nonparametric Theory. New York: Springer (1981)
40. Robinson, J.: The large sample power of permutation tests for randomization models. Ann Statist **1**, 291–296 (1973)
41. Rosenbaum, P.R.: The consequences of adjustment for a concomitant variable that has been affected by the treatment. J Roy Statist Soc A **147**, 656–666 (1984)
42. Rosenbaum, P.R.: Effects attributable to treatment: Inference in experiments and observational studies with a discrete pivot. Biometrika **88**, 219–231 (2001)
43. Rosenbaum, P.R.: Attributing effects to treatment in matched observational studies. J Am Statist Assoc **97**, 183–192 (2002)
44. Rosenbaum, P.R.: Observational Studies (2nd ed.). New York: Springer (2002)
45. Rosenbaum, P.R.: Covariance adjustment in randomized experiments and observational studies (with Discussion). Statist Sci **17**, 286–327 (2002)
46. Rosenbaum, P.R.: Exact confidence intervals for nonconstant effects by inverting the signed rank test. Am Statistician **57**, 132–138 (2003)
47. Rosenbaum, P.R.: Interference between units in randomized experiments. J Am Statist Assoc **102**, 191–200 (2007)
48. Rosenbaum, P.R.: Sensitivity analysis for m-estimates, tests, and confidence intervals in matched observational studies. Biometrics **63**, 456-464 (2007)
49. Rosenbaum, P.R.: Confidence intervals for uncommon but dramatic responses to treatment. Biometrics **63**, 1164–1171 (2007)
50. Rubin, D.B.: Estimating causal effects of treatments in randomized and nonrandomized studies. J Educ Psychol **66**, 688–701 (1974)
51. Rubin, D.B., Stuart, E.A., Zanutto, E.L.: A potential outcomes view of value-added assessment in education. J Educ Behav Statist **29**, 103–116 (2004)
52. Salsburg, D.S.: Alternative hypotheses for the effects of drugs in small-scale clinical studies. Biometrics **42**, 671–674 (1986)
53. Salsburg, D.S.: The Use of Restricted Significance Tests in Clinical Trials. New York: Springer (1992)
54. Shadish, W. R., Cook, T. D., Campbell, D. T.: Experimental and Quasi-Experimental Designs for Generalized Causal Inference. Boston: Houghton-Mifflin (2002)
55. Shaffer, J.P.: Bidirectional unbiased procedures. J Am Statist Assoc **69**, 437–439 (1974)
56. Small, D., Ten Have, T.R., Rosenbaum, P.R.: Randomization inference in a group-randomized trial of treatments for depression: covariate adjustment, noncompliance and quantile effects. J Am Statist Assoc **103**, 271-279 (2008)
57. Smith, J.A., Todd, P.E.: Does matching overcome LaLonde's critique of nonexperimental estimators? J Econometrics **125**, 305–353 (2005)
58. Sobel, M.E.: What do randomized studies of housing mobility demonstrate?: Causal inference in the face of interference. J Am Statist Assoc **476**, 1398–1407 (2006)
59. Stephenson, W.R.: A general class of one-sample nonparametric test statistics based on subsamples. J Am Statist Assoc **76**, 960–966 (1981)
60. Tobin, J.: Estimation of relationships for limited dependent variables. Econometrica **26**, 24–36 (1958)
61. Weiss, L.: A note on confidence sets for random variables. Ann Math Statist **26**, 142–144 (1955)
62. Welch, B.L.: On the z-test in randomized blocks and Latin squares. Biometrika **29**, 21–52 (1937)
63. Wilcoxon, F.: Individual comparisons by ranking methods. Biometrics **1**, 80–83 (1945)
64. Wilk, M.B.: The randomization analysis of a generalized randomized block design. Biometrika **42**, 70–79 (1955)

Chapter 3
Two Simple Models for Observational Studies

Abstract Observational studies differ from experiments in that randomization is not used to assign treatments. How were treatments assigned? This chapter introduces two simple models for treatment assignment in observational studies. The first model is useful but naïve: it says that people who look comparable are comparable. The second model speaks to a central concern in observational studies: people who look comparable in the observed data may not actually be comparable; they may differ in ways we did not observe.

3.1 The Population Before Matching

In the population before matching, there are L subjects, $\ell = 1, 2, \ldots, L$. Like the subjects in the randomized experiment in Chapter 2, each of these L subjects has an observed covariate $\mathbf{x}_\ell$, an unobserved covariate u_ℓ, an indicator of treatment assignment, Z_ℓ, where $Z_\ell = 1$ if the subject receives treatment or $Z_\ell = 0$ if the subject receives the control, a potential response to treatment, $r_{T\ell}$, which is seen if the subject receives treatment, $Z_\ell = 1$, a potential response to control, $r_{C\ell}$, which is seen if the subject receives the control, $Z_\ell = 0$, and an observed response, $R_\ell = Z_\ell r_{T\ell} + (1 - Z_\ell) r_{C\ell}$. Now, however, treatments Z_ℓ are not assigned by the equitable flip of a fair coin.

In the population before matching, we imagine that subject ℓ received treatment with probability π_ℓ, independently of other subjects, where π_ℓ may vary from one person to the next and is not known. More precisely,

$$\pi_\ell = \Pr(Z_\ell = 1 \mid r_{T\ell}, r_{C\ell}, \mathbf{x}_\ell, u_\ell), \tag{3.1}$$

$$\Pr(Z_1 = z_1, \ldots, Z_L = z_L \mid r_{T1}, r_{C1}, \mathbf{x}_1, u_1, \ldots, r_{TL}, r_{CL}, \mathbf{x}_L, u_L)$$
$$= \pi_1^{z_1} (1 - \pi_1)^{1-z_1} \cdots \pi_L^{z_L} (1 - \pi_L)^{1-z_L} = \Pi_{\ell=1}^{L} \pi_\ell^{z_\ell} (1 - \pi_\ell)^{1-z_\ell}.$$

It is natural to ask whether very little or a great deal has just been assumed. One might reasonably worry if a great deal has just been assumed, for perhaps the central problems in observational studies have just been assumed away. Addressing difficult problems by assuming they are not there is one of the less reliable tactics, so it is reasonable to seek some assurance that this has not just happened. Indeed, it has not. Actually, because of the delicate nature of unobserved variables, nothing at all has been assumed: there is always an unobserved u_ℓ that makes all of this true, and true in a fairly trivial way. Specifically, if we set $u_\ell = Z_\ell$, then $\pi_\ell = \Pr(Z_\ell = 1 \mid r_{T\ell}, r_{C\ell}, \mathbf{x}_\ell, u_\ell)$ is $\Pr(Z_\ell = 1 \mid r_{T\ell}, r_{C\ell}, \mathbf{x}_\ell, Z_\ell)$ which is, of course, $\pi_\ell = 1$ if $Z_\ell = 1$ and $\pi_\ell = 0$ if $Z_\ell = 0$; moreover, with π_ℓ so defined, $\Pi_{\ell=1}^{L} \pi_\ell^{z_\ell} (1 - \pi_\ell)^{1-z_\ell}$ is 1 for the observed treatment assignment and 0 for all other treatment assignments, and

$$\Pr(Z_1 = z_1, \ldots, Z_L = z_L \mid r_{T1}, r_{C1}, \mathbf{x}_1, u_1, \ldots, r_{TL}, r_{CL}, \mathbf{x}_L, u_L)$$

$$= \Pr(Z_1 = z_1, \ldots, Z_L = z_L \mid r_{T1}, r_{C1}, \mathbf{x}_1, Z_1, \ldots, r_{TL}, r_{CL}, \mathbf{x}_L, Z_L)$$

is also 1 for the observed treatment assignment and 0 for all other treatment assignments; so it's all true, in a trivial way, with $u_\ell = Z_\ell$. Until something is done to restrict the behavior of the unobserved covariate u_ℓ, nothing has been assumed: (3.1) is a representation, not a model, because there is always some u_ℓ that makes (3.1) true. A representation, such as (3.1), is a manner of speaking; it does not become a model until it says something that may be false.

3.2 The Ideal Matching

Imagine that we could find two subjects, say k and ℓ, such that exactly one was treated, $Z_k + Z_\ell = 1$, but they had the same probability of treatment, $\pi_k = \pi_\ell$, and we made those two subjects into a matched pair.[1] Obviously, it is something of a challenge to create this matched pair because we do not observe u_k and u_ℓ, and we observe either r_{Tk} or r_{Ck} but not both, and either $r_{T\ell}$ or $r_{C\ell}$ but not both. It is a challenge because we cannot create this pair by matching on observable quantities. We could create this pair by matching on the observable $\mathbf{x}$, so $\mathbf{x}_k = \mathbf{x}_\ell$, and then flipping a fair coin to determine (Z_k, Z_ℓ), assigning one member of the pair to treatment, the other to control; that is, we could create this pair by conducting a randomized paired experiment, as in Chapter 2. Indeed, with the aid of random assignment, matching on $\mathbf{x}$ may be prudent [27], but is not necessary to achieve $\pi_k = \pi_\ell$ because the π's are created by the experimenter. If randomization is infeasible or unethical, then

---

[1] The alert reader will notice that the supposition that there could be two subjects, k and ℓ, with $Z_k + Z_\ell = 1$ but $\pi_k = \pi_\ell$ is already a step beyond the mere representation in (3.1), because it precludes setting $u_i = Z_i$ for all i and instead requires $0 < \pi_i < 1$ for at least two subjects.

3.2 The Ideal Matching

how might one attempt to find a treated-control pair such that the paired units have the same probability of treatment, $\pi_k = \pi_\ell$?

Return for a moment to the study by Joshua Angrist and Victor Lavy [3] of class size and educational test performance, and in particular to the $I = 86$ pairs of Israeli schools in §1.3, one with between 31 and 40 students in the fifth grade ($Z = 0$), the other with between 41 and 50 students in the fifth grade ($Z = 1$), the pairs being matched for a covariate **x**, namely the percentage of disadvantaged students. Strict adherence to Maimonides' rule would require the slightly larger fifth grade cohorts to be taught in two small classes, and the slightly smaller fifth grade cohorts to be taught in one large class. As seen in Figure 1.1, adherence to Maimonides' rule was imperfect but strict enough to produce a wide separation in typical class size. In their study, (r_T, r_C) are average test scores in fifth grade if the fifth grade cohort was a little larger ($Z = 1$) or a little smaller ($Z = 0$). What separates a school with a slightly smaller fifth grade cohort ($Z = 0$, 31–40 students) and a school with a slightly larger fifth grade cohort ($Z = 1$, 41–50 students)? Well, what separates them is the enrollment of a handful of students in grade 5. It seems reasonably plausible that whether or not a few more students enroll in the fifth grade is a relatively haphazard event, an event not strongly related to the average test performance (r_T, r_C) that the fifth grade would exhibit with a larger or smaller cohort. That is, building a study in the way that Angrist and Lavy did, it does seem reasonably plausible that π_k and π_ℓ are fairly close in the $I = 86$ matched pairs of two schools.

Properly understood, a 'natural experiment' is an attempt to find in the world some rare circumstance such that a consequential treatment was handed to some people and denied to others for no particularly good reason at all, that is, haphazardly [2, 7, 31, 47, 58, 83, 92, 96, 102, 106]. The word 'natural' has various connotations, but a 'natural experiment' is a 'wild experiment' not a 'wholesome experiment,' natural in the way that a tiger is natural, not in the way that oatmeal is natural. To express the same thought differently: to say that 'it does seem reasonably plausible that π_k and π_ℓ are fairly close' is to say much less than 'definitely, $\pi_k = \pi_\ell$ by the way treatments were randomly assigned.' Haphazard is a far cry from randomized, a point given proper emphasis by Mark Rosenzweig and Ken Wolpin [92] and by Timothy Besley and Anne Case [7] in reviews of several natural experiments.

Nonetheless, it does seem reasonable to think that π_k and π_ℓ are fairly close for the 86 paired Israeli schools in Figure 1.1. This might be implausible in some other context, say in a national survey in the United States where schools are funded by local governments, so that class size might be predicted by the wealth or poverty of the local community. Even in the Israeli schools, the claim that π_k and π_ℓ are plausibly close depended upon: (i) matching for the percentage of students who were disadvantaged, $\mathbf{x}_k$, and (ii) defining Z_k in terms of cohort size rather than class size. In the pair of boxplots on the upper right of Figure 1.1 depicting class size and cohort size, there is considerable adherence to Maimonides' rule, but there are a few deviations: in particular, two large fifth grade cohorts that strict adherence to the rule would have divided, and several small cohorts that were taught in smaller classes than strict adherence to the rule would have permitted. Small variations in the size of the fifth grade cohort are plausibly haphazard, but deviations from Maimonides' rule may have been deliberate, considered responses to special circumstances, and hence not haphazard. As defined in Figure 1.1, Z_k reflects the haphazard element, namely susceptibility to Maimonides' rule based on cohort size, rather than realized class size. Defining Z_k in this way is analogous to an 'intention-to-treat' analysis (§1.2.6) in a randomized experiment.[2]

Suppose indeed we could find two subjects, k and ℓ, with the same probability of treatment, $\pi_k = \pi_\ell$, where the common value of π_k and π_ℓ is typically unknown. What would follow from this? From (3.1),

$$\Pr(Z_k = z_k, Z_\ell = z_\ell | r_{Tk}, r_{Ck}, \mathbf{x}_k, u_k, r_{T\ell}, r_{C\ell}, \mathbf{x}_\ell, u_\ell) \tag{3.2}$$

$$= \pi_k^{z_k} (1 - \pi_k)^{1-z_k} \pi_\ell^{z_\ell} (1 - \pi_\ell)^{1-z_\ell} \tag{3.3}$$

$$= \pi_\ell^{z_k + z_\ell} (1 - \pi_\ell)^{(1-z_k)+(1-z_\ell)} \text{ because } \pi_k = \pi_\ell, \tag{3.4}$$

which is, so far, of limited use, because π_ℓ is unknown. Suppose, however, that we found two subjects, k and ℓ, with the same probability of treatment, $\pi_k = \pi_\ell$, with precisely the additional fact that exactly one was treated, $Z_k + Z_\ell = 1$. Given that one of these subjects was treated, the chance that it was k and not ℓ would be

$$\Pr(Z_k = 1, Z_\ell = 0 | r_{Tk}, r_{Ck}, \mathbf{x}_k, u_k, r_{T\ell}, r_{C\ell}, \mathbf{x}_\ell, u_\ell, Z_k + Z_\ell = 1)$$

$$= \frac{\Pr(Z_k = 1, Z_\ell = 0 | r_{Tk}, r_{Ck}, \mathbf{x}_k, u_k, r_{T\ell}, r_{C\ell}, \mathbf{x}_\ell, u_\ell)}{\Pr(Z_k + Z_\ell = 1 | r_{Tk}, r_{Ck}, \mathbf{x}_k, u_k, r_{T\ell}, r_{C\ell}, \mathbf{x}_\ell, u_\ell)}$$

$$= \frac{\pi_\ell^{1+0} (1 - \pi_\ell)^{(1-1)+(1-0)}}{\pi_\ell^{1+0} (1 - \pi_\ell)^{(1-1)+(1-0)} + \pi_\ell^{0+1} (1 - \pi_\ell)^{(1-0)+(1-1)}}$$

$$= \frac{\pi_\ell (1 - \pi_\ell)}{\pi_\ell (1 - \pi_\ell) + \pi_\ell (1 - \pi_\ell)} = \frac{1}{2}$$

which, unlike (3.2), would be immensely useful. If we could find among the L subjects in the population a total of $2I$ different subjects matched in I pairs in just this way, then we would have reconstructed the distribution of treatment assignments $\mathbf{Z}$ in the paired, randomized experiment, and inference about the effects caused by treatments would be straightforward; indeed, the methods of Chapter 2 would be applicable.[3] Keep in mind that we are speaking here about mere imaginings, suppositions, hopes, aspirations and attempts, as distinct from simple, undeniable facts created by randomization in Chapter 2. In the words of the proverb: "If wishes were horses, beggars would ride."

---

[2] As in §1.2.6, this definition of Z_k leaves open the possibility of using cohort size Z_k as an instrument for the actual size of the class, and indeed Angrist and Lavy [3] reported instrumental variable analyses.

[3] The argument just presented is the simplest version of a general argument developed in [69].

3.2 The Ideal Matching

Where do things stand? As has been seen, the ideal matching would pair individuals, k and ℓ, with different treatments, $Z_k + Z_\ell = 1$, but the same probability of treatment, $\pi_k = \pi_\ell$. If we could do this, we could reconstruct a randomized experiment from observational data; however, we cannot do this. The wise attempt to find 'natural experiments' is the attempt to move towards this ideal. That is what you do with ideals: you attempt to move towards them. No doubt, many if not most 'natural experiments' fail to produce this ideal pairing [7, 92]; moreover, even if one 'natural experiment' once succeeded, there would be no way of knowing for sure that it had succeeded. A second attempt to approach the ideal is to find a context in which the covariates used to determine treatment assignment are measured covariates, $\mathbf{x}_k$, not unobserved covariates, u_ℓ, and then to match closely for the measured covariates, $\mathbf{x}_k$.[4] Matching for observed covariates is the topic of Part II of this book. Although we cannot see π_k and π_ℓ and so cannot see whether they are fairly close, perhaps there are things we can see that provide an incomplete and partial check on the closeness of π_k and π_ℓ. This leads to the devices of 'quasi-experiments' [10, 12, 58, 97] mentioned in §1.2.4, such as multiple control groups [51, 75, 78], 'multiple operationalism' or coherence [12, 36, 63, 82, 105], 'control constructs' and known effects [57, 77, 78, 97, 110], and one or more pretreatment measures of what will become the response.[5] Even when it 'seems reasonably plausible that π_k

---

[4] See, for instance, the argument made by David Gross and Nicholas Souleles, quoted in §4.2 from their paper [29]. The example in Chapter 7 attempts to use both approaches, although it is, perhaps, not entirely convincing in either attempt. In Chapter 7, the dose of chemotherapy received by patients with ovarian cancer is varied by comparing two types of providers of chemotherapy, gynecological oncologists (GO) and medical oncologists (MO), where the latter treat more intensively. The most important clinical covariates are measured, including clinical stage, tumor grade, histology, and comorbid conditions. Despite these two good attempts, it is far from clear that the GO-vs-MO assignment is haphazard, even after adjustments, and it quite likely that the oncologist knew considerably more about the patient than is recorded in electronic form.

[5] Figure 1.2 is an example of quasi-experimental reasoning. Instead of comparing college attendance in two groups, say the children of deceased fathers in 1979–1981, when the benefit program was available, and the children of deceased fathers in 1982–1983, after the program was withdrawn, Figure 1.2 follows Susan Dynarski's [20] study in comparing four groups. Figure 1.2 shows at most a small shift in college attendance by the group of children without deceased fathers, who were never eligible for the program. This is a small check, but a useful one: it suggests that the comparison of children of deceased fathers in 1979–1981 and 1982–1983, though totally confounded with time, may not be seriously biased by this confounding, because a substantial shift in college attendance by unaffected groups over this time is not evident. The quasi-experimental design in Figure 1.2 is widely used, and is known by a variety of not entirely helpful names, including 'pretest-posttest nonequivalent control group design' and 'difference-in-differences design.' Kevin Volpp and colleagues [108] used this design in a medical context to study the possible effects on patient mortality of a reduction, to 80 hours per week, of the maximum number of hours that a medical resident may work in a hospital. See [1, 5, 6, 11, 84] for discussion in general terms.

The study by Garen Wintemute and colleagues [112], as discussed in §4.3, contains another example of quasi-experimental reasoning using a 'control construct,' that is, an outcome thought to be unaffected by the treatment. They were interested in the possible effects of a change in gun control laws on violent crimes using guns. One would not expect a change in gun control laws to have a substantial effect on nonviolent, nongun crimes, and so the rate of such crimes is their control construct or unaffected outcome. See §4.3. An example with two control groups subject to different biases is discussed in §11.3.

and π_ℓ are fairly close,' we will need to ask about the consequences of being close rather than equal, about the possibility they are somewhat less close than we had hoped, about the possibility that they are disappointingly far apart. Answers are provided by sensitivity analysis; see §3.4.

3.3 A Naïve Model: People Who Look Comparable Are Comparable

Assigning treatments by flipping biased coins whose unknown biases are determined by observed covariates

The first of the two models in the title of this chapter asserts, rather naïvely, that people who look comparable in terms of measured covariates, x_k, actually are comparable, and to a nontrivial extent, an element of luck determined every treatment assignment. Most observational studies report, perhaps briefly or perhaps at length, at least one analysis that tentatively assumes that the naïve model is true. Most of the controversy and skepticism associated with observational studies is a reference of one sort or another to this model's naïveté.

The naïve model restricts the representation (3.1) in two quite severe ways. First, it says that treatment assignment probabilities, $\pi_\ell = \Pr(Z_\ell = 1 \mid r_{T\ell}, r_{C\ell}, \mathbf{x}_\ell, u_\ell)$, that condition on $(r_{T\ell}, r_{C\ell}, \mathbf{x}_\ell, u_\ell)$ may depend upon the observed covariates $\mathbf{x}_\ell$ but not upon the potential responses $(r_{T\ell}, r_{C\ell})$ or the unobserved covariate u_ℓ. Second, the model says $0 < \pi_\ell < 1$ for all ℓ, so that every person ℓ has a chance of receiving either treatment, $Z_\ell = 1$, or control, $Z_\ell = 0$. Putting this together with (3.1), the naïve model says:

$$\pi_\ell = \Pr(Z_\ell = 1 \mid r_{T\ell}, r_{C\ell}, \mathbf{x}_\ell, u_\ell) = \Pr(Z_\ell = 1 \mid \mathbf{x}_\ell) \tag{3.5}$$

and

$$0 < \pi_\ell < 1, \ \ell = 1, 2, \ldots, L, \tag{3.6}$$

with

$$\Pr(Z_1 = z_1, \ldots, Z_L = z_L \mid r_{T1}, r_{C1}, \mathbf{x}_1, u_1, \ldots, r_{TL}, r_{CL}, \mathbf{x}_L, u_L) \tag{3.7}$$

$$= \prod_{\ell=1}^{L} \pi_\ell^{z_\ell} (1 - \pi_\ell)^{1-z_\ell}. \tag{3.8}$$

The naïve model would be true if treatments Z_ℓ were assigned by independent flips of a fair coin: then $\pi_\ell = \frac{1}{2}$ for $\ell = 1, 2, \ldots, L$, and all of (3.5)–(3.8) is true. Indeed, the naïve model would be true if treatments Z_ℓ were assigned by independent flips of a group of biased coins, where the same biased coin is used whenever two people, k and ℓ, have the same observed covariate, $\mathbf{x}_k = \mathbf{x}_\ell$, and no coin has probability 1 or 0 of a head; then all of (3.5)–(3.8) is true. For (3.5)–(3.8) to be true, it is not necessary that the biases of these coins be known, but it is necessary that the bias

3.3 A Naïve Model: People Who Look Comparable Are Comparable

depend on $\mathbf{x}_\ell$ alone. The first feature of the model (3.5)–(3.8) says treatment assignment Z is conditionally independent of (r_T, r_C, u) given $\mathbf{x}$, or in Phillip Dawid's [17] notation, $Z \perp\!\!\!\perp (r_T, r_C, u) \,|\, \mathbf{x}$. In [67], Donald Rubin and I gave the name 'strongly ignorable treatment assignment given $\mathbf{x}$' to this naïve model (3.5)–(3.8).[6]

As will be seen in a moment, it would be very convenient if the naïve model (3.5)–(3.8) were true. If (3.5)–(3.8) were true, causal inference in an observational study would become a mechanical problem, so that if the sample size were large, and you followed certain steps, and you did not make mistakes in following the steps, then a correct causal inference would result; a computer could do it. So it would be convenient if (3.5)–(3.8) were true. To believe that something is true because it would be convenient if it were true is a fair definition of naïveté.

The ideal match and the naïve model

If the naïve model were true, then the ideal match of §3.2 could be produced simply by matching treated and control subjects with the same value of the observed covariate $\mathbf{x}$. This follows immediately from (3.5)–(3.8) and the discussion in §3.2. If the naïve model were true, one could reconstruct the distribution of treatment assignments $\mathbf{Z}$ in a randomized paired experiment from observational data simply by matching for the observed covariates, $\mathbf{x}$. Conventional methods of statistical analysis — the paired t-test, Wilcoxon's signed rank test, m-estimates, etc. — would work if (3.5)–(3.8) were true, and are quite likely to fail if (3.5)–(3.8) is false in a nontrivial way.

---

[6] Techincally, strongly ignorable treatment assigment given $\mathbf{x}$ is $Z \perp\!\!\!\perp (r_T, r_C) \,|\, \mathbf{x}$ and $0 < \Pr(Z = 1 \,|\, \mathbf{x}) < 1$ for all $\mathbf{x}$, or in a single expression, $0 < \Pr(Z = 1 \,|\, r_T, r_C, \mathbf{x}) = \Pr(Z = 1 \,|\, \mathbf{x}) < 1$ for all $\mathbf{x}$; that is, no explicit reference is made to an unobserved covariate u. I introduce u here, rather than in §3.4 where it is essential, to simplify the transition between §3.3 and §3.4. The remainder of this footnote is a slightly technical discussion of the formal relationship between (3.5) and the condition as given in [67]; it will interest only the insanely curious.

In a straightforward way [17, Lemma 3], condition (3.5) implies $Z \perp\!\!\!\perp (r_T, r_C) \,|\, \mathbf{x}$ and $0 < \Pr(Z = 1 \,|\, \mathbf{x}) < 1$ for all $\mathbf{x}$. Also, in a straightforward way [17, Lemma 3], if (3.5) is true then (i) $Z \perp\!\!\!\perp (r_T, r_C) \,|\, (\mathbf{x}, u)$, (ii) $0 < \Pr(Z = 1 \,|\, \mathbf{x}) < 1$ for all $\mathbf{x}$, and (iii) $Z \perp\!\!\!\perp u \,|\, \mathbf{x}$ are true. Moreover, (ii) and (iii) together imply (iv) $0 < \Pr(Z = 1 \,|\, \mathbf{x}, u) < 1$ for all $(\mathbf{x}, u)$. In words, condition (3.5) implies both strong ignorability given $\mathbf{x}$ and also strong ignorability given $(\mathbf{x}, u)$. Now, conditions (i) and (iv), without condition (iii), are the key elements of the sensitivity model in §3.4: they say, that treatment assignment would have been strongly ignorable if u had been measured and included with $\mathbf{x}$; so the failure to measure u is the source of our problems. Moreover, if condition (iii) were true in addition to (i) and (iv), then (3.5) and strong ignorability given $\mathbf{x}$ follow. In brief, strong ignorability given $\mathbf{x}$ is implied by the addition of condition (iii) to strong ignorability given $(\mathbf{x}, u)$. In words, one can reasonably think of the naive model §3.3 as the sensitivity model of §3.4 together with the irrelevance of u as expressed by $Z \perp\!\!\!\perp u \,|\, \mathbf{x}$ in condition (iii).

What is the propensity score?

So far, the talk has been of matching exactly for **x** as if that were easy to do. If **x** contains many covariates, then matching treated and control subjects with the same or similar **x** will be difficult if not impossible. Imagine, for instance, that **x** contains 20 covariates. Then any one subject, ℓ, might be above or below the median on the first covariate (two possibilities), above or below the median on the second covariate (two more possibilities, making $2 \times 2 = 4$ possibilities so far), and so on. Just in terms of being above or below the median on each covariate, there are $2 \times 2 \times \cdots \times 2 = 2^{20} = 1,048,576$ possibilities, or more than a million possibilities. With thousands of subjects, it will often be difficult to find two subjects who even match in the limited way of being on the same side of the median for all 20 covariates. This turns out to be less of a problem than it might seem at first, because of a device called the 'propensity score' [67].

In a sample of L subjects from an infinite population, the propensity score [67] is the conditional probability of treatment $Z = 1$ given the observed covariates **x**, or $e(\mathbf{x}) = \Pr(Z = 1 | \mathbf{x})$. Several aspects of this definition deserve immediate emphasis. The propensity score is defined in terms of the observed covariates, **x**, whether or not the naïve model (3.5)–(3.8) is true, that is, whether or not the 'ideal match' of §3.2 can be produced by matching on the observed covariates **x**. The form of (3.5)–(3.8) correctly suggests that the propensity score $e(\mathbf{x}) = \Pr(Z = 1 | \mathbf{x})$ will be more useful when (3.5)–(3.8) is true; however, the propensity score is defined in terms of the observable quantities, Z and **x**, whether or not (3.5)–(3.8) is true. In a randomized experiment, the propensity score is known because of random assignment. In contrast, in an observational study the propensity score is typically unknown; however, because $e(\mathbf{x}) = \Pr(Z = 1 | \mathbf{x})$ is defined in terms of observable quantities, namely treatment assignment Z and observed covariates, **x**, it is straightforward to estimate the propensity score in an observational study. In a randomized experiment, π_ℓ in (3.1) is known and in an observational study π_ℓ is unknown, but because π_ℓ depends upon the unobservable (r_T, r_C, u) as well as **x**, it is not possible to estimate π_ℓ in an observational study. Of course, if the naïve model (3.5)–(3.8) were true, then $\pi_\ell = e(\mathbf{x}_\ell)$; indeed, this is virtually the defining feature of that model.

Balancing property of the propensity score

The propensity score has several useful properties. The first property, the balancing property, is always true, whether or not the naïve model (3.5)–(3.8) is true. The balancing property [67] says that treated ($Z = 1$) and control ($Z = 0$) subjects with the same propensity score $e(\mathbf{x})$ have the same distribution of the observed covariates, **x**,

$$\Pr\{\mathbf{x} | Z = 1, e(\mathbf{x})\} = \Pr\{\mathbf{x} | Z = 0, e(\mathbf{x})\} \qquad (3.9)$$

or equivalently

$$Z \perp\!\!\!\perp \mathbf{x} \,\big|\, e(\mathbf{x}), \qquad (3.10)$$

3.3 A Naïve Model: People Who Look Comparable Are Comparable

so treatment Z and observed covariates $\mathbf{x}$ are conditionally independent given the propensity score.[7]

Because of the balancing property (3.10), if you pair two people, k and ℓ, one of whom is treated, $Z_k + Z_\ell = 1$, so that they have the same value of the propensity score, $e(\mathbf{x}_k) = e(\mathbf{x}_\ell)$, then they may have different values of the observed covariate, $\mathbf{x}_k \neq \mathbf{x}_\ell$, but in this pair, the specific values of the observed covariate $(\mathbf{x}_k, \mathbf{x}_\ell)$ will be unrelated to the treatment assignment (Z_k, Z_ℓ). If you form many pairs in this way, then the distribution of the observed covariates $\mathbf{x}$ will look about the same in the treated ($Z = 1$) and control ($Z = 0$) groups, even though individuals in matched pairs will typically have different values of $\mathbf{x}$. Although it is difficult to match on 20 covariates at once, it is easy to match on one covariate, the propensity score $e(\mathbf{x})$, and matching on $e(\mathbf{x})$ will tend to balance all 20 covariates.

In the randomized NSW experiment in Chapter 2, Table 2.1 displayed balance on observed covariates, $\mathbf{x}$. Randomization alone tends to balance observed covariates, $\mathbf{x}$, and Table 2.1 combined randomization and matching. In an observational study, it is often possible to match on an estimate $\widehat{e}(\mathbf{x})$ of the propensity score $e(\mathbf{x})$ and to produce balance on observed covariates $\mathbf{x}$ similar to the balance in Table 2.1; see, for example, Chapter 7.

Randomization is a much more powerful tool for balancing covariates than matching on an estimate of the propensity score. The difference is that the propensity score balances the observed covariates, $\mathbf{x}$, whereas randomization balances observed covariates, unobserved covariates, and potential responses, $(r_T, r_C, \mathbf{x}, u)$. This difference is not something you can see: randomization makes a promise about something you cannot see. Stated formally, if you match on $e(\mathbf{x})$, then $Z \perp\!\!\!\perp \mathbf{x} \mid e(\mathbf{x})$, but if you assign treatments by flipping fair coins, then $Z \perp\!\!\!\perp (r_T, r_C, \mathbf{x}, u)$. Randomization provides a basis for believing that an unobserved covariate u is balanced, but matching on propensity scores provides no basis for believing that.

Part II of this book discusses the practical aspects of matching, and an estimate $\widehat{e}(\mathbf{x})$ of the propensity score $e(\mathbf{x})$ is one of the tools used; see §8.2.[8] In common practice, the estimate of the propensity score, $\widehat{e}(\mathbf{x})$, is based on a model, such as a logit model, relating treatment assignment, Z, and observed covariates, $\mathbf{x}$. The logit model may include interactions, polynomials, and transformations of the covariates in $\mathbf{x}$, so it need not be a linear logit model narrowly conceived. Logit models for $\widehat{e}(\mathbf{x})$ are used for convenience; other methods can and have been used instead [56].

---

[7] The proof is easy [67]. Recall the following basic facts about conditional expectations: if A, B and C are random variables, then $E(A) = E\{E(A \mid B)\}$ and $E(A \mid C) = E\{E(A \mid B, C) \mid C\}$. From the definition of conditional independence, to prove (3.10), it suffices to prove $\Pr\{Z = 1 \mid \mathbf{x}, e(\mathbf{x})\} = \Pr\{Z = 1 \mid e(\mathbf{x})\}$. Because $e(\mathbf{x})$ is a function of $\mathbf{x}$, conditioning on $\mathbf{x}$ fixes $e(\mathbf{x})$, so $\Pr\{Z = 1 \mid \mathbf{x}, e(\mathbf{x})\} = \Pr(Z = 1 \mid \mathbf{x}) = e(\mathbf{x})$. Because $Z = 0$ or $Z = 1$, for any random variable D, we have $\Pr(Z = 1 \mid D) = E(Z \mid D)$. With $A = Z$, $B = \mathbf{x}$, $C = e(\mathbf{x})$, we have $\Pr\{Z = 1 \mid e(\mathbf{x})\} = E\{Z \mid e(\mathbf{x})\} = E\{E\{Z \mid \mathbf{x}, e(\mathbf{x})\} \mid e(\mathbf{x})\} = E\{\Pr\{Z = 1 \mid \mathbf{x}, e(\mathbf{x})\} \mid e(\mathbf{x})\} = E\{e(\mathbf{x}) \mid e(\mathbf{x})\} = e(\mathbf{x}) = \Pr(Z = 1 \mid \mathbf{x})$, which is what we needed to prove.

[8] Part II makes use of a stronger version of the balancing property than stated in (3.10). Specifically, if you match on $e(\mathbf{x})$ and any other aspect of $\mathbf{x}$, say $h(\mathbf{x})$, you still balance $\mathbf{x}$, that is, $Z \perp\!\!\!\perp \mathbf{x} \mid \{e(\mathbf{x}), h(\mathbf{x})\}$ for any $h(\mathbf{x})$. See [67] for the small adjustments required in the proof.

After matching, balance on observed covariates is checked by comparing the distributions of observed covariates in treated and control groups, as in Tables 2.1 and 7.1; for general discussion, see §9.1. Because (3.10) is known to be true for the true propensity score, $e(\mathbf{x})$, a check on covariate balance, as in Table 7.1, is a diagnostic check of the model that produced the estimated propensity score, $\widehat{e}(\mathbf{x})$, and may lead to a revision of that model.

In most but not all cases, when you substitute an estimate, say $\widehat{e}(\mathbf{x})$, for the unknown true parameter, $e(\mathbf{x})$, the estimate performs somewhat less well than the true parameter would perform. In fact, this is not true for propensity scores: estimated scores, $\widehat{e}(\mathbf{x})$, tend to work slightly better than true scores, $e(\mathbf{x})$. Estimated scores tend to slightly overfit, producing slightly better than chance balance on observed covariates, $\mathbf{x}$, in the data set used to construct the score. Estimated scores produce 'too much covariate balance,' but because propensity scores are used to balance covariates, 'too much covariate balance' is just fine.[9]

Propensity scores and ignorable treatment assignment

The balancing property of propensity scores (3.10) is always true, but a second property of propensity scores would follow if the naïve model (3.5)–(3.8) were true. Recall that if (3.5)–(3.8) were true, then the 'ideal match' of §3.2 could be produced simply by matching for the observed covariate $\mathbf{x}$. Recall also that it may be difficult to match closely for every one of the many covariates in $\mathbf{x}$, but it is easy to match on one variable, the propensity score, $e(\mathbf{x})$, and doing that balances all of $\mathbf{x}$. The second property closes this loop. It says: if (3.5)–(3.8) were true, then the 'ideal match' of §3.2 could be produced by matching on the propensity score, $e(\mathbf{x})$, alone [67]. The proof is very short: the 'ideal match' is to match on $\pi_\ell = \Pr(Z_\ell = 1 \mid r_{T\ell}, r_{C\ell}, \mathbf{x}_\ell, u_\ell)$, but if the naïve model (3.5)–(3.8) were true, then $\pi_\ell = e(\mathbf{x}_\ell)$, so matching on the propensity score is matching on π_ℓ. In words, if it suffices to match for the observed covariates, $\mathbf{x}$, then it suffices to match for the propensity score, $e(\mathbf{x})$.

The issue just discussed concerns a second property of propensity scores that holds when the naïve model (3.5)–(3.8) is true. This second property may be expressed in a different way. The naïve model (3.5)–(3.8) assumes

$$Z \perp\!\!\!\perp (r_T, r_C, u) \mid \mathbf{x}, \qquad (3.11)$$

and the second property says that (3.11) implies

$$Z \perp\!\!\!\perp (r_T, r_C, u) \mid e(\mathbf{x}) \qquad (3.12)$$

---

[9] Theoretical arguments also show that estimated propensity scores can have better properties than true propensity scores; see [69, 76].

3.3 A Naïve Model: People Who Look Comparable Are Comparable

so the single variable $e(\mathbf{x})$ may be used in place of the many observed covariates in $\mathbf{x}$.[10]

Summary: Separating two tasks, one mechanical, the other scientific

If the naïve model (3.5)–(3.8) were true, then the distribution of treatment assignments $\mathbf{Z}$ in a paired randomized experiment could be produced in an observational study by simply matching on observed covariates, $\mathbf{x}$. That might be a challenge if $\mathbf{x}$ contains many observed covariates, but in fact, matching on one covariate, the propensity score, $e(\mathbf{x})$, will balance all of the observed covariates, $\mathbf{x}$, and if (3.5)–(3.8) were true, this match would also produce the distribution of treatment assignments $\mathbf{Z}$ in a paired randomized experiment. Better still, it is straightforward to check whether matching has balanced the observed covariates; one simply checks whether matched treated and control groups are similar in terms of $\mathbf{x}$.

The naïve model (3.5)–(3.8) is important, but not because it is plausible; it is not plausible. The controversy and skepticism that almost invariably attends an observational study almost invariably refers back to this model's naïveté. Rather, the naïve model (3.5)–(3.8) is important because it cleanly divides inference in observational studies into two separable tasks. One is a fairly mechanical task that typically can be completed successfully, and can be seen to have been completed successfully, before the second task is engaged. This first task can take the form of matching treated and control subjects so that observed covariates are seen to be balanced. The first task is to compare treated and control subjects who look comparable prior to treatment. The second task engages the concern that people who look comparable may not be comparable. If people were not randomly assigned to treatments, then perhaps there are reasons they received the treatments they did, but those reasons are not visible to us because the observed covariates $\mathbf{x}$ provide an incomplete picture of the situation prior to treatment. The second task is not a mechanical but rather a scientific task, one that can be controversial and difficult to bring to a rapid and definitive closure; this task is, therefore, more challenging, and hence more interesting. The clever opportunities of natural experiments, the subtle devices of quasi-experiments, the technical tools of sensitivity analysis that extract information from nonidentified models — each of these is an attempt to engage the second task.

---

[10] In technical terms, if treatment assignment is strongly ignorable given $\mathbf{x}$, then it is strongly ignorable given $e(\mathbf{x})$. The proof of this version is also short [67]. In Note 7, we saw that $\Pr\{Z=1 \mid e(\mathbf{x})\} = \Pr(Z=1 \mid \mathbf{x})$. Condition (3.11) says $\Pr(Z=1 \mid r_T, r_C, \mathbf{x}, u) = \Pr(Z=1 \mid \mathbf{x})$. Together they say $\Pr(Z=1 \mid r_T, r_C, \mathbf{x}, u) = \Pr\{Z=1 \mid e(\mathbf{x})\}$, which implies (3.12).

3.4 Sensitivity Analysis: People Who Look Comparable May Differ

What is sensitivity analysis?

If the naïve model (3.5)–(3.8) were true, the distribution of treatment assignments **Z** in a randomized paired experiment could be reconstructed by matching for the observed covariate, **x**. It is common for a critic to argue that, in a particular study, the naïve model may be false. Indeed, it may be false. Typically, the critic accepts that the investigators matched for the observed covariates, **x**, so treated and control subjects are seen to be comparable in terms of **x**, but the critic points out that the investigators did not measure a specific covariate u, did not match for u, and so are in no position to assert that treated and control groups are comparable in terms of u. This criticism could be dismissed in a randomized experiment — randomization does tend to balance unobserved covariates — but the criticism cannot be dismissed in an observational study. This difference in the unobserved covariate u, the critic continues, is the real reason outcomes differ in the treated and control groups: it is not an effect caused by the treatment, but rather a failure on the part of the investigators to measure and control imbalances in u. Although not strictly necessary, the critic is usually aided by an air of superiority: "This would never happen in my laboratory."

It is important to recognize at the outset that our critic may be, but need not be, on the side of the angels. The tobacco industry and its (sometimes distinguished) consultants criticized, in precisely this way, observational studies linking smoking with lung cancer [103]. In this instance, the criticism was wrong. Investigators and their critics stand on level ground [8].

It is difficult if not impossible to give form to arguments of this sort until one has a way of speaking about the degree to which the naïve model is false. In an observational study, one could never assert with warranted conviction that the naïve model is precisely true. Trivially small deviations from the naïve model will have a trivially small impact on the study's conclusions. Sufficiently large deviations from the naïve model will overturn the results of any study. Because these two facts are always true, they quickly exhaust their usefulness. Therefore, the magnitude of the deviation is all-important. The sensitivity of an observational study to bias from an unmeasured covariate u is the magnitude of the departure from the naïve model that would need to be present to materially alter the study's conclusions.[11]

The first sensitivity analysis in an observational study concerned smoking and lung cancer. In 1959, Jerry Cornfield and his colleagues [15] asked about the magnitude of the bias from an unobserved covariate u needed to alter the conclusion

---

[11] In general, a sensitivity analysis asks how the conclusion of an argument dependent upon assumptions would change if the assumptions were relaxed. The term is sometimes misused to refer to performing several parallel statistical analyses without regard to the assumptions upon which they depend. If several statistical analyses all depend upon the same assumption — for instance, the naïve model (3.5) — then performing several such analyses provides no insight into consequences of the failure of that assumption.

3.4 Sensitivity Analysis: People Who Look Comparable May Differ

from observational studies that heavy smoking causes lung cancer. They concluded that the magnitude of the bias would need to be enormous.

The sensitivity analysis model: Quantitative deviation from random assignment

The naïve model (3.5)–(3.8) said that two people, k and ℓ, with the same observed covariates, $\mathbf{x}_k = \mathbf{x}_\ell$, have the same probability of treatment given $(r_T, r_C, \mathbf{x}, u)$, i.e., $\pi_k = \pi_\ell$, where $\pi_k = \Pr(Z_k = 1 \mid r_{Tk}, r_{Ck}, \mathbf{x}_k, u_k)$ and $\pi_\ell = \Pr(Z_\ell = 1 \mid r_{T\ell}, r_{C\ell}, \mathbf{x}_\ell, u_\ell)$. The sensitivity analysis model speaks about the same probabilities in (3.1), saying that the naïve model (3.5)–(3.8) may be false, but to an extent controlled by a parameter, $\Gamma \geq 1$. Specifically, it says that two people, k and ℓ, with the same observed covariates, $\mathbf{x}_k = \mathbf{x}_\ell$, have odds[12] of treatment, $\pi_k/(1-\pi_k)$ and $\pi_\ell/(1-\pi_\ell)$, that differ by at most a multiplier of Γ; that is, in (3.1),

$$\frac{1}{\Gamma} \leq \frac{\pi_k/(1-\pi_k)}{\pi_\ell/(1-\pi_\ell)} \leq \Gamma \quad \text{whenever } \mathbf{x}_k = \mathbf{x}_\ell. \tag{3.13}$$

If $\Gamma = 1$ in (3.13), then $\pi_k = \pi_\ell$, so (3.5)–(3.8) is true; that is, $\Gamma = 1$ corresponds with the naïve model. In §3.1, expression (3.1) was seen to be a representation and not a model — something that is always true for suitably defined u_ℓ — but that representation took $\pi_\ell = 0$ or $\pi_\ell = 1$, which implies $\Gamma = \infty$ in (3.13). In other words, numeric values of Γ between $\Gamma = 1$ and $\Gamma = \infty$ define a spectrum that begins with the naïve model (3.5)–(3.8) and ends with something that is hollow in the sense that it is always true, namely (3.1). The hollow statement that is always true, namely (3.1), is the statement that 'association does not imply causation,' that is, a sufficiently large departure from the naïve model can explain away as noncausal any observed association.

If $\Gamma = 2$, and if you, k, and I, ℓ, look the same, in the sense that we have the same observed covariates, $\mathbf{x}_k = \mathbf{x}_\ell$, then you might be twice as likely as I to receive the treatment because we differ in ways that have not been measured. For instance, if your $\pi_k = 2/3$ and my $\pi_\ell = 1/2$, then your odds of treatment rather than control are $\pi_k/(1-\pi_k) = 2$ or 2-to-1, whereas my odds of treatment rather than control are $\pi_\ell/(1-\pi_\ell) = 1$ or 1-to-1, and you are twice as likely as I to receive treatment, $\{\pi_k/(1-\pi_k)\}/\{\pi_\ell/(1-\pi_\ell)\} = 2$ in (3.13).[13]

---

[12] Odds are an alternative way of expressing probabilities. Probabilities and odds carry the same information in different forms. A probability of $\pi_k = 2/3$ is an odds of $\pi_k/(1-\pi_k) = 2$ or 2-to-1. Gamblers prefer odds to probabilities because odds express the chance of an event in terms of fair betting odds, the price of a fair bet. It is easy to move from probability π_k to odds $\omega_k = \pi_k/(1-\pi_k)$ and back again from odds ω_k to probability $\pi_k = \omega_k/(1+\omega_k)$.

[13] Implicitly, the critic is saying that the failure to measure u is the source of the problem, or that (3.5) would be true with $(\mathbf{x}, u)$ in place of $\mathbf{x}$, but is untrue with $\mathbf{x}$ alone. That is, the critic is saying $\pi_\ell = \Pr(Z_\ell = 1 \mid r_{T\ell}, r_{C\ell}, \mathbf{x}_\ell, u_\ell) = \Pr(Z_\ell = 1 \mid \mathbf{x}_\ell, u_\ell)$. As in §3.1, because of the delicate nature of unobserved variables, this is a manner of speaking rather than a tangible distinction. If the formalities are understood to refer to $\pi_\ell = \Pr(Z_\ell = 1 \mid r_{T\ell}, r_{C\ell}, \mathbf{x}_\ell, u_\ell)$, then it is not necessary to

Sensitivity analysis model when pairs are matched for observed covariates

The sensitivity analysis model (3.13) is quite general in its applicability [85, Chapter 4], but here its implications for matched pairs are developed [74]. Suppose that two subjects, k and ℓ, with the same observed covariates, $\mathbf{x}_k = \mathbf{x}_\ell$, are paired, with precisely the additional fact that one of them is treated and the other control, $Z_k + Z_\ell = 1$. Then in the representation (3.1), the chance that k is treated and ℓ is control is:

$$\Pr(Z_k = 1, Z_\ell = 0 \mid r_{Tk}, r_{Ck}, \mathbf{x}_k, u_k, r_{T\ell}, r_{C\ell}, \mathbf{x}_\ell, u_\ell, Z_k + Z_\ell = 1) = \frac{\pi_k}{\pi_k + \pi_\ell}. \quad (3.14)$$

If in addition the sensitivity model (3.13) were true in (3.1), then simple algebra yields

$$\frac{1}{1+\Gamma} \leq \frac{\pi_k}{\pi_k + \pi_\ell} \leq \frac{\Gamma}{1+\Gamma}. \quad (3.15)$$

In words, the condition (3.13) becomes a new condition (3.15) on paired individuals where one is treated and the other control, $Z_k + Z_\ell = 1$. If $\Gamma = 1$, then all three terms in (3.15) equal $\frac{1}{2}$, as in the randomized experiment in Chapter 2. As $\Gamma \to \infty$, the lower bound in (3.13) tends to zero and the upper bound tends to one.

Instead of pairing just two individuals, k and ℓ, suppose we pair $2I$ distinct individuals of the L individuals in the population in just this way, insisting that within each pair the two subjects have the same observed covariates and different treatments. Renumber these paired subjects into I pairs of two subjects, $i = 1, 2, \ldots, I$, $j = 1, 2$, so $\mathbf{x}_{i1} = \mathbf{x}_{i2}$, $Z_{i1} = 1 - Z_{i2}$ in each of the I pairs.[14] If (3.1) and (3.13) are true, then the distribution of treatment assignments in the I pairs satisfies

$$Z_{i1}, i = 1, \ldots, I \text{ are mutually independent,} \quad (3.16)$$

$$Z_{i2} = 1 - Z_{i1}, i = 1, \ldots, I, \quad (3.17)$$

$$\frac{1}{1+\Gamma} \leq \frac{\pi_{i1}}{\pi_{i1} + \pi_{i2}} \leq \frac{\Gamma}{1+\Gamma}, i = 1, \ldots, I. \quad (3.18)$$

This is very similar in form to the distribution of treatment assignments in a randomized paired experiment in Chapter 2, except that in the experiment (3.14) was $\frac{1}{2}$

---

insist that $\pi_\ell = \Pr(Z_\ell = 1 \mid \mathbf{x}_\ell, u_\ell)$. Conversely, if (3.1) and (3.13) were true as they stand, then there is an unobserved covariate $\tilde{u}_\ell$ such that (3.1) and (3.13) are true with $\pi_\ell = \Pr(Z_\ell = 1 \mid \mathbf{x}_\ell, \tilde{u}_\ell)$; simply take $\tilde{u}_\ell = \pi_\ell = \Pr(Z_\ell = 1 \mid r_{T\ell}, r_{C\ell}, \mathbf{x}_\ell, u_\ell)$.

[14] In a fussy technical sense, the numbering of pairs and people within pairs is supposed to convey nothing about these people, except that they were eligible to be paired, that is, they have the same observed covariates, different treatments, with $2I$ distinct people. Information about people is supposed to be recorded in variables that describe them, such as Z, $\mathbf{x}$, u, r_T, r_C, not in their position in the data set. You can't put your brother-in-law in the last pair just because of that remark he made last Thanksgiving; you have to code him in an explicit brother-in-law variable. Obviously, it is easy to make up subscripts that meet this fussy requirement: number the pairs at random, then number the people in a pair at random. The fussy technical point is that, in going from the L people in (3.1) to the $2I$ paired people, no information has been added and tucked away into the subject numbers — the criteria for pairs are precisely $\mathbf{x}_{i1} = \mathbf{x}_{i2}$, $Z_{i1} + Z_{i2} = 1$ with $2I$ distinct individuals.

3.5 Welding Fumes and DNA Damage

for $i = 1, \ldots, I$, whereas in (3.16)–(3.18) the treatment assignment probabilities may vary from pair to pair, are unknown, but are bounded by $1/(1+\Gamma)$ and $\Gamma/(1+\Gamma)$. If $\Gamma = 1.0001$, then (3.14) would differ trivially from a randomized paired experiment, but as $\Gamma \to \infty$ the difference can become arbitrarily large.

Suppose that we had calculated a P-value or a point estimate or confidence interval from a paired observational study matched for observed covariates **x**, by simply applying conventional statistical methods, that is, the methods in Chapter 2 for a randomized paired experiment. Those inferences would have their usual properties if the naïve model (3.5)–(3.8) were true, that is, if $\Gamma = 1$. How might those inferences change if Γ were some specific number larger than 1, indicating some bias due to failure to control for u? Using (3.16)–(3.18) and a few calculations, we can often deduce the range of possible P-values or point estimates or confidence intervals for a specified Γ. Consider, for instance, the P-value for testing the null hypothesis of no treatment effect. If the naïve model, $\Gamma = 1$, led to a P-value of, say, 0.001, and if $\Gamma = 2$ yields a range of possible P-values from 0.0001 to 0.02, then a bias of magnitude $\Gamma = 2$ creates greater uncertainty but does not alter the qualitative conclusion that the null hypothesis of no effect is not plausible. If the critic is thinking in terms of a moderately large deviation from a randomized trial, in which similar looking people may differ by a factor of $\Gamma = 2$ in their odds of treatment, then the critic is simply mistaken: the bias would have to be considerably larger than $\Gamma = 2$ to make no treatment effect plausible.

Every study is sensitive to sufficiently large biases. There is always a value of Γ such that, for that value and larger values of Γ, the interval of possible P-values includes small values, perhaps 0.0001, and large values, perhaps 0.1. A sensitivity analysis simply displays how the inference changes with Γ. For smoking and lung cancer, the bias would have to be enormous, $\Gamma = 6$; see [85, Chapter 4]. The question answered by a sensitivity analysis is: how large does Γ have to be before one must concede that the critic's criticism might be correct?

It is time to consider an example.

3.5 Welding Fumes and DNA Damage

Sensitivity analysis when testing the hypothesis of no treatment effect

The fumes produced by electric welding contain chromium and nickel and have been judged genotoxic in laboratory tests [39]. Werfel and colleagues [111] looked for evidence of DNA damage in humans by comparing 39 male welders to 39 male controls matched for age and smoking habits. Table 3.1 displays the comparability of the two groups with respect to the three covariates used in matching. Clearly, Table 3.1 is a rather limited demonstration of comparability.

Werfel et al. [111] presented several measures of genetic damage, including the measurement of DNA single strand breakage and DNA-protein cross-links using elution rates through polycarbonate filters with proteinase K. Broken strands are

Table 3.1 Covariate balance in 39 matched welder-control pairs. Covariates are gender, smoking and age.

		Welders	Controls
Male		100%	100%
Smokers		69%	69%
Age	Mean	39	39
	Minimum	23	23
	Lower Quartile	34	32
	Median	38	36
	Upper Quartile	46	46
	Maximum	56	59

expected to pass through filters more quickly, at a higher rate. Figure 3.1 depicts the elution rates and their matched pair differences. The differences are mostly positive, with higher elution rates for welders, and the differences are fairly symmetric about their median, with longer tails than the Normal distribution.

Table 3.2 is the sensitivity analysis for the one-sided P-value using Wilcoxon's signed rank statistic to test the null hypothesis of no treatment effect against the alternative that exposure to welding fumes caused an increase in DNA damage. The first row, $\Gamma = 1$, is the usual randomization inference, which would be appropriate if the 78 men had been paired for age and smoking and randomly assigned to their careers as a welder or a nonwelder. In the first row, the range of possible P-values is a single number, 3.1×10^{-7}, because there would be no uncertainty about the distribution of treatment assignments, $\mathbf{Z}$, in a randomized experiment. The naïve model (3.5)–(3.8) would also lead to $\Gamma = 1$ and the single P-value in the first row of Table 3.2. If this had been a randomized experiment, there would have been strong evidence against the null hypothesis of no effect. However, it was not a randomized experiment. The P-value in the first row of Table 3.2 says that it is implausible that the difference seen in Figure 3.1 is due to chance, the flip of a coin that assigned one man to treatment, another to control. The P-value in the first row of Table 3.2 does not speak to the critic's concern that the difference seen in Figure 3.1 is neither due to chance nor due to an effect caused by welding, but reflects instead some way that the matched welders and controls are not comparable. A small P-value, here 3.1×10^{-7}, computed assuming either randomization or equivalently the naïve model (3.5)–(3.8) does nothing to address the critic's concern. It is, however, possible to speak to that concern.

The second row permits a substantial departure from random treatment assignment or (3.5)–(3.8). It says that two men of the same age and smoking status — the same $\mathbf{x}$ — may not have the same chance of a career as a welder: one such man may be twice as likely as another to choose a career as a welder, $\Gamma = 2$, because they differ in terms of a covariate u that was not measured. This introduces a new source of uncertainty beyond chance. Using (3.16)–(3.18), we may determine every possible P-value that could be produced when $\Gamma = 2$, and it turns out that the smallest possible P-value is 3.4×10^{-12} and the largest possible P-value is 0.00064. Although

3.5 Welding Fumes and DNA Damage

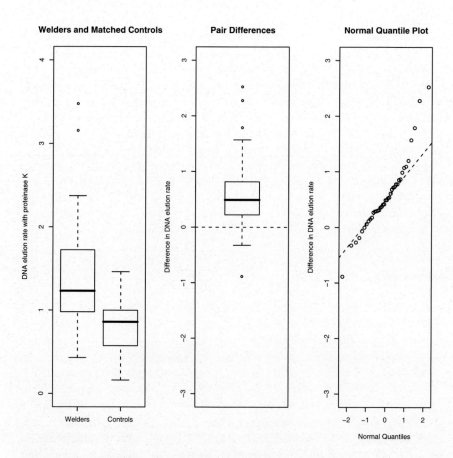

Fig. 3.1 DNA elution rates through polycarbonate filters with proteinase K for 39 male welders and 39 male controls matched for age and smoking. This assay is a measure of DNA single strand breakage and DNA-protein cross-links. In the boxplot of differences, there is a line at zero. In the Normal quantile plot, the line is fitted to the median and quartiles.

Table 3.2 Sensitivity analysis for the one-tailed P-value for testing the null hypothesis of no treatment effect on DNA elution rates with proteinase K in 39 pairs of a male welder and a male control matched for age and smoking. The table gives the lower (min) and upper (max) bounds on the one-sided P-value for departures from random assignment of various magnitudes, Γ. For $\Gamma = 1$, the two P-values are equal to each other and equal to the randomization P-value from Chapter 2. For $\Gamma > 1$, there is a range $[P_{\min},\ P_{\max}]$ of possible P-values. This study is sensitive only to very large biases, for instance $\Gamma = 5$, because at this point the range includes both small and large, significant and insignificant, P-values.

Γ	$P_{\min}$	$P_{\max}$
1	3.1×10^{-7}	3.1×10^{-7}
2	3.4×10^{-12}	0.00064
3	$< 10^{-15}$	0.011
4	$< 10^{-15}$	0.047
5	$< 10^{-15}$	0.108

a bias of magnitude $\Gamma = 2$ would introduce greater uncertainty, there would be no real doubt that the null hypothesis of no treatment effect is not plausible.

As seen in Table 3.2, all possible P-values are less than 0.05 for departures from randomization as large as $\Gamma = 4$. A bias of $\Gamma = 4$ is a very large departure from a randomized experiment. In a randomized experiment, each man in each pair has probability $\frac{1}{2}$ of receiving treatment. If $\Gamma = 4$, then in a matched pair, one man might have probability $\Gamma/(1+\Gamma) = 4/5$ of treatment and the other might have probability $1/(1+\Gamma) = 1/5$; however, even such a large departure from a randomized experiment is quite unlikely to produce the difference seen in Figure 3.1.

By $\Gamma = 5$, the situation has changed. Now the range of possible P-values includes some that are much smaller than the conventional 0.05 level and others that are considerably higher, the range being from $< 10^{-15}$ to 0.108. A very large departure from random assignment of magnitude $\Gamma = 5$ could produce the difference seen in Figure 3.1 even if welding has no effect on DNA elution rates.

Computations

The P-values in Table 3.2 are exact: in principle, they could be produced by direct enumeration analogous to those in §2.3.3, except that, now, different treatment assignments **Z** have different probabilities constrained by (3.16)–(3.18). The actual exact computations are done more efficiently; see Appendix §3.9.

Look back at (3.16)–(3.18) and consider how to set the probabilities to make T as large or as small as possible. In the absence of ties, the upper bound in Table 3.2 is obtained by comparing Wilcoxon's signed rank statistic to the distribution of a random variable, $\overline{\overline{T}}$, which is the sum of I independent random variables, $i = 1, 2, \ldots, I$, that take the value i with probability $\Gamma/(1+\Gamma)$ or the value 0 with probability $1/(1+\Gamma)$. In parallel, the lower bound in Table 3.2 is obtained by comparing Wilcoxon's signed rank statistic to the distribution of a random variable, $\overline{T}$, which is the sum of I independent random variables, $i = 1, 2, \ldots, I$, that take the value i with probability $1/(1+\Gamma)$ or the value 0 with probability $\Gamma/(1+\Gamma)$.

Although exact computation is quite feasible for moderate sample sizes, I, a large sample approximation is easier and typically adequate. In Table 3.2, Wilcoxon's signed rank statistic is $T = 715$ and there are no ties. In the absence of ties, for a specified Γ, the largest null distribution[15] of T subject to (3.16)–(3.18) has expec-

---

[15] What does it mean to speak of the 'largest distribution'? One random variable, A, is said to be stochastically larger than another random variable, B, if $\Pr(A \geq k) \geq \Pr(B \geq k)$ for every k. That is, A is more likely than B to jump over a bar at height k, no matter how high k the bar is set. Because this must be true *for every* k, it is a rather special relationship between random variables. For instance, it might happen that $\Pr(A \geq -1.65) = 0.95$, $\Pr(A \geq 1.65) = 0.05$ while $\Pr(B \geq -1.65) = 0.80$, $\Pr(B \geq 1.65) = 0.20$, so neither A nor B is stochastically larger than the other. For instance, this is true if A is Normal with mean zero and standard deviation 1, and B is Normal with mean zero and standard deviation 2. The intuition is overwhelmingly strong that the largest null distribution of Wilcoxon's signed rank statistic is obtained by driving the chance of a positive difference up to its maximum, namely $\Gamma/(1+\Gamma)$ in (3.18), and this intuition turns out to be correct. It takes a small amount of effort in this case to show the distribution is

3.5 Welding Fumes and DNA Damage

tation

$$E\left(\overline{\overline{T}}\middle|\mathscr{F},\mathscr{Z}\right) = \frac{\Gamma}{1+\Gamma}\cdot\frac{I(I+1)}{2} \qquad (3.19)$$

and variance

$$\text{var}\left(\overline{\overline{T}}\middle|\mathscr{F},\mathscr{Z}\right) = \frac{\Gamma}{(1+\Gamma)^2}\frac{I(I+1)(2I+1)}{6}. \qquad (3.20)$$

For $\Gamma = 1$, formulas (3.19) and (3.20) reduce to the formulas for randomization inference in §2.3.3, namely $E(T|\mathscr{F},\mathscr{Z}) = I(I+1)/4$ and $\text{var}(T|\mathscr{F},\mathscr{Z}) = I(I+1)(2I+1)/24$. For $\Gamma = 3$, the expectation is

$$E\left(\overline{\overline{T}}\middle|\mathscr{F},\mathscr{Z}\right) = \frac{3}{(1+3)}\cdot\frac{39(39+1)}{2} = 585 \qquad (3.21)$$

in (3.19) and the variance is

$$\text{var}\left(\overline{\overline{T}}\middle|\mathscr{F},\mathscr{Z}\right) = \frac{3}{(1+3)^2}\frac{39(39+1)(2\cdot 39+1)}{6} = 3851.25 \qquad (3.22)$$

in (3.20). For large I, the standardized deviate,

$$\frac{T - E\left(\overline{\overline{T}}\middle|\mathscr{F},\mathscr{Z}\right)}{\sqrt{\text{var}\left(\overline{\overline{T}}\middle|\mathscr{F},\mathscr{Z}\right)}} = \frac{715 - 585}{\sqrt{3851.25}} = 2.0948 \qquad (3.23)$$

is compared to the standard Normal cumulative distribution, $\Phi(\cdot)$, to yield the approximate upper bound on the one-sided P-value, $1 - \Phi(2.0948) = 0.018$, which is close to the exact value of 0.011 in Table 3.2. The lower bound is obtained in a parallel manner, with

$$E\left(\overline{T}\middle|\mathscr{F},\mathscr{Z}\right) = \frac{1}{1+\Gamma}\cdot\frac{I(I+1)}{2} \qquad (3.24)$$

and $\text{var}\left(\overline{T}\middle|\mathscr{F},\mathscr{Z}\right) = \text{var}\left(\overline{\overline{T}}\middle|\mathscr{F},\mathscr{Z}\right)$ given again by (3.20).

Ties are a minor inconvenience. In the notation of §2.3.3, the expectations become

$$E\left(\overline{\overline{T}}\middle|\mathscr{F},\mathscr{Z}\right) = \frac{\Gamma}{1+\Gamma}\sum_{i=1}^{I}s_i q_i, \qquad (3.25)$$

---

actually stochastically largest; see [85, Chapter 4]. In other cases, the same result is available with somewhat more effort. In still other cases, one can speak of a 'largest distribution' only asymptotically, that is, only in large samples; see [26] or [85, Chapter 4]. This presents no problem in practice, because the asymptotic results are quite adequate and easy to use [85, Chapter 4]; however, it does make the theory of the paired case simpler than, say, the theory of matching with several controls matched to each treated subject. To see a parallel discussion of the paired case and the case of several controls, see [89].

Table 3.3 Sensitivity analysis for the one-sided 95% confidence interval for a constant, additive treatment effect τ on DNA elution rates. As usual, the hypothesis of a constant effect $H_0 : \tau = \tau_0$ is tested by testing no effect on $Y_i - \tau_0$ for the given value of Γ. The one-sided 95% confidence interval is the set of values of τ_0 not rejected in the one-sided, 0.05 level test. As Γ increases, there is greater potential deviation from random treatment assignment in (3.13), and the confidence interval grows longer. For instance, a treatment effect of $\tau_0 = 0.30$ would be implausible in a randomized experiment, $\Gamma = 1$, but not in an observational study with $\Gamma = 2$.

Γ	1	2	3
95% Interval	$[0.37, \infty)$	$[0.21, \infty)$	$[0.094, \infty)$

$$E\left(\overline{T} \mid \mathscr{F}, \mathscr{Z}\right) = \frac{1}{1+\Gamma} \sum_{i=1}^{I} s_i q_i, \qquad (3.26)$$

while the variance becomes

$$\operatorname{var}\left(\overline{T} \mid \mathscr{F}, \mathscr{Z}\right) = \operatorname{var}\left(\overline{\overline{T}} \mid \mathscr{F}, \mathscr{Z}\right) = \frac{\Gamma}{(1+\Gamma)^2} \sum_{i=1}^{I} (s_i q_i)^2. \qquad (3.27)$$

The remaining calculations are unchanged.

Sensitivity analysis for a confidence interval

Table 3.3 is the sensitivity analysis for the one-sided 95% confidence interval for an additive, constant treatment effect discussed in §2.4.2. As in a randomized experiment, the hypothesis that $H_0 : r_{Tij} = r_{Cij} + \tau_0$ is tested by testing the null hypothesis of no treatment effect on the adjusted responses, $R_{ij} - \tau_0 Z_{ij}$, or equivalently on the adjusted, treated-minus-control pair differences, $Y_i - \tau_0$. The one-sided 95% confidence interval is the set of values of τ_0 not rejected by a one-sided, 0.05 level test.

From Table 3.2, the hypothesis $H_0 : \tau = \tau_0$ for $\tau_0 = 0$ is barely rejected for $\Gamma = 4$ because the maximum possible one-sided P-value is 0.047. For $\Gamma = 3$, the maximum possible one-sided P-value is 0.04859 for $\tau_0 = .0935$ and is 0.05055 for $\tau_0 = .0936$, so after rounding to two significant digits, the one-sided 95% confidence interval is $[0.094, \infty)$.

Sensitivity analysis for point estimates

For each value of $\Gamma \geq 1$, a sensitivity analysis replaces a single point estimate, say $\widehat{\tau}$, by an interval of point estimates, say $[\widehat{\tau}_{\min}, \widehat{\tau}_{\max}]$ that are the minimum and maximum point estimates for all distributions of treatment assignments satisfying (3.16)–(3.18). Unlike a test or a confidence interval, and like a point estimate, this interval $[\widehat{\tau}_{\min}, \widehat{\tau}_{\max}]$ does not reflect sampling uncertainty; however, it does reflect uncertainty introduced by departures from random treatment assignment in (3.13) or (3.16)–(3.18).

3.6 Bias Due to Incomplete Matching

Table 3.4 Sensitivity analysis for the Hodges-Lehmann (HL) point estimate for a constant, additive treatment effect τ on DNA elution rates.

Γ	1	2	3
HL Estimate	[0.51, 0.51]	[0.36, 0.69]	[0.27, 0.81]

In a randomized experiment in §2.4.3, the Hodges-Lehmann point estimate of a constant, additive treatment effect, τ, was obtained by computing Wilcoxon's signed rank statistic T from $Y_i - \tau_0$ and solving for the estimate, $\hat{\tau}$, as the value of τ_0 that brings T as close as possible to its null expectation, $I(I+1)/4$. It is not difficult to show[16] the interval of point estimates, $[\hat{\tau}_{\min}, \hat{\tau}_{\max}]$, is obtained by finding the value $\hat{\tau}_{\min}$ such that when Wilcoxon's T is computed from $Y_i - \hat{\tau}_{\min}$ it is as close as possible to (3.19), and the value $\hat{\tau}_{\max}$ such that when T is computed from $Y_i - \hat{\tau}_{\max}$ it is as close as possible to (3.24). For example, with $\Gamma = 2$, the maximum expectation in (3.19) is $\{\Gamma/(1+\Gamma)\}I(I+1)/2$ or $\{2/(1+2)\}39(39+1)/2 = 520$. When Wilcoxon's T is computed from $Y_i - .35550001$ it is $T = 519$, but when it is computed from $Y_i - .35549999$ it is $T = 521$, so $\hat{\tau}_{\min} = 0.3555$.

Table 3.4 displays $[\hat{\tau}_{\min}, \hat{\tau}_{\max}]$ for three values of Γ. For $\Gamma = 1$, there is a single point estimate, $\hat{\tau} = 0.51$, the value that would be obtained in a randomized experiment. A large departure from random assignment of magnitude $\Gamma = 3$ could reduce that by almost half, to $\hat{\tau}_{\min} = 0.27$.

3.6 Bias Due to Incomplete Matching

Chapter 3 has followed Chapter 2 in focusing on internal validity, as discussed in §2.6. That is, the focus has been on inference about treatment effects for the $2I$ matched individuals, not on whether the same treatment effects would be found in the population of L individuals. When treatment effects vary from person to person, as seemed to be the case in §2.5, changing the individuals under study may change the magnitude of the effect.

Although not required for internal validity, it is common in practice to match all of the treated subjects in the population, so that the number of matched pairs, I, equals the number of treated subjects $\sum Z_\ell$ in the available population of L individuals. The goal here is an aspect of external validity, specifically the ability to speak about treatment effects in the original population of L individuals. If the population of L individuals were itself a random sample from an infinite population, and if all treated subjects were matched, then under the naïve model (3.5)–(3.8), the average treated-minus-control difference in observed responses, $(1/I)\sum Y_i$, would be unbiased for the expected treatment effect on people who typically receive the treatment, namely $E(r_T - r_C | Z = 1)$; however, this is typically untrue if some treated subjects are deleted, creating what is known as the 'bias due to incomplete matching' [72].

---

[16] See [80] or [85, Chapter 4].

Unless explicitly instructed to do otherwise, the matching methods in Part II will match all treated subjects.

If some treated subjects, ℓ, in the population have propensity scores near 1, $e(\mathbf{x}_\ell) \approx 1$, they will be very difficult to match. Virtually everyone with this $\mathbf{x}_\ell$ will receive treatment. Rather than delete individuals one at a time based on extreme propensity scores, $e(\mathbf{x}_\ell)$, it is usually better to go back to the covariates themselves, $\mathbf{x}_\ell$, perhaps redefining the population under study to be a subpopulation of the original population of L subjects. A population defined in terms of $e(\mathbf{x}_\ell)$ is likely to have little meaning to other investigators, whereas a population defined in terms of one or two familiar covariates from $\mathbf{x}_\ell$ will have a clear meaning. In this redefined population, all treated subjects are matched. For instance, in a study of the effects of joining a gang at age 14 on subsequent violence [33, 34], a handful of the most extremely violent, chronically violent boys at age 13 all joined gangs at age 14. These few extremely violent boys had no plausible controls — all of the potential controls were substantially less violent than these few boys prior to age 14. Disappointing as this may be, there is no plausible way to estimate the effect of a treatment in a subpopulation that always receives the treatment. Using violence prior to age 14, the study population was redefined to exclude the extremely chronically violent subpopulation, with no claim that similar effects would be found in that subpopulation. See §12.4 for further discussion of the gang study.

3.7 Summary

Two simple models for treatment assignment in observational studies have been discussed. The two models define and divide the two tasks faced by the investigator.

The first model is naïve: it says that two people who look comparable in terms of observed covariates $\mathbf{x}_\ell$ are comparable. People who look comparable in terms of $\mathbf{x}_\ell$ are said to be 'ostensibly comparable'; on the surface, they appear to be comparable, but they may not be so. If the naïve model were true, it would suffice to match treated and control subjects for the observed covariates $\mathbf{x}_\ell$. More precisely, if the naïve model were true, and treated and control subjects were matched so that $\mathbf{x}_{ij} = \mathbf{x}_{ik}$ for different subjects j and k in the same matched set i, then this alone would reproduce the distribution of treatment assignments in a randomized experiment. The key difficulty in an observational study is that there is usually little or no reason to believe the naïve model is true.

The sensitivity analysis model says people who look comparable in terms of observed covariates $\mathbf{x}_\ell$ may differ in terms of one or more covariates u_ℓ that were not measured. The sensitivity analysis model says that two subjects, j and k, who look comparable in terms of observed covariates, and so might be placed in the same matched set i — that is, two subjects with $\mathbf{x}_{ij} = \mathbf{x}_{ik}$ — may differ in their odds of treatment by a factor of $\Gamma \geq 1$. When $\Gamma = 1$, the sensitivity analysis model reduces to the naïve model, yielding the distribution of treatment assignments in a randomized experiment. When $\Gamma > 1$, the treatment assignment probabilities are unknown, but unknown to a bounded degree. For each fixed value of $\Gamma \geq 1$, there is a range of possible inferences, for instance an interval of possible P-values or point estimates. For $\Gamma = 1$, the interval is a single point, namely the randomization inference. As $\Gamma \to \infty$, the interval widens until, at some point, it is so long as to be uninformative, for instance, including both small and large P-values. The sensitivity analysis determines the magnitude of bias, measured by Γ, that would need to be present to qualitatively alter the conclusions of the study, that is, to produce an interval so long that it is uninformative. There is always such a Γ, but the numerical value of Γ varies dramatically from one observational study to the next. The naïve model assumes $\Gamma = 1$. The phrase 'association does not imply causation' refers to letting $\Gamma \to \infty$. The sensitivity analysis determines the value of Γ that is relevant in light of the data at hand.

The two models define and divide the two tasks in an observational study. The first task is to compare people who look comparable. The first task can be done somewhat mechanically and completed: we may reach a stage where we are all forced to agree that the people who are being compared under alternative treatments do indeed look comparable in terms of observed covariates **x**. At this point, the first task is done. Part II of this book discusses the first task, namely matching for observed covariates. The second task is to address the possibility that differing outcomes in treated and control groups are not effects caused by the treatment, but instead reflect some way in which treated and control groups are not comparable in terms of covariates that were not measured. The second task is not a mechanical task, not one that could be handed over to a computer. The second task, being attended by controversy, is more challenging, and hence more interesting. Sensitivity analysis, the tactics in Chapter 5, and the concepts in Parts III and IV are aimed at the second task.

3.8 Further Reading

In this book, see Chapter 8 for further discussion of propensity scores, and see Chapter 5 and Parts III and IV for further discussion of sensitivity analysis. Propensity scores originate in [67] and are discussed in [16, 18, 30, 37, 35, 41, 56, 52, 54, 64, 65, 69, 70, 71, 76, 86, 93, 94, 95, 107] and [85, §3]. The method of sensitivity analysis described here is not restricted to matched pairs and is discussed in [74, 79, 80, 81, 26, 86, 89] and [85, §4]; see also [48]. That one-parameter sensitivity analysis may be reexpressed in terms of two parameters, one describing the relationship between Z and u, the other describing the relationship between r_C and u; see [91]. The one-parameter and two-parameter analyses are numerically the same, but the latter may aid in interpreting the former. Other approaches to sensitivity analysis are discussed in [13, 14, 15, 24, 25, 28, 42, 68, 53, 54, 55, 50, 66, 68, 73, 113, 99, 59]. For a few applications of sensitivity analysis, see [60, 19, 49, 100].

3.9 Appendix: Exact Computations for Sensitivity Analysis

Exact sensitivity distribution of Wilcoxon's statistic

For small to moderate I, the exact upper bound on the distribution of Wilcoxon's signed rank statistic may be obtained quickly in R as the convolution of I probability generating functions. Only the situation without ties is considered. Pagano and Tritchler [62] observed that permutation distributions may often be obtained in polynomial time by applying the fast Fourier transform to the convolution of characteristic functions or generating functions. Here, probability generating functions are used along with the R function `convolve`.

The distribution of $\overline{\overline{T}}$ is the distribution of the sum of I independent random variables, $i = 1, 2, \ldots, I$, taking the value i with probability $\Gamma/(1+\Gamma)$ and the value 0 with probability $1/(1+\Gamma)$. The ith random variable has probability generating function

$$h_i(x) = \frac{1}{1+\Gamma} + \frac{\Gamma x^i}{1+\Gamma}, \tag{3.28}$$

and $\overline{\overline{T}}$ has generating function $\Pi_{i=1}^{I} h_i(x)$. In R, the generating function of a random variable taking integer values $0, 1, \ldots, B$ is represented by a vector of dimension $B+1$ whose $b+1$ coordinate gives the probability that the random variable equals b. For instance, $h_3(x)$ is represented by

$$\left(\frac{1}{1+\Gamma}, 0, 0, \frac{\Gamma}{1+\Gamma} \right). \tag{3.29}$$

The distribution of $\overline{\overline{T}}$ is obtained by convolution as a vector with $1 + I(I+1)/2$ coordinates representing $\Pi_{i=1}^{I} h_i(x)$ and giving the $\Pr\left(\overline{\overline{T}} = b\right)$ for $b = 0, 1, \ldots, I(I+1)/2$ where $I(I+1)/2 = \sum i$.

The 39 matched pair differences are
```
> dif
 [1]  0.148  0.358  1.572  2.526  0.287 -0.271  0.494
 [8]  0.716  0.411  0.988  1.073  1.097  0.491  0.294
[15]  0.062  0.417 -0.886  0.314  0.178  0.867  0.539
[22]  1.791 -0.001 -0.067  0.779  0.113  0.729  0.374
[29]  0.610  2.277  0.303 -0.326  0.527  1.203  0.854
[36]  0.269  0.683 -0.190  0.778
```

For $\Gamma = 2$ in Table 3.2, the upper bound on the one sided P-value is
```
> senWilcoxExact(dif,gamma=2)
      pval   T ties
[1,] 0.0006376274 715 0
```

The R functions that compute this exact P-value follow.

3.9 Appendix: Exact Computations for Sensitivity Analysis

```
> senWilcoxExact
function(d,gamma=1){
 a<-abs(d)
 rk<-rank(a)
 s<-(d>0)*1
 tie<-1*(d==0)
 sgn<-s+tie/2
 wt<-sum(sgn*rk)
 out<-matrix(NA,1,3)
 colnames(out)<-c("pval","T","ties")
 out[1,3]<-sum(tie)
 out[1,2]<-wt
 out[1,1]<-wilcsenexacttail(floor(wt),length(d),gamma=gamma)
 out
}

> wilcsenexacttail
function(k,n,gamma=1){
#Upper tail probability for Wilcoxon's signed rank statistic
#Prob(T>=k) in n pairs with specified gamma
1-sum(wilcsenexact(n,gamma=gamma)[1:k])}

> wilcsenexact
function(n,gamma=1){
#Computes the upper bound distribution of
#Wilcoxon's signed rank statistic with n pairs
#Returns a vector g of length 1+sum(1:n)
#where g[k+1] is the probability the statistic
#equals k, for k in 0:sum(1:n)
#Uses gconv
 p<-gamma/(1+gamma)
 g<-c(1-p,p)
 for (i in 2:n){
 gi<-rep(0,i+1)
 gi[1]<-1-p
 gi[i+1]<-p
 g<-gconv(g,gi)
 }
 g
}

> gconv
function(g1,g2){
#convolution of g1 and g2
convolve(g1,rev(g2),type="o")}
```

References

1. Abadie, A.: Semiparametric difference-in-differences estimators. Rev Econ Stud **72**, 1–19 (2005)
2. Angrist, J.D., Krueger, A.B.: Empirical strategies in labor economics. In: Ashenfelter, O., Card, D. (eds.) Handbook of Labor Economics, Volume 3, pp. 1277–1366. New York: Elsevier (1999)
3. Angrist, J. D., Lavy, V.: Using Maimonides' rule to estimate the effect of class size on scholastic achievement. Q J Econ **114**, 533–575 (1999)
4. Angrist, J., Hahn, J.: When to control for covariates? Panel asymptotics for estimates of treatment effects. Rev Econ Statist **86**, 58–72 (2004)
5. Athey, S., Imbens, G.W.: Identification and inference in nonlinear difference-in-differences models. Econometrica **74**, 431–497 (2006)
6. Bertrand, M., Duflo, E., Mullainathan, S.: How much should we trust difference-in-differences estimates? Q J Econ **119**, 249–275 (2004)
7. Besley, T., Case, A.: Unnatural experiments? Estimating the incidence of endogenous policies. Econ J **110**, 672–694 (2000)
8. Bross, I.D.J.: Statistical criticism. Cancer **13**, 394–400 (1961)
9. Bross, I.D.J.: Spurious effects from an extraneous variable. J Chron Dis **19**, 637–647 (1966)
10. Campbell, D.T.: Factors relevant to the validity of experiments in social settings. Psych Bull **54**, 297–312 (1957)
11. Campbell, D.T.: Reforms as experiments. Am Psychol 409–429 (1969)
12. Campbell, D.T.: Methodology and Epistemology for Social Science: Selected Papers. Chicago: University of Chicago Press (1988)
13. Copas, J.B., Li, H.G.: Inference for non-random samples. J Roy Statist Soc B **59**, 55–77 (1997)
14. Copas, J.B., Eguchi, S.: Local sensitivity approximations for selectivity bias. J Roy Statist Soc B **63**, 871–896 (2001)
15. Cornfield, J., Haenszel, W., Hammond, E., Lilienfeld, A., Shimkin, M., Wynder, E.: Smoking and lung cancer: Recent evidence and a discussion of some questions. J Natl Cancer Instit **22**, 173–203 (1959)
16. D'Agostino, R.B.: Propensity score methods for bias reduction in the comparison of a treatment to a non-randomized control group. Statist Med **17**, 2265–2281 (1998)
17. Dawid, A.P.: Conditional independence in statistical theory (with Discussion). J Roy Statist Soc B **41**, 1–31 (1979)
18. Dehejia, R.H., Wahba, S.: Propensity score-matching methods for nonexperimental causal studies. Rev Econ Statist **84**, 151–161 (2002)
19. Diprete, T.A., Gangl, M.: Assessing bias in the estimation of causal effects. Sociolog Method **34**, 271–310 (2004)
20. Dynarski, S.M.: Does aid matter? Measuring the effect of student aid on college attendance and completion. Am Econ Rev **93**, 279–288 (2003)
21. Fenech, M., Changb, W.P., Kirsch-Voldersc, M., Holland, N., Bonassie, S., Zeiger, E.: HUMN project: Detailed description of the scoring criteria for the cytokinesis-block micronucleus assay using isolated human lymphocyte cultures. Mutat Res **534**, 65–75 (2003)
22. Foster, E.M., Bickman, L.: Old wine in new skins: The sensitivity of established findings to new methods. Eval Rev **33**, 281–306 (2009)
23. Frangakis, C.E., Rubin, D.B.: Principal stratification in causal inference. Biometrics **58**, 21–29 (2002)
24. Gastwirth, J.L.: Methods for assessing the sensitivity of comparisons in Title VII cases to omitted variables. Jurimetrics J **33**, 19–34 (1992)
25. Gastwirth, J.L., Krieger, A.M., Rosenbaum, P.R.: Dual and simultaneous sensitivity analysis for matched pairs. Biometrika **85**, 907–920 (1998)
26. Gastwirth, J.L., Krieger, A.M., Rosenbaum, P.R.: Asymptotic separability in sensitivity analysis. J Roy Statist Soc B **62**, 545–555 (2000)

27. Greevy, R., Lu, B., Silber, J.H., Rosenbaum, P.R.: Optimal matching before randomization. Biostatistics **5**, 263–275 (2004)
28. Greenland, S.: Basic methods of sensitivity analysis. Int J Epidemiol **25**, 1107–1116 (1996)
29. Gross, D.B., Souleles, N.S.: Do liquidity constraints and interest rates matter for consumer behavior? Evidence from credit card data. Q J Econ **117**, 149–185 (2002)
30. Hahn, J.Y.: On the role of the propensity score in efficient semiparametric estimation of average treatment effects. Econometrica **66**, 315–331 (1998)
31. Hamermesh, D.S.: The craft of labormetrics. Indust Labor Relat Rev **53**, 363–380 (2000)
32. Hansen, B.B.: The prognostic analogue of the propensity score. Biometrika **95**, 481–488 (2008)
33. Haviland, A., Nagin, D.S., Rosenbaum, P.R.: Combining propensity score matching and group-based trajectory analysis in an observational study. Psych Methods **12**, 247–267 (2007)
34. Haviland, A.M., Nagin, D.S., Rosenbaum, P.R., Tremblay, R.E.: Combining group-based trajectory modeling and propensity score matching for causal inferences in nonexperimental longitudinal data. Devel Psych **44**, 422–436 (2008)
35. Hirano, K., Imbens, G.W., Ridder, G.: Efficient estimation of average treatment effects using the estimated propensity score. Econometrica **71**, 1161–1189 (2003)
36. Hill, A.B.: The environment and disease: Association or causation? Proc Roy Soc Med **58**, 295–300 (1965)
37. Hill, J.L., Waldfogel, J., Brooks-Gunn, J., Han, W.J.: Maternal employment and child development: A fresh look using newer methods. Devel Psychol **41**, 833–850 (2005)
38. Ho, D.E., Imai, K., King, G., Stuart, E.A.: Matching as nonparametric preprocessing for reducing model dependence in parametric causal inference. Polit Anal **15**, 199–236 (2007)
39. International Agency for Research on Cancer: IARC Monographs on the Valuation of Carcinogenic Risks of Chemicals to Humans: Chromium, Nickel and Welding, Volume 49, pp. 447–525. Lyon: IARC (1990)
40. Imai, K.: Statistical analysis of randomized experiments with non-ignorable missing binary outcomes: an application to a voting experiment. Appl Statist **58**, 83–104 (2009)
41. Imbens, G.W.: The role of the propensity score in estimating dose response functions. Biometrika **87**, 706–710 (2000)
42. Imbens, G.W.: Sensitivity to exogeneity assumptions in program evaluation. Am Econ Rev **93**, 126–132 (2003)
43. Imbens, G.W.: Nonparametric estimation of average treatment effects under exogeneity: A review. Rev Econ Statist **86**, 4–29 (2004)
44. Imbens, G.W., Wooldridge, J.M.: Recent developments in the econometrics of program evaluation. J Econ Lit **47**, 5–86 (2009)
45. Joffe, M.M., Ten Have, T.R., Feldman, H.I., Kimmel, S.E.: Model selection, confounder control, and marginal structural models: Review and new applications. Am Statistician **58**, 272–279 (2004)
46. Johnson, B.A., Tsiatis, A.A.: Estimating mean response as a function of treatment duration in an observational study, where duration may be informatively censored. Biometrics **60**, 315–323 (2004)
47. Katan, M.B.: Commentary: Mendelian randomization, 18 years on. Int J Epidemiol **33**, 10–11 (2004)
48. Keele, L.J.: Rbounds: An R Package for Sensitivity Analysis with Matched Data. http://www.polisci.ohio-state.edu/faculty/lkeele/rbounds.html
49. Lee, M.J., Lee, S.J.: Sensitivity analysis of job-training effects on reemployment for Korean women. Empiric Econ **36**, 81–107 (2009)
50. Lin, D.Y., Psaty, B.M., Kronmal, R.A.: Assessing sensitivity of regression to unmeasured confounders in observational studies. Biometrics **54**, 948–963 (1998)
51. Lu, B., Rosenbaum, P.R.: Optimal matching with two control groups. J Comput Graph Statist **13**, 422–434 (2004)
52. Manski, C.: Nonparametric bounds on treatment effects. Am Econ Rev **80**, 319–323 (1990)

53. Manski, C.F.: Identification Problems in the Social Sciences. Cambridge: Harvard University Press (1995)
54. Manski, C.F., Nagin, D.S.: Bounding disagreements about treatment effects: a case study of sentencing and recidivism. Sociol Method **28**, 99–137 (1998)
55. Marcus, S.M.: Using omitted variable bias to assess uncertainty in the estimation of an AIDS education treatment effect. J Educ Behav Statist **22**, 193–201 (1997)
56. McCaffrey, D.F., Ridgeway, G., Morral, A.R.: Propensity score estimation with boosted regression for evaluating causal effects in observational studies. Psych Meth **9**, 403–425 (2004)
57. McKillip, J.: Research without control groups: A control construct design. In: Methodological Issues in Applied Social Psychology, F. B. Bryant, et al., eds., New York: Plenum Press, pp. 159–175 (1992)
58. Meyer, B.D.: Natural and quasi-experiments in economics. J Business Econ Statist **13**, 151–161 (1995)
59. Mitra, N. Heitjan, D.F.: Sensitivity of the hazard ratio to nonignorable treatment assignment in an observational study. Statist Med **26**, 1398–1414 (2007)
60. Normand, S-L., Landrum, M.B., Guadagnoli, E., Ayanian, J.Z., Ryan, T.J., Cleary, P.D., McNeil, B.J.: Validating recommendations for coronary angiography following acute myocardial infarction in the elderly: A matched analysis using propensity scores. J Clin Epidemiol **54**, 387–398 (2001)
61. Normand, S-L., Sykora, K., Li, P., Mamdani, M., Rochon, P.A., Anderson, G.M.: Readers guide to critical appraisal of cohort studies: 3. Analytical strategies to reduce confounding. Brit Med J **330**, 1021–1023 (2005)
62. Pagano, M., Tritchler, D.: On obtaining permutation distributions in polynomial time. J Am Statist Assoc **78**, 435–440 (1983)
63. Reynolds, K.D., West, S.G.: A multiplist strategy for strengthening nonequivalent control group designs. Eval Rev **11**, 691–714 (1987)
64. Robins, J.M., Mark, S.D., Newey, W.K.: Estimating exposure effects by modeling the expectation of exposure conditional on confounders. Biometrics **48**, 479–495 (1992)
65. Robins, J.M., Ritov, Y.: Toward a curse of dimensionality appropriate (CODA) asymptotic theory for semi-parametric models. Statist Med **16**, 285–319 (1997)
66. Robins, J.M., Rotnitzky, A., Scharfstein, D.: Sensitivity analysis for selection bias and unmeasured confounding in missing data and causal inference models. In: Statistical Models in Epidemiology, E. Halloran and D. Berry, eds., pp. 1–94. New York: Springer (1999)
67. Rosenbaum, P.R., Rubin, D.B.: The central role of the propensity score in observational studies for causal effects. Biometrika **70**, 41–55 (1983)
68. Rosenbaum, P. R., Rubin, D.B.: Assessing sensitivity to an unobserved binary covariate in an observational study with binary outcome. J Roy Statist Soc B **45**, 212–218 (1983)
69. Rosenbaum, P.R.: Conditional permutation tests and the propensity score in observational studies. J Am Statist Assoc **79**, 565–574 (1984)
70. Rosenbaum, P.R., Rubin, D.B.: Reducing bias in observational studies using subclassification on the propensity score. J Am Statist Assoc **79**, 516–524 (1984)
71. Rosenbaum, P.R., Rubin, D.B.: Constructing a control group by multivariate matched sampling methods that incorporate the propensity score. Am Statistician **39**, 33–38 (1985)
72. Rosenbaum, P. R., Rubin, D.B.: The bias due to incomplete matching. Biometrics **41**, 106–116 (1985)
73. Rosenbaum, P.R.: Dropping out of high school in the United States: An observational study. J Educ Statist **11**, 207–224 (1986)
74. Rosenbaum, P.R.: Sensitivity analysis for certain permutation inferences in matched observational studies. Biometrika **74**, 13–26 (1987)
75. Rosenbaum, P.R.: The role of a second control group in an observational study (with Discussion). Statist Sci **2**, 292–316 (1987)
76. Rosenbaum, P.R.: Model-based direct adjustment. J Am Statist Assoc **82**, 387–394 (1987)
77. Rosenbaum, P.R.: The role of known effects in observational studies. Biometrics **45**, 557–569 (1989)

References

78. Rosenbaum, P.R.: On permutation tests for hidden biases in observational studies. Ann Statist **17**, 643–653 (1989)
79. Rosenbaum, P.R., Krieger, A.M.: Sensitivity analysis for two-sample permutation inferences in observational studies. J Am Statist Assoc **85**, 493–498 (1990)
80. Rosenbaum, P.R.: Hodges-Lehmann point estimates in observational studies. J Am Statist Assoc **88**, 1250–1253 (1993)
81. Rosenbaum, P.R.: Quantiles in nonrandom samples and observational studies. J Am Statist Assoc **90**, 1424–1431 (1995)
82. Rosenbaum, P.R.: Signed rank statistics for coherent predictions. Biometrics **53**, 556–566 (1997)
83. Rosenbaum, P.R.: Choice as an alternative to control in observational studies (with Discussion). Statist Sci **14**, 259–304 (1999)
84. Rosenbaum, P.R.: Stability in the absence of treatment. J Am Statist Assoc **96**, 210–219 (2001)
85. Rosenbaum, P.R.: Observational Studies (2nd ed.). New York: Springer (2002)
86. Rosenbaum, P.R.: Covariance adjustment in randomized experiments and observational studies (with Discussion). Statist Sci **17**, 286–327 (2002)
87. Rosenbaum, P.R.: Design sensitivity in observational studies. Biometrika **91**, 153–164 (2004)
88. Rosenbaum, P. R.: Heterogeneity and causality: Unit heterogeneity and design sensitivity in observational studies. Am Statistician **59**, 147–152 (2005)
89. Rosenbaum, P.R.: Sensitivity analysis for m-estimates, tests, and confidence intervals in matched observational studies. Biometrics **63**, 456–464 (2007)
90. Rosenbaum, P.R., Silber, J.H.: Sensitivity analysis for equivalence and difference in an observational study of neonatal intensive care units. J Am Statist Assoc **104**, 501–511 (2009)
91. Rosenbaum, P.R., Silber, J.H.: Amplification of sensitivity analysis in observational studies. J Am Statist Assoc, to appear.
92. Rosenzweig, M.R., Wolpin, K.I.: Natural 'natural experiments' in economics. J Econ Lit **38**, 827–874 (2000)
93. Rotnitzky, A., Robins, J.M.: Semiparametric regression estimation in the presence of dependent censoring. Biometrika **82**, 805–820 (1995)
94. Rubin, D.B., Thomas, N.: Characterizing the effect of matching using linear propensity score methods with normal distribution. Biometrika **79**, 797–809 (1992)
95. Rubin, D.B., Thomas, N.: Combining propensity score matching with additional adjustments for prognostic covariates. J Am Statist Assoc **95**, 573–585 (2000)
96. Rutter, M.: Identifying the Environmental Causes of Disease: How Do We Decide What to Believe and When to Take Action? London: Academy of Medical Sciences (2007)
97. Shadish, W.R., Cook, T.D., Campbell, D. T.: Experimental and Quasi-Experimental Designs for Generalized Causal Inference. Boston: Houghton-Mifflin (2002)
98. Shadish, W. R., Cook, T. D.: The renaissance of field experimentation in evaluating interventions. Annu Rev Psychol **60**, 607–629 (2009)
99. Shepherd, B.E., Gilbert, P.B., Mehrotra, D.V.: Eliciting a counterfactual sensitivity parameter. Am Statistician **61**, 56–63 (2007)
100. Silber, J.H., Rosenbaum, P. R., Trudeau, M.E., Chen, W., Zhang, X., Lorch, S.L., Rapaport-Kelz, R., Mosher, R.E., Even-Shoshan, O.: Preoperative antibiotics and mortality in the elderly. Ann Surg **242**, 107–114 (2005)
101. Small, D., Rosenbaum, P.R.: War and wages: The strength of instrumental variables and their sensitivity to unobserved biases. J Am Statist Assoc **103**, 924–933 (2008)
102. Spielman, R.S., Ewens, W.J.: A sibship test for linkage in the presence of association: The sib transmission/disequilibrium test. Am J Hum Genet **62**, 450–458 (1998)
103. Stolley, P.D.: When genius errs — R.A. Fisher and the lung cancer controversy. Am J Epidemiol **133**, 416–425 (1991)
104. Stone, R.: The assumptions on which causal inferences rest. J Roy Statist Soc B **55**, 455–466 (1993)

105. Trochim, W.M.K.: Pattern matching, validity and conceptualization in program evaluation. Eval Rev **9**, 575–604 (1985)
106. Vandenbroucke, J.P.: When are observational studies as credible as randomized trials? Lancet **363**, 1728–1731 (2004)
107. VanderWeele, T.: The use of propensity score methods in psychiatric research. Int J Methods Psychol Res **15**, 95–103 (2006)
108. Volpp, K.G., Rosen, A.K., Rosenbaum, P.R., Romano, P.S., Even-Shoshan, O., Wang, Y., Bellini, L., Behringer, T., Silber, J.H.: Mortality among hospitalized Medicare beneficiaries in the first 2 years following ACGME resident duty hour reform. J Am Med Assoc **298**, 975–983 (2007)
109. Wang, L.S., Krieger, A.M.: Causal conclusions are most sensitive to unobserved binary covariates. Statist Med **25**, 2257–2271 (2006)
110. Weiss, N.S.: Can the "specificity" of an association be rehabilitated as a basis for supporting a causal hypothesis? Epidemiology **13**, 6–8 (2002)
111. Werfel, U., Langen, V., Eickhoff, I., Schoonbrood, J., Vahrenholz, C., Brauksiepe, A., Popp, W., Norpoth, K.: Elevated DNA single-strand breakage frequencies in lymphocytes of welders. Carcinogenesis **19**, 413–418 (1998)
112. Wintemute, G.J., Wright, M.A., Drake, C.M., Beaumont, J.J.: Subsequent criminal activity among violent misdemeanants who seek to purchase handguns: risk factors and effectiveness of denying handgun purchase. J Am Med Assoc **285**, 1019–1026 (2001)
113. Yu, B.B., Gastwirth, J.L.: Sensitivity analysis for trend tests: Application to the risk of radiation exposure. Biostatistics **6**, 201–209 (2005)
114. Zanutto, E., Lu, B., Hornik, R.: Using propensity score subclassification for multiple treatment doses to evaluate a national antidrug media campaign. J Educ Behav Statist **30**, 59–73 (2005)

Chapter 4
Competing Theories Structure Design

Abstract In a well designed experiment or observational study, competing theories make conflicting predictions. Several examples, some quite old, are used to illustrate. Also discussed are: the goals of replication, empirical studies of reasons for effects, and the importance of systemic knowledge in eliminating errors.

> About thirty years ago there was much talk that geologists ought only to observe and not theorise; and I well remember some one saying that at this rate a man might as well go into a gravel-pit and count the pebbles and describe the colours. How odd it is that anyone should not see that all observation must be for or against some view if it is to be of any service.
>
> Charles Darwin 1861 [16]
Letter to Henry Fawcett

> What goes on in science is not that we try to have theories that accommodate our experiences; it's closer that we try to have experiences that adjudicate among our theories.
>
> Jerry Fodor [21, pages 202–203]

> It takes a theory to kill a theory ... given our need to have a systematic way of thinking about complicated reality.
>
> Paul A. Samuelson [55, page 304]

4.1 How Stones Fall

In his *Physics*, Aristotle claimed a heavy object falls faster than a light one. Much in daily experience confirms this, or seems to. Stones fall faster than feathers, for instance. Everyone has seen this. How can you doubt what you and everyone else have seen?

Galileo doubted that Aristotle had it right. In his dialogue[1] *Two New Sciences* [25], Galileo proposed a thought experiment, perhaps the most famous of all thought experiments. Suppose Aristotle were correct, says Galileo, and suppose we connected a large stone to a small one. Would the two connected stones fall faster or slower than the large stone falling alone? In Galileo's words [25, pages 66–67]:

> *Salviati*: But without other experiences, by a short and conclusive demonstration, we can prove clearly that it is not true that a heavier moveable is moved more swiftly than another, less heavy, these being of the same material, and in a word, those of which Aristotle speaks. ... if we had two moveables whose natural speeds were unequal, it is evident that were we to connect the slower to the faster, the latter would be partly retarded by the slower, and this would be partly speeded up by the faster. Do you not agree with me in this opinion?
> *Simplicio*: It seems to me that this would undoubtedly follow.
> *Salviati*: But if this is so, and it is also true that a large stone is moved with eight degrees of speed, for example, and a smaller one with four degrees, then joining both together, their composite will be moved with a speed less than eight degrees. But the two stones joined together make a larger stone than that first one which was moved with eight degrees of speed; therefore this greater stone is moved less swiftly than the lesser one. But this is contrary to your assumption. So you see how, from the supposition that the heavier body is moved more swiftly than the less heavy, I concluded that the heavier moves less swiftly.
> *Simplicio*: I find myself in a tangle ...

Galileo develops his own theory involving a 'law of uniformly accelerated motion:' [25, page 166]:

> Proposition II. Theorem II: If a moveable descends from rest in uniformly accelerated motion, the spaces run through in any times whatever are ... as the squares of those times.

In later Newtonian terms, if an object is acted upon by a single, constant force, namely gravity, its acceleration will be constant in time, its velocity will increase linearly with time, and the distance it travels will increase with the square of the time spent traveling.[2] Galileo's proposition concerns a theory that asserts that the instantaneous velocity of a falling stone is ever increasing at a constant rate of increase, is never the same for two distinct instants. Instantaneous velocity is not measurable, but the proposition says the theory has a measurable consequence, a relationship between distance traveled and time spent traveling.

Naturally, Galileo set up an experiment to test the testable consequence of his theory. Or, at least, it seems natural to us, though it was something of a new idea at the time. Things fall quickly, making measurement difficult, so Galileo began by slowing the speed of the fall. Harré [29, pages 79–81] writes:

> The experiment involved cutting and polishing a groove in a wooden beam and lining the groove with parchment. A polished bronze ball was let roll down the groove when the

---

[1] The full title is "Discourses and Mathematical Demonstrations Concerning Two New Sciences Pertaining to Mechanics and Local Motions" by Galileo Galilei, chief philosopher and mathematician to the most serene Grand Duke of Tuscany. It was written while Galileo was under house arrest for teaching the Copernican theory. Galileo could not publish it in Italy, and it was published by Elzevir in Leyden in 1638.

[2] This is, of course, a simple calculus problem. Then again, calculus was invented, in part, with a view to making this a simple problem.

4.1 How Stones Fall

beam was set on an incline ... Variations in time for many runs of the same descent were very small. The theoretically derived relation between distances and times for uniformly accelerating motion was tested by letting the ball roll a quarter, then half, then two-thirds and so on, of the length of the groove, measuring the times for the journey in each case. The ball did indeed take half the time required for a full descent to reach the quarter way point.

Much is instructive in Galileo's approach.

- Galileo develops his theory in dialogue with, in opposition to, existing theories.[3]
- Everyday impressions that might casually be taken to support Aristotle's theory are immediately challenged, not as false impressions, but as providing no support to Aristotle's theory. Galileo turns attention to a "heavier moveable ... of the same material." If stones fall faster than feathers, but heavy, larger stones fall no faster than lighter, smaller stones of the same material, then that is evidence against, not evidence in favor of Aristotle's theory, for it says that something besides weight causes stones and feathers to fall at different rates.[4] Repeatedly, either by abstraction in argument or by experimental procedure — "polishing a groove ... lining the groove with parchment ... a polished bronze ball" — disturbing influences are removed so that weight and weight alone varies.[5]
- The case Galileo makes is neither strictly theoretical nor strictly empirical. Using a thought experiment, he argues that Aristotle's theory contradicts itself, that it is necessarily false without reference to any particular experimental observation. The thought experiment is not totally convincing: if an actual experiment failed to reproduce the thought experiment, we would be puzzled rather than certain that the actual experiment erred. Nonetheless, Galileo's theoretical argument creates a space for competing theories and their experimental evaluation.[6]
- Galileo introduces a competing theory, a beautiful theory, in terms of constant acceleration. In point of fact, this theory speaks about things that cannot be seen or measured, namely instantaneous velocity and how it changes. Galileo develops an observable, testable consequence of this theory. The testable consequence involves very precise predictions concerning the relationship between distance traveled and time spent traveling.

---

[3] The importance of working with several theories at once is often stressed. Paul Feyerabend writes: "You can be a good empiricist only if you are prepared to work with many alternative theories rather than with a single point of view and 'experience.' ... Theoretical pluralism is assumed to be an essential feature of all knowledge that claims to be objective ... [19, pages 14–15] [T]he evidence that might refute a theory can often be unearthed only with the help of an incompatible alternative [20, page 29]." See also [11, 45].

[4] Robert Nozick [42, pages 261–263], Peter Achinstein [1], and Kent Staley [59] argue that whether or not E consitutes evidence for T is a matter itself open to empirical challenge and investigation.

[5] The systematic exclusion of sources of variation besides the one cause under study is familiar in every scientific laboratory, and it was named the 'method of difference' by John Stuart Mill [38]; see Chapter 15.

[6] Thomas Kuhn [34] and J.R. Brown [5] present perspectives on the role of thought experiments in scientific work. In particular, Kuhn writes [34, page 264]: "By transforming felt anomaly to concrete contradiction, the thought experiment [... provided the...] first clear view of the misfit between experience and implicit expectation ...."

- The experiment is limited in scope and is atypical of situations in which objects fall. The descent is slowed by a beam set on an incline, permitting precise measurements to be compared with a theory that makes precise predictions. The object, a polished bronze ball, is atypical of falling objects. And so on. No attempt is made to survey the falls of all of the world's falling objects, because those comparisons would be subject to innumerable disturbing influences that would obscure the issue under examination. Laura Fermi and Gilberto Bernardini [18, page 20] write that Galileo's experiments "reproduced the essential elements of the phenomenon under controlled and simplified conditions."
- Galileo conducted a severe test of his theory, but did little to demonstrate that it is always true: perhaps the theory is correct only for one bronze ball, on one inclined beam, with suitable parchment, in Italy, in the 1600s.[7]

4.2 The Permanent-Debt Hypothesis

In his permanent income hypothesis, Milton Friedman argued that personal consumption today is guided by expected long-term income, not current income or cash on hand.[8] Much in daily experience confirms this, or seems to. At a business school, doctoral students receive tuition and a stipend but spend modestly, while MBAs pay tuition with no stipend, yet spend less modestly. This is consistent with the permanent income hypothesis: MBAs expect to earn more in the long run. Whatever may be the ultimate fate of the permanent income hypothesis, it embodies a plausible claim: a rational person would anticipate future income in deciding about current consumption.[9]

The permanent income hypothesis has been challenged in various ways. David Gross and Nicholas Souleles [27, page 149] write:

> The canonical Permanent-Income Hypothesis (PIH) assumes that consumers have certainty-equivalent preferences and do not face any liquidity constraints. Under these assumptions the marginal propensity to consume (MPC) out of liquid wealth depends on model parameters, but generally averages less than 0.1. The MPC out of predictable income or 'liquidity' (e.g., increases in credit limits), which do not entail wealth effects, should be zero. The leading alternative view of the world is that liquidity constraints are pervasive. Even when

---

[7] The notion that scientific theories are testable but not demonstrable is thematic in the work of Sir Karl Popper [48]. He writes: "Theories are not verifiable, but they can be 'corroborated' ... [W]e should try to assess what tests, what trials, [the theory] has withstood [48, page 251] ... [I]t is not so much the number of corroborating instances which determines the degree of corroboration as the severity of the various tests to which hypothesis can be, and has been, subjected [48, page 267]." Obviously, if you doubted Galileo, you could polish a bronze ball ...

[8] Friedman discusses his permanent income hypothesis in his *A Theory of the Consumption Function* [23]. A concise, simplified version of the permanent income hypothesis is given in a couple of pages by Romer [51, Chapter 7]. Zellner [72, III.C] and Friedman [24, Chapter 12] reprint sections of Friedman's *A Theory of the Consumption Function*.

[9] Perhaps there is a certain similarity between a rational person and a polished bronze ball: both are uncommon in nature, but their behavior is of interest nonetheless.

4.2 The Permanent-Debt Hypothesis

they do not currently bind, they can be reinforced by precautionary motives concerning the possibility that they bind in the future. Under this view the MPC out of liquidity can equal one over a range of levels for "cash-on-hand," defined to include available credit.

As in §4.1, the starting point is not one theory, but the contrast between two theories. Contrast must become conflict: one must find a quiet undisturbed location where contrasting theories make conflicting predictions. Gross and Souleles [27, pages 150–151] continue:

> To test whether liquidity constraints and interest rates really matter in practice, this paper uses a unique new data set containing a panel of thousands of individual credit card accounts from several different card issuers. The data set ... includes essentially everything that the issuers know about their accounts, including information from people's credit applications, monthly statements, and credit bureau reports. In particular, it separately records credit limits and credit balances, allowing us to distinguish credit supply and demand, as well as account specific interest rates. These data allow us to analyze the response of debt to changes in credit limits and thereby estimate the MPC out of liquidity, both on average and across different types of consumers. The analysis generates clean tests distinguishing the PIH, liquidity constraints, precautionary saving, and behavioral models of consumption ...

The permanent income hypothesis predicts that an offer of credit unrelated to permanent income should not prompt an increase in spending. Gross and Souleles are seeking a circumstance in which there is a change in available credit with no change in expected long-term income. There is a great deal of activity in these credit card accounts that obscures the intended contrast of theories, and efforts are needed to remove these irrelevant disturbances. Gross and Souleles adopt two strategies: "first, we use an unusually rich set of control variables..." [27, page 154]. For instance, a person might be offered more credit because of increased credit scores or debts that appear to be under control, and perhaps these are related to income, but Gross and Souleles know what the card issuers know about these matters, so variations in these quantities can be controlled. Second, they exploit "'timing rules' built into the credit supply functions ... many issuers will not consider ... an account for a line change if it has been less than six months or less than one year since the last line change" [27, page 155]. Their point is that these timing rules generate small jumps in available credit that are not much related to an individual's financial circumstances: if your credit account is entirely unchanged in all its aspects (or more accurately, if this is the case after adjustment for the many control variables), you will be offered more credit after a lapse of time. Gross and Souleles are trying, perhaps successfully, to isolate a change in credit availability accompanied by no other consequential change. They conclude [27, page 181]: "We found that increases in credit limits generate an immediate and significant rise in debt, counter to the PIH."

The goal here is no different than the goal in §4.1: to display an effect that contrasts two theories as sharply as possible, removing to the greatest degree possible all disturbances that might obscure that contrast. There is no attempt to generalize from a sample to a population. The two theories began as general theories; there is no need to generalize further. The goal is to contrast conflicting predictions of two general theories in the stillness of a laboratory.

4.3 Guns and Misdemeanors

The 1968 Federal Gun Control Act prohibits the purchase of guns by felons. Beginning in 1991, California prohibited the purchase of handguns by individuals convicted of violent misdemeanors, including assault or brandishing a firearm. It is not clear whether such prohibitions are effective at reducing gun-related violence. On one theory, it is fairly easy to purchase a handgun illegally, for instance by purchasing the gun from someone who can buy it legally, so the prohibitions deter only those who wish to avoid illegal activity, perhaps not the best target for deterrence. On a different theory, legal prohibitions backed by punishment deter individuals who wish to avoid punishment.

The two theories are plain enough, but contrasting them is not so easy. Under what circumstances do the two theories make conflicting predictions? Under both theories, it would not be surprising if violent men remain violent, with or without restrictions on handgun purchases. The laws restrict purchases by violent individuals; these individuals cannot be compared with unrestricted, nonviolent individuals.

The effects of California's law were investigated by Garen Wintemute, Mona Wright, Christiana Drake, and James Beaumont [71] in the following way. They compared two groups of Californians who had been convicted of violent misdemeanors of the sort that would have prevented a legal purchase of a handgun beginning in 1991. One group consisted of individuals who had applied to purchase a handgun in 1989 or 1990, before the law took effect. The second group consisted of individuals who applied to purchase a handgun in 1991 and whose application was denied. These two groups may not be perfectly comparable, but at least both groups had been convicted of violent misdemeanors and both groups sought to purchase a handgun. In terms of demographics and previous convictions, the two groups looked fairly similar [71, Table 1].

If California's law were effective, one expects to see a lower rate of gun and violent crime in the group denied a handgun purchase, with no difference in nongun or nonviolent crime. If the law were ineffective, one expects to see similar rates of gun and violent crime in both groups. With or without adjustments for demographics and previous convictions, Wintemute et al. [71, Tables 2 and 3] found lower rates of gun and violent crime in the group denied a handgun purchase, with little indication of a difference in the rate of other crimes.

The reasoning here is much the same as in §4.1 and §4.2. Admittedly, the two theories are quite plain: one says the policy works, the other denies this. And yet, because the policy is targeted at offenders, care is needed to identify circumstances in which these theories make conflicting predictions.

4.4 The Dutch Famine of 1944–1945

Does malnutrition in utero reduce mental performance at age 19? The two theories are straightforward: it does or it doesn't. It is not straightforward, however, to identify circumstances in which these theories lead to different predictions. Reliable information is needed about mental performance at age 19 and about mother's diet 20 years prior to that. Controls are needed whose mothers were not malnourished but who were otherwise similar in background, education, social class, etc. Neither such information nor such controls are commonly available.

Zena Stein, Mervyn Susser, Gerhart Saenger and Francis Marolla [61] found what was needed in the Dutch famine of 1944–1945. They wrote [61, page 708]:

> On 17 September 1944 British paratroops landed at Arnhem in an effort to force a bridgehead across the Rhine. At the same time, in response to a call from the Dutch government-in-exile in London, Dutch rail workers went on strike. The effort to take the bridgehead failed, and the Nazis in reprisal imposed a transport embargo on western Holland. A severe winter froze the barges in the canals, and soon no food was reaching the large cities ... At their lowest point the official food rations reached 450 calories per day, a quarter of the minimum standard. In cities outside the famine area, rations almost never fell below 1300 calories per day. ... The Dutch famine was remarkable in three respects: (i) Famine has seldom if ever struck where extensive, reliable and valid data allow the effects to be analyzed within specified conditions of social environment. (ii) The famine was sharply circumscribed in both time and place. (iii) The type and degree of nutritional deprivation during the famine were known with a precision unequaled in any large human population before or since.

To know a child's date and place of birth is to know whether the mother was affected by the famine. So Stein et al. defined famine and control regions in Holland, as well as a cohort of children born just before the start of the famine, several cohorts exposed to the famine at various times in utero, and a cohort conceived after the famine ended. In these cohorts, virtually all males received medical examinations and psychological testing, including an IQ test, in connection with induction into the military. For 1700 births at two hospitals, one in a famine region (Rotterdam) and one in a control region (Heerlan), birth weights were available. Stein et al. [61, page 712] concluded: "Starvation during pregnancy had no detectable effects on the mental performance of surviving male offspring."

Subsequent studies used the Dutch famine to examine other outcomes, including schizophrenia, affective disorders, obesity, and breast cancer; see [35, 63] for reviews of these studies. For studies that have taken a similar approach in other situations, see [3, 57].

The Dutch famine was unrepresentative both of typical in utero development and typical famines, and it was useful for studies of development precisely in the ways it was unrepresentative. In representative situations, famine is confounded with other factors that obscure the effects caused by malnutrition in utero.

4.5 Replicating Effects and Biases

The randomized experiment in its idealized, perhaps unattainable, form is subject to a single source of uncertainty — that stemming from a finite sample size. In this ideal, biases of every sort have been eliminated by blocking and randomization;

therefore, the sole function of replication is to increase the sample size. Actual experiments are a step or two removed from the ideal, and observational studies are several steps removed from that. In an observational study, even an excellent one, there is the possibility that can never be entirely eliminated: treated and control subjects who look comparable may not actually be comparable, so differing outcomes may not be treatment effects.[10] In observational studies, the principal and important function of replication is to shed some light on biases of this sort.

Here, too, there are two theories. The first theory asserts that previous studies have produced estimates of treatment effects without much bias. The competing theory denies this, asserting instead that previous studies were biased in certain particular ways, and if those biases were removed, then the ostensible treatment effect would be removed with them. The replication does not repeat the original study; rather, it studies the same treatment effect in a context that is not subject to particular biases claimed by the competing theory. Replication is not repetition: the issue at hand is whether the ostensible effect can be reproduced if a conjectured bias is removed.

An example is David Card and Alan Krueger's [9, 10] replication of their earlier study [8] of the effects of the minimum wage on employment[11]; see also §11.3. On 1 April 1992, New Jersey raised its minimum wage by about 20%, whereas neighboring Pennsylvania left its minimum wage unchanged. In their initial study, Card and Krueger [8] used survey data to examine changes in employment at fast food restaurants, such as Burger King and Wendy's, in New Jersey and eastern Pennsylvania, before and after the increase in New Jersey's minimum wage. They found "no evidence that the rise in New Jersey's minimum wage reduced employment at fast food restaurants in the state." Their careful and interesting study received some critical commentary; e.g., [41]. One of several issues was the quality of the employment data obtained by a telephone survey, and another was simply that New Jersey and Pennsylvania differ in many ways relevant to employment change, not just in their approach to the minimum wage. In 1996, the minimum wage as set by the U.S. Federal Government was raised, forcing up the minimum wage in Pennsylvania, without forcing changes in New Jersey whose state minimum wage was already above the new Federal minimum wage. Card and Krueger [9, 10] then replicated the original study with the roles of New Jersey and Pennsylvania now reversed, using employment data from payroll records for Unemployment Insurance made available by the U.S. Bureau of Labor Statistics. Their findings were similar. The replication does not eliminate all concerns raised about the original study, but it does make two specific concerns less plausible.

---

[10] In an observational study, an increase in sample size ensures only that the estimator is closer to whatever the estimator estimates. Typically, the estimator estimates the sum of a treatment effect and a bias whose unknown magnitude does not decrease as the sample size increases. When a bias of this sort is not zero, as the sample size increases, confidence intervals shrink in length to exclude the true treatment effect, and hypothesis tests are ever more likely to reject the true hypothesis. It is a mistake to attach great importance to an increase in the sample size (or to an increase in statistical efficiency) when substantial biases of fixed size are present or likely.

[11] For a brief survey of the economics of the minimum wage, see [6, §12.1].

4.5 Replicating Effects and Biases

Mervyn Susser writes:

> The epidemiologist [...seeks...] consistency of results in a variety of repeated tests ... Consistency is present if the result is not dislodged in the face of diversity in times, places, circumstances, and people, as well as of research design [65, page 88] ... The strength of the argument rests on the fact that diverse approaches produce similar results [64, page 148].

Susser's careful statement is easily, perhaps typically, misread. Mere consistency upon replication is not the goal. Mere consistency means very little. Rather, the goal is "consistency ... in a variety of ... tests" and "consistency ... not dislodged in the face of diversity." The mere reappearance of an association between treatment and response does not convince us that the association is causal — whatever produced the association before has produced it again. It is the tenacity of the association — its ability to resist determined challenges — that is ultimately convincing. For a similar view with examples from physics, see Allan Franklin and Colin Howson [22]. For a similar view with examples from mathematics in a Bayesian formulation, see Georg Polya [47, especially pages 463–464].

To illustrate, consider two sequences of studies, one sequence involving the effectiveness of treatments for drug addiction, the other involving the effects of advertising on prices. The first sequence increases the sample size, but the bias of greatest concern remains the same. The second sequence asks the same question in very different contexts.

Several nonexperimental studies of the effects of treatments for heroin or cocaine addiction have found that people who remain in treatment for at least three months are more likely to remain drug-free than people who drop out of treatment before three months. These studies had involved the Drug Abuse Reporting Program (DARP) and the Treatment Outcomes Prospective Study (TOPS). In a third study using a different source of data, Hubbard et al. [31, page 268] wrote:

> The general finding from DARP and TOPS that treatment duration of at least three months is associated statistically and clinically with more positive outcomes supports the inference of treatment effectiveness. The following analysis retests this hypothesis...

In their new study, Hubbard et al. [31] found, again, that people who dropped out of treatment went on to use more illegal drugs than people who remained in treatment. Is this an effect caused by treatment? Or is it simply that a person who has little enthusiasm for ending his addiction is both more likely to drop out of treatment and more likely to use illegal drugs? A summary report of the National Academy of Sciences [36, page 17] was understandably skeptical:

> Depending on the process by which users are selected into treatment programs and the determinants of dropout from such programs ... the data may make treatment programs seem more or less cost-effective than they actually are.

These three studies are all open to the same question: Is a comparison of people who remain in treatment and people who drop out revealing an effect caused by the treatment, or is it revealing something about the people who stay in treatment and those who drop out? Because the same question can be asked of each study, the replications increase the sample size — which was large from the start, and never much at issue — but they do not make progress towards answering a basic concern.

What is the effect of advertising on prices? More precisely, what is the effect on prices of imposing or removing a restriction on the advertising of prices? One study by Amihai Glazer [26] made use of a strike from 10 August 1978 to 5 October 1978 that shut down three daily New York City newspapers, the New York Times, the New York Post, and the Daily News. The strike reduced advertising of retail food prices in New York City. Glazer looked at changes in retail food prices in Queens, the eastern borough of New York City, comparing Queens with adjacent Nassau County, outside the City, where the major newspaper, Newsday, was unaffected by the strike. A second study by Jeffrey Milyo and Joel Waldfogel [39] examined changes in liquor prices in Rhode Island before and after the U.S. Supreme Court struck down Rhode Island's ban on liquor advertising, comparing Rhode Island with adjacent areas of Massachusetts that were unaffected by the Court decision. In a third study, C. Robert Clark [12] examined the effect on the prices of children's breakfast cereals of Quebec's ban on advertising aimed at children under the age of 13. Clark compared the prices of cereals for children and cereals for adults in Quebec and in the rest of Canada. Each situation — the newspaper strike in New York, the Supreme Court decision in Rhode Island, the ban on advertising to children in Quebec — has idiosyncrasies that might be mistaken for an effect of advertising on prices, but there is no obvious reason why these idiosyncrasies should align to reproduce the same association between advertising and prices. Unlike the studies of treatments for drug addiction, where one alternative explanation suffices for three studies, in the advertising studies, three unrelated idiosyncrasies would have to produce similar associations. That could happen, but it seems less and less plausible as more and more studies are designed to replicate an effect, if it is a real effect, but to avoid replicating biases.

The goal in replication is to study the same treatment effect in a new context that removes some plausible bias that may have affected previous studies [53].

4.6 Reasons for Effects

> To suppose universal laws of nature capable of being apprehended by the mind and yet having no reason for their special forms, but standing inexplicable and irrational, is hardly a justifiable position. Uniformities are precisely the sort of facts that need to be accounted for. ... Law is *par excellence* the thing that wants a reason.
>
> Charles Sanders Peirce [44]
> The Architecture of Theories

There is an obvious distinction between empirical evidence in support of a theory and a reasoned argument in support of a theory, but as is often the case with obvious distinctions, this one becomes less obvious upon close inspection. It is, after all, common to argue that what was taken as empirical evidence in support of a theory is

4.6 Reasons for Effects

not evidence at all.[12] It is equally common to cast doubt upon a reasoned argument with the aid of empirical evidence. To offer a reasoned argument in support of the claim that a treatment does or should or would have a certain effect is to create a new object for empirical investigation, namely the argument itself.[13] Consider an example.

In an effort to reduce gun violence, various municipalities have created programs to buy handguns from their owners. For instance, Milwaukee did this in 1994-1996, offering $50 for a workable gun. Buyback programs are popular because they are voluntary: they do not meet the same resistance that a coercive program might meet. However, voluntary programs may also be less effective than coercive programs. Do voluntary buyback programs work to reduce gun violence?

One could argue that buybacks have to work: whenever a gun is destroyed, that gun cannot be used in a future homicide or suicide. One could argue that the program can't really work: a rational person would turn in a gun for $50 only if he does not plan to use it, so the program is buying guns that would otherwise rest harmlessly in the attic. One could argue that buybacks work because people aren't rational: many gun deaths are accidents, and others stem from impulsive responses to rage or depression, so buying guns from people who do not plan to use them helps people to adhere to their peaceful plans. (See Ainslie [2] or Rachlin [50] for cogent general discussions of rational control of irrational impulses.) One could argue that buybacks can't really work because many violent individuals do not have peaceful plans and will not sell their guns, intending to thwart their own violent inclinations: "[T]he self-images of violent criminals are always congruent with their violent criminal actions," wrote Athens [4, page 68] in a careful, perceptive study. In short, it is not difficult to offer a variety of reasons in support of a variety of possible effects. Whatever the strengths or weakness of these arguments as arguments for or against particular effects, the arguments create objects for empirical investigation, namely the validity of the arguments themselves.

Basic to each of these arguments is the question: Do buyback programs buy the guns that would otherwise be used in gun violence? One aspect of this question is: Do buyback programs typically buy the types of guns typically used in gun violence? A second aspect is: Do buyback programs buy guns in sufficient quantity to affect gun violence? Evelyn Kuhn and colleagues [33] addressed the first aspect and a report of the U.S. National Academy of Sciences [69, pages 95–96] addressed the second.

Table 4.1 is Kuhn et al.'s [33] comparison of the caliber of repurchased guns and homicide guns. To a dramatic degree, with an odds ratio of ten or more, re-

---

[12] See §4.1 and Note 4.

[13] Dretske [17, pages 20–22] writes: "More often than not a reason, or the giving of reasons, supplies a recipe ... for the possible falsification of the statement or statements for which the reason is given. By this I mean that although some statements, considered in isolation, may appear irrefutable, they lose this invulnerability when taken in the context provided by evidential support. ... If Q is a reason for believing P true, then although P need not be true, Q must be. We can, and often do, give reasons — sometimes very good reasons — for believing something that is not true ... [A]ny challenge to the truth of Q is simultaneously a challenge to the acceptability of Q as a reason."

Table 4.1 Homicide and buyback handguns in Milwaukee by caliber. As the odds ratios indicate, homicide and buyback guns are quite different.

Caliber	Buyback	Homicide	Odds Ratio
Small: .22, .25, .32	719	75	1.0
Medium: .357, .38, 9mm	182	202	10.6
Large: .40, .44, .45	20	40	19.2
Total	941	369	

purchased guns are small caliber and homicide guns are large caliber. Kuhn et al. make several similar comparisons, finding that repurchased guns tended to be small, obsolete revolvers, whereas homicide guns tended to be large, inexpensive, semiautomatic pistols. They found similar but not identical results for suicide guns. The buyback programs are buying types of guns infrequently used in gun violence.

The report of the U.S. National Academy of Sciences makes this same point and two others [69, pages 95–96]:

> The theory on which gun buy-back programs is based is flawed in three respects. First, the guns that are typically surrendered in gun buy-backs are those that are least likely to be used in criminal activities ... Second, because replacement guns are relatively easily obtained, the actual decline in the number of guns on the street may be smaller than the number of guns that are turned in. Third, the likelihood that any particular gun will be used in a crime in a given year is low. In 1999, approximately 6,500 homicides were committed with handguns. There are approximately 70 million handguns in the United States. Thus, if a different handgun were used in each homicide, the likelihood that a particular handgun would be used to kill an individual in a particular year is 1 in 10,000. The typical buy-back program yields less than 1,000 guns.

In brief, their third point claims that the scale of buyback programs is too small to have a meaningful effect.

Table 4.1, related results in [33], and the arguments in the quoted paragraph from [69, pages 95–96] are intended to undermine the reasoning that justifies gun buyback programs, but they provide no direct evidence about the effects of such programs on violence. In [33], violence is not measured. Table 4.1 and the quoted paragraph attack the idea that gun buyback programs could work by buying the guns used in gun violence, because the programs buy the wrong types of guns and too few of them. It is, nonetheless, not entirely inconceivable that gun buyback programs do work to reduce violence, but do this in some other way. Imagine a highly publicized buyback program, in which, each night, the television news interviews someone who turned in a gun, reporting the accumulated total of guns repurchased, accompanied by comments from civic leaders about how the city is "getting rid of its guns and ridding itself of violence." Even if the program bought the wrong guns and too few of them, it is not entirely inconceivable that the publicity, fanciful though it may be, would affect public sentiment in a way that ultimately affected the level of gun violence. Of course, in arguing for a new policy, one might hope to say more than: 'it is not entirely inconceivable that the policy might work.'

In the discussion here, a precise distinction is intended between 'direct evidence of a treatment effect' and a 'reason for a treatment effect.' The distinction does not

4.6 Reasons for Effects

refer to the presence or absence of empirical evidence: empirical evidence is present in some form in most scientific work. Nor does the distinction refer to the strength or quality of the evidence: in the case of gun buybacks, the evidence against the reasons for an effect may be more compelling than the direct evidence of ineffectiveness [69, pages 95–96]. 'Direct evidence of a treatment effect' means a study of the relevant treatment, on the relevant subjects or units, measured in terms of the relevant outcome. A randomized experiment or an observational study may provide direct evidence of an effect in this sense, and the evidence is likely to be stronger if it comes from a well-conducted randomized experiment. In contrast, an empirical study of a 'reason for an effect' is a study that is a step removed; it provides a reason for thinking the relevant treatment would have a particular effect were it to be applied to the relevant subjects measured in terms of the relevant response, but one or more of these elements is not actually present. In the case of gun buybacks in Table 4.1 and in [33], the relevant outcome is the level of gun violence, which is not measured; instead, the results refer to the types of guns that are repurchased. In the paragraph above quoted from [69, pages 95–96], the relevant outcome is again gun violence, but the results refer to the number of guns repurchased. Dafna Kanny and colleagues [32] studied the effects on helmet use of a law requiring bicyclists to wear helmets: this provides a plausible reason for thinking the law reduced accidents, but it is a step removed, because helmet use rather than accident reduction is the outcome. In the 1964 U.S. Surgeon General's Report, *Smoking and Health*, it is noted that [60, page 143] "there is evidence from numerous laboratories that tobacco smoke condensates and extracts of tobacco are carcinogenic for several animal species." This is a reason for thinking that tobacco is carcinogenic in humans, but it is a step removed, because the subjects are not human. Much laboratory work in the biomedical sciences studies reasons for effects in humans by examining effects on animals or cell cultures or molecules. The use of surrogate outcomes in medicine is the study of a reason for an effect in the sense defined here; for example, the use of colonic polyps as a surrogate for mortality from colorectal cancer [56]. Laboratory style experimentation in economics studies reasons for effects when small experimental incentives are proxies for incentives faced in actual economic decisions [46]. A theoretical argument about what effect would appear in a tidy and simple theoretical world may constitute a reason for an effect in the untidy and complex world we actually inhabit; see, for instance, §11.3.

There is no substitute for direct evidence of the effect of a treatment. And yet, evidence about reasons for an effect remains important. Evidence about reasons for an effect may either strengthen or undermine direct evidence of an effect, and either outcome is constructive [54, 67].

4.7 The Drive for System

> A good scientific theory is under tension from two opposing forces: the drive for evidence and the drive for system. Theoretical terms should be subject to observable criteria, the more the better; and they should lend themselves to systematic laws, the simpler the better, other things being equal. If either of these drives were unchecked by the other, it would issue in something unworthy of the name of scientific theory: in the one case a mere record of observations, and in the other a myth without foundation.
>
> W.V.O. Quine [49, page 90]

> [I]mpersonal knowledge ... has a special commitment to reasons. Bodies of knowledge are essentially, if to varying degrees with different subjects, systematic. There is both a pure and an applied reason for this. Pure, because the aim is not just to know but to understand, and in scientific cases at least understanding necessarily implies organization and economy. Applied, because a body of knowledge will only be freely extensible and open to criticism if rationally organized. ... So knowledge, in this sense, must have reasons.
>
> Bernard Williams [70, page 56]

The first paragraph of Chapter 1 recalled Borgman's political cartoon in which a TV newsman presents "Today's Random Medical News" that "according to a report released today ... coffee can cause depression in twins." The cartoon accurately depicts a popular misconception. It is often thought, incorrectly, that a scientific study has a clear and stable interpretation on the day it is published, and that interpretation is 'news.' In fact, the considered judgment about a scientific study hardens slowly, like cement, as the study is subjected to critical commentary, reconciled with other studies, both past and future, and integrated into a systemic understanding of the larger topic to which the study is a contribution. The science of the undergraduate classroom — that is, the science that is presented simply as the way things are — is not a list of findings from a sequence of individually decisive studies, but rather the systemic reconciliation and integration of many, perhaps conflicting, studies. The study of 'reasons for effects' in §4.6 is part of this systemic reconciliation. Gaps and conflicts in systemic knowledge provoke new investigations aimed at closing gaps, resolving conflicts. In this sense, there is no 'science news.' On the day of its release, the new report faces a fate that is far too uncertain to constitute news. By the time a systemic understanding has hardened, by the time today's report has found a resting place in that understanding, both the report and the systemic understanding are far too old to constitute news.

Consider, for instance, Steven Clinton and Edward Giovannucci's [13] review in 1998 of the possibility that diet affects the risk of prostate cancer. Their review attempted to organize, reconcile, and lend systematic form to 193 scientific reports. The following excerpts indicate the flavor of their review.

> Studies of prostate cancer have yielded inconsistent results concerning the relationship between various measurements of adult body mass or obesity and risk [of prostate cancer, citing 11 studies, which are then contrasted in detail] [13, page 422]. ... An association between prostate cancer incidence and tobacco use has been inconsistent ... Additional studies focusing on the critical timing of smoking during the life cycle ... are necessary [13,

page 419]. ... The relationship between alcohol consumption and risk of prostate cancer has been evaluated in a series of studies. No strong evidence has emerged for an association [citing 4 studies] [13, page 424]. ... Reports of a correlation between diets rich in meat or dairy products and risk of prostate cancer are frequent [citing 12 studies] [13, page 425]. ... [Citing Giovannucci's own study, in which a nutrient, lycopene, found in tomatoes, was studied:] men in the highest quintile of lycopene intake experienced a 21% lower risk [... but ...] investigators should use caution in assuming that lycopene mediates a relationship between consumption of tomato products and lower risk of prostate cancer [13, page 429]. ... A recent study was designed to examine the effects of selenium supplementation on recurrence of skin cancer in a high-risk population [citing a randomized clinical trial]. Selenium treatment did not influence the risk of skin cancer, although selenium-treated patients had a nonsignificant reduction in total cancer mortality, as well as a lower risk of prostate cancer. Additional studies are necessary ... [13, pages 429–30]

The serious, evenhanded, cautious tone of Clinton and Giovannucci's review is in marked contrast to the presentation of diet and health in the popular press. Clinton and Giovannucci's calls for additional studies are often cited by subsequent studies [40, 43]. An updated review six years later reports some progress but many questions still open [37].

Most scientific fields seek a systemic understanding that: (i) contrasts and reconciles past studies, (ii) guides future studies towards promising gaps in the current understanding, and (iii) integrates theoretical and empirical work. For an example from economics, see Jonathan Gruber's [28] review of the economic issues related to providing health insurance to the uninsured. For an example from criminology, see the report by Charles Wellford and colleagues [69] concerning firearms and violence. For an example from the epidemiology of dementia emphasizing methodology as the origin for inconsistent results, see the review by Nicola Coley, Sandrine Andrieu, Virginie Gardette, Sophie Gillette-Guyonnet, Caroline Sanz, Bruno Vellas, and Alain Grand [14].

The drive for system plays a key role in eliminating scientific error, but it also plays a key role in deciding what to study next. A new experiment or observational study contributes to an extant understanding, one that is partly systematic, partly incomplete, partly conflicting, and doubtless partly, perhaps substantially, in error. A new study is judged by its contribution to organizing, completing, reconciling, or correcting that understanding.

4.8 Further Reading

Karl Popper [48], Georg Polya [47], John Platt [45], and Paul Feyerabend [19] offer abstract discussions relevant to the current chapter; see also [52]. The material in §4.5 and §4.6 is discussed in greater detail in [53] and [54], respectively.

References

1. Achinstein, P.: Are empirical evidence claims a priori? Brit J Philos Sci **46**, 447–473 (1995)
2. Ainslie, G.: Breakdown of Will. New York: Cambridge University Press (2002)
3. Almond, D.: Is the 1918 influenza pandemic over? Long-term effects of in utero influenza exposure in the post-1940 US population. J Polit Econ **114**, 672–712 (2006)
4. Athens, L.: Violent Criminal Acts and Actors Revisited. Urbana: University of Illinois Press (1997)
5. Brown, J. R.: The Laboratory of the Mind: Thought Experiments in the Natural Sciences. New York: Routledge (1991)
6. Cahuc, P., Zylberberg, A.: Labor Economics. Cambridge, MA: MIT Press (2004)
7. Card, D.: The causal effect of education. In: Handbook of Labor Economics, O. Ashenfelter and D. Card, eds. New York: North-Holland (2001)
8. Card, D., Krueger, A.: Minimum wages and employment: A case study of the fast-food industry in New Jersey and Pennsylvania. Am Econ Rev **84**, 772–793 (1994)
9. Card, D., Krueger, A.: A reanalysis of the effect of the New Jersey minimum wage increase on the fast-food industry with representative payroll data. NBER Working Paper 6386. Boston: National Bureau of Economic Research (1998)
10. Card, D., Krueger, A.: Minimum wages and employment: A case study of the fast-food industry in New Jersey and Pennsylvania: Reply. Am Econ Rev **90**, 1397–1420 (2000)
11. Chamberlin, T.C.: The method of multiple working hypotheses. Science **15**, 92 (1890), reprinted Science **148**, 754–759 (1965)
12. Clark, C.R.: Advertising restrictions and competition in the children's breakfast cereal industry. J Law Econ **50**, 757–780 (2007)
13. Clinton, S.K., Giovannucci, E.: Diet, nutrition, and prostate cancer. Annu Rev Nutr **18**, 413–440 (1998)
14. Coley, N., Andrieu, S., Gardette. V., Gillette-Guyonnet, S., Sanz, C., Vellas, B., Grand, A.: Dementia prevention: Methodological explanations for inconsistent results. Epidemiol Rev 35–66 (2008)
15. Coulibaly, B., Li, G.: Do homeowners increase consumption after the last mortgage payment? An alternative test of the permanent income hypothesis. Rev Econ Statist **88**, 10–19 (2006)
16. Darwin, C.: Letter to Henry Fawcett, 18 September 1861. (1861)
17. Dretske, F.I.: Reasons and falsification. Philos Q **15**, 20–34 (1965)
18. Fermi, L., Bernardini, G.: Galileo and the Scientific Revolution. New York: Basic Books (1961), reprinted New York: Dover (2003)
19. Feyerabend, P.: How to be a good empiricist — a plea for tolerance in matters epistemological. In: Nidditch, P.H., ed. The Philosophy of Science, New York: Oxford University Press (1968)
20. Feyerabend, P.: Against Method. London: Verso. (1975)
21. Fodor, J.A.: The dogma that didn't bark. Mind **100**, 201-220 (1991)
22. Franklin, A., Howson, C.: Why do scientists prefer to vary their experiments? Stud Hist Philos Sci **15**, 51-62 (1984)
23. Friedman, M.: A Theory of the Consumption Function. Princeton, NJ: Princeton University Press (1957)
24. Friedman, M.: The Essence of Friedman. Stanford, CA: Hoover Institution Press (1987)
25. Galileo, G.: Two New Sciences (1638). Madison: University of Wisconsin Press (1974)
26. Glazer, A.: Advertising, information and prices. Econ Inquiry **19**, 661-671 (1981)
27. Gross, D.B., Souleles, N.S.: Do liquidity constraints and interest rates matter for consumer behavior? Evidence from credit card data. Q J Econ **117**, 149–185 (2002)
28. Gruber, J.: Covering the uninsured in the United States. J Econ Lit **46**, 571-606 (2008)
29. Harré, R.: Great Scientific Experiments. London: Phaidon Press (1981), reprinted New York: Dover (2002)
30. Ho, D.E., Imai, K.: Estimating causal effects of ballot order from a randomized natural experiment: The California alphabet lottery, 1978–2002. Public Opin Q **72**, 216–240 (2008)

References

31. Hubbard, R.L., Craddock, S.G., Flynn, P.M., Anderson, J., Etheridge, R.M.: Overview of 1-year follow-up outcomes in the Drug Abuse Treatment Outcome Study (DATOS). Psychol Addict Behav **11**, 261-278 (1997)
32. Kanny, D., Schieber, R.A., Pryor, V., Kresnow, M.: Effectiveness of a state law mandating use of bicycle helmets among children. Am J Epidemiol **154**, 1072–1076 (2001)
33. Kuhn, E.M., Nie, C.L., O'Brien, M.E., Withers, R.L., Wintemute, G.J., Hargarten, S.W.: Missing the target: A comparison of buyback and fatality related guns. Inj Prev **8**, 143–146 (2002)
34. Kuhn, T.: A function for thought experiments. In: Kuhn, T., The Essential Tension: Selected Studies in Scientific Tradition and Change, pages 240–265. Chicago: University of Chicago Press. (1979)
35. Lumey, L.H., Stein, A.D., Kahn, H.S., van der Pal-de Bruin, K.M., Blauw, G.J., Zybert, P.A., Susser, E.S.: Cohort profile: The Dutch hunger winter families study. Int J Epidemiol **36**, 1196–1204 (2007)
36. Manski, C.F., Pepper, J.V., Thomas, Y.F. (eds): Assessment of Two Cost-Effectiveness Studies on Cocaine Control Policy. Washington, DC: National Academies Press (1999)
37. McCullough, M.L., Giovannucci, E.L.: Diet and cancer prevention. Oncogene **23**, 6349–6364 (2004)
38. Mill, J.S.: A System of Logic: The Principles of Evidence and the Methods of Scientific Investigation. Indianapolis: Liberty Fund (1867)
39. Milyo, J., Waldfogel, J.: The effect of price advertising on prices: Evidence in the wake of 44 Liquormart. Am Econ Rev **89**, 1081-1096 (1999)
40. Mitrou, P.N., Albanes, D., Weinstein, S.J., Pietinen, P., Taylor, P.R., Virtamo, J., Leitzmann, M.F.: A prospective study of dietary calcium, dairy products and prostate cancer risk. Int J Cancer **120**, 2466–2473 (2007)
41. Neumark, D., Wascher, W.: Minimum wages and employment: A case study of the fast-food industry in New Jersey and Pennsylvania: Comment. Am Econ Rev **90**, 1362–1396 (2000)
42. Nozick, R.: Philosophical Explanations. Cambridge, MA: Harvard University Press (1981)
43. Park, S.Y., Murphy, S.P., Wilkens, L.R., Stram, D.O., Henderson, B.E., Kolonel, L.N.: Calcium, vitamin D, and dairy product intake and prostate cancer risk — The multiethnic cohort study. Am J Epidemiol **166**, 1259–1269 (2007)
44. Peirce, C.S.: The architecture of theories. In: Charles Sanders Peirce: Selected Writings. New York: Dover (1958)
45. Platt, J.: Strong inference. Science **146**, 347–353 (1964)
46. Plott, C.R., Smith, V.L. (eds): Handbook of Experimental Economics Results. New York: North Holland (2008)
47. Polya, G.: Heuristic reasoning and the theory of probability. Am Math Month **48**, 450–465 (1941)
48. Popper, K.R.: The Logic of Scientific Discovery. New York: Harper and Row (1935 in German, 1959 in English)
49. Quine, W.V.O.: What price bivalence? J Philos **78**, 90–95 (1981)
50. Rachlin, H.: Self-control: Beyond commitment (with Discussion). Behav Brain Sci **18**, 109–159 (1995)
51. Romer, D.: Advanced Macroeconomics. New York: McGraw-Hill/Irwin (2005)
52. Rosenbaum, P.R.: Choice as an alternative to control in observational studies (with Discussion). Statist Sci **14**, 259–304 (1999)
53. Rosenbaum, P.R.: Replicating effects and biases. Am Statistician **55**, 223–227 (2001)
54. Rosenbaum, P.R.: Reasons for effects. Chance 5–10 (2005)
55. Samuelson, P.A.: Schumpeter as economic theorist. In: P.A. Samuelson, Collected Papers, Volume 5, pages 201-327. Cambridge, MA: MIT Press (1982, 1986)
56. Schatzkin, A., Gail, M.: The promise and peril of surrogate end points in cancer research. Nature Rev Cancer **2**, 19–27 (2002)
57. Sparen, P., Vagero, D., Shestov, D.B., Plavinskaja, S., Parfenova, N., Hoptiar, V., Paturot, D., Galanti, M.R.: Long term mortality after severe starvation during the siege of Leningrad: prospective cohort study. Br Med J **328**, 11-14 (2004)

58. Souleles, N.S.: The response of household consumption to income tax refunds. Am Econ Rev **89**, 947–958 (1999)
59. Staley, K.W.: Robust evidence and secure evidence claims. Philos Sci **71**, 467–488. (2004)
60. Surgeon General's Advisory Committee: Smoking and Health. Princeton, NJ: Van Nostrand (1964)
61. Stein, Z., Susser, M., Saenger, G., Marolla, F.: Nutrition and mental performance. Science **178**, 708–713 (1972)
62. Stigler, G.J.: The economics of minimum wage legislation. Am Econ Rev **36**, 358–365 (1946)
63. Susser, E., Hoek, H.W., Brown, A.: Neurodevelopmental disorders after prenatal famine: the story of the Dutch famine study. Am J Epidemiol **147**, 213–216 (1998)
64. Susser, M.: Causal Thinking in the Health Sciences: Concepts and Strategies in Epidemiology. New York: Oxford University Press (1973)
65. Susser, M.: Falsification, verification and causal inference in epidemiology: Reconsideration in the light of Sir Karl Popper's philosophy. In: Susser, M., ed., Epidemiology, Health and Society: Selected Papers, pp. 82–93, New York: Oxford University Press (1987)
66. Urmson, J.O.: Aristotle's Ethics. Oxford: Blackwell (1988)
67. Weed, D.L., Hursting, S.D.: Biologic plausibility in causal inference: current method and practice. Am J Epidemiol **147**, 415–425 (1998)
68. Weiss, N.S.: Can the "specificity" of an association be rehabilitated as a basis for supporting a causal hypothesis? Epidemiology **13**, 6–8 (2002)
69. Wellford, C.F., Pepper, J.V., Petrie, C.V., and the Committee to Improve Research and Data on Firearms: Firearms and Violence. Washington, DC: National Academies Press (2004)
70. Williams, B.: Knowledge and reasons. In: Williams, B., Philosophy as a Humanistic Discipline, pages 47–56. Princeton, NJ: Princeton University Press. (2006)
71. Wintemute, G.J., Wright, M.A., Drake, C.M., Beaumont, J.J.: Subsequent criminal activity among violent misdemeanants who seek to purchase handguns: Risk factors and effectiveness of denying handgun purchase. J Am Med Assoc **285**, 1019–1026 (2001)
72. Zellner, A. Readings in Economic Statistics and Econometrics. Boston: Little, Brown (1968)

Chapter 5
Opportunities, Devices, and Instruments

Abstract What features of the design of an observational study affect its ability to distinguish a treatment effect from bias due to an unmeasured covariate u_{ij}? This topic, which is the focus of Part III of the book, is sketched in informal terms in the current chapter. An opportunity is an unusual setting in which there is less confounding with unobserved covariates than occurs in common settings. One opportunity may be the base on which one or more natural experiments are built. A device is information collected in an effort to disambiguate an association that might otherwise be thought to reflect either an effect or a bias. Typical devices include: multiple control groups, outcomes thought to be unaffected by the treatment, coherence among several outcomes, and varied doses of treatment. An instrument is a relatively haphazard nudge towards acceptance of treatment where the nudge itself can affect the outcome only if it prompts acceptance of the treatment. Although competing theories structure design, opportunities, devices, and instruments are ingredients from which designs are built.

5.1 Opportunities

The well-ordered world

> ... a nostalgia for caprice and chaos ...
> E.M. Cioran [23, Page 2]

In a well-ordered world, a rationally ordered world, each person would have no choice but to receive the best treatment for that person, the decision being based on undisputed knowledge and expert deliberation. That well-ordered world would be the antithesis of a randomized experiment, in which people receive treatments for no reason at all, simply the flip of a fair coin, the experimental design reflecting our acknowledged ignorance about the best treatment and our determined efforts to reduce that ignorance. In that well-ordered world, it would be difficult to learn anything new. In that world, policy has settled into the "deep slumber of decided opinion" [59, page 42]. Happily for the investigator planning an observational study, we do not live in that well-ordered world.

An observational study may begin with an opportunity, an arbitrary, capricious, chaotic disruption of the ordered world of everyday. The investigator may ask: How can this disruption be put to use? Is there a pair of scientific theories that on an average day do not yield conflicting predictions, but today, just because of this disruption, do yield conflicting predictions? In Chapter 2, randomness was found to be useful. Following Richard Sennett [93], we may ask: Is disorder useful?

Questions

I. Does a low level of ambient light cause auto accidents? It would not do to compare shoppers driving at 11:00 am, commuters driving at dusk, party-goers driving home at 3:00 am; their rates of accidents might differ for reasons other than the difference in ambient light. It would not do to compare the rates of accidents at 5:00 pm in Helsinki and Tel Aviv, because a difference in ambient light is one difference among many others. What would you compare?

II. If murderers, rapists, and armed robbers condemned to death were not executed, would they do greater injury to other inmates and prison staff than is done by others incarcerated for violent crimes? What would you compare?

III. At present in the United States, married couples pay taxes jointly on the sum of their incomes, where the total income matters, but the division of income between husband and wife does not. Also, the tax rates are progressive, with higher tax rates at higher incomes. Under such a system, if a husband and wife have very different incomes, the tax disincentives to work may be quite different than if they were not married. It is natural to ask: What effect does joint taxation have on the labor supply of husbands and wives? How would you study the effects of joint taxation on labor supply given that the whole country is subject to joint taxation?

IV. Does growing up in a poor neighborhood depress later adult earnings? The question asks for a separation of the effects of the neighborhood from any attributes of the individual's own family circumstances. What would you compare?

Solutions

I. You are looking for a change in ambient light with no other consequential change. When does that happen? It happens twice a year with the switch into and out of daylight savings time. John Sullivan and Michael Flannagan [102] looked at change in fatal auto accident rates at dawn and dusk in the weeks before and after the discontinuity in ambient light produced by the switch into and out of daylight savings time.

5.1 Opportunities

Does increased or reduced sleep affect auto accident rates? In a different use of the change into or out of daylight savings time, Mats Lambe and Peter Cummings [51] looked at Mondays before and after the switch, noting that on the Monday immediately after the switch, people often have added or subtracted an hour of sleep.

The use of daylight savings time to vary either ambient light or sleep duration is an example of a discontinuity design, in which a treatment shifts abruptly and discontinuously on some dimension, here time, while the sources of bias that are of greatest concern are likely to shift gradually and continuously, so the bias in estimating the treatment effect is small near the discontinuity. Donald Thistlethwaite and Donald Campbell [105] were the first to see discontinuities as opportunities for observational studies; see also [11, 12, 15, 25, 34, 46].

II. Generally, it is not easy to study the behavior that a person would have exhibited had they not been executed. In the case of Furman v. Georgia in 1972, the U.S. Supreme Court found that the then current methods of implementing the death penalty were 'cruel and unusual' and therefore in violation of the U.S. Constitution, invalidating the death sentences of more than 600 prisoners facing execution. Marquart and Sorensen [56] looked at 47 Texas inmates whose sentences were commuted to life imprisonment, examining their behavior from 1973 to 1986, compared with a cohort of similar violent offenders not on death row.

Supreme Court decisions are not infrequently seen as opportunities; e.g., [60]. For present purposes, it makes little difference whether we view the Supreme Court as capricious in striking down legislation or legislators as capricious in ignoring the strictures of the Constitution; not wisdom, but sharp abrupt change is sought.

III. Today, the U.S. tax code is well-ordered in the limited sense that the whole country is subject to joint taxation, but it was not always so. Prior to the Revenue Act of 1948, married couples paid joint taxes on their combined incomes in states with community property laws and separate taxes on their individual incomes in states without such laws. Sara LaLumia [50] compared the change in the labor supplied by husbands and wives, from before the Revenue Act of 1948 to after, in states with and without community property laws.

IV. The public housing program in Toronto assigned families to locations based on vacancies at the time a family reached the top of the waiting list, and Philip Oreopoulos [62] used this relatively haphazard arrangement to study neighborhood effects on adult earnings. He wrote:

> This paper is the first to examine the effects of the neighborhood on the long-run labor market outcomes of adults who were assigned as children to different residential housing projects in Toronto... All families in the Toronto program were assigned to various housing projects throughout the city at the time they reach the top of the waiting list... [F]amilies cannot specify a preference... The Toronto housing program also permits comparison across a wide variety of subsidized housing projects... [some located] in central downtown, while others are in middle-income areas in the suburbs.

The Toronto housing program is an opportunity to the extent that position on a waiting list predicts earnings later in life only because it affects the assigned neighborhood. The Toronto housing program is an opportunity to the extent that something consequential (your neighborhood) was determined by something inconsequential (your position on a waiting list).

5.2 Devices

5.2.1 Disambiguation

In his President's Address to the Royal Society of Medicine, Sir Austin Bradford Hill [38] asked:

> Our observations reveal an association between two variables, perfectly clear-cut and beyond what we would care to attribute to the play of chance. What aspects of that association should we especially consider before deciding that the most likely interpretation of it is causation?

Association does not imply causation: an association between treatment and outcome is ambiguous, possibly an effect caused by the treatment, possibly a bias from comparing people who are not comparable despite looking comparable. 'Disambiguate' is a word with an ugly sound but the right attitude. Ambiguity is opposed, perhaps ultimately defeated, by activity specifically targeted at resolving ambiguity.[1] Devices enlarge the set of considered associations with a view to disambiguating the association between treatment and outcome. Donald Campbell's paper [18] of 1957 was one of the first to systematically consider the role of devices.

5.2.2 Multiple control groups

In the absence of random assignment, the mere fact that controls escaped treatment is prima facie evidence that they are not comparable to treated subjects. How did they escape treatment? Perhaps the controls were denied treatment, or declined the offer of treatment, or lived too far away to receive treatment, or were studied before treatment became available. Depending upon the context, these reasons for escaping treatment may appear consequential or innocuous. How might one obtain relevant evidence to supplement mere appearances?

If there are several ways to escape treatment, then several control groups are possible. Direct evidence about bias from unmeasured covariates is sometimes available, depending upon the context, by using more than one control group in an observational study. For instance, one control group might consist of people who declined treatment and another control group of people who were denied treatment.

For example, in 2005, Majan Bilban and Cvetka Jakopin [14] asked whether lead and zinc miners exposed to radon gas and heavy metals were suffering from ex-

---

[1] Citing Bentham, the Oxford English Dictionary writes: "'disambiguate': verb, to remove ambiguity from."

5.2 Devices

Table 5.1 Comparison of micronuclei frequency among lead-zinc miners and two control groups. The two-sample unpooled t-statistics are given together with the standard errors (SE) of the means.

Group	Label	n	Mean	SE	t vs LP	t vs SR
Local population	LP	57	6.005	0.377	–	−0.98
Slovene residents	SR	61	6.400	0.143	0.98	–
Mine workers	MW	67	14.456	0.479	13.87	16.13

cess levels of genetic damage. They compared 70 Slovene lead-zinc miners at one mine with two control groups, one consisting of local residents close to the mine, the other consisting of other Slovene residents who lived at a considerable distance from the mine. The mine itself was a source of lead pollution in the local environment. Bilban and Jakopin looked at several standard measures in genetic toxicology, including the frequency of micronuclei. In this assay [29], a blood sample is drawn and blood lymphocytes are cultured. In a normal cell division, the genetic material in one cell divides to become the genetic material in the two nuclei of two new separate cells. If this process does not go well, each new cell may not have a single nucleus, but may have bits of genetic material scattered in several micronuclei. The frequency of micronuclei (MN) is the measure of genetic damage; specifically, Bilban and Jakopin looked at 500 binuclear cells per person and recorded the total number of micronuclei in the 500 cells.

Table 5.1 displays Bilban and Jakopin's [14, Table IV] data for micronuclei among miners and the two control groups. It is clear that the miners have higher frequencies of micronuclei than both control groups, which do not differ greatly from each other.[2] Another example using two control groups is discussed in §11.3.

The advantages and limitations of multiple control groups may be developed in formal statistical terms, but several issues are fairly clear without formalities. First, if two control groups have differing outcomes after adjustment for observed covariates, then that cannot be due to an effect caused by the treatment, and it must indicate that at least one and perhaps both control groups are not suitable as control groups, perhaps that they differ with respect to unmeasured covariates. If the two control groups differ from each other with respect to relevant unmeasured covariates, then at least one of the two control groups must differ from the treated group with respect to these covariates.

Second, two control groups can be useful only if they differ in some useful way. If having two control groups were of value merely because there are two instead

---

[2] The analysis as I have done it in Table 5.1 is unproblematic because the situation is so dramatic, with t's either above 10 or below 1. In less dramatic situations, this analysis is not appropriate, for several reasons. First, a more powerful test for effect would use both control groups at once. Second, Table 5.1 performs several tests with no effort to control the rate of misstatements by these tests. Third, Table 5.1 takes the absence of a difference between the two control groups as supporting their comparability, but the failure to reject a null hypothesis is not evidence in favor of that hypothesis. For instance, the two control groups might not differ significantly because of limited power, or else they might differ significantly but the differences might be too small to invalidate their usefulness in the main comparison with the treated group. These issues will be discussed with greater care in §19.3.

of one, then any one control group could be divided in half at random to produce two groups, but of course that would provide no new information about unmeasured covariates. The two control groups in Table 5.1 help to indicate that there is nothing special about the region immediately around the mine; only miners, not local residents, have elevated MN levels. Donald Campbell [19] argued that it is wise to select two control groups so they differ with respect to a covariate that, though unmeasured, is known to differ substantially in the two groups. Similar outcomes in two such control groups provide evidence that imbalances in the unmeasured covariate are not responsible for treatment-vs-control differences in outcomes; see [69, 71] for discussion of this in terms of the power and unbiasedness of tests for bias from unmeasured covariates. Following M. Bitterman, Campbell refers to this as 'control by systematic variation:' the unmeasured covariate is systematically varied without producing substantial variation in the outcome.

Third, similar outcomes in two control groups do not ensure that treatment-versus-control comparisons are unbiased. The two control groups may both be biased in the same way. For instance, if a willingness to work in a lead-zinc mine is associated with a general lack of concern about health hazards, one might discover that miners are more prone to tolerate other health hazards, say cigarette smoke and excessive alcohol, than are other Slovenes living near or far from the mine. A comparison of two control groups may have considerable power to detect certain specific biases — biases from covariates that have been systematically varied — and virtually no power to detect other biases — biases that are present in a similar way in both control groups [69, 71]. This is a very simple case of a general issue discussed carefully by Dylan Small [96].

5.2.3 Coherence among several outcomes

Austin Bradford Hill [38] in the quotation in §5.2.1 argued that there were aspects of the observed association that might aid in distinguishing an actual treatment effect from mere failure to adjust for some covariate that was not measured. Section 5.2.2 considered multiple control groups. In §3.4, it was seen that studies vary in their sensitivity to bias from unmeasured covariates: small biases, measured by Γ, could explain the observed association between treatment and outcome in some studies, but only large biases do so in other studies. So the degree of sensitivity to unmeasured biases is another aspect of the observed association that is relevant to distinguishing treatment effects and biases. Are there additional aspects to consider?

Table 5.2 displays data for the fifth grade for four of the 86 pairs of Israeli schools in Figure 1.1 from Angrist and Lavy's [3] study of academic test performance and class size manipulated by Maimonides' rule. Recall from §1.3 that schools were paired for the percentage of disadvantaged students in the school, x, and the outcomes were the average math and verbal test scores for the fifth grade. The middle portion of Table 5.2 shows Maimonides' rule in imperfect operation. School

5.2 Devices

Table 5.2 Four of the 86 pairs of two Israeli schools in Angrist and Lavy's study of test performance and class size manipulated using Maimonides' rule. Pair $j = 3$ violated Maimonides' rule, dividing the cohort for school $j = 2$ which had a cohort size of 40.

Pair i	School j	Percentage Disadvantaged x	Cohort Size	Number of Classes	Class Size	Average Math Score	Average Verbal Score
1	1	9	46	2	23.0	72.1	81.1
1	2	8	40	1	40.0	63.1	79.4
2	1	1	45	2	22.5	78.5	85.5
2	2	1	32	1	32.0	68.1	75.7
3	1	0	47	2	23.5	78.1	80.0
3	2	0	40	2	20.0	64.4	80.4
4	1	1	45	2	22.5	76.4	82.9
4	2	1	33	1	33.0	67.0	84.0

$(i, j) = (1, 1)$ had 46 students in the fifth grade and was divided to form two classes with an average size of 23, while school $(i, j) = (1, 2)$ had 40 students in the fifth grade and was not divided, yielding one class with 40 students. Pair $i = 3$ violates Maimonides' rule because the cohort of size 40 in school $(i, j) = (3, 2)$ was divided.

Table 5.2 suggests several questions. The first question, discussed in the current section, concerns 'coherence.'[3] As is often the case, the most compelling argument in support of the importance of 'coherence' comes from considering 'incoherence.' Suppose that larger cohorts divided into smaller classes had decidedly superior results in mathematics when compared with smaller cohorts taught in larger classes, but also decidedly inferior results in the verbal test. That result would be incoherent. Faced with an incoherent result of this sort, the investigator would be hard-pressed to argue that smaller classes produce superior academic achievement. In fact, Figure 1.1 suggests gains in both math and verbal test scores, a coherent result. If incoherence presents a substantial obstacle to a claim that the treatment caused its ostensible effects, then the absence of incoherence — that is, coherence — should entail some strengthening of that claim. Can this intuition be formalized?

Recall from §2.4.1 the hypothesis of an additive treatment effect. If indeed the treatment had additive effects, τ_{math} on the math test, τ_{verb} on the verbal test, then a claim that smaller class sizes improve test performance is the claim that $\tau_{math} \geq 0$ and $\tau_{verb} \geq 0$ with at least one strict inequality. To say that smaller classes cause a decisive improvement in both mathematics and verbal test performance is to say that both τ_{math} and τ_{verb} are substantially positive. A gain in math and a loss in verbal test performance, $\tau_{math} > 0$ and $\tau_{verb} < 0$, is not logically impossible, but it would not be compatible with the anticipated benefit from smaller class sizes. The policy question is whether smaller classes confer benefits commensurate with their unambiguously higher costs, and for that the relevant question is whether both τ_{math} and τ_{verb} are substantially positive. Viewed geometrically, a gain in mathematics

---

[3] Hill [38] used the attractive term 'coherence' but did not give it a precise meaning. Campbell [20] used the term 'multiple operationalism' in a more technical sense, one that is quite consistent with the discussion in the current section. Trochim [106] uses the term 'pattern matching' in a similar way. Reynolds and West [66] present a compelling application.

is half of the line, $\tau_{math} > 0$, but a gain in both mathematics and verbal scores is a quarter of the plane, $(\tau_{math}, \tau_{verb}) > (0,0)$, so the latter is a more focused hypothesis.

In the previous paragraph I have made an argument that a relevant treatment effect has a certain specific form. You may or may not have found that argument compelling. No matter. For present purposes, the issue is not whether the argument in the previous paragraph is compelling to you in this particular case, but rather that any claim of coherence or incoherence depends, explicitly or implicitly, on an argument of this sort. A claim is made that an actual treatment effect, or a policy-relevant treatment effect, or a useful treatment effect must have a certain form. Coherence then means a pattern of observed associations compatible with this anticipated form, and incoherence means a pattern of observed associations incompatible with this form. Claims of coherence or incoherence are arguable to the extent that the anticipated form of treatment effect is arguable.

A simple way to look for an increase in both math and verbal scores is to add together the two signed rank statistics for the math and verbal scores;[4] this is one instance of the 'coherent signed rank statistic' [76] and [81, §9]. Although discussed in technical terms in §17.2, the intuition behind the coherent signed rank statistic is clear without technical details. If larger cohorts divided in half tend to have better math and verbal scores than smaller cohorts, both signed rank statistics will tend to be large, and the coherent statistic will be extremely large. If gains in math scores are offset by losses in verbal scores, the sum of the two signed rank statistics will not be especially large.

We may conduct a sensitivity analysis for the coherent signed rank statistic. See [76] and [81, §9]. The intuition here is that a coherent result will be judged less sensitive to unobserved bias if the coherent signed rank statistic is used, whereas an incoherent result will be more sensitive to unobserved bias. Section 17.2 will develop some of the formalities that justify this intuition. The computations are very similar in form to those in §3.5 and equally simple to perform; see [76] and [81, §9].

Table 5.3 shows the sensitivity analysis for the signed rank statistic applied to the math score alone, to the verbal score alone, and to the coherent signed rank statistic that combines them. Table 5.3 presents the upper bound on the one-sided P-value for testing the null hypothesis of no treatment effect against an increase in test performance. The lower bound on the P-value is highly significant in all cases and is not presented. The math score becomes sensitive to unobserved bias at about $\Gamma = 1.45$, the verbal score at a little more than $\Gamma = 1.55$ and the coherent statistic at about $\Gamma = 1.70$. In brief, a noticeably larger bias, Γ, would need to be present to explain away the coherent pattern of associations than would be needed to explain

---

[4] In an obvious way, in adding the two signed rank statistics, one is committing oneself to a particular direction of effect for the two outcomes. If one anticipated gains in math scores together with declines in verbal scores, one might replace the verbal score by their negation before summing the signed rank statistics. Because the two outcomes are ranked separately, approximately equal weight is being given to each of the outcomes. The coherent signed rank statistic may be used with more than two oriented outcomes. It may also be adjusted to include varied doses of treatment [76].

5.2 Devices

Table 5.3 Sensitivity analysis and coherence in Angrist and Lavy's study of academic test performance and class size manipulated by Maimonides' rule. The table gives the upper bound on the one-sided P-value for the math test score, the verbal test score and their coherent combination. In this instance, a larger bias, say $\Gamma = 1.7$, is needed to explain away the coherent association, whereas the association with the math test score could be explained by $\Gamma = 1.45$.

Γ	Math	Verbal	Coherent
1.00	0.0012	0.00037	0.00018
1.40	0.043	0.020	0.011
1.45	0.057	0.027	0.015
1.55	0.092	0.047	0.026
1.65	0.138	0.075	0.043
1.70	0.164	0.092	0.054

away the ostensible effect on either math or verbal scores taken in isolation. See §19.4 for further discussion of this analysis.

The results in Table 5.3 for Israeli schools are more sensitive to unobserved biases than the results in Table 3.2 for welders. That is an unambiguous fact clearly visible in the two sets of data. However, the contexts are different, and the contexts are relevant to thinking about what biases are plausible. Angrist and Lavy [3] worked hard to prevent unmeasured biases: the treated and control schools in Figure 1.1 and Table 5.2 were selected to be similar in terms of percentage disadvantaged, x, and to differ slightly in fifth grade cohort size on opposite sides of Maimonides' cutpoint of 40 students. Could this comparison omit a covariate u strongly predictive of both math and verbal test performance and about 1.7 times more common among schools with a slightly larger fifth grade cohort? Certainly, that remains a logical possibility, but Angrist and Lavy's results in Figure 1.1 are not extremely fragile given the magnitude of bias that seems plausible here.

5.2.4 Known effects

Unaffected outcomes or control outcomes

In §5.2.3, coherence referred to the possibility that the treatment was thought to affect several outcomes in known directions. It may happen, however, that we think a treatment will affect one outcome and not affect another, and we wish to exploit the anticipated absence of effect to provide information about unmeasured biases [57, 66, 67, 70, 71, 110, 25]. A control group is a group of subjects known to be unaffected by the treatment; by analogy, an outcome known to be unaffected by the treatment is sometimes called a 'control outcome' or a 'control construct.' A control outcome is sometimes linked to Hill's [38] notion of the 'specificity' of a treatment effect [70, 110].[5] In particular, McKillip [57] argues that under some circumstances, a control outcome might suffice in the absence of a control group.

An argument involving a control outcome appeared in §4.3 in the study by Wintemute and colleagues [112] of the effects in California of the introduction of a law in 1991 denying handguns to individuals convicted of violent misdemeanors. Recall that they compared individuals who had convictions for violent misdemeanors and who applied to purchase a handgun in either 1989-1990, before the law, or 1991, after the law had taken effect. Also, recall that they found lower rates of gun and violent crime when the law was in effect. Is this an effect caused by the law? Over time, there could be shifts in crime rates with, say, shifts in demographics or economic conditions. It is not clear how or why restrictions on handgun purchases would affect nonviolent crimes. Wintemute et al. [112] also looked at rates of nonviolent and nongun crime, finding little change in these rates following the change in law. In their study, where an effect is plausible, treatment and outcome are associated; where an effect is not particularly plausible, treatment and outcome are not associated.

In a similar manner, Sadik Khuder and colleagues [49] study the impact of a smoking ban in workplaces and in public areas on the rate of hospital admission for coronary heart disease. As a control outcome, they used admissions for non-smoking related diseases.

Although the role of known effects may be developed in formal terms, the main issues are fairly clear without technical detail. To be useful, an unaffected outcome must have certain properties. If an unaffected outcome were useful merely because it is unaffected, then such an outcome could always be created artificially using random numbers; however, that could not provide insight into unmeasured biases. To be useful, an unaffected outcome must be associated with some unmeasured covariate. Plausibly, trends in nonviolent crime or nongun crime are associated with general trends in lawless behavior, and therefore are useful in distinguishing an effect of the law from a general increase or decrease in crime. It is not difficult to show that an unaffected outcome can provide a consistent and unbiased test[6] of

---

[5] Specificity of a treatment effect in Hill's sense [38] is sometimes understood as referring to the number of outcomes associated with the treatment, but more recent work has emphasized the absence of associations with outcomes the treatment is not expected to affect [70, 110].

[6] Consistency and unbiasedness are two concepts of minimal competence for a test of a null hypothesis H_0 against an alternative hypothesis H_A. Consistency says the test would work if the sample size were large enough. Unbiasedness says the test is oriented in the correct direction in samples of all sizes. One would be hard pressed to say the test is actually a test of H_0 against H_A if consistency and unbiasedness failed in a material way. To be a 5% *level* test of H_0, the chance of a P-value less than 0.05 must be at most 5% when H_0 is true. The *power of a test* of a null hypothesis, H_0, against an alternative hypothesis, H_A, is the probability that H_0 will be rejected when H_A is true. If the test is performed at the 5% level, then the power of the test is the probability of a P-value less than or equal to 0.05 when H_0 is false and H_A is true instead. We would like the power to be high. The test is *consistent* against H_A if the power increases to 1 as the sample size increases — that is, rejection of H_0 in favor of H_A is nearly certain if H_A is true and the sample size is large enough. The test is an *unbiased* test of H_0 against H_A if the power is at least equal to the level whenever H_A is true. If the test is performed at the 5% level, then it is unbiased against H_A if the power is at least 5% when H_A is true.

5.2 Devices

imbalance in an unmeasured covariate with which it is associated; see [70, 71] and [81, §6]. A substantial imbalance in an unmeasured covariate may yield only a faint echo in an unaffected outcome if the association between them is weak.[7] An unaffected outcome may guide the scope of a sensitivity analysis [74].

Bias of known direction

Claims to know the direction of the bias induced by an unmeasured covariate are common in discussions of the findings of observational studies. To the extent that such claims are warranted, they may disambiguate certain associations between treatment and outcome.

Did adherence to the gold standard lengthen the Great Depression? Building upon work by Milton Friedman and Anna Schwartz [30] about the United States, studies by E.U. Choudhri and L.A. Kochin [22] and B. Eichengreen and J. Sachs [26] compared nations that were not on the gold standard, that left the gold standard at various times, or that remained on the gold standard. Discussing these studies, Benjamin Bernanke [13] wrote:

> [B]y 1935 countries that had left gold relatively early had largely recovered from the Depression, while Gold Block countries remained at low levels of output and employment. ... If choices of exchange-rate regime were random, these results would leave little doubt ... Of course, in practice the decision about whether to leave the gold standard was endogenous to a degree ... In fact, these results are very unlikely to be spurious ... [A]ny bias created by endogeneity of the decision to leave gold would appear to go the wrong way, as it were, to explain the facts: The presumption is that economically weaker countries, or those suffering the deepest depressions, would be the first to devalue or abandon gold. Yet the evidence is that countries leaving gold recovered substantially more rapidly and vigorously than those who did not. Hence, any correction for endogeneity ... should tend to strengthen the association of economic expansion and the abandonment of gold.

Bernanke's claim is typical of claims to know the direction of a particular bias. The claim is not put forward in the language of absolute certainty. Rather the claim is put forward as a clarification of the logical and yet somewhat implausible consequences of the claim that the association between treatment and outcome was produced by a bias of a particular kind. To assert the claimed bias is to assert also its necessary but implausible implications.

A fairly risky but sometimes compelling study design exploits a claim to know that the most plausible bias runs counter to the claimed effects of the treatment. In this design, two groups are compared that are known to be incomparable, but incomparable in a direction that would tend to mask an actual effect rather than create a spurious one. The logic behind this design is valid: if the bias runs counter to the anticipated effect, and the bias is ignored, inferences about the effect will be conservative, so the bias will not lead to spurious rejection of no effect in favor of the anticipated effect; see [70, 71] and [81, §6]. Nonetheless, the design is risky

---

[7] More precisely, the Kullback-Leibler information in the unaffected outcome is never greater, and is typically much smaller, than the information in the unmeasured covariate itself [70].

because a bias in one direction combined with an effect in the other may cancel, so an actual effect is missed. Moreover, the claim to know the direction of the bias may be disputed.

Does disability insurance discourage work? In one study exploiting unmeasured biases of known direction, John Bound [16] compared successful and unsuccessful applicants for U.S. Social Security disability insurance. His premise was that successful applicants were often more severely disabled than unsuccessful applicants, that is, that the U.S. Social Security Administration was employing some reasonable criteria in sorting the applications into two piles. In this case, the incentive to not work was given to individuals who were, presumably, less able to work. Presumably, the difference in work behavior between successful and unsuccessful applicants overstates the effect of the incentive, because it combines the impact of the incentive and greater disability. In fact, Bound [16] found that relatively few of the unsuccessful applicants returned to work, and on that basis claimed that the incentive effects were not extremely large.

Do laws that prevent handgun purchases by convicted felons prevent violent crimes? What would be a suitable control group for comparison with convicted felons? Mona Wright, Garen Wintemute, and Frederick Rivara [113] conducted a study exploiting a bias of known direction. They compared convicted felons whose attempt to purchase a handgun was denied with persons arrested for felonies but not convicted whose purchase was permitted. Presumably, some of those arrested but not convicted were innocent, so presumably the group of purchasers in their study contains fewer individuals who have previously committed felony offenses. Though the most plausible bias works in the opposite direction, Wright et al. found that, adjusting for age, purchasers were 13% more likely than denied felons to be arrested for gun crimes in the following three years [113, relative risk 1.13 in Table 1].

5.2.5 Doses of treatment

Does a dose-response relationship strengthen causal inference?

Much has been written about doses of treatment and their relationship to claims of cause and effect. Many such claims contradict other related claims, or appear to. Section 17.3 will attempt to sort out some of these claims in formal terms. Here, some of the claims are mentioned and an example (guided by §17.3) is presented. (A different aspect of doses is discussed in §11.3.)

The most familiar claim is due to Austin Bradford Hill in the paper quoted previously. Hill [38, page 298] wrote:

> [I]f the association is one which can reveal a biological gradient, or dose-response curve, then we should look most carefully for such evidence. For instance, the fact that the death rate from cancer of the lung rises linearly with the number of cigarettes smoked daily, adds a very great deal to the simpler evidence that cigarette smokers have a higher death rate than non-smokers.

5.2 Devices

Hill's claim is often disputed. Kenneth Rothman [90, page 18] wrote:

> Some causal associations, however, show no apparent trend of effect with dose; an example is the association between DES and adenocarcinoma of the vagina ... Associations that do show a dose-response trend are not necessarily causal; confounding can result in such a trend between a noncausal risk factor and disease if the confounding factor itself demonstrates a biologic gradient in its relation with disease.

In a subtle reinterpretation of Hill's notion of dose and response, Noel Weiss [109, page 488] says much the same:

> [O]ne or more confounding factors can be related closely enough to both exposure and disease to give rise to [a dose response relationship] in the absence of cause and effect.

These remarks speak of a dose-response relationship as something that is present or absent, with contributions to the strength of evidence that are correspondingly present or absent. Is this the best way to speak about either a dose-response relationship or the strength of evidence? We might ask: would an analysis that took account of doses of treatment exhibit less sensitivity to unmeasured biases? The answer is accessible by analytical means in any one study [83], and the answer varies from one study to the next. Hill makes a positive reference to smoking and cancer of the lung, and Rothman makes a negative reference to DES and cancer of the vagina, but the existence of both positive and negative examples is consistent with the notion that doses sometimes reduce sensitivity to unobserved biases.

Cochran [24] argued that observational studies should be patterned after simple experiments, and the good, standard advice about clinical trials is to compare two treatments that are as different as possible; see Peto, Pike, Armitage, Breslow, Cox, Howard, Mantel, McPherson, Peto, and Smith [64, page 590]. Of course, this entails widely separated doses with no attempt to discover a graduated continuum of dose and response. Is a graduated continuum the relevant issue? Or is it simply important to anticipate negligible effects among treated subjects receiving negligible doses? The issue turns out to be quite unambiguous; see §17.3 and [84].

Doses may be suspect in one sense or another. In Angrist and Lavy's [3] study of class size and educational test performance, cohort size was thought to be haphazard, but class size — the dose of treatment — was thought to be haphazard only to the extent that it was determined by cohort size using Maimonides' rule. The distinction between treated and control may be free of error, while the dose magnitudes may be subject to errors of measurement [86, 98]. In both instances, suspect doses require instruments if bias is to be avoided; see §5.3.

It is time to consider an example.

Genetic damage from paint and paint thinners

Professional painters are exposed to a variety of potential hazards in paint and paint thinners, including organic solvents and lead. Do such exposures cause genetic damage? Pinto, Ceballos, García and colleagues [65] compared male professional painters in Yucatan, Mexico, with male clerks, matched for age, with respect to the

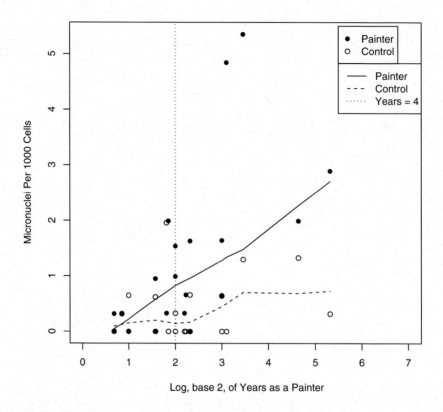

Fig. 5.1 Micronuclei in 22 male painters and 22 male controls matched for age plotted against $\log_2(\text{years})$ where years is the number of years of employment as a painter for the painter in each matched pair. The two curves are lowess smooths. The vertical line is at the median of 4 years or $\log_2(\text{years}) = 2$.

frequency of micronuclei in 3000 oral epithelial cells scraped from the cheek. See §5.2.2 for a discussion of the micronucleus assay. The painters were public building painters who worked without masks or gloves, and the dose d_i of their exposure[8] is the number of years of employment as a painter, recorded on a logarithmic scale, specifically $\log_2(\text{years})$, so $\log_2(2^k) = k$, and in particular, $\log_2(4) = 2$ and $\log_2(32) = 5$. Pinto et al. [65]'s data appear in Table 5.4 and Figure 5.1. Obviously, age and years as a painter are highly correlated, but the controls are closely matched

---

[8] When $\mathscr{F}$ was introduced in Chapter 2, treatment was applied at a single dose, and so doses were not mentioned. In general, if there are fixed doses, one dose d_i for each pair i, then the doses are also part of $\mathscr{F}$. Because previous discussions involving $\mathscr{F}$ had a single dose, we may adopt the new definition that includes doses in $\mathscr{F}$ without altering the content of those previous discussions.

5.2 Devices

Table 5.4 Micronnuclei per 1000 cells in 22 male painters and 22 male clerks matched for age. For painters, the number of years worked as a painter is recorded, along with its logarithm to the base of 2. Pairs are sorted by increasing years. A line separates painters with less than four years of work as a painter.

Pair	Painter Age	Control Age	Years Painting	$\log_2$(years)	Painter Micronuclei	Control Micronuclei
1	18	18	1.6	0.68	0.32	0.00
2	20	20	1.6	0.68	0.00	0.00
3	40	39	1.6	0.68	0.00	0.00
4	29	29	1.8	0.85	0.32	0.32
5	20	18	2.0	1.00	0.00	0.65
6	26	27	2.0	1.00	0.00	0.00
7	23	23	3.0	1.58	0.95	0.00
8	30	30	3.0	1.58	0.00	0.62
9	31	31	3.5	1.81	0.33	1.96
10	52	51	3.6	1.85	1.99	0.00
11	22	22	4.0	2.00	0.99	0.33
12	47	47	4.0	2.00	1.54	0.00
13	40	39	4.6	2.20	0.33	0.00
14	22	22	4.7	2.23	0.66	0.00
15	23	23	5.0	2.32	0.00	0.00
16	42	42	5.0	2.32	1.63	0.66
17	35	36	8.0	3.00	0.65	0.00
18	48	49	8.0	3.00	1.64	0.64
19	60	58	8.6	3.10	4.84	0.00
20	62	64	11.0	3.46	5.35	1.30
21	41	40	25.0	4.64	1.99	1.33
22	60	63	40.0	5.32	2.89	0.32

for age. In Figure 5.1, painters have higher frequencies of micronuclei, and the difference between painters and matched controls increases with $\log_2$ (years).

Recall from §2.3.3 that Wilcoxon's signed rank statistic is $T = \sum_{i=1}^{I} \text{sgn}(Y_i) \cdot q_i$ where $\text{sgn}(a) = 1$ if $a > 0$, $\text{sgn}(a) = 0$ if $a \leq 0$, and q_i is the rank of $|Y_i|$. The dose-weighted signed rank statistic [76, 83, 108] with doses d_i, $i = 1, \ldots, I$, is similar except it weights the pairs i by the doses d_i, $T_{\text{dose}} = \sum_{i=1}^{I} \text{sgn}(Y_i) \cdot q_i \cdot d_i$. An alternative statistic is the signed rank statistic applied to the subset of pairs with large doses, say $d_i \geq \tilde{d}$, that is,

$$T_{\text{high}} = \sum_{\{i: d_i \geq \tilde{d}\}} \text{sgn}(Y_i) \cdot q_i. \tag{5.1}$$

In the example, $\tilde{d}$ is set at the median dose, here, four years or $\tilde{d} = \log_2(4) = 2$; see the vertical line in Figure 5.1. The sensitivity analysis for these statistics is easy to perform and closely parallels the sensitivity analysis for the signed rank statistic in §3.5; see [76, 83].

Table 5.5 asks whether the dose-response pattern in Figure 5.1 reduced sensitivity to bias from failure to adjust for an unmeasured covariate u_{ij}. In this instance, weighting by doses, T_{dose}, is marginally less sensitive to unmeasured bias than ig-

Table 5.5 Did a dose-response relationship reduce sensitivity to unmeasured covariates? Tests of the null hypothesis of no treatment effect on micronuclei frequency in 22 male painters and 22 male clerks matched for age. Comparison of the upper bounds on the one-sided P-value using Wilcoxon's signed rank statistic, the dose-weighted signed rank statistic, and the signed rank statistic applied to high dose pairs. The dose, d_i, is $\log_2(\text{years})$. The signed rank statistic applied to the high-dose pairs used 12 of the 22 pairs in which the painter had worked for at least four years as a painter, $d_i = \log_2(\text{years}) \geq 2$, discarding ten pairs.

Γ	Ignoring doses, T	Weighting by doses, T_{dose}	Using only the 12 high-dose pairs, T_{high}
1	0.0032	0.0025	0.0012
2	0.064	0.038	0.016
2.5	0.12	0.067	0.028
3.3	0.22	0.12	0.048

noring doses, T, but using only the 12 pairs with high doses, T_{high}, is substantially less sensitive to unmeasured bias. Specifically, a bias of $\Gamma = 2$ could just explain the association between professional painting and micronuclei frequency if that association is measured ignoring doses using Wilcoxon's signed rank statistic, T, because the upper bound on the one-sided P-value is 0.064. A bias of $\Gamma = 2.5$ could just explain the association between treatment and response if the dose-weighted signed rank statistic, T_{dose}, is used. In contrast, even a bias of $\Gamma = 3.3$ could not explain the high levels of micronuclei in the 12 pairs with at least four years of work as a painter, because the upper bound on the one-sided significance level is 0.048.

A dose-response relationship reduces sensitivity to bias from unmeasured covariates in some studies but not in others [83]. A mere upward trend in a plot or table does not decide the issue, whereas Table 5.5 is decisive. The relative performance of T, T_{dose}, and T_{high} are examined further in §17.3 and [88]; see also [84].

5.2.6 Differential effects and generic biases

What are differential effects?

If there are two treatments, say A and B, each of which may be applied or withheld, then there is a 2×2 factorial arrangement of treatments in four cells. Extending the notation of Chapter 3, person ℓ falls in one of four cells: (i) neither A nor B ($Z_{A\ell} = 0$, $Z_{B\ell} = 0$), (ii) A but not B, ($Z_{A\ell} = 1$, $Z_{B\ell} = 0$), (iii) B but not A, ($Z_{A\ell} = 0$, $Z_{B\ell} = 1$), and (iv) both A and B, ($Z_{A\ell} = 1$, $Z_{B\ell} = 1$). If subjects ℓ were randomized to the cells of this 2×2 design, with $Z_{A\ell}$ and $Z_{B\ell}$ independently determined by the flips of a fair coin, then there are several comparisons or contrasts that may be examined, and because randomization would prevent bias in all such comparisons, the choice of comparison would reflect only our interests or considerations of parsimony. When randomization is not used, the situation is different. Depending upon the context,

5.2 Devices

it may happen that certain comparisons are less susceptible than others to certain types of biases.

The differential effect [87] of the two treatments is the effect of giving one treatment in lieu of the other. It is a comparison of two of the cells of the 2×2 factorial design, but not one of the most familiar comparisons. Specifically, the differential effect compares cells (ii) and (iii) above, that is, A but not B, ($Z_{A\ell} = 1, Z_{B\ell} = 0$) versus B but not A, ($Z_{A\ell} = 0, Z_{B\ell} = 1$). In a randomized experiment, there is no more interest in this comparison than in other comparisons. In an observational study, it may happen that an unmeasured bias that promotes A also promotes B, and in this case the differential effect may be less biased, or even unbiased, when other factorial contrasts are severely biased.

Whether the differential effect is of interest will depend upon the context. Obviously, the differential effect of A versus B is not the effect of A versus not-A; it may happen that A and B have substantial but similar effects, so their differential effect is small; in this case, the differential effect of A versus B reveals little about the effect of A. In other cases, the differential effect of A versus B may reveal quite a bit about the effect A versus not-A.

Examples of differential effects in observational studies

The study by Leonard Evans [14] in §1.4 is a skillful and fairly compelling use of differential effects to disambiguate treatment effects and unmeasured biases. Table 1.1 describes fatal accidents in which precisely one person in the front seat wore seat belts and precisely one of these two people died. There are, of course, many fatal accidents in which both people in the front seat wore seat belts and many others in which neither person wore seat belts. In other words, Table 1.1 describes the differential comparison, belted-driver-with-unbelted-passenger versus unbelted-driver-with-belted-passenger, which comprises two cells of a 2×2 factorial. As discussed in §1.4, the worry is that wearing seat belts is a precaution that may be accompanied by other precautions, driving at slower speeds or driving at a greater distance from the car ahead. A comparison that used the other cells of the 2×2 factorial — twice belted or twice unbelted — might be extremely biased, with use of seat belts taking the credit for cautious driving that occurred in unmeasured ways. In contrast, Table 1.1 compares two people in the same car, in the same accident, one belted, the other not, with the unbelted person at greater risk regardless of position. In a strict sense, Table 1.1 refers to the differential effect of belting with an unbelted companion, and that effect might be different from, say, belting with a belted companion: an unbelted companion may be hurled about the car and become a hazard to others. And yet, in this context, the beneficial differential effect of belting is strongly suggestive of a beneficial main effect of belting, and the differential effect is plausibly less affected by unmeasured biases.

Does regular use of nonsteroidal anti-inflammatory drugs (NSAIDs), such as the pain reliever ibuprofen, reduce the risk of Alzheimer disease? Several studies had found a negative association, but in a review of the subject, in 't Veld et al. [47] raised

Table 5.6 Attributes of the mothers who used marijuana, crack, neither or both in the U.S. National Institute on Drug Abuse's study of live births in the District of Columbia. Mothers who used either marijuana or crack but not both are more similar to each other than either group is to mothers who used neither drug.

	Neither	Marijuana	Crack	Both
Number of Babies	931	11	39	5
Cigarette use during pregnancy (%)	22	100	92	100
Alcohol use during pregnancy (%)	23	64	74	100
Late or no prenatal care (%)	10	27	48	40
Less than high school education (%)	23	54	49	60
Married (%)	34	0	3	0
Did not want to get pregnant (%)	32	46	59	80

the possibility that people who are in the early stages of cognitive impairment might be less aware of pain or less active in seeking treatment for pain, thereby depressing their use of pain relievers. That is, perhaps the early stages of Alzheimer disease cause a reduction in the use of NSAIDs, rather than NSAIDs reducing the risk of Alzheimer disease. James Anthony and colleagues [6] addressed this possibility in the following way. Acetaminophen is a pain reliever that is not an NSAID. Anthony et al. [6, Table 2] found that the risk of Alzheimer disease among users of just NSAID pain relievers was about half that of users of just analgesic compounds that were not NSAIDs. A generic tendency to be less aware of pain could explain a negative association between Alzheimer disease and NSAIDs, but it does not explain the differential association that a preference for NSAIDs over other pain relievers is negatively associated with Alzheimer disease. Plausibly, Alzheimer disease causes people to be passive in the face of pain, but it is far less plausible that Alzheimer disease causes people to reach for acetaminophen rather than ibuprofen. The differential effect of acetaminophen versus ibuprofen is not plausibly affected by certain biases that plausibly affect the main effect of ibuprofen.

In practical cases it is not possible to demonstrate that differential effects are less affected than main effects by certain unobserved biases, precisely because the biases are not observed. It is possible, in an impractical way, to illustrate this phenomenon by pretending for a moment that certain observed covariates are not observed. Table 5.6 compares four groups of mothers in the National Institute on Drug Abuse [61] study of babies born alive in eight hospitals in Washington, D.C. None of these mothers reported using heroin during pregnancy, and the mothers are classified by whether they used marijuana, crack, neither or both during pregnancy. In terms of several variables, such as cigarette and alcohol use and prenatal care, mothers who used either drug are more similar to each other than they are to mothers who used neither drug. The differential effect of marijuana versus cocaine on the health of newborn babies may or may not be of interest; that depends upon the context. What Table 5.6 suggests is that this differential effect is likely to be less biased than either main effect formed by comparing with mothers who abstained from narcotics; see [87] for further discussion.

What are generic biases?

A 'generic unobserved bias' is one that promotes the use of both treatments [87]. The following definition from [87] makes this precise.

Definition 5.1. There is only generic unobserved bias in the assignment of two treatments, A and B, if

$$\frac{\Pr(Z_{A\ell}=1, Z_{B\ell}=0 \mid r_{T\ell}, r_{C\ell}, \mathbf{x}_\ell, u_\ell)}{\Pr(Z_{A\ell}=0, Z_{B\ell}=1 \mid r_{T\ell}, r_{C\ell}, \mathbf{x}_\ell, u_\ell)} = \vartheta(\mathbf{x}_\ell), \quad \ell = 1, \ldots, L, \quad (5.2)$$

for some (typically unknown) function $\vartheta(\mathbf{x}_\ell)$.

Condition (5.2) says that the odds of receiving treatment A in lieu of treatment B may depend on observed covariates, $\mathbf{x}_\ell$, but not on $(r_{T\ell}, r_{C\ell}, u_\ell)$. Condition (5.2) is considerably weaker than the naïve model's condition (3.5). That is, (5.2) may be true when quantities such as $\Pr(Z_{A\ell} = 1 \mid r_{T\ell}, r_{C\ell}, \mathbf{x}_\ell, u_\ell)$ do depend upon $(r_{T\ell}, r_{C\ell}, u_\ell)$. For instance, in [6], (5.2) says that the odds of taking acetaminophen rather than ibuprofen does not depend upon the unobservable $(r_{T\ell}, r_{C\ell}, u_\ell)$, but the chance of taking ibuprofen at all may depend upon $(r_{T\ell}, r_{C\ell}, u_\ell)$. In brief, (5.2) may be true when the naïve model is false.

It is possible to show [87] that a generic bias of the form (5.2) does not bias the differential effect of treatment A versus treatment B, although it may bias the main effects of A and of B. Specifically, suppose (5.2) is true with $0 < \vartheta(\mathbf{x}_\ell) < \infty$, and I matched pairs are formed, matching exactly for $\mathbf{x}_\ell$, in such a way that each pair contains one subject who received treatment A but not B and one subject who received treatment B but not A. Then the distribution of assignments to A in lieu of B within these matched pairs is the randomization distribution in Chapter 2; see [87] for proof. If (5.2) is false, then it is possible to conduct a sensitivity analysis that is concerned only with the degree of violation of (5.2) and not with other violations of the naïve model, and this sensitivity analysis closely resembles the sensitivity analysis in Chapter 3; see [87].

5.3 Instruments

What is an instrument?

An instrument is a random nudge to accept a treatment, where the nudge may or may not induce acceptance of the treatment, and the nudge can affect the outcome only if it does succeed in inducing acceptance of the treatment. Paul Holland [40] offered the 'encouragement design' as a prototype of an instrument; see also the important paper by Joshua Angrist, Guido Imbens, and Donald Rubin [2]. In a paired, randomized encouragement design, individuals are paired based on measured pretreatment covariates, and for each pair, a coin is flipped to decide which member of the pair will be encouraged to accept the treatment. In an encouragement design, the experimenter hopes that encouraged individuals will accept the treatment and individuals who are not encouraged will decline the treatment; however, some individuals may in fact decline the treatment despite encouragement, and others may take the treatment in the absence of encouragement. A typical context would be encouragement to diet or exercise, or other similar settings in which a voluntary change in behavior is essential to the treatment.

There is, however, more to an instrument than randomly assigned encouragement. The encouragement itself must have no effect on the outcome besides its effect on acceptance of the treatment. This is often called the 'exclusion restriction.' Consider encouragement to exercise with weight loss as an outcome. To be an instrument, randomized encouragement to exercise must affect weight loss only by inducing exercise. Suppose you are indeed randomly selected and strongly encouraged to exercise. At the end of each week, you are told yet again that you are a failure and embarrassment for not exercising during the week, but despite this, you will not exercise; instead, with all that good 'encouragement,' you fall into a deep depression, stop eating, and your weight drops. In this case, encouragement to exercise is not an instrument for exercise: it changed your weight, but not by inducing you to exercise.

Violations of the exclusion restriction are no small matter. If violation of the exclusion restriction is ignored in the exercise example, your weight loss will be attributed to exercise, though it was produced by depression instead.

Instruments are rare, but valuable when they exist.

Example: Noncompliance in a double-blind randomized trial

Randomized encouragement to accept either an active drug or a double-blind placebo is the experimental design that most closely approximates an instrument. In this case, encouragement is actually randomized. Also, because of the double-blind placebo, neither the subject nor the investigator knows what the subject is being encouraged to do. In this case, there are few opportunities for encouragement to affect a clinical outcome without shifting the amount of active drug consumed; that is, the exclusion restriction is likely to be satisfied. Even this situation is imperfect to the extent that side effects of the active drug alert the investigator or the subject to the identity of the treatment.

An example is from Jeffrey Silber and colleagues' [95] randomized, double-blind clinical trial of the drug enalapril to protect the hearts of children who had received an anthracycline as part of cancer chemotherapy. Briefly, this is the 'AAA trial' for the 'ACE-inhibitor after anthracycline randomized trial.' Anthracyclines are fairly effective at curing certain childhood cancers, but they may damage a child's heart. The study focused on children, teens mostly, whose cancer appeared to have been cured, but who showed signs of cardiac decline. The AAA trial randomized 135 children to enalapril or a double-blind placebo and measured cardiac function over several years. Although the children were encouraged to take a specific dose of

5.3 Instruments

drug or placebo, some children took less than that dose. It is likely that neither the child, nor the child's parents, nor the attending physician knew whether the child was taking less than a full dose of enalapril or less than a full dose of placebo [95], so it is likely that any effect of encouragement on cardiac function is a consequence of biological effects of the quantity of enalapril consumed. In this context, randomized encouragement is an instrument for the amount of enalapril actually consumed. Because it is an instrument, it provides a principled solution to the problem of noncompliance, that is, a principled answer to the question: How effective is the prescribed dose of drug on subjects willing to take that dose, despite the refusal of some subjects to take the prescribed dose? An instrumental variable analysis of noncompliance in the AAA trial is discussed by Robert Greevy, Jeffrey Silber, Avital Cnaan, and me [33].

Example: Maimonides' rule

Recall from §1.3 and §5.2.3 the study by Angrist and Lavy [3] of academic test performance and class size manipulated by Maimonides' rule. In that study, small variations in the size of the fifth grade cohort often produced large variations in class size for cohorts whose size was close to the cutpoint of 40 students; see Figure 1.1. In Figure 1.1, some schools had class sizes in defiance of Maimonides' rule; there is some noncompliance. Focus on schools with cohort sizes near 40, that is, the schools in Figure 1.1. In this case, small variations in cohort size are presumably haphazard, while class size, not cohort size, is presumably what affects academic performance. If these two presumptions were actually correct for the schools in Figure 1.1, then whether or not the cohort size exceeds 40 is an instrument for class size. In Table 5.2, in pair #3, there is noncompliance: in violation of Maimonides' rule, both schools in this pair had two small classes.

Notation for an instrument in a paired encouragement design

In a paired, randomized encouragement design, one person in pair i is picked at random and encouraged, denoted $Z_{ij} = 1$; the other person in the pair is not encouraged, denoted $Z_{ij} = 0$, so $Z_{i1} + Z_{i2} = 1$. The jth subject in pair i has two potential doses of treatment, dose d_{Tij} if encouraged to accept the treatment, $Z_{ij} = 1$, or dose d_{Cij} if not encouraged to accept the treatment, $Z_{ij} = 0$. We observed either d_{Tij} if $Z_{ij} = 1$ or d_{Cij} if $Z_{ij} = 0$, but the pair (d_{Tij}, d_{Cij}) is not jointly observed for any subject ij. The dose of treatment actually received is d_{Tij} if $Z_{ij} = 1$ or d_{Cij} if $Z_{ij} = 0$, so it is $D_{ij} = Z_{ij} d_{Tij} + (1 - Z_{ij}) d_{Cij}$ in either case. The doses may be continuous, ordinal, or binary.[9] In Angrist and Lavy's [3] study, D_{ij} is the observed average class size, so $D_{11} = 23$ and $D_{12} = 40$ for the first pair in Table 5.2.

---

[9] As in Note 8, when $\mathscr{F}$ was defined in Chapter 2, the potential doses (d_{Tij}, d_{Cij}) were always equal to $(1,0)$ and so were not mentioned. In general, if there are potential doses, (d_{Tij}, d_{Cij}), then they are part of $\mathscr{F}$. Because previous discussions involving $\mathscr{F}$ had a single dose, we may

In Silber's AAA trial [33, 95], while on-study, a child assigned to placebo, $Z_{ij} = 0$, took no enalapril, $d_{Cij} = 0$, and a compliant child assigned to enalapril, $Z_{ij} = 1$, took the full assigned dose of enalapril, $d_{Tij} = 1$; however, many children were somewhat noncompliant and took somewhat less than the full dose, so $d_{Tij} < 1$ for many children ij. The decision by the experimenter to assign a child to enalapril or placebo, Z_{ij}, is randomized. The decision by the child to be compliant, $(d_{Tij} = 1, d_{Cij} = 0)$, or not compliant, $(d_{Tij} < 1, d_{Cij} = 0)$, is the child's decision, doubtless influenced by the child's parents. Quite possibly, compliant children, $(d_{Tij} = 1, d_{Cij} = 0)$, differ from noncompliant children, $(d_{Tij} < 1, d_{Cij} = 0)$. Most of the children in the experiment were teens, newly engaging or rejecting sports, alcohol, tobacco and narcotics. Perhaps compliant and noncompliant children are equally likely to smoke or engage in binge drinking, but perhaps not. Randomization has ensured that encouraged children are similar to unencouraged children, that the assignment to enalapril or placebo is equitable, but there is nothing to ensure that compliant children are comparable to less compliant children.

The paired randomized experiment in Chapter 2 is the special case in which encouragement is always perfectly decisive for a binary dose. Specifically, the paired randomized experiment is the special case with $(d_{Tij}, d_{Cij}) = (1, 0)$ for all i and j, so $D_{ij} = Z_{ij}$. In other words, people take the full dose of treatment, $D_{ij} = 1$, if encouraged to do so, $Z_{ij} = 1$, and they take none of the treatment, $D_{ij} = 0$, if not encouraged to accept the treatment, $Z_{ij} = 0$.

An instrument is said to be 'strong' if d_{Tij} is much larger than d_{Cij} for most or all individuals ij; in this case, encouragement strongly shifts the dose received by most subjects. Figure 1.1 depicts a strong instrument: in many schools, average class size shifted decisively downward when enrollment passed 40 students. An instrument is said to be 'weak' if d_{Tij} is close to or equal to d_{Cij} for most or all individuals ij; in this case, most individuals ignore encouragement. The most popular method of analysis using an instrument, namely two-stage least squares, tends to give incorrect inferences when an instrument is weak [17]: confidence intervals that claim 95% coverage may cover only 85% of the time. The confidence intervals and tests using an instrument in this book do not have this problem [45]: 95% intervals cover 95% of the time with weak and strong instruments.

An 'intention-to-treat analysis' compares the encouraged group, $Z_{ij} = 1$, to the unencouraged group, $Z_{ij} = 0$, ignoring compliance behavior. It compares the intended treated group to the intended control group, ignoring actual treatment. The advantage of such an analysis is that it is fully justified by the random assignment of encouragement, Z_{ij}; that is, the groups being compared are comparable. The disadvantage is that 'intention-to-treat analysis' estimates the effect of encouragement to accept the treatment, not the effect of the treatment itself. Both effects are interesting, but they can be quite different. Quitting smoking might be highly beneficial to your health, but difficult to do. Encouragement to quit smoking might be highly ineffective, because most people do not comply by quitting, but quitting itself might be highly effective. Both effects are important — the effects of encouragement to

---

adopt the new definition that includes doses in $\mathscr{F}$ without altering the content of those previous discussions.

5.3 Instruments

quit, the effects of quitting — but they are different things. In Silber's AAA trial [33, 95], if enalapril is beneficial, then enalapril consumed is likely to be more effective than enalapril in the bottle, and the intention-to-treat analysis estimates an idiosyncratic mixture of these two effects. Despite its limitations, and because of its strengths, the intention-to-treat analysis is one of the basic analyses that should be reported in any randomized trial with noncompliance. The analyses in Table 5.3 in §5.2.3 were intention-to-treat analyses: they ignored actual class size, D_{ij}, which sometimes violated Maimonides' rule. Is there a principled way to take account of noncompliance?

The hypothesis that effect is proportional to dose

One hypothesis that satisfies the exclusion restriction asserts that the effect of encouragement, Z_{ij}, on the response, (r_{Tij}, r_{Cij}), is proportional to its effect on the dose, (d_{Tij}, d_{Cij}):

$$r_{Tij} - r_{Cij} = \beta \left(d_{Tij} - d_{Cij} \right) \quad \text{for } i = 1, \ldots, I, \ j = 1, 2. \tag{5.3}$$

In (5.3), if encouragement Z_{ij} does not affect your dose, $d_{Tij} = d_{Cij}$, then it does not affect your response, $r_{Tij} = r_{Cij}$. For instance, in Silber's AAA trial [33, 95], (5.3) implies that if assignment to enalapril, $Z_{ij} = 1$, does not actually induce a child to take enalapril, so $d_{Tij} = d_{Cij} = 0$, then it does not affect cardiac function, $r_{Tij} - r_{Cij} = 0$. In Silber's AAA trial, a compliant child has $(d_{Tij}, d_{Cij}) = (1, 0)$, so (5.3) would imply that such a child has $r_{Tij} - r_{Cij} = \beta(1 - 0) = \beta$, and β is the effect of the full dose, $d_{Tij} = 1$, for a compliant child. Similarly, for a less than fully compliant child who would take half the assigned dose, $d_{Tij} = \frac{1}{2}$, the hypothesis (5.3) implies an effect on cardiac function of $r_{Tij} - r_{Cij} = \beta \left(\frac{1}{2} - 0 \right) = \beta/2$.

In Angrist and Lavy's [3] study, hypothesis (5.3) asserts that cohort size matters for test performance (r_{Tij}, r_{Cij}) in direct proportion to its impact on class size, (d_{Tij}, d_{Cij}). Expressed differently, we expect passing the cutpoint of 40 students to matter in pair $i = 1$ of Table 5.2, but not much in pair $i = 3$. In (5.3), β is the point gain in test performance, $r_{Tij} - r_{Cij}$, caused by a change in class size of $d_{Tij} - d_{Cij}$, and Figure 1.1 suggests β is negative, with a decrease in class size increasing performance.

Inference about β

In an elementary but useful way, (5.3) may be rearranged as

$$r_{Tij} - \beta d_{Tij} = r_{Cij} - \beta d_{Cij} = a_{ij}, \text{ say, for } i = 1, \ldots, I, \ j = 1, 2. \tag{5.4}$$

Suppose that we wish to test the hypothesis $H_0 : \beta = \beta_0$ in (5.3). From this hypothesis and the observable data, we may calculate the adjusted response, $R_{ij} - \beta_0 D_{ij}$. If the hypothesis $H_0 : \beta = \beta_0$ were true, then $R_{ij} - \beta_0 D_{ij} = r_{Tij} - \beta d_{Tij}$ if the jth subject in pair i is encouraged, $Z_{ij} = 1$, and $R_{ij} - \beta_0 D_{ij} = r_{Cij} - \beta d_{Cij}$ if this subject is not encouraged, so that in either case $R_{ij} - \beta_0 D_{ij} = a_{ij}$ using (5.4).[10] That is, if $H_0 : r_{Tij} - r_{Cij} = \beta_0 (d_{Tij} - d_{Cij})$, $i = 1,\ldots,I$, $j = 1,2$, were true, the adjusted responses, $R_{ij} - \beta_0 D_{ij} = a_{ij}$, would satisfy the null hypothesis of no treatment effect. In light of this, the hypothesis $H_0 : \beta = \beta_0$ in (5.3) may be tested in a randomized encouragement design by calculating a statistic, such as Wilcoxon's signed rank statistic T, from the adjusted responses, and comparing T with its usual randomization distribution in §2.3.3; see [75]. Specifically, the signed rank statistic is computed from the encouraged ($Z_{ij} = 1$) minus unencouraged ($Z_{ij} = 0$) difference in adjusted responses, $Y_i^{(\beta_0)} = (Z_{i1} - Z_{i2}) \{(R_{i1} - \beta_0 D_{i1}) - (R_{i2} - \beta_0 D_{i2})\}$. If $H_0 : \beta = \beta_0$ in (5.3) is true, then $Y_i^{(\beta_0)} = (Z_{i1} - Z_{i2})(a_{i1} - a_{i2})$ is $\pm |a_{i1} - a_{i2}|$; moreover, if $H_0 : \beta = \beta_0$ is true and encouragement, Z_{ij}, is randomized, then $Y_i^{(\beta_0)}$ is $a_{i1} - a_{i2}$ or $-(a_{i1} - a_{i2})$ each with probability $\frac{1}{2}$, so that, in particular, $Y_i^{(\beta_0)}$ is symmetrically distributed about zero. Notice that this reasoning is exactly parallel to §2.4; only the form of the hypothesis has changed [33, 45, 75, 78, 97].

The hypothesis $H_0 : \beta = 0$ in (5.3) is the hypothesis $H_0 : r_{Tij} = r_{Cij}$ for $i = 1,\ldots,I$, $j = 1,2$; that is, $H_0 : \beta = 0$ in (5.3) is Fisher's sharp null hypothesis of no treatment effect. To reject either hypothesis is to reject the other: it is the same hypothesis, tested using the same test statistic, computed in the same way, compared with the same null distribution, yielding identical P-values. Because these two hypotheses are the same, the intention-to-treat analysis will reject Fisher's hypothesis of no treatment effect if and only if the instrumental variable (IV) analysis rejects $H_0 : \beta = 0$. As developed here, the IV analysis cannot find a treatment effect if the intention-to-treat analysis does not find one; indeed, they yield identical significance levels in testing the hypothesis of no effect.

Also, in strict parallel with §2.4, a $1 - \alpha$ confidence set for β is obtained by testing every hypothesis $H_0 : \beta = \beta_0$ and retaining for the confidence set the values not rejected by the test at level α; then, a $1 - \alpha$ confidence interval is the shortest interval containing the confidence set. This confidence set has the correct coverage rate even if the instrument is weak [45, 75]. If the instrument is extremely weak, the interval may compensate by becoming longer, and in extreme cases it may become infinite in length. A long confidence interval is a warning that the instrument is providing little information about the effect caused by the treatment. Better warned than misled.

Finally, in strict parallel with §2.4, a point estimate $\widehat{\beta}$ of β may be obtained by the method of Hodges and Lehmann; that is, by equating T computed from $R_{ij} - \beta_0 D_{ij}$ to its null expectation, $I(I+1)/4$, and solving the equation for the estimate, $\widehat{\beta}$.

The sensitivity of IV inferences to departures from randomly assigned encouragement applies the method of §3.4 to T when computed from the adjusted re-

---

[10] Review Note 9. If the hypothesis $H_0 : r_{Tij} - r_{Cij} = \beta_0 (d_{Tij} - d_{Cij})$ is true, then $R_{ij} - \beta_0 D_{ij} = a_{ij}$ is fixed, not varying with Z_{ij}. In other words, because the $(r_{Tij}, r_{Cij}, d_{Tij}, d_{Cij})$'s are part of $\mathscr{F}$, if H_0 is true, then $\beta_0 = (r_{Tij} - r_{Cij})/(d_{Tij} - d_{Cij})$ is determined by $\mathscr{F}$, so using (5.4), the quantity a_{ij} may be calculated from $\mathscr{F}$.

5.3 Instruments

sponses $R_{ij} - \beta_0 D_{ij}$; see [75, 78, 97]. Weak instruments are invariably sensitive to small departures from random assignment of encouragement [97].

Example: IV analysis for Maimonides' rule

In Angrist and Lavy's [3] study of academic test performance and class size manipulated by Maimonides' rule, the hypothesis $H_0 : \beta = \beta_0$ in (5.3) with $\beta_0 = -0.1$ for the math test asserts that a one-student increase in class size produces a decrease of $\frac{1}{10}$ point in the average math score. At the cutpoint of 40 students, Maimonides' rule reduces the average class size by about 20 students, so if $H_0 : \beta = -0.1$ were true, that reduction would increase average math test scores by about 2 points. Is that hypothesis plausible?

To test this hypothesis, apply Wilcoxon's signed rank test to

$$Y_i^{(\beta_0)} = (Z_{i1} - Z_{i2})\{(R_{i1} - \beta_0 D_{i1}) - (R_{i2} - \beta_0 D_{i2})\} \tag{5.5}$$

with $\beta_0 = -0.1$. For instance, in Table 5.2, in pair $i = 1$,

$$Y_1^{(-0.1)} = (Z_{11} - Z_{12})\{(R_{11} - \beta_0 D_{11}) - (R_{12} - \beta_0 D_{12})\} \tag{5.6}$$

$$= (1 - 0)[\{72.1 - (-0.1) \cdot 23.0\} - \{63.1 - (-0.1) \cdot 40.0\}] = 7.3. \tag{5.7}$$

Applying Wilcoxon's signed rank test to the $I = 86$ adjusted differences, $Y_i^{(-0.1)}$, yields a one-sided P-value of 0.00909, suggesting that $\beta < \beta_0 = -0.1$, and yields a two-sided P-value of $2 \times 0.00909 = 0.01818$. In Figure 1.1, if cohort size were an instrument for class size — that is, if cohort size were random and influenced math scores only by influencing class size through (5.3) — then it would be quite clear that a one-student increase in class size depresses average math scores by more than $\frac{1}{10}$ of a point.

A confidence interval is built by inverting the hypothesis test; see §2.4. Testing every hypothesis $H_0 : \beta = \beta_0$ in (5.3) in this way, retaining those not rejected by the two-sided 0.05 level test, yields the 95% confidence interval for β of $[-0.812, -0.151]$. For instance, the one-sided P-value for testing $H_0 : \beta = -0.1515$ is 0.0252, so this hypothesis is barely not rejected in a two-sided, 0.05 level test, whereas the one-sided P-value for testing $H_0 : \beta = -1.51$ is 0.0249, so this hypothesis is barely rejected. Under Maimonides' rule, for a 20 student increase in class size as the cohort size goes from 41 to 40, the confidence interval for the change in average math scores 20β is $20 \times [-0.812, -0.151]$ or $[-16.24, -3.02]$ points.

A point estimate is obtained from the test by the method of Hodges and Lehmann; see §2.4. With $I = 86$ pairs of schools, the null expectation of Wilcoxon's signed rank statistic T is $I(I+1)/4 = 86(86+1)/4 = 1870.5$. If T is computed from $Y_i^{(-0.4518)}$, it yields $T = 1871$, a little too high, but if T is computed from $Y_i^{(-0.4519)}$, it yields $T = 1870$, a little too low, so upon rounding to three digits, the Hodges-

Lehmann point estimate $\widehat{\beta}$ of β in (5.3) is $\widehat{\beta} = -0.452$ or $20\widehat{\beta} = -9.04$ math score points for a 20 student increase in class size.

If cohort size were truly an instrument for class size and if (5.3) were true as well, then $Y_i^{(\beta)}$ would be symmetric about zero. Figure 5.2 plots the 'residuals,' that is the differences in test performance adjusted for the estimated effect of class size, $Y_i^{(\widehat{\beta})} = Y_i^{(-0.452)}$. The residuals $Y_i^{(\widehat{\beta})}$ do not deviate noticeably from symmetry about zero.

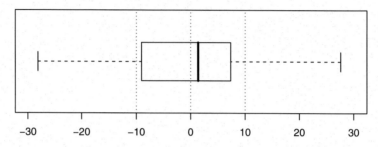

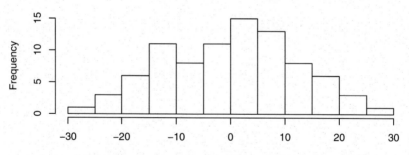

Fig. 5.2 IV residuals of math test scores, $Y_i^{(\beta_0)}$ with $\beta_0 = -0.452$, in $I = 86$ pairs of two Israeli schools. If cohort size were in fact an instrument for class size in these schools, and if $\beta = -0.452$, then these residuals would be symmetric about zero. The residuals look approximately symmetric about zero, with a mean of -0.03.

As discussed in §5.2.3, for the 86 pairs of schools in Figure 1.1 with fifth grade cohorts between 30 and 50, cohort size may not be completely random, so that cohort size is not actually an instrument for class size. Because of the nature of the comparison, very large biases may seem implausible, but no matter how carefully a nonrandomized comparison is structured, it is never possible to completely rule out the possibility of small biases. As in §3.4 and §5.2.3, a sensitivity analysis may be conducted; see [75, 78] and [81, §5]. In fact, the math test results in Table 5.3 are the sensitivity analysis for the hypothesis $H_0 : \beta = 0$. For $\Gamma = 1.1$ or $\Gamma = 1.2$, the maximum one sided P-values for testing $H_0 : \beta = -0.1$ are, respectively, 0.0238 and 0.0508. For $\Gamma \geq 1$, there is an interval $\left[\widehat{\beta}_{\min}, \widehat{\beta}_{\max}\right]$ of possible point estimates of β where, as we have seen, for $\Gamma = 1$, the single estimate is $\widehat{\beta} = \widehat{\beta}_{\min} = \widehat{\beta}_{\max} = -0.452$. As Γ increases, $\widehat{\beta}_{\min}$ decreases and $\widehat{\beta}_{\max}$ increases. For $\Gamma = 1.83$, $\widehat{\beta}_{\max} = -0.1$, whereas for $\Gamma = 2.21$, $\widehat{\beta}_{\max} = 0$. In brief, from Table 5.3, rejection of no effect of class size is insensitive to small biases, $\Gamma \leq 1.4$, and a point estimate of no effect, $\widehat{\beta}_{\max} = 0$, requires $\Gamma = 2.21$. Again, the qualitative conclusions reached by Angrist and Lavy [3] are not extremely fragile; small deviations from a truly random instrument would not alter their qualitative conclusions.

When and why are instruments valuable?

At first glance, the use of instruments seems unattractive. A natural experiment seeks a consequential treatment to which people are exposed in a haphazard way. In contrast, an instrument requires both a haphazard exposure and additionally the exclusion restriction — that is, the exposure must affect the outcome only by affecting the treatment. If a natural experiment rests on one doubtful assumption, a study using an instrument rests on two doubtful assumptions. At first glance, the step towards instruments seems to be a step in the wrong direction.

This first impression is reinforced by a second. In reading the scientific literature, many claimed instruments seem implausible as instruments, perhaps because they are clearly not random, perhaps because the exclusion restriction seems implausible, or perhaps because the authors treat these central issues in a glib and superficial manner. The examples from Angrist and Lavy [3] and Silber and colleagues [33, 95] are atypical in that they are at least plausible as instruments.

Given these two impressions, why seek instruments? When are they particularly valuable? When are they not needed? Instruments are of greatest value when any straightforward comparison of treated and control subjects is almost inevitably affected by the same bias, a bias that will recur whenever and wherever a straightforward comparison is made [45, §1]. A typical example is the effect of additional education on earnings. Almost inevitably, students who obtain post-secondary degrees have performed well in school during the ages of compulsory education, have greater motivation, better standardized test scores, typically greater financial resources, and so on, than students who leave school at an earlier stage. Repeatedly finding that people with more education earn more does little to isolate the effects actually caused by education; that is, the biases replicate with the same consistency that — perhaps with greater consistency than — the effects replicate [79]. In such a situation, it may be possible to find haphazard nudges that, at the margin, enable or discourage additional education, such as discontinuities in admission or aid criteria, proximity to affordable education, shifts in the cost or availability of aid. These nudges may be biased in various ways, but there may be no reason for them to be consistently biased in the same direction, so similar estimates of effect from studies subject to different potential biases gradually reduce ambiguity about what part is effect and what part is bias [79]. In this vein, David Card [21] surveys the many varied techniques used to estimate the economic returns actually caused by additional education.

5.4 Summary

Although structured by competing theories that make conflicting predictions, observational studies are built from opportunities, devices, and instruments.

Opportunities are special circumstances in which competing theories may be contrasted with unusual clarity. In these circumstances, the theories make sharply different predictions and the most common or plausible unmeasured biases are reduced or absent.

The association between a treatment and an outcome is ambiguous when treatments are not assigned at random. The association may be produced by an effect caused by the treatment or by a comparison of people who are not truly comparable. Devices are the tools of disambiguation, the active effort to reduce ambiguity.

Certain treatments are consistently assigned in the same biased way, so that straightforward comparisons of treated and control groups inevitably reproduce the same bias. There is no direct opportunity to separate the bias from the treatment effect. In these cases, it may be possible to find haphazard nudges towards treatment, pushes that tilt but do not control treatment assignment. To the extent that these nudges are assigned at random and affect outcomes only through acceptance of the treatment, they form instruments that may permit the isolation of treatment effects. Instruments are rare, but useful when they exist.

5.5 Further Reading

Part III of this book and Chapters 6 through 9 of [81] are 'further reading' for this chapter. Part III discusses 'design sensitivity' which measures the degree to which different designs or data generating processes produce results that are insensitive to unmeasured biases. In contrast, Chapters 6 through 8 of [81] discuss the circumstances under which known effects or multiple control groups can detect unmeasured biases, presenting technical results about these topics, while Chapter 9 of [81] discusses some technical aspects of coherence; alternatively, for articles covering similar material, see [69, 70, 71, 72, 76]. Multiple control groups are discussed in [9, 10, 19, 41, 53, 63, 69, 71] and [81, §8]; see also §11.3 and §19.3.

Known effects are discussed in [57, 66, 70, 71, 110, 25] and [81, §6]. Differential effects and generic biases are discussed in [87] where further results and examples may be found; see also the related, interesting but different design used by Rahul Roychoudhuri and colleagues [91]. The view of instruments as a bridge between randomized experiments and observational studies is developed in various ways in [2, 7, 8, 28, 33, 31]. The link between instruments and randomization inference is developed in [33, 45, 75, 78, 82]. General discussions of 'opportunities' are found in [4, 35, 58, 77, 89, 92, 103, 107]. Lotteries often present 'opportunities;' see, for instance, [5, 8, 39, 43].

References

1. Abadie, A., Gardeazabal, J.: Economic costs of conflict: A case study of the Basque Country. Am Econ Rev **93**, 113–132 (2003)
2. Angrist, J.D., Imbens, G.W., Rubin, D.B.: Identification of causal effects using instrumental variables (with Discussion). J Am Statist Assoc **91**, 444–455 (1996)
3. Angrist, J.D., Lavy, V.: Using Maimonides' rule to estimate the effect of class size on scholastic achievement. Quart J Econ **114**, 533–575 (1999)
4. Angrist, J.D., Krueger, A.B.: Empirical strategies in labor economics. In: Ashenfelter, O., Card, D. (eds.), Handbook of Labor Economics, Volume 3, pp. 1277–1366. New York: Elsevier (1999)
5. Angrist, J., Lavy, V.: New evidence on classroom computers and pupil learning. Econ J **112**, 735–765 (2002)
6. Anthony, J.C., Breitner, J.C., Zandi, P.P., Meyer, M.R., Jurasova, I., Norton, M.C., Stone, S.V.: Reduced prevalence of AD in users of NSAIDs and H_2 receptor antagonists. Neurology **54**, 2066–2071 (2000)
7. Barnard, J., Du, J.T., Hill, J.L., Rubin, D.B.: A broader template for analyzing broken randomized experiments. Sociol Methods Res **27**, 285–317 (1998)
8. Barnard, J., Frangakis, C.E., Hill, J.L., Rubin, D.B.: Principal stratification approach to broken randomized experiments: A case study of School Choice vouchers in New York City. J Am Statist Assoc **98**, 299–311 (2003)
9. Battistin, E., Rettore, E.: Ineligibles and eligible non-participants as a double comparison group in regression-discontinuity designs. J Econometrics **142**, 715–730 (2008)
10. Behrman, J.R., Cheng, Y., Todd, P.E.: Evaluating preschool programs when length of exposure to the program varies: A nonparametric approach. Rev Econ Statist **86**, 108-132 (2004)
11. Berk, R.A., Rauma, D.: Capitalizing on nonrandom assignment to treatments: A regression-discontinuity evaluation of a crime-control program. J Am Statist Assoc **78**, 21–27 (1983)
12. Berk, R.A., de Leuuw, J.: An evaluation of California's inmate classification system using a regression discontinuity design. J Am Statist Assoc **94**, 1045–1052 (1999)
13. Bernanke, B.S.: The macroeconomics of the Great Depression: a comparative approach. J Money Cred Bank **27**, 1–28 (1995) Reprinted: Bernanke, B.S. Essays on the Great Depression, Princeton, NJ: Princeton University Press (2000)
14. Bilban, M., Jakopin, C.B.: Incidence of cytogenetic damage in lead-zinc mine workers exposed to radon. Mutagenesis **20**, 187–191 (2005)
15. Black, S.: Do better schools matter? Parental valuation of elementary education. Q J Econ **114**, 577–599 (1999)
16. Bound, J.: The health and earnings of rejected disability insurance applicants. Am Econ Rev **79**, 482–503 (1989)
17. Bound, J., Jaeger, D.A., Baker, R.M.: Problems with instrumental variables estimation when the correlation between the instruments and the endogenous explanatory variable is weak. J Am Statist Assoc **90**, 443–450 (1995)

18. Campbell, D.T.: Factors relevant to the validity of experiments in social settings. Psychol Bull **54**, 297–312 (1957)
19. Campbell, D.T.: Prospective: artifact and control. In: Rosenthal, R. and Rosnow, R. (eds.), Artifact in Behavioral Research, pp. 351–382. New York: Academic Press (1969)
20. Campbell, D.T.: Methodology and Epistemology for Social Science: Selected Papers. Chicago: University of Chicago Press (1988)
21. Card, D.: The causal effect of education. In: Ashenfelter, O., Card, D., (eds.) Handbook of Labor Economics. New York: North Holland (2001)
22. Choudhri, E.U., Kochin, L.A.: The exchange rate and the international transmission of business cycle disturbances: Some evidence from the Great Depression. J Money Cred Bank **12**, 565–574 (1980)
23. Cioran, E.M.: History and Utopia. Chicago: University of Chicago Press (1998)
24. Cochran, W.G.: The planning of observational studies of human populations (with Discussion). J Roy Statist Soc A **128**, 234–265 (1965)
25. Cook, T.D.: Waiting for life to arrive: a history of the regression-discontinuity designs in psychology, statistics and economics. J Econometrics **142**, 636–654 (2007)
26. Eichengreen, B., Sachs, J.: Exchange rates and economic recovery in the 1930's. J Econ Hist **45**, 925–946 (1985)
27. Evans, L.: The effectiveness of safety belts in preventing fatalities. Accid Anal Prev **18**, 229–241 (1986)
28. Frangakis, C.E., Rubin, D.B.: Addressing complications of intention-to-treat analysis in the combined presence of all-or-none treatment noncompliance and subsequent missing outcomes. Biometrika **86**, 365–379 (1999)
29. Fenech, M., Changb, W.P., Kirsch-Voldersc, M., Holland, N., Bonassie, S., Zeiger, E.: HUMN project: Detailed description of the scoring criteria for the cytokinesis-block micronucleus assay using isolated human lymphocyte cultures. Mutat Res **534**, 65–75 (2003)
30. Friedman, M., Schwartz, A.J.: A Monetary History of the United States. Princeton, NJ: Princeton University Press (1963)
31. Goetghebeur, E., Loeys, T.: Beyond intent to treat. Epidemiol Rev **24**, 85–90 (2002)
32. Gould, E. D., Lavy, V., Paserman, M. D.: Immigrating to opportunity: Estimating the effect of school quality using a natural experiment on Ethiopians in Israel. Q J Econ **119**, 489–526 (2004)
33. Greevy, R., Silber, J.H., Cnaan, A., Rosenbaum, P.R.: Randomization inference with imperfect compliance in the ACE-inhibitor after anthracycline randomized trial. J Am Statist Assoc **99**, 7–15 (2004)
34. Hahn, J., Todd, P., Van der Klaauw, W.: Identification and estimation of treatment effects with a regression-discontinuity design. Econometrica **69**, 201–209 (2001)
35. Hamermesh, D.S.: The craft of labormetrics. Indust Labor Relat Rev **53**, 363–380 (2000)
36. Hawkins, N.G., Sanson-Fisher, R.W., Shakeshaft, A., D'Este, C., Green, L.W.: The multiple baseline design for evaluating population based research. Am J Prev Med **33**, 162–168 (2007)
37. Heckman, J., Navarro-Lozano, S.: Using matching, instrumental variables, and control functions to estimate economic choice models. Rev Econ Statist **86**, 30–57 (2004)
38. Hill, A.B.: The environment and disease: Association or causation? Proc Roy Soc Med **58**, 295–300 (1965)
39. Ho, D.E., Imai, K.: Estimating the causal effects of ballot order from a randomized natural experiment: California alphabet lottery, 1978-2002. Public Opin Q **72**, 216–240 (2008)
40. Holland, P.W.: Causal Inference, path analysis, and recursive structural equations models. Sociolog Method **18**, 449–484 (1988)
41. Holland, P.W.: Choosing among alternative nonexperimental methods for estimating the impact of social programs: comment. J Am Statist Assoc **84**, 875–877 (1989)
42. Imbens, G.W.: The role of the propensity score in estimating dose response functions. Biometrika **87**, 706–710 (2000)

43. Imbens, G.W., Rubin, D.B., Sacerdote, B.I.: Estimating the effect of unearned income on labor earnings, savings, and consumption: Evidence from a survey of lottery players. Am Econ Rev **91**, 778-794 (2001)
44. Imbens, G.W.: Nonparametric estimation of average treatment effects under exogeneity: A review. Rev Econ Statist **86**, 4–29 (2004)
45. Imbens, G., Rosenbaum, P.R.: Robust, accurate confidence intervals with a weak instrument: Quarter of birth and education. J Roy Statist Soc A **168**, 109–126 (2005)
46. Imbens, G.W., Lemieux, T.: Regression discontinuity designs: A guide to practice. J Econometrics **142**, 615–635 (2008)
47. in 't Veld, B.A., Launer, L.J., Breteler, M.M.B., Hofman, A., Stricker, B.H.C.: Pharmacologic agents associated with a preventive effect on Alzheimer's disease. Epidemiol Rev **2**, 248-268 (2002)
48. Joffe, M.M., Colditz, G.A.: Restriction as a method for reducing bias in the estimation of direct effects. Statist Med **17**, 2233–2249 (1998)
49. Khuder, S.A., Milz, S., Jordan, T., Price, J., Silvestri, K., Butler, P.: The impact of a smoking ban on hospital admissions for coronary heart disease. Prev Med **45**, 3–8 (2007)
50. LaLumia, S.: The effects of joint taxation of married couples on labor supply and non-wage income. J Public Econ **92**, 1698-1719 (2008)
51. Lambe, M., Cummings, P.: The shift to and from daylight savings time and motor vehicle crashes. Accid Anal Prev **32**, 609–611 (2002)
52. Li, F., Frangakis, C.E.: Polydesigns and causal inference. Biometrics **62**, 343–351 (2006)
53. Lu, B., Rosenbaum, P.R.: Optimal matching with two control groups. J Comput Graph Statist **13**, 422–434 (2004)
54. Ludwig, J., Miller, D.L.: Does Head Start improve children's life chances? Evidence from a regression discontinuity design. Q J Econ **122**, 159–208 (2007)
55. Manski, C.: Nonparametric bounds on treatment effects. Am Econ Rev 319–323 (1990)
56. Marguart, J.W., Sorensen, J.R.: Institutional and postrelease behavior of Furman-commuted inmates in Texas. Criminology **26**, 677–693 (1988)
57. McKillip, J.: Research without control groups: A control construct design. In: Bryant, F.B., et al., (eds.), Methodological Issues in Applied Social Psychology, pp. 159–175. New York: Plenum Press (1992)
58. Meyer, B.D.: Natural and quasi-experiments in economics. J Business Econ Statist **13**, 151–161 (1995)
59. Mill, J.S.: On Liberty. New York: Barnes and Nobel (1859, reprinted 2004)
60. Milyo, J., Waldfogel, J.: The effect of price advertising on prices: evidence in the wake of 44 Liquormart. Am Econ Rev **89**, 1081–1096 (1999)
61. NIDA: Washington DC Metropolitan Area Drug Study (DC*MADS), 1992. U.S. National Institute on Drug Abuse: ICPSR Study No. 2347. http://www.icpsr.umich.edu (1999)
62. Oreopoulos, P.: Long-run consequences of living in a poor neighborhood. Q J Econ **118**, 1533–1575 (2003)
63. Origo, F.: Flexible pay, firm performance and the role of unions: New evidence from Italy. Labour Econ **16**, 64–78 (2009)
64. Peto, R., Pike, M., Armitage, P., Breslow, N., Cox, D., Howard, S., Mantel, N., McPherson, K., Peto, J., Smith, P.: Design and analysis of randomised clinical trials requiring prolonged observation of each patient, I. Br J Cancer **34**, 585–612 (1976)
65. Pinto, D., Ceballos, J.M., García, G., Guzmán, P., Del Razo, L.M., Gómez, E.V.H., García, A., Gonsebatt, M.E.: Increased cytogenetic damage in outdoor painters. Mutat Res **467**, 105–111 (2000)
66. Reynolds, K.D., West, S.G.: A multiplist strategy for strengthening nonequivalent control group designs. Eval Rev **11**, 691–714 (1987)
67. Rosenbaum, P.R.: From association to causation in observational studies. J Am Statist Assoc **79**, 41-48 (1984)
68. Rosenbaum, P.R.: Sensitivity analysis for certain permutation inferences in matched observational studies. Biometrika **74**, 13–26 (1987)

69. Rosenbaum, P.R.: The role of a second control group in an observational study (with Discussion). Statist Sci **2**, 292–316 (1987)
70. Rosenbaum, P.R.: The role of known effects in observational studies. Biometrics **45**, 557–569 (1989)
71. Rosenbaum, P.R.: On permutation tests for hidden biases in observational studies. Ann Statist **17**, 643–653 (1989)
72. Rosenbaum, P.R.: Some poset statistics. Ann Statist **19**, 1091–1097 (1991)
73. Rosenbaum, P.R.: Hodges-Lehmann point estimates in observational studies. J Am Statist Assoc **88**, 1250–1253 (1993)
74. Rosenbaum, P.R.: Detecting bias with confidence in observational studies. Biometrika **79**, 367–374 (1992)
75. Rosenbaum, P.R.: Comment on a paper by Angrist, Imbens, and Rubin. J Am Statist Assoc **91**, 465–468 (1996)
76. Rosenbaum, P.R.: Signed rank statistics for coherent predictions. Biometrics **53**, 556–566 (1997)
77. Rosenbaum, P.R.: Choice as an alternative to control in observational studies (with Discussion). Statist Sci **14**, 259–304 (1999)
78. Rosenbaum, P.R.: Using quantile averages in matched observational studies. Appl Statist **48**, 63–78 (1999)
79. Rosenbaum, P.R.: Replicating effects and biases. Am Statistician **55**, 223–227 (2001)
80. Rosenbaum, P.R.: Stability in the absence of treatment. J Am Statist Assoc **96**, 210–219 (2001)
81. Rosenbaum, P.R.: Observational Studies (2nd ed.). New York: Springer (2002)
82. Rosenbaum, P.R.: Covariance adjustment in randomized experiments and observational studies (with Discussion). Statist Sci **17**, 286–327 (2002)
83. Rosenbaum, P.R.: Does a dose-response relationship reduce sensitivity to hidden bias? Biostatistics **4**, 1–10 (2003)
84. Rosenbaum, P.R.: Design sensitivity in observational studies. Biometrika **91**, 153–164 (2004)
85. Rosenbaum, P. R.: Heterogeneity and causality: Unit heterogeneity and design sensitivity in observational studies. Am Statist **59**, 147–152 (2005)
86. Rosenbaum, P.R.: Exact, nonparametric inference when doses are measured with random errors. J Am Statist Assoc **100**, 511–518 (2005)
87. Rosenbaum, P.R.: Differential effects and generic biases in observational studies. Biometrika **93**, 573–586 (2006)
88. Rosenbaum, P.R.: What aspects of the design of an observational study affect its sensitivity to bias from covariates that were not observed? Festschrift for Paul W. Holland. Princeton, NJ: ETS (2009)
89. Rosenzweig, M.R., Wolpin, K.I.: Natural 'natural experiments' in economics. J Econ Lit **38**, 827–874 (2000)
90. Rothman, K.J.: Modern Epidemiology. Boston: Little, Brown (1986)
91. Roychoudhuri, R., Robinson, D., Putcha, V., Cuzick, J., Darby, S., Møller, H.: Increased cardiovascular mortality more than fifteen years after radiotherapy for breast cancer: a population-based study. BMC Cancer, Jan 15, 7–9 (2007)
92. Rutter, M.: Identifying the Environmental Causes of Disease: How Do We Decide What to Believe and When to Take Action? London: Academy of Medical Sciences. (2007)
93. Sennett, R.: The Uses of Disorder. New Haven, CT: Yale University Press (1971, 2008)
94. Shadish, W.R., Cook, T.D.: The renaissance of field experimentation in evaluating interventions. Annu Rev Psychol **60**, 607–629 (2009)
95. Silber, J.H., Cnaan, A., Clark, B.J., Paridon, S.M., Chin, A.J., et al.: Enalapril to prevent cardiac function decline in long-term survivors of pediatric cancer exposed to anthracyclines. J Clin Oncol **5**, 820–828 (2004)
96. Small, D.S.: Sensitivity analysis for instrumental variables regression with overidentifying restrictions. J Am Statist Assoc **102**, 1049–1058 (2007)

References

97. Small, D.S., Rosenbaum, P.R.: War and wages: the strength of instrumental variables and their sensitivity to unobserved biases. J Am Statist Assoc **103**, 924–933 (2008)
98. Small, D.S., Rosenbaum, P.R.: Error-free milestones in error-prone measurements. Ann Appl Statist, to appear (2009)
99. Sobel, M.E.: An introduction to causal inference. Sociol Methods Res **24**, 353–379 (1996)
100. Sommer, A., Zeger, S.L.: On estimating efficacy from clinical trials. Statist Med **10**, 45–52 (1991)
101. Stuart, E.A., Rubin, D.B.: Matching with multiple control groups with adjustment for group differences. J Educ Behav Statist **33**, 279–306 (2008)
102. Sullivan, J.M., Flannagan, M.J.: The role of ambient light level in fatal crashes: Inferences from daylight saving time transitions. Accid Anal Prev **34**, 487–498 (2002)
103. Summers, L.H.: The scientific illusion in empirical macroeconomics (with Discussion). Scand J Econ **93**, pp. 129–148 (1991)
104. Tan, Z.: Regression and weighting methods for causal inference using instrumental variables. J Am Statist Assoc **101**, 1607–1618 (2006)
105. Thistlethwaite, D.L., Campbell, D.T.: Regression-discontinuity analysis. J Educ Psychol **51**, 309–317 (1960)
106. Trochim, W.M.K.: Pattern matching, validity and conceptualization in program evaluation. Eval Rev **9**, 575–604 (1985)
107. Vandenbroucke, J.P.: When are observational studies as credible as randomized trials? Lancet **363**, 1728–1731 (2004)
108. van Eeden, C.: An analogue, for signed rank statistics, of Jureckova's asymptotic linearity theorem for rank statistics. Ann Math Statist **43**, 791–802 (1972)
109. Weiss, N.: Inferring causal relationships: Elaboration of the criterion of dose-response. Am J Epidemiol **113**, 487–490 (1981)
110. Weiss, N.: Can the 'specificity' of an association be rehabilitated as a basis for supporting a causal hypothesis? Epidemiol **13**, 6–8 (2002)
111. West, S.G., Duan, N., Pequegnat, W., Gaist, P., Des Jarlais, D.C., Holtgrave, D., Szapocznik, J., Fishbein, M., Rapkin, B., Clatts, M., Mullen, P.D.: Alternatives to the randomized controlled trial. Am J Public Health **98**, 1359–1366 (2008)
112. Wintemute, G.J., Wright, M.A., Drake, C.M., Beaumont, J.J.: Subsequent criminal activity among violent misdemeanants who seek to purchase handguns: risk factors and effectiveness of denying handgun purchase. J Am Med Assoc **285**, 1019–1026 (2001)
113. Wright, M.A., Wintemute, G.J., Rivara, F.P.: Effectiveness of denial of handgun purchase to persons believed to be at high risk for firearm violence. Am J Public Health **89**, 88–90 (1999)

Chapter 6
Transparency

Abstract Transparency means making evidence evident. An observational study that is not transparent may be overwhelming or intimidating, but it is unlikely to be convincing. Several aspects of transparency are briefly discussed.

> The beliefs we have most warrant for, have no safeguard to rest on, but a standing invitation to the whole world to prove them unfounded. If the challenge is not accepted, or is accepted and the attempt fails, we are far enough from certainty still; but we have done the best the existing state of human reason admits of...' This is the amount of certainty attainable by a fallible being, and this is the sole way of attaining it.
>
> John Stuart Mill [4, page 21]

> The objectivity of all science, including mathematics, is inseparably linked with its criticizability.
>
> Karl R. Popper [6, page 137]

Transparency means making evidence evident. An experiment, and by analogy an observational study, is not a private experience, not the source of some private conviction. In the absence of transparency, evidence, argument, and conclusions are not fully available for critical evaluation. Critical discussion, the standing invitation to undermine evidence or argument, is, for John Stuart Mill, the sole safeguard of our most warranted beliefs. For Karl Popper, critical discussion is inseparably linked with the objectivity of science and mathematics. To the extent that transparency is needed for critical discussion, transparency is no small issue.

David Cox [2, page 8] writes:

> An important aspect of analysis, difficult to achieve with complex methods, is transparency. That is, in principle, the pathway between the data and the conclusions should be as clear as is feasible. This is partly for the self-education of the analyst and also is for protection against errors.... It is also important for presenting conclusions.... Transparency strongly encourages the use of the simplest methods that will be adequate.

See also Cox [1, page 11].

In addition to the "use of the simplest methods that will be adequate," transparency is aided by the following considerations. Introducing a felicitous phrase, Mervyn Susser [8, page 74] writes: "A main object of research design is thus to *simplify the conditions of observation*." In a chapter with that title, Susser [8, Chapter 7] discusses: (i) limiting observations to the relevant segment of the population to "isolate the hypothetical causal variable and permit study of its effects alone" [8, page 74], (ii) "selecting suitable situations" [8, page 76], and (iii) matching to remove bias from observed covariates. Concerning (i), see also Joffe and Colditz [3], who use the term 'restriction.' In discussing the design of clinical trials, Richard Peto et al. [5, page 590] write: "A positive result is more likely, and a null result is more informative, if the main comparison is of only 2 treatments, these being as different as possible."

Modularity aids transparency. If a study addresses several issues, each potentially contentious, but the study is composed of separable, simple modular components, then there is the realistic prospect that the scope of contention will be constructively focused. Modularity limits leakage: controversy remains contained within the region of contention. Here are three modular questions that arise in most observational studies. Do the treated and untreated populations overlap sufficiently with respect to observed covariates to permit the construction of a comparable control group, or do the populations need to be restricted before such a comparison is attempted? Has matching succeeded in balancing observed covariates, so matched treated and control groups are comparable with respect to these observed covariates? Is it plausible that differing outcomes in treated and control groups are produced by imbalances in a specific unmeasured covariate? If these questions are separated and addressed one at a time, the first two questions may be settled with little controversy before engaging the inevitably more contentious third question. If the three questions are addressed simultaneously in one grand analysis, little conviction or consensus may develop even about the first two questions.

To Cox's good list of reasons for seeking transparency, I would add one more [7, Chapter 12]. If smaller issues are permitted to become unnecessarily complex, then the analysis may collapse under the weight of these complexities before engaging larger issues. In observational studies, if the adjustment for observed covariates becomes unnecessarily complex, as sometimes happens, then the analysis may never engage the fundamental issue, namely possible biases from covariates that were not measured. Like a good card trick, unnecessarily complex adjustments for observed covariates may distract attention from the fundamental issue in observational studies, but they are unlikely to shed much light on it.

An empirical study that is not transparent may be published or cited, but it is less likely to undergo serious critical discussion and therefore less likely to receive the implicit endorsement of surviving such discussion largely unscathed. Conversely, if critical discussion of a transparent study reveals potential ambiguities or alternative interpretations, that discussion may stimulate a replication that eliminates the ambiguities (§4.5). If a study has unambiguous faults, they are more likely to be discovered if the study is transparent, and hence there is greater reason to trust the conclusions of a transparent study when no such faults are identified.

References

1. Cox, D.R.: Planning of Experiments. New York: Wiley (1958)
2. Cox, D.R.: Applied statistics: a review. Ann Appl Statist **1**, 1–16 (2007)
3. Joffe, M.M., Colditz, G.A.: Restriction as a method for reducing bias in the estimation of direct effects. Statist Med **17**, 2233–2249 (1998)
4. Mill, J.S.: On Liberty. New York: Barnes and Nobel (1859, reprinted 2004)
5. Peto, R., Pike, M., Armitage, P., Breslow, N., Cox, D., Howard, S., Mantel, N., McPherson, K., Peto, J., Smith, P. Design and analysis of randomised clinical trials requiring prolonged observation of each patient, I. Br J Cancer **34**, 585–612 (1976)
6. Popper, K.R.: Objective Knowledge. Oxford: Oxford University Press (1972)
7. Rosenbaum, P.R.: Observational Studies. New York: Springer (2002)
8. Susser, M.: Causal Thinking in the Health Sciences. New York: Oxford University Press (1973)

Part II
Matching

Chapter 7
A Matched Observational Study

Abstract As a prelude to several chapters describing the construction of a matched control group, the current chapter presents an example of a matched observational study as it might (and did) appear in a scientific journal. When reporting a matched observational study, the matching methods are described very briefly in the *Methods* section. In more detail, the *Results* section presents tables or figures showing that the matching has been effective in balancing certain observed covariates, so that treated and control groups are comparable with respect to these specific variables. The *Results* section then compares outcomes in treated and control groups. Because matching has arranged matters to compare ostensibly comparable groups, the comparison of outcomes is often both simpler in form and more detailed in content than it might be if separate adjustments were required for each aspect of each outcome. Treated and control groups that appear comparable in terms of a specific list of measured covariates – groups that are ostensibly comparable – may nonetheless differ in terms of covariates that were not measured. Though not discussed in the current chapter, the important issue of unmeasured covariates in this example is discussed in Part III.

7.1 Is More Chemotherapy More Effective?

Jeffrey Silber, Dan Polsky, Richard Ross, Orit Even-Shoshan, Sandy Schwartz, Katrina Armstrong, Tom Randall, and I [6, 8] asked how the intensity of chemotherapy for ovarian cancer affected patient outcomes. We thought that greater intensity might prolong survival, perhaps at the cost of increased toxicity. What evidence bears on this question?

There is a basic difficulty in studying the intended effects of medical treatments outside of randomized controlled clinical trials [9]. In virtually all areas of medicine, most of the variation in treatment occurs in thoughtful and deliberate response to variation in the health, prognosis or wishes of patients. That is, treatment assignment is very far from being determined 'at random.' Ovarian cancer is unusual in this regard because there is a meaningful source of variation in treatment that is not a response to the patient. Chemotherapy for ovarian cancer is provided by two distinct specialties, namely medical oncologists (MOs), who provide chemotherapy for cancers of all kinds, and gynecologic oncologists (GOs), who primarily treat cancers of the ovary, uterus and cervix. Medical oncologists typically have a residency in internal medicine, followed by a fellowship emphasizing the administration of chemotherapy and the management of its side effects. Gynecologic oncologists typically complete a residency in obstetrics and gynecology, followed by a fellowship in gynecologic oncology, that includes training in surgical oncology and chemotherapy administration for gynecologic cancers. Unlike gynecologic oncologists, who are trained in surgery, medical oncologists are almost invariably not surgeons, so medical oncologists provide chemotherapy after someone else has performed surgery. We anticipated, as turned out to be the case, that MOs would use chemotherapy more intensively than GOs, both at the time of initial diagnosis and several years later if the cancer has spread from its site of origin. Is the greater intensity of chemotherapy found in the practice of MOs of benefit to patients?

The study was based on data that linked the Surveillance, Epidemiology and End Results (SEER) program of the U.S. National Cancer Institute to Medicare claims. The SEER data are collected at SEER sites in the United States, where some SEER sites are cities (e.g., Detroit) and others are states (e.g., New Mexico). The SEER data include clinical stage and tumor grade. The study used data on women older than 65 with ovarian cancer who were diagnosed between 1991 and 1999 and had appropriate surgery and at least some chemotherapy; see [8] for details. There were 344 such women who received chemotherapy from a GO, and 2011 such women who received chemotherapy from an MO.

7.2 Matching for Observed Covariates

We matched each GO patient to an MO patient, producing 344 matched pairs of two similar patients. Table 7.1 describes several of the covariates used in the matching. Was the matching successful in producing GO and MO groups that were reasonably comparable in terms of observed covariates?

The first group of variables in Table 7.1 describes the type of surgeon. Gynecologic oncologists are specialists in surgery for gynecologic cancers, but surgery is often performed by gynecologists or general surgeons. Not surprisingly, this variable is substantially out of balance before matching: a patient who has surgery performed by a GO is more likely to have chemotherapy provided by a GO. After matching, the distributions of surgeon types are almost identical.

Clinical stage is usually thought to be the most important predictor of survival in ovarian cancer. Perhaps because there is no particular reason for patients to seek chemotherapy from an MO or a GO, stage was not greatly out of balance before matching. For stage, the balance after matching is quite close.

7.2 Matching for Observed Covariates

Table 7.1 Comparability at baseline for patients with ovarian cancer: all 344 patients of a gynecologic oncologist (GO), 344 matched patients of a medical oncologist (MO), and all 2011 patients of medical oncologists. Values are in percent, unless labeled as means. Also included in the matching, but not in this table, were additional comorbid conditions, including anemia, angina, arrhythmia, asthma, coagulation disorder, electrolyte abnormality, hepatic dysfunction, hyperthyroidism, peripheral vascular disease, and rheumatoid arthritis. Notice that surgeon type, SEER site, and year of diagnosis were substantially out of balance prior to matching, but were in reasonable balance after matching.

		GO $n=344$	matched-MO $n=344$	all-MO $n=2,011$
Surgeon	GO	76	75	33
Type	Gyn	15	16	39
	General	8	8	28
Stage	I	9	9	9
	II	11	9	9
	III	51	53	47
	IV	26	26	31
	Missing	3	2	3
Tumor	1	5	4	4
Grade	2	16	13	17
	3	52	55	47
	4	9	8	11
	Missing	18	20	21
Demographics	Age, mean	72.2	72.2	72.8
	White	91	94	94
	Black	8	5	3
Selected	COPD	15	12	13
Comorbid	Hypertension	48	46	42
Conditions	Diabetes	11	8	8
	CHF	2	2	4
SEER Site	Connecticut	18	18	15
	Detroit	26	26	12
	Iowa	17	17	17
	New Mexico	7	7	3
	Seattle	9	9	16
	Atlanta	9	9	7
	Los Angeles	12	12	19
	San Francisco	1	1	9
Year of Diagnosis	1991	4	4	9
	1992	7	7	14
	1993	10	9	14
	1994	11	11	12
	1995	11	13	12
	1996	10	9	12
	1997	16	15	10
	1998	13	15	9
	1999	18	17	9
Propensity Score	$\widehat{e}(\mathbf{x})$, mean	0.23	0.21	0.14

In fact, whenever possible, we matched exactly for surgeon type and stage. This means that, whenever possible, a stage III patient whose surgery was performed by a gynecologist was paired with a stage III patient whose surgery was performed by a gynecologist. As seen in the marginal distributions in Table 7.1, this was not always possible. Naive intuition suggests that one should match exactly for every variable, but a moment's thought reveals this to be impossible. If there were just 30 binary variables, there would be 2^{30} or about a billion types of patients, so it is highly unlikely that one would find exact matches on all 30 variables for almost all of the patients. Nonetheless, the balance on the covariates in Table 7.1 is quite good, even though, unlike stage and surgeon type, most variables in Table 7.1 are not exactly matched. Section 9.1 will discuss covariate balance in greater detail. In practice, exact matching, if used at all, is reserved for one or two critically important variables.

Surgeon type and clinical stage are connected. A surgeon who is trained specifically in cancer surgery, such as a GO, is likely to combine an effort at curative surgery with extensive node sampling. In consequence, a cancer surgeon may find more cancer than a general surgeon, and may assign a patient to a higher clinical stage [2]. That is, a patient who is classified as stage II by a general surgeon might have been classified as stage III by a GO. The very meaning of one variable, clinical stage, depends upon another variable, surgeon type. However, because almost all of the pairs are matched exactly for stage and surgeon type, in almost all pairs, the two patients received the same stage from the same category of surgeon.

Tumor grade, demographics, and selected comorbid conditions are seen to be reasonably well balanced, and the same is true for other comorbid conditions mentioned in the caption of Table 7.1. Comorbid conditions matter both because they may pose a direct risk to the patient's survival and may also complicate and limit the administration of chemotherapy.

Gynecologic oncologists are unevenly spread through the United States, and the profession has been growing over time. For this reason, before matching, both SEER site and year of diagnosis were often quite different for the patients of GOs and MOs. For patients of GOs at SEER sites, 26% were in Detroit and only 1% in San Francisco, while for all MO patients, 12% were in Detroit and 9% in San Francisco. Expressed in terms of odds ratios, a woman is almost 20 times as likely to be treated by a GO in Detroit than in San Francisco. Although ovarian cancer is, presumably, much the same disease in Detroit and San Francisco, there are differences in wealth, demographics, and health services that might affect patient outcomes. After matching, the distributions of SEER sites are the same. Similarly, GO patients were more likely to have been diagnosed recently, simply because there are more GOs recently. Although ovarian cancer itself may not have changed much through the 1990s, the chemotherapies did improve [3], so a substantial imbalance in year of diagnosis would compare patients who received different drugs. After matching, the proportion of patients diagnosed in three time intervals, 1991–1992, 1993–1996, and 1997–1999, were identical, with only small imbalances for individual years inside these intervals. In fact, although this is not visible in Table 7.1, the joint behavior of the three time intervals and the SEER site is exactly balanced;

7.3 Outcomes in Matched Pairs

Table 7.2 Weeks with chemotherapy, in year 1 and in years 1 through 5, for 344 matched pairs. Values are the mean, quartiles, minimum, and maximum. Two-sided P-values are from Wilcoxon's signed rank test.

Period	Group	Mean	Min	25%	50%	75%	Max	P-value
Year 1	GO	6.63	1	5	6	8	19	0.0022
Year 1	MO	7.74	1	5	6	10	42	
Years 1-5	GO	12.07	1	5	9	16	70	0.00045
Years 1-5	MO	16.47	1	6	11	21	103	

for instance, the number of GO patients in 1991–1992 from Iowa equals the number of matched MO patients in 1991–1992 from Iowa. The balance on SEER site and year of diagnosis was obtained using "fine balance" [5, 6], a technique discussed in Chapter 10.

The large imbalances before matching in surgeon type, year of diagnosis, and SEER site, and the much smaller imbalances before matching in clinical stage, grade and comorbid conditions are somewhat encouraging. They hint at the possibility that the chemotherapy provider type reflects the relative availability of MOs and GOs more than it reflects attributes of the patient's disease. In any event, the measured covariates are reasonably well balanced after matching.

The final variable in Table 7.1 is the propensity score. It is an estimate of the conditional probability of receiving chemotherapy from a GO rather than an MO given the observed covariates. Propensity scores are a basic tool when matching on many variables [4]. Propensity scores were discussed conceptually in §3.3, and their role in multivariate matching is discussed further in Chapter 8. In Table 7.1, the mean propensity score is fairly similar in the two matched groups; indeed, the balance on many covariates is produced by balancing the propensity score; see (3.10).

7.3 Outcomes in Matched Pairs

As anticipated, many MOs often used chemotherapy more intensely than did GOs, both in the first year following diagnosis and in the first five years following diagnosis; see Table 7.2 and Figure 7.1. The difference at the medians is not large, but it is quite noticeable at the upper quartiles. Was greater intensity of treatment by MOs beneficial?

Despite a difference in intensity, survival was virtually identical for the patients of MOs and GOs; see Table 7.3. The standard test comparing paired censored survival times is the Prentice-Wilcoxon test [7]; it gives a two-sided P-value of 0.45. Table 7.3 also lists the number of patients at risk at the beginning of each of the first five years, that is, the number of patients alive and uncensored at the start of the year. The number of patients at risk is also very similar for matched patients of MOs and GOs.

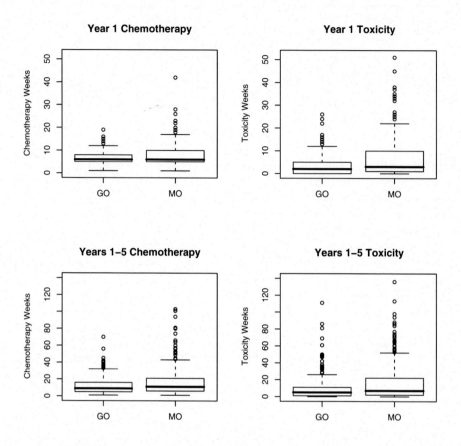

Fig. 7.1 Chemotherapy and toxicity in 344 pairs of patients with ovarian cancer, one treated by a GO, the other by an MO.

Patients of MOs had more weeks of toxicity than did patients of GOs; see Table 7.4 and Figure 7.1. In Table 7.4, both in year 1 and in years 1 to 5, the upper quartiles for MO patients are double those for GO patients.

The difference in intensity was associated with the difference in toxicity; see Figure 7.2. In Figure 7.2, the 344 matched pair differences in toxicity weeks, MO-minus-GO, are plotted against the 344 matched pair differences in chemotherapy weeks. Points project horizontally into the marginal boxplot of the differences in toxicity weeks. The curve is a lowess smooth [1, pages 168–180] as implemented in the statistical package R. Kendall's rank correlation between toxicity differences and chemotherapy differences is 0.39 and is significantly different from zero, with a two-sided significance level less than 10^{-10}.

In brief, it appears that MOs often treated more intensively than GOs, often with more toxicity, but survival was no different.

7.4 Summary

Table 7.3 Survival and number at risk in 344 matched pairs of one GO patient and one MO patient. The two-sided *P*-value of 0.45 is from the Prentice-Wilcoxon test [7] comparing paired survival times.

	GO Patients	MO Patients
Median (Years)	3.04	2.98
95% CI	[2.50, 3.40]	[2.69, 3.67]
1 Year Survival %	86.6	87.5
95% CI	[83.0, 90.2]	[84.0, 90.1]
2 Year Survival %	64.8	66.9
95% CI	[59.8, 69.9]	[61.9, 71.8]
5 Year Survival %	35.1	34.2
95% CI	[30.0, 40.2]	[29.2, 39.3]
Number at Risk Year 0	344	344
Number at Risk Year 1	298	301
Number at Risk Year 2	223	230
Number at Risk Year 3	173	172
Number at Risk Year 4	133	128

Table 7.4 Weeks with chemotherapy-associated toxicity, in year 1 and in years 1 through 5, for 344 matched pairs. Values are the mean, quartiles, minimum, and maximum. Two-sided *P*-values are from Wilcoxon's signed rank test. Chemotherapy-associated toxicity refers to inpatient or outpatient diagnoses for anemia, neutropenia, thrombocytopenia, diarrhea, dehydration or mucositis, and neuropathy.

Period	Group	Mean	Min	25%	50%	75%	Max	*P*-value
Year 1	GO	3.61	0	0	2	5	26	0.00000089
Year 1	MO	6.67	0	1	3	10	51	
Years 1-5	GO	8.89	0	1	5	11	111	0.000000026
Years 1-5	MO	16.29	0	2	7	22	136	

An obvious question is whether the patterns exhibited in Tables 7.2-7.4 and Figures 7.1 and 7.2 could be the result not of differences in intensity of chemotherapy applied to similar patients, but rather of unmeasured pretreatment differences between the patients themselves. Table 7.1 showed that, prior to matching, many covariates were substantially out of balance in the GO and MO groups. It is certainly possible that some other covariate, a covariate that was not measured, was also out of balance before matching, and because that variable was not controlled by matching, it is certainly possible that the variable is still out of balance in Tables 7.2–7.4 and Figures 7.1 and 7.2. This concern is central to Part III, where there will be further analysis of the ovarian data.

7.4 Summary

This chapter is an introduction to Part II. A matched comparison has been presented as it might be, and was, presented in a scientific journal [8], with little reference to the procedures used to construct the matched sample [6]. Matching procedures are

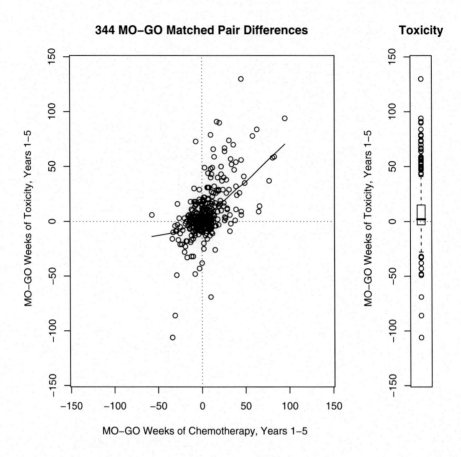

Fig. 7.2 Matched pair differences, MO−GO, in toxicity plotted against differences in chemotherapy. The curve is a lowess smooth.

the focus of Part II. The current chapter has made three basic points. First, matching on many observed covariates is often feasible; see Table 7.1. Second, the reader may examine the degree to which matched groups are comparable with respect to observed covariates, as well as which covariates are not among the observed covariates, without getting involved in the procedures used to construct the matched sample; again, see Table 7.1. Finally, straightforward analyses sufficed to take a close look at several outcomes — here, survival, chemotherapy, toxicity in Tables 7.2-7.4 and Figures 7.1-7.2 — because those analyses compared two matched groups that looked similar prior to treatment with respect to observed covariates. More precisely, analyses conducted under the naïve model for treatment assignment (§3.3) are straightforward. The possibility that the naïve model is not true is taken up in Part III.

7.5 Further Reading

The remainder of Part II is further reading for this chapter. The ovarian cancer study is discussed in detail by Jeffrey Silber and colleagues [6, 8].

References

1. Cleveland, W.S.: The Elements of Graphing Data. Summit, NJ: Hobart Press.
2. Mayer, A.R., Chambers, S.K., Graves, E., et al: Ovarian cancer staging: Does it require a gynecologic oncologist? Gynecol Oncol **47**, 223–337 (1992)
3. McGuire, W.P., Bundy, B., Wenzel, L., et al.: Cyclophosphamide and cisplatin compared with paclitaxel and cisplatin in patients with stage III and stage IV ovarian cancer. N Engl J Med **334**, 1–6 (1996)
4. Rosenbaum, P.R., Rubin, D.B.: The central role of the propensity score in observational studies for causal effects. Biometrika **70**, 41–55 (1983)
5. Rosenbaum, P.R.: Optimal matching in observational studies. J Am Statist Assoc **84**, 1024–1032 (1989)
6. Rosenbaum, P.R., Ross R.N., Silber, J.H.: Minimum distance matched sampling with fine balance in an observational study of treatment for ovarian cancer. J Am Statist Assoc **102**, 75–83. (2007)
7. O'Brien, P.C., Fleming, T.R.: A paired Prentice-Wilcoxon test for censored paired data. Biometrics **43**, 169–180 (1987)
8. Silber, J.H., Rosenbaum, P.R., Polsky, D., Ross, R.N., Even-Shoshan, O., Schwartz, S., Armstrong, K.A., Randall, T.C.: Does ovarian cancer treatment and survival differ by the specialty providing chemotherapy? J Clin Oncol **25**, 1169–1175 (2007); related editorial **25**, 1157–1159 (2007); related letters and rejoinders **25**, 3552–3558 (2007)
9. Vandenbroucke, J.P.: When are observational studies as credible as randomized trials? Lancet **363**, 1728–1731 (2004)

Chapter 8
Basic Tools of Multivariate Matching

Abstract The basic tools of multivariate matching are introduced, including the propensity score, distance matrices, calipers imposed using a penalty function, optimal matching, matching with multiple controls and full matching. The tools are illustrated with a tiny example from genetic toxicology ($n = 46$), an example that is so small that one can keep track of individuals as they are matched using different techniques.

8.1 A Small Example

The mechanics of matching are best illustrated with a very small example, not because such an example is representative but because it is possible to inspect the details of what goes on. The example considered here has 47 subjects and three covariates. Typical examples have many more subjects and many more covariates.

Welders are exposed to chromium and nickel, substances that can cause inappropriate links between DNA and proteins, which in turn may disrupt gene expression or interfere with replication of DNA. Costa, Zhitkovich, and Toniolo [10] measured DNA-protein cross-links in samples of white blood cells from 21 railroad arc welders exposed to chromium and nickel and 26 unexposed controls. All 47 subjects were male. In their data in Table 8.1, there are three covariates, namely age, race and current smoking behavior. The response is a measure of DNA-protein cross-links.

Summary measures describing the covariates appear at the bottom of Table 8.1. The welders are about five years younger than the controls on average, have relatively fewer African Americans, and more smokers. For what it is worth, the difference in age between welders and controls is significant as judged by the t-test, with a t of -2.25 and a two-sided significance level of 0.03. The two binary variables yield t's of -0.94 for race and 1.2 for smoking, and are neither significant by that standard nor by Fisher's exact test for a 2×2 table. In an experiment, random assignment leads to the expectation that one covariate in 20 will exhibit an imbalance

Table 8.1 Unmatched data for 21 railroad arc welders and 26 potential controls. Covariates are age, race (C=Caucasian, AA=African American), current smoker (Y=yes, N=no). Response is DPC = DNA-protein cross-links in percent in white blood cells. All 47 subjects are male.

Welders					Controls				
ID	Age	Race	Smoker	DPC	ID	Age	Race	Smoker	DPC
1	38	C	N	1.77	1	48	AA	N	1.08
2	44	C	N	1.02	2	63	C	N	1.09
3	39	C	Y	1.44	3	44	C	Y	1.10
4	33	AA	Y	0.65	4	40	C	N	1.10
5	35	C	Y	2.08	5	50	C	N	0.93
6	39	C	Y	0.61	6	52	C	N	1.11
7	27	C	N	2.86	7	56	C	N	0.98
8	43	C	Y	4.19	8	47	C	N	2.20
9	39	C	Y	4.88	9	38	C	N	0.88
10	43	AA	N	1.08	10	34	C	N	1.55
11	41	C	Y	2.03	11	42	C	N	0.55
12	36	C	N	2.81	12	36	C	Y	1.04
13	35	C	N	0.94	13	41	C	N	1.66
14	37	C	N	1.43	14	41	AA	Y	1.49
15	39	C	Y	1.25	15	31	AA	Y	1.36
16	34	C	N	2.97	16	56	AA	Y	1.02
17	35	C	Y	1.01	17	51	AA	N	0.99
18	53	C	N	2.07	18	36	C	Y	0.65
19	38	C	Y	1.15	19	44	C	N	0.42
20	37	C	N	1.07	20	35	C	N	2.33
21	38	C	Y	1.63	21	34	C	Y	0.97
					22	39	C	Y	0.62
					23	45	C	N	1.02
					24	42	C	N	1.78
					25	30	C	N	0.95
					26	35	C	Y	1.59
	Mean Age	AA %	Smoker %			Mean Age	AA %	Smoker %	
	38	10	52			43	19	35	

that is statistically significant at the 0.05 level, at least when the comparison is based on an appropriate randomization test, such as Fisher's exact test. [1]

A pair matching would form 21 pairs of a welder and a control, discarding five potential controls, in the process altering the marginal distribution of the three covariates in the control group. Pair matching is, perhaps, not the most attractive method in this particular example. In part, the removal of five controls can have only a moderate impact on the distribution of covariates. In part, in such a small example, it is not entirely comfortable to discard five controls. Then again, the oldest control, #2, is 63 years old, which is ten years older than the oldest welder, #18,

---

[1] Cochran [8] discusses the relationship between the magnitude of a t-statistic for covariate imbalance and the coverage rate of a confidence interval for a treatment effect when no adjustment is made for the covariate. He concludes that problems begin to occur before the conventional 0.05 level of significance is reached, and that attention should be given to covariates exhibiting a t-statistic of 1.5 in absolute value.

8.2 Propensity Score

Table 8.2 Limitations of pair matching for age in the welder data. Even the 21 youngest of the 26 controls are, on average, slightly older than the welders, so even a pair matching that focused exclusively on one covariate, age, could not entirely eliminate the age difference. The discrepancy is not large in this example but can be large in some studies. Other forms of matching besides pair matching can make further progress.

	Mean	Min	Q1	Median	Q3	Max
All 26 Controls	42.7	30	36	42	48	63
21 Youngest Controls	39.6	30	35	40	44	50
21 Welders	38.2	27	35	38	39	53

and at least 19 years older than all other welders; arguably, some controls are not suitable for comparison.

As Rubin [38] observes, there is a definite limit to what pair matching can accomplish. In many cases, what is needed falls well within this attainable limit; in other cases it does not. Importantly, to a large extent, this limit is not shared by more flexible matching strategies, such as 'full matching' [35]. One can see the limits of pair matching in Table 8.2. The mean age of the 21 welders is 38.2 years, while the mean age of the 26 controls is 42.7 years. Even if the only consideration were reducing the mean age in the control group, in this example, pair matching will fall slightly short of the goal, because the mean age of the 21 youngest of the 26 controls is 39.6 years. In parallel, for smoking, 11 of the 21 welders are smokers, while only 9 of the 26 potential controls are smokers, so pair matching can at most select all nine smoking controls, and cannot perfectly balance smoking. Moreover, the 21 youngest controls include only eight of the nine smokers. Furthermore, although pair matching can eliminate the imbalance in race, if it does this, it cannot use all nine smoking controls. The discrepancies are not large, and hence are perhaps tolerable in this example, but in other studies, a matching strategy more flexible than pair matching may be needed.

Before matching, there are $L = 47$ subjects in Table 8.1, $\ell = 1, 2, \ldots, L$. In the example, for subject ℓ, the observed covariate, $\mathbf{x}_\ell$, is three-dimensional, $\mathbf{x}_\ell = (x_{\ell 1}, x_{\ell 2}, x_{\ell 3})^T$, where (i) $x_{\ell 1}$ is the age of subject ℓ, (ii) $x_{\ell 2}$ encodes race, $x_{\ell 2} = 1$ if ℓ is African American, $x_{\ell 2} = 0$ if ℓ is Caucasian, (iii) $x_{\ell 3}$ encodes smoking, $x_{\ell 3} = 1$ if ℓ is a current smoker, $x_{\ell 3} = 0$ otherwise. For instance, $\mathbf{x}_1 = (38, 0, 0)^T$. The variable Z_ℓ distinguishes treated subjects from potential controls: $Z_\ell = 1$ for a treated subject, here a welder; $Z_\ell = 0$ for a potential control.

8.2 Propensity Score

As discussed in §3.3, the propensity score is the conditional probability of exposure to treatment given the observed covariates, $e(\mathbf{x}) = \Pr(Z = 1 \mid \mathbf{x})$. The propensity score is defined in terms of the observed covariates, $\mathbf{x}$, even though there is invariably concern about other covariates that were not measured. Properties of the propensity score are discussed in §3.3 and [30, 31, 33, 36].

In the simplest randomized experiment, with treatments assigned by independent flips of a fair coin, the propensity score is $e(\mathbf{x}) = \Pr(Z = 1 \mid \mathbf{x}) = \frac{1}{2}$ for all values of the covariate $\mathbf{x}$. In this case, the covariate $\mathbf{x}$ is not useful in predicting the treatment that a subject will receive. In such a completely randomized experiment, smokers are just as likely as nonsmokers to receive the treatment — the chance is $\frac{1}{2}$ for smokers and $\frac{1}{2}$ for nonsmokers — so smokers tend to show up in the treated group just about as often as they do in the control group, and any difference in smoking frequency is due to chance, the flip of a coin that assigned one subject to treatment, another to control.

Brief examination of Table 8.1 suggested that at least age, x_1, and possibly race and smoking, x_2 and x_3, could be used to predict treatment assignment Z, so that the propensity score is not constant. Saying that welders tend to be somewhat younger than controls is much the same as saying that the chance of being a welder is lower for someone who is older, that is, $e(\mathbf{x})$ is lower when x_1 is higher.

If two subjects have the same propensity score, $e(\mathbf{x})$, they may have different values of $\mathbf{x}$. For instance, a younger nonsmoker and an older smoker might have the same propensity score, $e(\mathbf{x})$, because welders are often younger smokers. Imagine that we have two individuals, a welder and a control, with the same propensity score $e(\mathbf{x})$ but different covariates $\mathbf{x}$. Although these two individuals differ in terms of $\mathbf{x}$, this difference in $\mathbf{x}$ will not be helpful in guessing which one is the welder, because $e(\mathbf{x})$ is the same. If subjects were matched for $e(\mathbf{x})$, they may be mismatched for $\mathbf{x}$, but the mismatches in $\mathbf{x}$ will be due to chance and will tend to balance, particularly in large samples. If young nonsmokers and old smokers have the same propensity score, then a match on the propensity score may pair a young nonsmoking welder to an old smoking control, but it will do this about as often as it pairs an old smoking welder to a young nonsmoking control.

In brief, matching on $e(\mathbf{x})$ tends to balance $\mathbf{x}$; see (3.10) in §3.3 or [30]. More precisely, treatment assignment Z is conditionally independent of observed covariates $\mathbf{x}$ given the propensity score $e(\mathbf{x})$. Moreover, the propensity score is the coarsest function of $\mathbf{x}$ with this balancing property, so if the propensity score $e(\mathbf{x})$ is not balanced, then the observed covariates $\mathbf{x}$ will continue to be useful in predicting treatment Z. Expressed differently, ignoring chance imbalances, as we might in sufficiently large samples, balancing the propensity score $e(\mathbf{x})$ is sufficient to balance the observed covariates $\mathbf{x}$ but also necessary to balance $\mathbf{x}$. It is important to understand that this is a true statement about observed covariates $\mathbf{x}$, and only about observed covariates, whether or not treatment assignment also depends on covariates that were not measured. On the negative side, success in balancing the observed covariates, $\mathbf{x}$, provides no assurance that unmeasured covariates are balanced. On the positive side, balancing observed covariates — comparing groups that, at least, look comparable in terms of the observed $\mathbf{x}$ — is a discrete task that can be successfully completed without reference to possible imbalances in covariates that were not measured. For the men in Table 8.1, imagine that having a father who worked as a welder is associated with a much higher chance that the son also will work as a welder, so father's occupation is an unmeasured covariate that could be used to predict treatment Z; then success in matching on the propensity score $e(\mathbf{x})$ would

8.2 Propensity Score

Table 8.3 Estimated propensity scores $\hat{e}(\mathbf{x})$ for 21 railroad arc welders and 26 potential controls. Covariates are age, race (C=Caucasian, AA=African American), current smoker (Y=yes, N=no).

	Welders				Controls			
ID	Age	Race	Smoker	$\hat{e}(\mathbf{x})$	ID	Age	Race	Smoker	$\hat{e}(\mathbf{x})$
1	38	C	N	0.46	1	48	AA	N	0.14
2	44	C	N	0.34	2	63	C	N	0.09
3	39	C	Y	0.57	3	44	C	Y	0.47
4	33	AA	Y	0.51	4	40	C	N	0.42
5	35	C	Y	0.65	5	50	C	N	0.23
6	39	C	Y	0.57	6	52	C	N	0.20
7	27	C	N	0.68	7	56	C	N	0.15
8	43	C	Y	0.49	8	47	C	N	0.28
9	39	C	Y	0.57	9	38	C	N	0.46
10	43	AA	N	0.20	10	34	C	N	0.54
11	41	C	Y	0.53	11	42	C	N	0.38
12	36	C	N	0.50	12	36	C	Y	0.64
13	35	C	N	0.52	13	41	C	N	0.40
14	37	C	N	0.48	14	41	AA	Y	0.35
15	39	C	Y	0.57	15	31	AA	Y	0.55
16	34	C	N	0.54	16	56	AA	Y	0.13
17	35	C	Y	0.65	17	51	AA	N	0.12
18	53	C	N	0.19	18	36	C	Y	0.64
19	38	C	Y	0.60	19	44	C	N	0.34
20	37	C	N	0.48	20	35	C	N	0.52
21	38	C	Y	0.60	21	34	C	Y	0.67
					22	39	C	Y	0.57
					23	45	C	N	0.32
					24	42	C	N	0.38
					25	30	C	N	0.63
					26	35	C	Y	0.65
Mean Age	AA %	Smoker %	Mean $\hat{e}(\mathbf{x})$		Mean Age	AA %	Smoker %	Mean $\hat{e}(\mathbf{x})$
38	10	52	0.51		43	19	35	0.39

tend to balance age, race and smoking, but there is no reason to expect it to balance father's occupation. In short, (i) matching on $e(\mathbf{x})$ is often practical even when there are many covariates in $\mathbf{x}$ because $e(\mathbf{x})$ is a single variable, (ii) matching on $e(\mathbf{x})$ tends to balance all of $\mathbf{x}$, and (iii) failure to balance $e(\mathbf{x})$ implies that $\mathbf{x}$ is not balanced.

The propensity score is unknown, but it can be estimated from the data at hand. In the example, the propensity score is estimated by a linear logit model

$$\log\left\{\frac{e(\mathbf{x}_\ell)}{1-e(\mathbf{x}_\ell)}\right\} = \zeta_0 + \zeta_1 x_{\ell 1} + \zeta_2 x_{\ell 2} + \zeta_3 x_{\ell 3}, \tag{8.1}$$

and the fitted values $\hat{e}(\mathbf{x}_\ell)$ from this model are the estimates of the propensity score. The estimates of the propensity score, $\hat{e}(\mathbf{x}_\ell)$, are displayed in Table 8.3. Control #2, a 63-year-old, Caucasian nonsmoker, has $\hat{e}(\mathbf{x}_\ell) = 0.09$, only a 9% estimated chance of being a welder. In contrast, control #12, a 36-year-old smoker, has $\hat{e}(\mathbf{x}_\ell) = 0.64$, a 64% chance of being a welder, so this control actually has covariates that are atypical of controls and more typical of welders. Welders #10 and #18 have similar estimated propensity scores $\widehat{e}(\mathbf{x}_\ell)$ but different patterns of covariates $\mathbf{x}_\ell$.

The limitations of pair matching, illustrated for age in §8.1, apply with any variable, including the propensity score. The mean of the 21 largest $\widehat{e}(\mathbf{x})$'s in the control group is 0.46, somewhat less than the mean of $\widehat{e}(\mathbf{x})$ in the treated group, namely 0.51, so no pair matching can completely close that gap.

8.3 Distance Matrices

In its simplest form, a distance matrix is a table with one row for each treated subject and one column for each potential control. For the welder data in Table 8.3, the distance matrix would have 21 rows and 26 columns; it would be 21×26. The value in row i and column j of the table is the 'distance' between the ith treated subject and the jth potential control. This 'distance' is a nonnegative number[2] or infinity, ∞, which measures the similarity of two individuals in terms of their covariates $\mathbf{x}$. Two individuals with the same value of $\mathbf{x}$ would have distance zero. An infinite distance in row i and column j is used to forbid matching the ith treated subject to the jth potential control.

For the welder data in Table 8.3, Table 8.4 displays all 21 rows and the first 6 of 26 columns on one such distance matrix. The distance is the squared difference in the estimated propensity score, $\widehat{e}(\mathbf{x})$. The first welder has $\widehat{e}(\mathbf{x}) = 0.46$ and the first control has $\widehat{e}(\mathbf{x}) = 0.14$, so the distance in the first row and first column of Table 8.4 is $(0.46 - 0.14)^2 = 0.10$. If the only concern were to obtain close matches on $\widehat{e}(\mathbf{x})$, then this might be a reasonable distance. The disadvantage is that two controls with the same propensity score, $\widehat{e}(\mathbf{x})$, may have different patterns of covariates, $\mathbf{x}$, and this is ignored in Table 8.4. For example, in the first row and third and fourth columns of Table 8.4, the distance is zero to two decimal places. Looking back at Table 8.3, the distances between welder #1 and potential controls #3 and #4 are, respectively, $(0.46 - 0.47)^2 = 0.0001$ and $(0.46 - 0.42)^2 = 0.0016$, so control #3 is ever so slightly closer to welder #1. However, in terms of the details of $\mathbf{x}$, control #4 looks to be the better match, a nonsmoker with a two-year difference in age, as opposed to control #3, a smoker with a six-year difference in age. Because younger smokers are more common in the welder group, the propensity score is indifferent between a younger nonsmoker and an older smoker, but the details of $\mathbf{x}$ suggest that control #4 is a better match for welder #1 than is control #4.

An alternative distance [32] insists that individuals be close on the propensity score, $\widehat{e}(\mathbf{x})$, but once this is achieved, the details of $\mathbf{x}$ affect the distance. With a caliper of width w, if two individuals, say k and ℓ, have propensity scores that differ by more than w — that is, if $|\widehat{e}(\mathbf{x}_k) - \widehat{e}(\mathbf{x}_\ell)| > w$ — then the distance between these

---

[2] The distance need not be, and typically is not, a distance in the sense that the word is used in metric space topology: it need not satisfy the triangle inequality.

8.3 Distance Matrices

Table 8.4 Squared differences in propensity scores between welders and controls. Rows are the 21 welders and columns are for the first 6 of 26 potential controls.

Welder	Control 1	Control 2	Control 3	Control 4	Control 5	Control 6
1	0.10	0.13	0.00	0.00	0.05	0.06
2	0.04	0.06	0.02	0.01	0.01	0.02
3	0.19	0.23	0.01	0.02	0.12	0.14
4	0.13	0.18	0.00	0.01	0.08	0.09
5	0.26	0.32	0.03	0.06	0.18	0.20
6	0.19	0.23	0.01	0.02	0.12	0.14
7	0.29	0.35	0.05	0.07	0.20	0.23
8	0.12	0.16	0.00	0.01	0.07	0.08
9	0.19	0.23	0.01	0.02	0.12	0.14
10	0.00	0.01	0.07	0.05	0.00	0.00
11	0.15	0.19	0.00	0.01	0.09	0.11
12	0.13	0.17	0.00	0.01	0.07	0.09
13	0.14	0.19	0.00	0.01	0.08	0.10
14	0.11	0.15	0.00	0.00	0.06	0.08
15	0.19	0.23	0.01	0.02	0.12	0.14
16	0.16	0.20	0.01	0.02	0.10	0.11
17	0.26	0.32	0.03	0.06	0.18	0.20
18	0.00	0.01	0.08	0.05	0.00	0.00
19	0.20	0.25	0.02	0.03	0.13	0.15
20	0.11	0.15	0.00	0.00	0.06	0.08
21	0.20	0.25	0.02	0.03	0.13	0.15

individuals is set to ∞, whereas, if $|\widehat{e}(\mathbf{x}_k) - \widehat{e}(\mathbf{x}_\ell)| \leq w$, the distance is a measure of proximity of $\mathbf{x}_k$ and $\mathbf{x}_\ell$. The caliper width, w, is often taken as a multiple of the standard deviation of the propensity score, $\widehat{e}(\mathbf{x})$, so that by varying the multiplier, one can vary the relative importance given to $\widehat{e}(\mathbf{x})$ and $\mathbf{x}$. In Table 8.3, the standard deviation of $\widehat{e}(\mathbf{x})$ is 0.172. Tables 8.5 and 8.6 illustrate two distance matrices using a caliper on the propensity score, in which the caliper is half of the standard deviation of the propensity score, or $0.172/2 = 0.086$.

In problems of practical size, a caliper of 20% of the standard deviation of the propensity score is more common, and even that may be too large. A reasonable strategy is to start with a caliper width of 20% of the standard deviation of the propensity score, adjusting the caliper if needed to obtain balance on the propensity score.

In both Tables 8.5 and 8.6, welder #1 has a propensity score that differs from potential controls #1 and #2 by more than 0.086 in absolute value, so the distance is infinite. There is a finite distance between welder #1 and potential controls #3 and #4, because their propensity scores are close, but in both Tables 8.5 and 8.6, the distance is much smaller for potential control #4, because this control is closer to welder #1 in age and smoking behavior. Control #2, who is 63 years old, is at infinite distance from all 21 welders. Experience [32] and simulation [13] suggest that matching on the propensity score $\widehat{e}(\mathbf{x})$ alone often suffices to balance the distribution of covariates, but the individual pairs may be quite different in terms of $\mathbf{x}$, as in the case of welder #1 and potential control #3; whereas, minimum distance

Table 8.5 Mahalanobis distances within propensity score calipers. Rows are the 21 welders and columns are for the first 6 of 26 potential controls. An ∞ signifies that the caliper is violated.

Welder	Control 1	Control 2	Control 3	Control 4	Control 5	Control 6
1	∞	∞	6.15	0.08	∞	∞
2	∞	∞	∞	0.33	∞	∞
3	∞	∞	∞	∞	∞	∞
4	∞	∞	12.29	∞	∞	∞
5	∞	∞	∞	∞	∞	∞
6	∞	∞	∞	∞	∞	∞
7	∞	∞	∞	∞	∞	∞
8	∞	∞	0.02	5.09	∞	∞
9	∞	∞	∞	∞	∞	∞
10	0.51	∞	∞	∞	10.20	11.17
11	∞	∞	0.18	∞	∞	∞
12	∞	∞	7.06	0.33	∞	∞
13	∞	∞	7.57	∞	∞	∞
14	∞	∞	6.58	0.18	∞	∞
15	∞	∞	∞	∞	∞	∞
16	∞	∞	8.13	∞	∞	∞
17	∞	∞	∞	∞	∞	∞
18	9.41	∞	∞	∞	0.18	0.02
19	∞	∞	∞	∞	∞	∞
20	∞	∞	6.58	0.18	∞	∞
21	∞	∞	∞	∞	∞	∞

matching within sufficiently narrow propensity score calipers also tends to balance the distribution of covariates but confers the added benefit of closer individual pairs, as in the case of welder #1 and potential control #4.

In Table 8.5, within the caliper, the distance is the Mahalanobis distance [24, 39], which generalizes to several variables the familiar notion of measuring distance in units of the standard deviation. Speaking very informally, in the Mahalanobis distance, a difference of one standard deviation counts the same for each covariate in $\mathbf{x}$. Even as an informal description, this is not quite correct. The Mahalanobis distance takes account of the correlations among variables. If one covariate in $\mathbf{x}$ were weight in pounds rounded to the nearest pound and another were weight in kilograms rounded to the nearest kilogram, then the Mahalanobis distance would come very close to counting those two covariates as a single covariate because of their high correlation.

If $\widehat{\Sigma}$ is the sample covariance matrix of $\mathbf{x}$, then the estimated Mahalanobis distance [24, 39] between $\mathbf{x}_k$ and $\mathbf{x}_\ell$ is $(\mathbf{x}_k - \mathbf{x}_\ell)^T \widehat{\Sigma}^{-1} (\mathbf{x}_k - \mathbf{x}_\ell)$. In the welder data in Table 8.3, $\mathbf{x}_\ell = (x_{\ell 1}, x_{\ell 2}, x_{\ell 3})^T$ where $x_{\ell 1}$ is age, $x_{\ell 2}$ is a binary indicator of race, and $x_{\ell 3}$ is a binary indicator of smoking, and the sample variance covariance matrix is

$$\widehat{\Sigma} = \begin{bmatrix} 54.04 & 0.39 & -0.94 \\ 0.39 & 0.13 & 0.02 \\ -0.94 & 0.02 & 0.25 \end{bmatrix} \qquad (8.2)$$

with inverse

8.3 Distance Matrices

$$\widehat{\Sigma}^{-1} = \begin{bmatrix} 0.021 & -0.077 & 0.084 \\ -0.077 & 8.127 & -1.009 \\ 0.084 & -1.009 & 4.407 \end{bmatrix}. \tag{8.3}$$

The Mahalanobis distance was originally developed for use with multivariate Normal data, and for data of that type it works fine. With data that are not Normal, the Mahalanobis distance can exhibit some rather odd behavior. If one covariate contains extreme outliers or has a long-tailed distribution, its standard deviation will be inflated, and the Mahalanobis distance will tend to ignore that covariate in matching. With binary indicators, the variance is largest for events that occur about half the time, and it is smallest for events with probabilities near zero and one. In consequence, the Mahalanobis distance gives greater weight to binary variables with probabilities near zero or one than to binary variables with probabilities closer to one half. In the welder data, two individuals with the same age and race but different smoking behavior have a Mahalanobis distance of 4.407, whereas two individuals with the same age and smoking behavior but different race have a Mahalanobis distance of 8.127, so a mismatch on race is counted as almost twice as bad as a mismatch on smoking. Two people who differed by 20 years in age with the same race and smoking would have a Mahalanobis distance of $0.021 \times 20^2 = 8.4$, so a difference in race counts about as much as a 20-year difference in age. In a different context, if there were binary indicators for the states of the United States, then the Mahalanobis distance would regard matching for Wyoming as vastly more important than matching for California, simply because fewer people live in Wyoming. In many contexts, rare binary covariates are not of overriding importance, and outliers do not make a covariate unimportant, so the Mahalanobis distance may not be appropriate with covariates of this kind.

A simple alternative to the Mahalanobis distance (i) replaces each of the covariates, one at a time, by its ranks, with average ranks for ties, (ii) premultiplies and postmultiplies the covariance matrix of the ranks by a diagonal matrix whose diagonal elements are the ratios of the standard deviation of untied ranks, $1, \ldots, L$, to the standard deviations of the tied ranks of the covariates, and (iii) computes the Mahalanobis distance using the ranks and this adjusted covariance matrix. Call this the 'rank-based Mahalanobis distance.' Step (i) limits the influence of outliers. After step (ii) is complete, the adjusted covariance matrix has a constant diagonal. Step (ii) prevents heavily tied covariates, such as rare binary variables, from having increased influence due to reduced variance. In the welder data, two individuals with the same age and race but different smoking behavior have a rank-based Mahalanobis distance of 3.2, whereas two individuals with the same age and smoking behavior but different race have a rank-based Mahalanobis distance of 3.1, so a mismatch on race is counted as about equal to a mismatch on smoking.

Tables 8.5 and 8.6 contrast the Mahalanobis distance and the rank-based Mahalanobis distance for the welder data. In both tables, in row #18 and column #1, there is the distance between a 53-year-old, Caucasian, nonsmoking welder and a 48-year-old, African American, nonsmoking control. In Table 8.5, the distance is 9.41, while in Table 8.6 the distance is 3.33. In fact, the four largest finite distances

Table 8.6 Rank-based Mahalanobis distances within propensity score calipers. Rows are the 21 welders and columns are for the first 6 of 26 potential controls. An ∞ signifies that the caliper is violated.

Welder	Control 1	Control 2	Control 3	Control 4	Control 5	Control 6
1	∞	∞	5.98	0.33	∞	∞
2	∞	∞	∞	0.47	∞	∞
3	∞	∞	∞	∞	∞	∞
4	∞	∞	10.43	∞	∞	∞
5	∞	∞	∞	∞	∞	∞
6	∞	∞	∞	∞	∞	∞
7	∞	∞	∞	∞	∞	∞
8	∞	∞	0.04	3.92	∞	∞
9	∞	∞	∞	∞	∞	∞
10	0.25	∞	∞	∞	3.72	4.01
11	∞	∞	0.28	∞	∞	∞
12	∞	∞	7.61	0.98	∞	∞
13	∞	∞	9.02	∞	∞	∞
14	∞	∞	6.83	0.64	∞	∞
15	∞	∞	∞	∞	∞	∞
16	∞	∞	10.61	∞	∞	∞
17	∞	∞	∞	∞	∞	∞
18	3.33	∞	∞	∞	0.05	0.01
19	∞	∞	∞	∞	∞	∞
20	∞	∞	6.83	0.64	∞	∞
21	∞	∞	∞	∞	∞	∞

in Table 8.5 are the four distances between Caucasians and African Americans, whereas race and smoking are about equally important in Table 8.6.

In short, a sturdy choice for a distance is the rank-based Mahalanobis distance within calipers on the propensity score, with the caliper width w adjusted to ensure good balance on the propensity score.

8.4 Optimal Pair Matching

An 'optimal pair matching' pairs each treated subject with a different control to minimize the total distance within matched pairs [34]. In the welder data in Table 8.3, this means forming 21 pairs using 21 different controls from the 26 potential controls so that the sum of the 21 distances within pairs is minimized. The problem is not trivial because the closest control to one treated subject may also be the closest control to another treated subject. For instance, in Table 8.4, many treated subjects are close to potential control #3, but control #3 will be paired to only one of them. As seen in Table 8.7, a best-first or greedy algorithm will not generally find an optimal pair matching.

Finding an optimal pair matching is known as the 'assignment problem' and was solved by Kuhn [22] in 1955. One of the fastest algorithms for the assignment problem was proposed by Bertsekas [4], who offers free Fortran code at his website

8.4 Optimal Pair Matching

Table 8.7 A small example showing that a greedy algorithm, or a best-first algorithm, does not solve the optimal matching problem. A greedy algorithm would pair treated #1 with control #1, and then would be forced to pair treated #2 with control #2 for a total distance of 0+1000, but it is possible to obtain a match with a total distance of 0.01+0.01.

	Control 1	Control 2
Treated 1	0.00	0.01
Treated 2	0.01	1000.00

Table 8.8 Optimal pair match using the squared difference in the propensity score. Covariates are age, race (C=Caucasian, AA=African American), current smoker (Y=yes, N=no), and the estimated propensity score $\widehat{e}(\mathbf{x})$.

	Welders				Matched Controls			
Pair	Age	Race	Smoker	$\widehat{e}(\mathbf{x})$	Age	Race	Smoker	$\widehat{e}(\mathbf{x})$
1	38	C	N	0.46	45	C	N	0.32
2	44	C	N	0.34	47	C	N	0.28
3	39	C	Y	0.57	39	C	Y	0.57
4	33	AA	Y	0.51	41	C	N	0.40
5	35	C	Y	0.65	34	C	Y	0.67
6	39	C	Y	0.57	31	AA	Y	0.55
7	27	C	N	0.68	35	C	Y	0.65
8	43	C	Y	0.49	41	AA	Y	0.35
9	39	C	Y	0.57	34	C	N	0.54
10	43	AA	N	0.20	50	C	N	0.23
11	41	C	Y	0.53	44	C	Y	0.47
12	36	C	N	0.50	42	C	N	0.38
13	35	C	N	0.52	40	C	N	0.42
14	37	C	N	0.48	44	C	N	0.34
15	39	C	Y	0.57	35	C	N	0.52
16	34	C	N	0.54	38	C	N	0.46
17	35	C	Y	0.65	36	C	Y	0.64
18	53	C	N	0.19	52	C	N	0.20
19	38	C	Y	0.60	36	C	Y	0.64
20	37	C	N	0.48	42	C	N	0.38
21	38	C	Y	0.60	30	C	N	0.63
	mean	%AA	%Y	mean	mean	%AA	%Y	mean
	38	10	52	0.51	40	10	38	0.46

at MIT. Hansen's [16] `pairmatch` function in his `optmatch` package makes Bertsekas' code available from inside the statistical package R; see [28] for free access to R, and see [25] for a general textbook about R. Dell'Amico and Toth [11] provide a recent review of the literature on the assignment problem. A Fortran implementation of Kuhn's Hungarian Method is available from the Association for Computing Machinery (ACM) as implemented by Carpaneto and Toth [7]. The SAS program `proc assign` also solves the assignment problem.

For the welder data, Table 8.8 displays an optimal match using the propensity score distances in Table 8.4. As the discussion in §8.3 led us to expect, the marginal means at the bottom of Table 8.8 are fairly well balanced, but the individual pairs are often not matched for race or smoking.

Table 8.9 Optimal pair match using the Mahalanobis distance within propensity score calipers. The caliper is half the standard deviation of the propensity score, or $0.172/2 = 0.086$. Covariates are age, race (C=caucasian, AA=African American), current smoker (Y=yes, N=no) and the estimated propensity score $\widehat{e}(\mathbf{x})$.

	Welders				Matched Controls			
Pair	Age	Race	Smoker	$\widehat{e}(\mathbf{x})$	Age	Race	Smoker	$\widehat{e}(\mathbf{x})$
1	38	C	N	0.46	44	C	N	0.34
2	44	C	N	0.34	47	C	N	0.28
3	39	C	Y	0.57	36	C	Y	0.64
4	33	AA	Y	0.51	41	AA	Y	0.35
5	35	C	Y	0.65	35	C	Y	0.65
6	39	C	Y	0.57	39	C	Y	0.57
7	27	C	N	0.68	30	C	N	0.63
8	43	C	Y	0.49	45	C	N	0.32
9	39	C	Y	0.57	36	C	Y	0.64
10	43	AA	N	0.20	48	AA	N	0.14
11	41	C	Y	0.53	44	C	Y	0.47
12	36	C	N	0.50	41	C	N	0.40
13	35	C	N	0.52	40	C	N	0.42
14	37	C	N	0.48	42	C	N	0.38
15	39	C	Y	0.57	35	C	N	0.52
16	34	C	N	0.54	38	C	N	0.46
17	35	C	Y	0.65	34	C	Y	0.67
18	53	C	N	0.19	52	C	N	0.20
19	38	C	Y	0.60	34	C	N	0.54
20	37	C	N	0.48	42	C	N	0.38
21	38	C	Y	0.60	31	AA	Y	0.55
	mean	%AA	%Y	mean	mean	%AA	%Y	mean
	38	10	52	0.51	40	14	38	0.45

Although it is not obvious from Tables 8.3, 8.5, and 8.6, because of competition for controls, there is no pair matching in which the caliper on the propensity score, $|\widehat{e}(\mathbf{x}_k) - \widehat{e}(\mathbf{x}_\ell)| \leq w$ with $w = 0.086$, is respected for all 21 matched pairs. This is true because of competition among welders for the same controls, despite the fact that each welder has a propensity score within $w = 0.086$ of at least one potential control. For this reason, the infinities in Tables 8.5 and 8.6 are not used; they are replaced by the addition of a 'penalty function,' which exacts a large but finite penalty for violations of the constraint; see, for instance, [2, pages 372–373] or [21, Chapter 6]. The penalty used here is $1000 \times \max{(0, |\widehat{e}(\mathbf{x}_k) - \widehat{e}(\mathbf{x}_\ell)| - w)}$, so if $|\widehat{e}(\mathbf{x}_k) - \widehat{e}(\mathbf{x}_\ell)| \leq w$ then the penalty is zero, but if $|\widehat{e}(\mathbf{x}_k) - \widehat{e}(\mathbf{x}_\ell)| > w$ then the penalty is $1000 \times (|\widehat{e}(\mathbf{x}_k) - \widehat{e}(\mathbf{x}_\ell)| - w)$. For instance, the penalty for matching welder #1 to potential control #1 is $1000 \times (|.4587 - 0.1437| - 0.0860) = 229$. This penalty is added to the Mahalanobis distance or rank-based Mahalanobis distance for the corresponding pair. Optimal matching will try to avoid the penalties by respecting the caliper, but when that is not possible, it will prefer to match so the caliper is only slightly violated for a few matched pairs.

Table 8.9 displays the optimal match using the Mahalanobis distances within propensity score calipers from Table 8.5. Again, the calipers are implemented not

8.5 Optimal Matching with Multiple Controls

Table 8.10 Optimal pair match using the rank-based Mahalanobis distance within propensity score calipers. The caliper is half the standard deviation of the propensity score, or $0.172/2 = 0.086$. Covariates are age, race (C=Caucasian, AA=African American), current smoker (Y=yes, N=no), and the estimated propensity score $\widehat{e}(\mathbf{x})$.

	Welders				Matched Controls			
Pair	Age	Race	Smoker	$\widehat{e}(\mathbf{x})$	Age	Race	Smoker	$\widehat{e}(\mathbf{x})$
1	38	C	N	0.46	44	C	N	0.34
2	44	C	N	0.34	47	C	N	0.28
3	39	C	Y	0.57	36	C	Y	0.64
4	33	AA	Y	0.51	41	AA	Y	0.35
5	35	C	Y	0.65	35	C	Y	0.65
6	39	C	Y	0.57	39	C	Y	0.57
7	27	C	N	0.68	30	C	N	0.63
8	43	C	Y	0.49	45	C	N	0.32
9	39	C	Y	0.57	35	C	N	0.52
10	43	AA	N	0.20	48	AA	N	0.14
11	41	C	Y	0.53	44	C	Y	0.47
12	36	C	N	0.50	41	C	N	0.40
13	35	C	N	0.52	40	C	N	0.42
14	37	C	N	0.48	42	C	N	0.38
15	39	C	Y	0.57	36	C	Y	0.64
16	34	C	N	0.54	38	C	N	0.46
17	35	C	Y	0.65	34	C	Y	0.67
18	53	C	N	0.19	52	C	N	0.20
19	38	C	Y	0.60	31	AA	Y	0.55
20	37	C	N	0.48	42	C	N	0.38
21	38	C	Y	0.60	34	C	N	0.54
	mean	%AA	%Y	mean	mean	%AA	%Y	mean
	38	10	52	0.51	40	14	38	0.45

with ∞'s but rather with a penalty function. Table 8.10 is similar, except the optimal match uses the rank-based Mahalanobis distance within propensity score calipers in Table 8.6, again using the penalty function to implement the calipers. The matches in Table 8.9 and Table 8.10 are similar to each other and better than Table 8.8 in the specific sense that inside individual pairs the matched individuals are closer.

8.5 Optimal Matching with Multiple Controls

In matching with multiple controls, each treated subject is matched to at least one, and possibly more than one, control. To match in a fixed ratio is to match each treated subject to the same number of controls; for instance, pair matching in §8.4 matches in a ratio of 1-to-1, while matching each treated subject to two controls is matching in a ratio of 1-to-2. To match in a variable ratio is to allow the number of controls to vary from one treated subject to another. In the welder data in Table 8.3, there are too few potential controls to match in a 1-to-2 fixed ratio, but it is possible to match with multiple controls in a variable ratio, and in particular to use all of the controls. The decision to match in fixed or variable ratio is a substantial one that affects both the quality of the matching and the analysis and presentation of results. These issues are sketched at the end of this section, following a variable match for the welder data.

Imagine for a moment that there were 42 or more potential controls in the welder data in Table 8.3; then it would be possible to match in a fixed 1-to-2 ratio, producing 21 matched sets, each containing one welder and two controls, where all 42 controls are different. An optimal matching with a fixed ratio of 1-to-2 would minimize the sum of the distances within matched sets between treated subjects and their matched controls, or the sum of 42 distances. Similar considerations apply to matching in a fixed ratio of 1-to-3, 1-to-4, and so on.

Matching with a variable number of controls is slightly more complex. The optimization algorithm decides not only who is matched to whom, but also how many controls to assign to each treated subject. So matching with a variable number of controls has more choices to make, but these choices must be constrained in some reasonable way to avoid trivial results. If all the distances were positive, and no constraints were imposed, the minimum distance matching with variable numbers of controls would always be a pair matching. A simple constraint is to insist that a certain number of controls be used. For instance, in the welder data, one might insist that all 26 controls be used. Alternatively, one might insist that 23 controls be used, which would permit, but not require, the algorithm to discard the three potential controls who are older than all of the welders. An attractive feature of fixing the total number of controls is that the total distance is then a sum of a fixed number of distances. In addition to constraining the total number of controls, a different type of constraint permits a treated subject to have, say, at least one but at most three controls. Given some set of constraints on the matching, an optimal matching with variable controls minimizes the total distance between treated subjects and controls inside matched sets.

For the welder data, Table 8.11 is an optimal matching with variable controls, using all 26 controls, based on the Mahalanobis distances with propensity score calipers. The calipers again use the penalty function in §8.4. The match consists of 19 pairs, one welder matched to three controls, and one welder matched to four controls. Matched set #10 consists of older African Americans. Matched set #18 consists of older Caucasian nonsmokers. Both groups were overrepresented in the control group. As might be expected from the discussion in §8.3, a variable match using the rank-based Mahalanobis distance (not shown) is generally similar, but gives about equal emphasis to race and smoking, as opposed to the emphasis on race in Table 8.11.

In a matched pair, a welder is compared with his matched control. For instance, in matched pair #1 in Table 8.11, the welder is 38 years old, the matched control is 44 years old, and the difference in age is $38 - 44 = -6$ years. In matched set #10, the welder is compared with the average of the three matched controls. In this way, the comparison in matched set #10 is just slightly less variable than it would be if one of the three controls were used and the other two were discarded. For instance, in set #10, the welder is 43 years old, the three matched controls have average age

8.5 Optimal Matching with Multiple Controls

Table 8.11 Optimal match with multiple controls using the Mahalanobis distance within propensity score calipers. The caliper is half the standard deviation of the propensity score, or $0.172/2 = 0.086$. Covariates are age, race (C=caucasian, AA=African American), current smoker (Y=yes, N=no) and the estimated propensity score $\widehat{e}(\mathbf{x})$.

Matched Set	Welders				Matched Controls			
	Age	Race	Smoker	$\widehat{e}(\mathbf{x})$	Age	Race	Smoker	$\widehat{e}(\mathbf{x})$
1	38	C	N	0.46	44	C	N	0.34
2	44	C	N	0.34	47	C	N	0.28
3	39	C	Y	0.57	36	C	Y	0.64
4	33	AA	Y	0.51	41	AA	Y	0.35
5	35	C	Y	0.65	35	C	Y	0.65
6	39	C	Y	0.57	36	C	Y	0.64
7	27	C	N	0.68	30	C	N	0.63
8	43	C	Y	0.49	45	C	N	0.32
9	39	C	Y	0.57	35	C	N	0.52
10	43	AA	N	0.20	51	AA	N	0.12
10					56	AA	Y	0.13
10					48	AA	N	0.14
11	41	C	Y	0.53	44	C	Y	0.47
12	36	C	N	0.50	41	C	N	0.40
13	35	C	N	0.52	40	C	N	0.42
14	37	C	N	0.48	42	C	N	0.38
15	39	C	Y	0.57	39	C	Y	0.57
16	34	C	N	0.54	38	C	N	0.46
17	35	C	Y	0.65	34	C	Y	0.67
18	53	C	N	0.19	63	C	N	0.09
18					56	C	N	0.15
18					52	C	N	0.20
18					50	C	N	0.23
19	38	C	Y	0.60	34	C	N	0.54
20	37	C	N	0.48	42	C	N	0.38
21	38	C	Y	0.60	31	AA	Y	0.55
	mean	%AA	%Y	mean	mean	%AA	%Y	mean
	38	10	52	0.51	40	14	40	0.45

$(51 + 56 + 48)/3 = 51.7$, so the difference in age in that set is $43 - 51.7 = -8.67$ years. Similarly, in set #18, the welder is 53 years old, the four matched controls have average age $(63 + 56 + 52 + 50)/4 = 55.25$, so the difference in age in that set is $53 - 55.25 = -2.25$ years. The 'average difference in age' is the average of these 21 differences. Similarly, the 'average age' in the control group is the average of these 21 average ages of controls. Implicitly, the control in matched set #1 counts as one person, but each of the controls in matched set #10 counts as $\frac{1}{3}$ of a person. In this way, although all controls are used, they are weighted to describe a younger, more Caucasian population with more smokers, that is, a population like the 21 welders. Weighted in this way, the mean age among the matched controls is 40, about two years older than the mean age of 38 for the welders. The same process is applied to a binary variable, such as smoking, to obtain a 40% smoking rate among controls. Although all controls are used in Table 8.11, the balance in the covariate means is similar to pair matching in §8.4 which discarded five controls. This process of weighting is known as 'direct adjustment,' and it is one of the basic differences between matching with a fixed ratio and matching with a variable ratio.[3] When matching with a fixed ratio — for instance, with pair matching in §8.4 — each matched control has the same weight.

The principal advantage of matching in a fixed ratio, such as 1-to-3 matching, is that summary statistics, including those that might be displayed in graphs, may be computed from treated and control groups in the usual way, without using direct adjustment to give unequal weights to observations. For instance, one might display boxplots or Kaplan-Meier survival curves for treated and control groups. This advantage will weigh heavily when potential controls are abundant and the biases that must be removed are not large. A smaller issue is statistical efficiency, which in nominal terms, though often not in practical terms, slightly favors matching in fixed ratio over matching with a variable ratio [26].

There are several advantages to matching with a variable ratio [26, 34]. First, the matched sets will be more closely matched, in the following precise sense. If one finds the minimum distance match with, say, an *average* of three controls per treated subject and two to four controls for each treated subject, the total distance within matched sets will never be larger, and will typically be quite a bit smaller, than in fixed ratio matching with three controls.[4] This is visible in Table 8.11: the two welders with low propensity scores were matched to the seven controls with low propensity scores, and trying to allocate these seven more evenly among welders would have produced a larger mismatch on these propensity scores. Second, matching in fixed ratio requires the number of controls to be an integer multiple of the number of treated subjects, and this restriction may be inconvenient or undesirable for any of a variety of reasons; for instance, for the welder data in Table 8.3, the only possible matching in fixed ratio is pair matching. Third, just as §8.1 discussed definite limits to what can be accomplished with pair matching, there are also definite limits to what can be accomplished by matching with multiple controls, but these limits are better when matching with a variable ratio [26].

Finding an optimal match with variable controls is equivalent to solving a particular minimum-cost flow problem in a network [34]. In the statistical package R, Hansen's [16] fullmatch function in his optmatch package will determine an optimal match with variable controls from a distance matrix by setting his parameter min.controls to equal 1 rather than its default value of 0; the user will often

---

[3] Direct adjustment can be applied to pretty much anything, not just means and proportions. In particular, an empirical distribution function is little more than a sequence of proportions, and it is clear how to apply direct adjustment to proportions. From the weighted empirical distribution function, pretty much anything else can be computed; for instance, medians and quartiles. In [17], medians and quartiles from directly adjusted empirical distribution functions are used to construct directly adjusted boxplots. In the weighted empirical distribution function, the 44-year-old control in matched set #1 has mass $1/21$, whereas the 63-year-old control in matched set #18 has mass $1/(4 \times 21)$.

[4] This is true because, when the number of controls matched is the same, optimal matching with variable controls solves a less constrained optimization problem than matching with a fixed ratio, yet the two problems have the same objective function, so the optimum is never worse.

8.6 Optimal Full Matching

Table 8.12 A distance matrix showing that the best full matching can be much better than the best pair matching or the best matching with variable controls. The best full matching is (1,a,b) and (2,3,c) with a distance of $0.01 + 0.01 + 0.01 + 0.01 = 0.04$. A best pair matching is (1,a), (2,b), (3,c), with a distance of $0.01 + 1000.00 + 0.01 = 1000.02$. Because the number of potential controls equals the number of treated subjects, every matching with variable controls is a pair matching, so variable matching can do no better than pair matching.

	Control a	Control b	Control c
Treated 1	0.01	0.01	1000.00
Treated 2	1000.00	1000.00	0.01
Treated 3	1000.00	1000.00	0.01

wish to adjust the parameter `max.controls`, which limits the number of controls matched to each treated subject, and the parameter `omit.fraction` which determines the number of controls to use. An alternative approach is to use the assignment algorithm in §8.4 but with an altered and enlarged distance matrix [27]; for instance, this can be done with `proc assign` in SAS. For a textbook discussion of network optimization, see [1, 5, 9].

Statistical analysis is not difficult when matching with multiple controls, whether in fixed or variable ratios; see [23, pages 132–145], [12, pages 384–387] or [37, pages 135–139]. For an illustration, in [17, 18], a study of gang violence among teenage boys is analyzed twice, first as a study matched with a fixed ratio of 1-to-2, and second as a study with two strata matched separately, one matched with between two and seven controls and an average of five controls, the other matched with between one and six controls with an average of three controls.

8.6 Optimal Full Matching

In matching with a variable number of controls in §8.5, a treated subject could be matched to one or more controls. In full matching, the reverse situation is permitted as well: one control may be matched to several treated subjects [35]. Table 8.12 shows that full matching can be vastly better than pair matching or matching with a variable number of controls. Because full matching includes as special cases all of the matching procedures considered in §8.4-§8.5, the optimal full matching will produce matched sets that are at least as close as these procedures.

Table 8.13 is an optimal full match for the welder data in Table 8.3, using all 26 controls, and the Mahalanobis distance within propensity score calipers, with the calipers implemented using the penalty function in §8.4. Matched set #1 is a pair. Matched set #2 has one welder and seven controls. Matched set #3 has one control and seven welders. Within each matched set, the welders and controls look similar.

As in matching with variable controls, summary statistics for the control group in full matching must be directly adjusted. In parallel with §8.5, the welder in matched set #2 is compared with the average of the seven controls, so these controls implicitly have weight $\frac{1}{7}$. In matched set #3, each of the seven welders is compared

Table 8.13 Optimal full match using Mahalanobis distance within propensity score calipers. The caliper is half the standard deviation of the propensity score, or $0.172/2 = 0.086$. Covariates are age, race (C=Caucasian, AA=African American), current smoker (Y=yes, N=no), and the estimated propensity score $\hat{e}(\mathbf{x})$.

	Welders				Matched Controls			
Matched Set	Age	Race	Smoker	$\hat{e}(\mathbf{x})$	Age	Race	Smoker	$\hat{e}(\mathbf{x})$
1	38	C	N	0.46	40	C	N	0.42
2	44	C	N	0.34	47	C	N	0.28
2					45	C	N	0.32
2					44	C	N	0.34
2					41	AA	Y	0.35
2					42	C	N	0.38
2					42	C	N	0.38
2					41	C	N	0.40
3	41	C	Y	0.53	39	C	Y	0.57
3	39	C	Y	0.57				
3	39	C	Y	0.57				
3	39	C	Y	0.57				
3	39	C	Y	0.57				
3	38	C	Y	0.60				
3	38	C	Y	0.60				
4	33	AA	Y	0.51	31	AA	Y	0.55
5	35	C	Y	0.65	35	C	Y	0.65
5					34	C	Y	0.67
6	27	C	N	0.68	30	C	N	0.63
7	43	C	Y	0.49	44	C	Y	0.47
8	43	AA	N	0.20	51	AA	N	0.12
8					56	AA	Y	0.13
8					48	AA	N	0.14
9	36	C	N	0.50	35	C	N	0.52
9	35	C	N	0.52				
10	37	C	N	0.48	38	C	N	0.46
10	37	C	N	0.48				
11	34	C	N	0.54	34	C	N	0.54
12	35	C	Y	0.65	36	C	Y	0.64
12					36	C	Y	0.64
13	53	C	N	0.19	63	C	N	0.09
13					56	C	N	0.15
13					52	C	N	0.20
13					50	C	N	0.23
	mean	%AA	%Y	mean	mean	%AA	%Y	mean
	38	10	52	0.51	39	10	55	0.50

with the same control, so this control implicitly has weight 7. As seen at the bottom of Table 8.13, after weighting or direct adjustment, the means in the control group are quite similar to the means in the welder group.

Although the matched sets in Table 8.13 are quite homogeneous, the quite unequal set sizes can lead to some inefficiency. Tables 8.14 and 8.15 are two of many possible variations on the same theme. In Table 8.14, the matched sets can be pairs or triples, and only 21 of the 26 controls are used, in parallel pair matchings in §8.4.

8.6 Optimal Full Matching

Table 8.14 Optimal full match with sets of size 2 or 3 and 21 controls using the Mahalanobis distance within propensity score calipers. The caliper is half the standard deviation of the propensity score, or $0.172/2 = 0.086$. Covariates are age, race (C=Caucasian, AA=African American), current smoker (Y=yes, N=no) and the estimated propensity score $\widehat{e}(\mathbf{x})$.

Matched Set	Welders				Matched Controls			
	Age	Race	Smoker	$\widehat{e}(\mathbf{x})$	Age	Race	Smoker	$\widehat{e}(\mathbf{x})$
1	38	C	N	0.46	42	C	N	0.38
1					42	C	N	0.38
2	44	C	N	0.34	45	C	N	0.32
2					44	C	N	0.34
3	39	C	Y	0.57	36	C	Y	0.64
3	38	C	Y	0.60				
4	33	AA	Y	0.51	31	AA	Y	0.55
5	35	C	Y	0.65	35	C	Y	0.65
6	39	C	Y	0.57	39	C	Y	0.57
6	39	C	Y	0.57				
7	27	C	N	0.68	30	C	N	0.63
8	43	C	Y	0.49	44	C	Y	0.47
8	41	C	Y	0.53				
9	43	AA	N	0.20	51	AA	N	0.12
9					48	AA	N	0.14
10	36	C	N	0.50	35	C	N	0.52
10	35	C	N	0.52				
11	37	C	N	0.48	41	C	N	0.40
11					38	C	N	0.46
12	39	C	Y	0.57	36	C	Y	0.64
12	38	C	Y	0.60				
13	34	C	N	0.54	34	C	N	0.54
14	35	C	Y	0.65	34	C	Y	0.67
15	53	C	N	0.19	56	C	N	0.15
15					52	C	N	0.20
16	37	C	N	0.48	40	C	N	0.42
	mean	%AA	%Y	mean	mean	%AA	%Y	mean
	38	10	52	0.51	39	10	52	0.50

In Table 8.15, all 26 controls are used, with at most two welders or at most three controls in any matched set. In Tables 8.14 and 8.15, welders and controls in the same matched set are reasonably similar, and the (weighted) means in the control group are closer to the welder means than for the several pair matchings in §8.4.

In one specific sense, an optimal full matching is an optimal design for an observational study [35]. Specifically, define a stratification to be a partitioning of the subjects into groups or strata based on the covariates with the one requirement that each stratum must contain at least one treated subject and at least one control. The quality of a stratification might reasonably be judged by a weighted average of all the within strata distances between treated subjects and controls. For instance, if a stratum contained two treated subjects and three controls, then the average distance between the treated and control subjects in this stratum is an average of the $6 = 2 \times 3$ distances between two treated and three control subjects. The stratum-specific av-

Table 8.15 Optimal full match using all 26 controls with sets containing at most two welders and at most three controls, using the Mahalanobis distance within propensity score calipers. The caliper is half the standard deviation of the propensity score, or $0.172/2 = 0.086$. Covariates are age, race (C=Caucasian, AA=African American), current smoker (Y=yes, N=no), and the estimated propensity score $\widehat{e}(\mathbf{x})$.

Matched Set	Welders				Matched Controls			
	Age	Race	Smoker	$\widehat{e}(\mathbf{x})$	Age	Race	Smoker	$\widehat{e}(\mathbf{x})$
1	38	C	N	0.46	44	C	N	0.34
1					41	AA	Y	0.35
1					42	C	N	0.38
2	44	C	N	0.34	50	C	N	0.23
2					47	C	N	0.28
2					45	C	N	0.32
3	39	C	Y	0.57	36	C	Y	0.64
3	39	C	Y	0.57				
4	33	AA	Y	0.51	31	AA	Y	0.55
5	35	C	Y	0.65	34	C	Y	0.67
6	27	C	N	0.68	30	C	N	0.63
7	43	C	Y	0.49	44	C	Y	0.47
7	41	C	Y	0.53				
8	39	C	Y	0.57	39	C	Y	0.57
8	39	C	Y	0.57				
9	43	AA	N	0.20	51	AA	N	0.12
9					56	AA	Y	0.13
9					48	AA	N	0.14
10	36	C	N	0.50	35	C	N	0.52
10	35	C	N	0.52				
11	37	C	N	0.48	42	C	N	0.38
11					38	C	N	0.46
12	34	C	N	0.54	34	C	N	0.54
13	35	C	Y	0.65	35	C	Y	0.65
14	53	C	N	0.19	63	C	N	0.09
14					56	C	N	0.15
14					52	C	N	0.20
15	38	C	Y	0.60	36	C	Y	0.64
15	38	C	Y	0.60				
16	37	C	N	0.48	41	C	N	0.40
16					40	C	N	0.42
	mean	%AA	%Y	mean	mean	%AA	%Y	mean
	38	10	52	0.51	39	11	56	0.50

erage distances are combined into a single number, specifically a weighted average distance, perhaps weighting by the number of treated subjects (here 2) or the total number of subjects (here $5 = 2 + 3$) or the number of control subjects (here 3). No matter which of these weightings is used, no matter what distance is used, there is always a full matching that minimizes this weighted average distance [35]. Moreover, with continuous covariates and reasonable distance functions, with probability 1, only a full matching will minimize the weighted average distance [35]. Stated informally but briefly, a stratification that makes treated subjects and controls as similar as possible is always a full matching.[5]

An optimal full match can be found as a minimum cost flow problem in a network [35, 15]. In the statistical package R, starting with a distance matrix, Hansen's [16] fullmatch function in his optmatch package will find an optimal full matching. In SAS, proc netflow can find an optimal full matching, although it requires quite a bit of coaxing. Bertsekas [5] provides Fortran code for network optimization at his web page at MIT; this code is called by Hansen's optmatch package in R. Implementation in R is illustrated in Chapter 13.

8.7 Efficiency

Efficiency is a secondary concern in observational studies; the primary concern is biases that do not diminish as the sample size increases [8]. If there is a bias of fixed size, then as the sample size increases it quickly dominates the mean square error,[6] so one has a highly efficient estimate of the wrong answer. Despite its secondary role, there is an important fact about efficiency with multiple controls that deserves close attention.

Imagine that we will match in fixed ratio, k controls matched to each treated subject in each of I matched sets. Imagine further that the situation satisfies the assumptions associated with the paired t-test. That is: (i) each matched set has a pair parameter that is eliminated by differencing, (ii) there is an additive treatment effect, τ, (iii) beyond the pair effect, there are independent Normally distributed errors with expectation zero, constant variance ω^2. If the matching were perfect, exactly controlling all bias from observed and unobserved covariates, then the treated-minus-control difference in means is unbiased for τ with variance $(1+1/k)\omega^2/I$. Table 8.16 displays the variance multiplier, $(1+1/k)$, for several choices of k. Matching with $k=2$ substantially reduces the variance of the mean difference, from 2 to 1.5, taking it halfway to $k=\infty$. The gains after that are much smaller, from 1.5 to 1.25 by going from $k=2$ to $k=4$, from 1.1 to 1 by going from $k=10$ to $k=\infty$. See [18] for informal presentation of related results. Detailed efficiency results are found in [17, 26].

There is abundant evidence from theory [26] and from case studies [45] suggesting that Table 8.16 exaggerates the gain from additional controls. The reason is simple: the best $k=2$ controls will be more closely matched for observed covariates than the best $k=10$ controls, and the quality of the match affects both bias and

---

[5] Although the proofs of these claims require some attention to detail, the underlying technique may be described briefly. Specifically, if a stratification is not a full match, then some stratum can be subdivided without increasing, and possibly decreasing, its average distance. Subdividing repeatedly terminates in a full match.

[6] There is a sense in which Pitman's asymptotic relative efficiency isn't well-defined (or isn't meaningful) when there is a bias whose magnitude does not diminish with increasing sample size. See Chapter 15 where this subject is developed in a precise sense.

Table 8.16 Variance ratio $1 + 1/k$ when matching k controls to each treated subject. Here, $k = \infty$ is only slightly better than $k = 6$, which is slightly better than $k = 4$.

Number of Controls	1	2	4	6	10	∞
Variance Multiplier	2	1.50	1.25	1.17	1.10	1

variance. The premise of Table 8.16 is that the quality of the match does not change as more controls are used, and that premise is false.

If large numbers of close potential controls are available without cost, use of $k = 2$ controls is definitely worthwhile, and $k = 4$ or $k = 6$ may yield some further improvement. In a precise, quantifiable sense, the issue of design sensitivity in Part III is far more important than using every available control, and the considerations discussed there will often encourage use of some controls and not others; see in particular Chapter 15 and §17.3.

8.8 Summary

Matching on one variable, the propensity score, tends to produce treated and control groups that, in aggregate, are balanced with respect to observed covariates; however, individual pairs that are close on the propensity score may differ widely on specific covariates. To form closer pairs, a distance is used that penalizes large differences on the propensity score, and then finds individual pairs that are as close as possible. Pairs or matched sets are constructed using an optimization algorithm. Matching with variable controls and full matching combine elements of matching with elements of direct adjustment. Full matching can often produce closer matches than pair matching.

8.9 Further Reading

Matching for propensity scores is discussed in [30, 32]. Optimal matching is discussed in [34]. Matching with variable controls is discussed in [26], where it is shown that matching with variable controls can remove more bias than matching in fixed ratio. The optimality of full matching is proved in [35]. See the paper by Ben Hansen and Stephanie Olsen Klopfer [15] for the best algorithmic results on full matching. For examples of optimal pair matching, see [41, 43, 44]; for matching in fixed ratio, see [45, 18]; for variable matching, see [17], and for full matching, see the papers by Ben Hansen [14], Elizabeth Stuart and K.M. Green [46], and Ruth Heller, Elisabetta Manduchi, and Dylan Small [19]. For a detailed result on the statistical efficiency of alternative designs, see the appendix of [17].

References

1. Ahuja, R.K., Magnanti, T.L., Orlin, J.B.: Network Flows: Theory, Algorithms, and Applications. Upper Saddle River, NJ: Prentice Hall (1993)
2. Avriel, M.: Nonlinear Programming. Upper Saddle River, New Jersey: Prentice Hall. (1976)
3. Bergstralh, E.J., Kosanke, J.L., Jacobsen, S.L.: Software for optimal matching in observational studies. Epidemiology **7**, 331–332 (1996)
4. Bertsekas, D.P.: A new algorithm for the assignment problem. Math Program **21**, 152–171 (1981)
5. Bertsekas, D.P.: Linear Network Optimization. Cambridge, MA: MIT Press (1991)
6. Braitman, L.E., Rosenbaum, P.R.: Rare outcomes, common treatments: Analytic strategies using propensity scores. Ann Intern Med **137**, 693–695 (2002)
7. Carpaneto, G., Toth, P.: Algorithm 548: Solution of the Assignment Problem [H]. ACM Trans Math Software **6**, 104–111 (1980)
8. Cochran, W.G.: The planning of observational studies of human populations (with Discussion). J Roy Statist Soc A **128**, 234–265.
9. Cook, W.J., Cunningham, W.H., Pulleyblank, W.R., Schrijver, A.: Combinatorial Optimization. New York: Wiley (1998)
10. Costa, M., Zhitkovich, A., Toniolo, P.: DNA-protein cross-links in welders: Molecular implications. Cancer Res **53**, 460–463 (1993)
11. Dell'Amico, M. and Toth, P.: Algorithms and codes for dense assignment problems: the state of the art. Discrete App Math **100**, 17–48 (2000)
12. Fleiss, J.L., Levin, B., Paik, M.C.: Statistical Methods for Rates and Proportions. New York: Wiley (2001)
13. Gu, X.S., Rosenbaum, P.R.: Comparison of multivariate matching methods: Structures, distances, and algorithms. J Comput Graph Statist **2**, 405–420 (1993)
14. Hansen, B.B.: Full matching in an observational study of coaching for the SAT. J Am Statist Assoc **99**, 609–618 (2004)
15. Hansen, B.B., Klopfer, S.O.: Optimal full matching and related designs via network flows. J Comp Graph Statist **15**, 609–627 (2006)
16. Hansen, B.B.: Optmatch: Flexible, optimal matching for observational studies. R News **7**, 18–24 (2007)
17. Haviland, A.M., Nagin, D.S., Rosenbaum, P.R.: Combining propensity score matching and group-based trajectory analysis in an observational study. Psychol Methods **12**, 247–267 (2007)
18. Haviland, A.M., Nagin, D.S., Rosenbaum, P.R., Tremblay, R.: Combining group-based trajectory modeling and propensity score matching for causal inferences in nonexperimental longitudinal data. Dev Psychol **44**, 422–436 (2008)
19. Heller, R., Manduchi, E., Small, D.: Matching methods for observational microarray studies. Bioinformatics **25**, 904–909 (2009)
20. Ho, D., Imai, K., King, G., Stuart, E.A.: Matching as nonparametric preprocessing for reducing model dependence in parametric causal inference. Polit Anal **15**, 199–236 (2007)
21. Karmanov, V.G.: Mathematical Programming. Moscow: Mir.
22. Kuhn, H.W.: The Hungarian method for the assignment problem. Naval Res Logist Quart **2**, 83–97 (1955)
23. Lehmann, E.L.: Nonparametrics. San Francisco: Holden Day (1975)
24. Mahalanobis, P.C.: On the generalized distance in statistics. Proc Natl Inst Sci India **12**, 49–55 (1936)
25. Maindonald, J., Braun, J.: Data Analysis and Graphics Using R. New York: Cambridge University Press (2005)
26. Ming, K., Rosenbaum, P. R. Substantial gains in bias reduction from matching with a variable number of controls. Biometrics **56**, 118–124 (2000)
27. Ming, K., Rosenbaum, P.R.: A note on optimal matching with variable controls using the assignment algorithm. J Comp Graph Statist **10**, 455–463 (2001)

28. R Development Core Team.: R: A Language and Environment for Statistical Computing. Vienna: R Foundation, http://www.R-project.org (2007)
29. Papadimitriou, C.H., Steiglitz, K.: Combinatorial Optimization: Algorithms and Complexity. Englewood Cliffs, NJ: Prentice-Hall (1982)
30. Rosenbaum, P.R., Rubin, D.B.: The central role of the propensity score in observational studies for causal effects. Biometrika **70**, 41–55 (1983)
31. Rosenbaum, P.R.: Conditional permutation tests and the propensity score in observational studies. J Am Statist Assoc **79**, 565–574 (1984)
32. Rosenbaum, P.R., Rubin, D.B.: Constructing a control group by multivariate matched sampling methods that incorporate the propensity score. Am Statistician **39**, 33–38 (1985)
33. Rosenbaum, P.R.: Model-based direct adjustment. J Am Statist Assoc **82**, 387–394 (1987)
34. Rosenbaum, P.R.: Optimal matching in observational studies. J Am Statist Assoc **84**, 1024–1032 (1989)
35. Rosenbaum, P.R.: A characterization of optimal designs for observational studies. J Roy Statist Soc B **53**, 597–610 (1991)
36. Rosenbaum, P.R.: Covariance adjustment in randomized experiments and observational studies (with Discussion). Statist Sci **17**, 286–327 (2002)
37. Rosenbaum, P.R.: Observational Studies. New York: Springer (2002)
38. Rubin, D.B.: Matching to remove bias in observational studies. Biometrics **29**,159–183 (1973)
39. Rubin D.B.: Bias reduction using Mahalanobis metric matching. Biometrics **36**, 293–298 (1980)
40. Sekhon, J.S.: Opiates for the matches: Matching methods for causal inference. Annu Rev Pol Sci **12**, 487-508 (2009)
41. Silber, J.H., Rosenbaum, P.R., Trudeau, M.E., Even-Shoshan, O., Chen, W., Zhang, X., Mosher, R.E.: Multivariate matching and bias reduction in the surgical outcomes study. Medical Care **39**, 1048–1064 (2001)
42. Silber, J.H., Rosenbaum, P.R., Trudeau, M.E., Chen, W., Zhang, X., Lorch, S.L., Rapaport-Kelz, R., Mosher, R.E, Even-Shoshan, O.: Preoperative antibiotics and mortality in the elderly, Ann Surg **242**, 107–114 (2005)
43. Silber, J.H., Rosenbaum, P.R., Polsky, D., Ross, R.N., Even-Shoshan, O., Schwartz, S., Armstrong, K.A., Randall, T.C.: Does ovarian cancer treatment and survival differ by the specialty providing chemotherapy? J Clin Oncol **25**, 1169–1175 (2007)
44. Silber, J.H., Lorch, S.L., Rosenbaum, P.R., Medoff-Cooper, B., Bakewell-Sachs, S., Millman, A., Mi, L., Even-Shoshan, O., Escobar, G.E.: Additional maturity at discharge and subsequent health care costs. Health Serv Res **44**, 444–463 (2009)
45. Smith, H.L.: Matching with multiple controls to estimate treatment effects in observational studies. Sociol Method **27**, 325–353 (1997)
46. Stuart, E.A., Green, K.M.: Using full matching to estimate causal effects in nonexperimental studies: Examining the relationship between adolescent marijuana use and adult outcomes. Dev Psychol **44**, 395–406 (2008)

Chapter 9
Various Practical Issues in Matching

Abstract Having constructed a matched control group, one must check that it is satisfactory, in the sense of balancing the observed covariates. If some covariates are not balanced, then adjustments are made to bring them into balance. Three adjustments are almost exact matching, exact matching, and the use of small penalties. Exact matching has a special role in extremely large problems, where it can be used to accelerate computation. Matching when some covariates have missing values is discussed.

9.1 Checking Covariate Balance

Although a table such as Table 7.1 in Chapter 7 may suffice to describe covariate balance in a scientific paper [15], somewhat more is typically needed when constructing a matched sample. Checks on covariate balance in matching are informal diagnostics, not unlike residuals in a regression. They aid in thinking about whether the treated and control groups overlap sufficiently to be matched, and whether the match currently under consideration has achieved reasonable balance or whether some refinements are required. The study of ovarian cancer [15] in Chapter 7 is used to illustrate.

One common measure of covariate imbalance is a slightly unusual version of an absolute standardized difference in means [12]. The numerator of the standardized difference is simply the treated-minus-control difference in covariate means or proportions, and it is computed before and after matching. The first variable, surgeon-type-GO, in Table 7.1 is a binary indicator, which is 1 if a GO performed the surgery and is 0 if someone else performed the surgery. Before matching, the difference in means is $0.76 - 0.33 = 0.43$, or 43%, and after matching it is $0.76 - 0.75 = 0.01$ or 1%. Write $\bar{x}_{tk}$, $\bar{x}_{ck}$, $\bar{x}_{cmk}$ for the means of covariate k in, respectively, the treated group, the control group before matching, and the matched control group, so $\bar{x}_{t1} = 0.76$, $\bar{x}_{c1} = 0.33$, and $\bar{x}_{cm1} = 0.75$ for the first covariate in Table 7.1. It is the denominator that is just slightly unusual in two ways. First, the denominator always describes the standard deviation before matching, even when measuring imbalance after matching. We are asking whether the means or proportions are close; we do not want the answer to be hidden by a simultaneous change in the standard deviation. Second, the standard deviation before matching is calculated in a way that gives equal weight to the standard deviation in the treated and control groups before matching. In many problems, the potential control group is much larger than the treated group, but we do not want to give much more weight to the standard deviation in the control group. Write s_{tk} and s_{ck} for the standard deviations of covariate k in the treated group and in the control group before matching. The pooled standard deviation for covariate k is $\sqrt{\left(s_{tk}^2 + s_{ck}^2\right)/2}$. The absolute standardized difference before matching is $sd_{bk} = |\bar{x}_{tk} - \bar{x}_{ck}| / \sqrt{\left(s_{tk}^2 + s_{ck}^2\right)/2}$ and the absolute standardized difference after matching is $sd_{mk} = |\bar{x}_{tk} - \bar{x}_{cmk}| / \sqrt{\left(s_{tk}^2 + s_{ck}^2\right)/2}$; notice that they are identical except that $\bar{x}_{cmk}$ replaces $\bar{x}_{ck}$. For the first covariate in Table 7.1, surgeon-type-GO, the absolute standardized difference is 0.95 before matching and 0.02 after matching, that is, almost a full standard deviation before matching, and about 2% of a standard deviation after matching.

The boxplots in Figure 9.1 display 67 absolute standardized differences before and after matching. The list of 67 covariates is slightly redundant; for instance, all three categories of surgeon type appear as three binary variables, even though the value of one of these variables is determined by the values of the other two. The imbalances before matching are quite large: there are four covariates with differences of more than half a standard deviation. After matching, the median absolute standardized difference is 0.03 or 3% of a standard deviation, and the maximum is 0.14. In fact, because fine balancing was used to construct this matched sample, 18 of the 67 absolute standardized differences equal zero exactly.

The principal advantage of an absolute standardized difference over an unstandardized difference, say $\bar{x}_{tk} - \bar{x}_{cmk}$, is that variables on different scales, such as age and hypertension, can be plotted in a single graph for quick inspection. The disadvantage is that a covariate such as age means more in terms of years than in terms of standard deviations. In practice, it is helpful to examine an unstandardized table such as Table 7.1 in addition to graphs of standardized differences.

In matching with variable controls, as in §8.5, or in full matching, as in §8.6, the mean in the matched control group, $\bar{x}_{cmk}$, is a weighted mean, as described in §8.5 and §8.6. Then the absolute standardized difference is computed using this weighted mean, $|\bar{x}_{tk} - \bar{x}_{cmk}| / \sqrt{\left(s_{tk}^2 + s_{ck}^2\right)/2}$.

It might seem desirable that all of the absolute standardized differences equal zero, but this would not happen even in a completely randomized experiment. How does the imbalance in the boxplots in Figure 9.1 compare with the imbalance expected in a completely randomized experiment?

Imagine a completely randomized experiment. This means that 688 unmatched patients are randomly divided into two groups, each with 344 patients. If a randomization test were applied to one covariate to compare the distribution of the covariate in these randomly formed groups, it would produce a P-value less than or equal to

9.1 Checking Covariate Balance

0.05 with probability 0.05; in fact, it would produce a *P*-value less than or equal to α with probability α for every α between zero and one.[1] Fisher's exact test for a 2×2 table is one such randomization test, and Wilcoxon's rank sum test is another. With 688 patients, it suffices to use the large sample approximations to these tests.

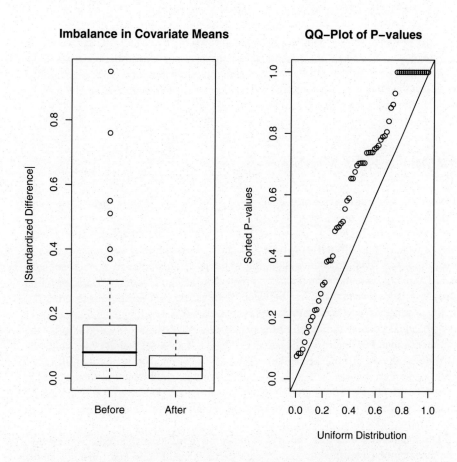

Fig. 9.1 Balance checks for 67 covariates in the study of ovarian cancer. The boxplots show absolute values of standardized differences in means between the GO and MO groups. The quantile-quantile plot compares the 67 two-sample *P*-values with the uniform distribution, with the line of equality. The boxplot shows that the imbalances in covariates were greatly reduced by matching, while the quantile-quantile plot shows the imbalance on observed covariates after matching is somewhat better than expected for a completely randomized experiment with patients randomly assigned to GO or MO. Balance on observed covariates does not imply balance on covariates that were not observed.

---

[1] This statement is not exactly correct, owing to the discreteness of a randomization distribution, but it is close enough.

The quantile-quantile plot in Figure 9.1 compares the 67 two-sample P-values with the uniform distribution. A quantile-quantile plot compares the quantiles in a sample with the quantiles in a probability distribution to see how they compare [2, pages 143–149]. If the sample looks like the distribution, the points will fall close to the line of equality. In Figure 9.1 the points fall above the line of equality. This means that the two-sample p-values for the 688 patients tend to be larger than they would be if they came from the uniform distribution. In other words, matching on these 67 covariates has produced more balance on these 67 covariates than we would have expected if we had not matched these patients, but instead had assigned them to treatment or control at random. Of course, randomization has the key benefit that it tends to balance variables that were not measured, whereas matching can only be expected to balance observed covariates.

9.2 Almost Exact Matching

If there are a few covariates of overriding importance, each taking just a few values, then one may wish to match exactly on these covariates whenever this is possible. For instance, in Table 7.1, for reasons discussed in §7.2, an effort was made to match exactly on both surgeon type and stage, although this was not possible in every instance. The procedure is similar to the use of penalties in §8.4. If subjects k and ℓ do not have the same values on these few key covariates, then a substantial penalty is added to the distance between them as recorded in the distance matrix. If the penalty is large enough, optimal matching will avoid the penalties whenever possible, and when avoiding all of the penalties is not possible, it will minimize the number of matches that incur the penalty. That is, if an exact match is possible, one will be produced; if not, the match will be as close to exact as possible. In either case, once the match is as exact as possible, the best such match is selected using other considerations represented in the distance matrix. The approach is flexible: it may be used with pair matching (§8.4), matching with variable controls (§8.5), or full matching (§8.6).

Obviously, as the distance matrix increasingly emphasizes these key covariates, it also de-emphasizes the other covariates, perhaps resulting in poor matches on the other covariates. This is the principal disadvantage of all methods that emphasize exact matching: gains in one area are paid for by losses in another, and sometimes the price is high. In contrast, methods that emphasize balancing covariates, such as propensity scores (§8.2) and fine balance (Chapter 10), may produce gains in one area without losses in another.

To illustrate, in the welder data in Table 8.3, there were 11 smokers among the 21 welders and 9 smokers among the 26 potential controls. In pair matching, an exact match on smoking is not possible: at most 9 of the 11 smoking welders can be matched to smoking controls. In fact, none of the three pair matches in Tables 8.8-8.10 had used all nine smoking controls. Table 9.1 is an optimal pair match using the Mahalanobis distance within propensity score calipers with almost exact matching

9.2 Almost Exact Matching

Table 9.1 Optimal pair match using the Mahalanobis distance within propensity score calipers with almost exact matching for smoking. The caliper is half the standard deviation of the propensity score, or $0.172/2 = 0.086$. Covariates are age, race (C=Caucasian, AA=African American), current smoker (Y=yes, N=no), and the estimated propensity score $\widehat{e}(\mathbf{x})$.

	Welders				Matched Controls			
Pair	Age	Race	Smoker	$\widehat{e}(\mathbf{x})$	Age	Race	Smoker	$\widehat{e}(\mathbf{x})$
1	38	C	N	0.46	44	C	N	0.34
2	44	C	N	0.34	45	C	N	0.32
3	39	C	Y	0.57	36	C	Y	0.64
4	33	AA	Y	0.51	41	AA	Y	0.35
5	35	C	Y	0.65	35	C	Y	0.65
6	39	C	Y	0.57	39	C	Y	0.57
7	27	C	N	0.68	30	C	N	0.63
8	43	C	Y	0.49	56	AA	Y	0.13
9	39	C	Y	0.57	36	C	Y	0.64
10	43	AA	N	0.20	48	AA	N	0.14
11	41	C	Y	0.53	44	C	Y	0.47
12	36	C	N	0.50	41	C	N	0.40
13	35	C	N	0.52	40	C	N	0.42
14	37	C	N	0.48	42	C	N	0.38
15	39	C	Y	0.57	35	C	N	0.52
16	34	C	N	0.54	38	C	N	0.46
17	35	C	Y	0.65	34	C	Y	0.67
18	53	C	N	0.19	52	C	N	0.20
19	38	C	Y	0.60	34	C	N	0.54
20	37	C	N	0.48	42	C	N	0.38
21	38	C	Y	0.60	31	AA	Y	0.55
	mean	%AA	%Y	mean	mean	%AA	%Y	mean
	38	10	52	0.51	40	19	43	0.45

for smoking. Specifically, the distance matrix for the match in Table 8.5 was used; however, if there was a mismatch on smoking between a welder and a potential control, a penalty was added to the distance between them. This penalty, 5322, was ten times the largest distance in the previous distance matrix used for Table 8.5, where the previous distance matrix already contained penalties to implement the caliper on the propensity score, as described in §8.4.

In Table 9.1, all nine smoking controls are used, unlike Tables 8.8–8.10. Each of the nine smoking controls is matched to a welder who smokes. Inevitably, two smoking welders are matched to nonsmoking controls; they are #15 and #19. In the means at the bottom of Table 9.1, the balance on smoking is slightly better than in Tables 8.8–8.10, but the balance on race is worse.

Almost exact matching is sometimes useful when there is interest in estimating the treatment effect separately in several subgroups, say for smokers and nonsmokers in Table 9.1. For the ovarian cancer study in Chapter 7, in [15, Table 5], results were reported separately for the 263 pairs in which both patients had stage III or stage IV cancer. Because the matching was 'almost exact' for stage, only 10 of 344 pairs had a stage III or IV cancer matched to another stage or missing stage, and so could not be used straightforwardly in such a subgroup analysis.

Table 9.2 Is exact matching feasible in the welder data? The 2x2 contingency tables show that exact matching is feasible for race, but not for smoking. Covariates are race (C=Caucasian, AA=African American), and current smoker (Y=yes, N=no)

	Smoker	Nonsmoker
Welder	11	10
Potential Control	9	17

	AA	C
Welder	2	19
Potential Control	5	21

Almost exact matching adds large penalties to a distance matrix when treated and control subjects differ on a key covariate. Sometimes a covariate is uncomfortably out of balance but not of overriding importance. In this case, small penalties for mismatches are sometimes used. There is, at present, little theory to guide the use of small penalties; nonetheless, they often work. The Mahalanobis distance between two randomly selected individuals from a single multivariate Normal population has expectation equal to twice the number of covariates, so adding a penalty of 2 to a Mahalanobis distance may be viewed informally as doubling the importance of that one covariate.

9.3 Exact Matching

Almost exact matching, as in §9.2, will result in an exact match whenever one is available, so for many purposes a separate discussion of exact matching is not needed. Nonetheless, there are computational reasons to distinguish exact and almost exact matching.

A simple contingency table will indicate whether an exact match is possible. For the welder data, Table 9.2 shows that exact matching is feasible for race (because $5 \geq 2$ and $21 \geq 19$) but not feasible for smoking (because $9 < 11$), although 'almost exact matching' is feasible for smoking, as illustrated in §9.2.

When feasible, an exact match can be found by subdividing the matching problem into several smaller problems and piecing the answers together. For instance, knowing from Table 9.2 that an exact match for race is feasible, the match can be divided into two separate problems, namely matching the two African American welders and matching the 19 Caucasian welders.

In small problems, there is no advantage to such a subdivision, but in extremely large matching problems, there can be a useful reduction in computational effort, both in required memory storage and in speed of computation. For instance, suppose that there is a nominal covariate with five categories, each containing m treated subjects and n potential controls, so that there are $5m + 5n$ subjects in total. A distance matrix for all $5m + 5n$ subjects is $5m \times 5n$ with $25mn$ entries, whereas each of the five subdivided matching problems has a distance matrix that is $m \times n$ with mn entries, so it is 1/25th as large. The worst-case time bound for the assignment problem grows as the cube of the number of subjects [4, page 147], here $5m+5n$, so the reduction to a problem of size $m+n$ is expected to substantially accelerate computation.

If a critically important covariate is continuous, it may enter the matching in two versions, one continuous, the other discrete. For instance, a nominal variable can be formed by cutting at quantiles of the continuous variable, and this nominal variable may be used to subdivide the problem for exact matching. The only reasons to do this is to save computer memory or accelerate computation. Within each subdivision, the continuous variable may be used in a distance, such as the Mahalanobis distance, or with a caliper, as described in §8.4.

A large sample size should be a luxury, not a hindrance. To enjoy that luxury when matching, consider exact matching on one or two important covariates, subdividing the matching problem into several smaller problems.

9.4 Missing Covariate Values

A covariate has missing values if it is recorded for many people but not for some people. The pattern of missing data refers to which values are missing and which values are present, so you observe the pattern of missing data, though you do not observe the missing values. The constant worry with missing values is that what you cannot see may differ in important ways from what you can see. If a manuscript is missing certain words, it makes all the difference whether the manuscript was damaged by rain or edited by a censor. If edited by a censor, and if you know what motivates the censor, then the fact that a word is missing speaks volumes about what the word must have been, even though you still do not know the word. If a high school student reports his mother's level of education in a survey but not his father's, you might wonder whether he ever met his father. The pattern of missing data may or may not say something about the values that are missing, but knowing the pattern does not tell you the missing values.

The propensity score $e(\mathbf{x})$ is the conditional probability of treatment given the observed covariates, $e(\mathbf{x}) = \Pr(Z=1 \mid \mathbf{x})$, and matching on $e(\mathbf{x})$ tends to balance $\mathbf{x}$; see §3.3 or [10, Theorems 1 and 2]. If a coordinate of $\mathbf{x}$ is sometimes missing, think of that coordinate as being a number if it is not missing or * if it is missing. When some covariates have missing values, the propensity score remains well-defined: it refers to the conditional probability of treatment given $\mathbf{x}$, that is $e(\mathbf{x}) = \Pr(Z=1 \mid \mathbf{x})$, where now some of the coordinates may equal *. It only takes a moment to review the proof of the balancing property of the propensity score to realize that the proof cares not a bit whether the coordinates of $\mathbf{x}$ take only numeric values or sometimes take the value *; see [11, Appendix B] for an explicit and detailed discussion. This says something useful, but alas something less than one might hope. It says that matching on $e(\mathbf{x})$ will tend to balance the observed covariates and the pattern of missing covariates; however, the missing values themselves may or may not be balanced. The subject is easiest to discuss in the context of an example, which will happen in §13.4.

The usual way to estimate $e(\mathbf{x})$ when some covariates are missing is to (i) append indicator variables for the pattern of missing data, (ii) for each covariate, plug in an arbitrary but fixed value for each *, and (iii) fit a logit model with these variables, taking the fitted probabilities from the model as the estimates $\widehat{e}(\mathbf{x})$. It is not difficult to verify that the presence of the missing value indicators means that the arbitrary values in step (ii) do not affect $\widehat{e}(\mathbf{x})$, though they do affect the coefficients of the model.[2] Such a logit model is a specific parametric form for a conditional probability $e(\mathbf{x}) = \Pr(Z=1 \mid \mathbf{x})$ relating two observed quantities, namely Z and $\mathbf{x}$ with its *'s. As is always true with any such model, it may or may not correctly represent the form of $\Pr(Z=1 \mid \mathbf{x})$ — perhaps there should be an interaction linking the missing value indicator for variable #7 with the indicator for variable #19 — however, the issues here are no different from any other attempt to model a conditional probability of one observable given another. Again, see [11, Appendix B] for additional detail.

Further discussion is aided by an example; see §13.4.

9.5 Further Reading

For further discussion of checks on covariate balance, see the brief discussion in Cochran's paper [3], and the longer discussions in the papers by Ben Hansen and Jake Bowers [5], Ben Hansen [6], Kosuke Imai, Gary King, and Elizabeth Stuart [7] , and Sue Marcus and colleagues [9]. Propensity scores with missing data are discussed in [11, Appendix B]. The use of penalty functions in optimization is fairly standard; see [1, 8].

References

1. Avriel, M.: Nonlinear Programming. Englewood Cliffs, NJ: Prentice Hall (1976)
2. Cleveland, W.S.: The Elements of Graphing Data. Summit, NJ: Hobart Press.
3. Cochran, W.G.: The planning of observational studies of human populations (with Discussion). J Roy Statist Soc A **128**, 234–265.
4. Cook, W.J., Cunningham, W.H., Pulleyblank, W.R., Schrijver, A.: Combinatorial Optimization. New York: Wiley (1998)
5. Hansen, B.B., Bowers, J.: Covariate balance in simple, stratified and clustered comparative studies. Statist Sci **23**, 219–236 (2008)
6. Hansen, B.B.: The essential role of balance tests in propensity-matched observational studies. Statist Med **12**, 2050–2054 (2008)
7. Imai, K., King, G., Stuart, E.A.: Misunderstandings between experimentalists and observationalists about causal inference. J Roy Statist Soc A **171**, 481–502 (2008)

---

[2] Technically, the fitted probabilities in logit regression are invariant under affine transformations of the predictors.

References

8. Karmanov, V.G.: Mathematical Programming. Moscow: Mir.
9. Marcus, S.M., Siddique, J., Ten Have, T.R., Gibbons, R.D., Stuart, E., Normand, S-L. T. Balancing treatment comparisons in longitudinal studies. Psychiatric Ann **38**, 12 (2008)
10. Rosenbaum, P.R., Rubin, D.B.: The central role of the propensity score in observational studies for causal effects. Biometrika **70**, 41–55 (1983)
11. Rosenbaum, P.R., Rubin, D.B.: Reducing bias in observational studies using subclassification on the propensity score. J Am Statist Assoc **79**, 516–524 (1984)
12. Rosenbaum, P.R., Rubin, D.B.: Constructing a control group by multivariate matched sampling methods that incorporate the propensity score. Am Statist **39**, 33–38 (1985)
13. Rosenbaum, P.R.: Optimal matching in observational studies. J Am Statist Assoc **84**, 1024–1032 (1989)
14. Rosenbaum, P.R., Ross R.N., Silber, J.H.: Minimum distance matched sampling with fine balance in an observational study of treatment for ovarian cancer. J Am Statist Assoc **102**, 75–83. (2007)
15. Silber, J.H., Rosenbaum, P.R., Polsky, D., Ross, R.N., Even-Shoshan, O., Schwartz, S., Armstrong, K.A., Randall, T.C.: Does ovarian cancer treatment and survival differ by the specialty providing chemotherapy? J Clin Oncol **25**, 1169–1175 (2007)

Chapter 10
Fine Balance

Abstract Fine balance means constraining a match to balance a nominal variable, without restricting who is matched to whom, when matching to minimize a distance between treated and control subjects. It may be applied to: (i) a nominal variable with many levels that is difficult to balance using propensity scores, (ii) a rare binary variable that is difficult to control using a distance, or (iii) the interaction of several nominal variables. The fine balance constraint and the distance can emphasize different covariates. When exact balance is unobtainable, fine balance can be used to obtain a specific pattern of imbalances.

10.1 What Is Fine Balance?

Fine balance is a constraint on an optimal matching that forces a nominal variable to be balanced [5, 6]. Fine balance is explicitly aimed at covariate balance, and it makes no effort to match exactly for this nominal variable. The pairing can focus on other covariates with the knowledge that this nominal variable will be balanced.

In the study of ovarian cancer in Chapter 7 and [7], the eight SEER sites and three intervals 1991–1992, 1993–1996, and 1997–1999 for year of diagnosis were finely balanced, as was their $24 = 8 \times 3$ category interaction. In other words, the number of patients diagnosed in 1991–1992 in Connecticut, for instance, was exactly the same for GO patients and the matched control group of MO patients, although these patients were not typically matched to each other. As you may recall from Table 7.1, SEER site and year of diagnosis were substantially out of balance prior to matching, but they are not usually viewed as extremely important by clinicians. For this reason, the distance and various penalties used in this match sought individual pairs with the same clinical stage from the same surgeon type with similar propensity scores, and paid no explicit attention to SEER site, yet the fine balance constraint ensured SEER site was balanced in aggregate.

Propensity scores use probability to balance covariates [4], appealing to the law of large numbers [2, Chapter 10]; in effect, if a coin is flipped sufficiently often, the number of heads tends to balance the number of tails. Although they work well in many contexts, propensity scores may not balance covariates when the number of trials (or flips) is effectively small. For instance, if the data record little or no information about income, education, pollution, and so on, but do record postal zip code, then one might think to match for zip code as a proxy for various unmeasured variables. In many contexts, there will be only a handful of people from any one zip code, so matching exactly for Zip Code will yield poor matches for other covariates; moreover, the law of large numbers is of no help. In such a context, fine balancing can balance zip code, without matching people from the same zip code to one another. Similar considerations apply if there are several binary covariates in which one category is rarely seen; these can be combined into one nominal variable with several rare levels and one common level, and this amalgamated covariate can be finely balanced. More generally, if several important nominal variables are thought to interact with each other, fine balancing may be used to exactly balance the many categories formed from their interaction (i.e., their direct product).

10.2 Constructing an Exactly Balanced Control Group

As in Chapter 8, fine balance is best understood with the aid of a small example. For the welder data [1] in Chapter 8, it was clear from Tables 8.3 and 9.2 that it was possible to balance race but not smoking. Section 10.2 finds a match that finely balances race, while §10.3 finds a match that controls the nature of the joint imbalance in race and smoking.

In Tables 8.3 and 9.2, among the potential controls, there are five African Americans where two are needed for balance, and there are 21 Caucasians, where 19 are needed for balance, so three African Americans and two Caucasians must be removed to obtain balance. This is done by expanding the distance matrix, adding five rows. Table 10.1 adds five rows to Table 8.5, which recorded the Mahalanobis distances within propensity score calipers for the first 6 of 26 potential controls. The five extra rows are E1 to E5, making the distance matrix square, 26×26. The first two of these extra rows, E1 and E2, will take away two Caucasians, while the last three of the extra rows, E3, E4, and E5, will take away three African Americans. Because control #1 in column 1 of Table 10.1 is African American, he is at infinite distance from the two Caucasian extras and at zero distance from the three African American extras. Because controls #2–#6 in columns 2 to 6 of Table 10.1 are Caucasian, they are at zero distance from the two Caucasian extras and at infinite distance from the three African American extras. A similar pattern occurs in the remaining 20 columns of the distance matrix.

With a distance matrix patterned as in Table 10.1, an optimal match is found, and the extras and their matched controls are discarded. To avoid the infinite distances, three African American and two Caucasian controls must be matched to the extras, and the match that remains is balanced.

10.2 Constructing an Exactly Balanced Control Group

Table 10.1 Mahalanobis distances within propensity score calipers with extra rows to finely balance race. Rows are the 21 welders plus five extras (E), and columns are for the first 6 of 26 potential controls. An ∞ signifies that the caliper is violated.

Welder	Control 1	Control 2	Control 3	Control 4	Control 5	Control 6
1	∞	∞	6.15	0.08	∞	∞
2	∞	∞	∞	0.33	∞	∞
3	∞	∞	∞	∞	∞	∞
4	∞	∞	12.29	∞	∞	∞
5	∞	∞	∞	∞	∞	∞
6	∞	∞	∞	∞	∞	∞
7	∞	∞	∞	∞	∞	∞
8	∞	∞	0.02	5.09	∞	∞
9	∞	∞	∞	∞	∞	∞
10	0.51	∞	∞	∞	10.20	11.17
11	∞	∞	0.18	∞	∞	∞
12	∞	∞	7.06	0.33	∞	∞
13	∞	∞	7.57	∞	∞	∞
14	∞	∞	6.58	0.18	∞	∞
15	∞	∞	∞	∞	∞	∞
16	∞	∞	8.13	∞	∞	∞
17	∞	∞	∞	∞	∞	∞
18	9.41	∞	∞	∞	0.18	0.02
19	∞	∞	∞	∞	∞	∞
20	∞	∞	6.58	0.18	∞	∞
21	∞	∞	∞	∞	∞	∞
E1	∞	0	0	0	0	0
E2	∞	0	0	0	0	0
E3	0	∞	∞	∞	∞	∞
E4	0	∞	∞	∞	∞	∞
E5	0	∞	∞	∞	∞	∞

Table 10.2 displays the minimum distance finely balanced match. Race is perfectly balanced, but the pairs are not exactly matched for race. In one sense, Table 10.2 is a slight improvement over Table 8.9 because the control group means for age and smoking are the same, while the means for race and the propensity score are slightly closer to the welder means. Notice that the finely balanced match selected two African American smokers, even though a nonsmoker was available as a control for pair #10.

The general procedure is essentially the same. If there is a nominal covariate for which balance is required, then (i) cross-tabulate that covariate with treatment, as in Table 9.2; (ii) determine the number of controls that must be removed from each category of the covariate to achieve balance; (iii) add one row for each control that must be removed, with zero distance to its own category and infinite distance to other categories; (iv) find an optimal match for this new, square distance matrix; (v) discard the extra rows and their matched controls. The nominal covariate that is balanced may be formed by combining several other nominal covariates or intervals that cut a continuous variable.

Table 10.2 Optimal pair match using the Mahalanobis distance within propensity score calipers, with fine balance for race. This match is constrained to balance race. The caliper is half the standard deviation of the propensity score, or $0.172/2 = 0.086$. Covariates are age, race (C=Caucasian, AA=African American), current smoker (Y=yes, N=no), and the estimated propensity score $\widehat{e}(\mathbf{x})$.

	Welders				Matched Controls			
Pair	Age	Race	Smoker	$\widehat{e}(\mathbf{x})$	Age	Race	Smoker	$\widehat{e}(\mathbf{x})$
1	38	C	N	0.46	44	C	N	0.34
2	44	C	N	0.34	47	C	N	0.28
3	39	C	Y	0.57	36	C	Y	0.64
4	33	AA	Y	0.51	41	AA	Y	0.35
5	35	C	Y	0.65	35	C	Y	0.65
6	39	C	Y	0.57	36	C	Y	0.64
7	27	C	N	0.68	30	C	N	0.63
8	43	C	Y	0.49	45	C	N	0.32
9	39	C	Y	0.57	39	C	Y	0.57
10	43	AA	N	0.20	50	C	N	0.23
11	41	C	Y	0.53	44	C	Y	0.47
12	36	C	N	0.50	41	C	N	0.40
13	35	C	N	0.52	40	C	N	0.42
14	37	C	N	0.48	42	C	N	0.38
15	39	C	Y	0.57	35	C	N	0.52
16	34	C	N	0.54	38	C	N	0.46
17	35	C	Y	0.65	34	C	Y	0.67
18	53	C	N	0.19	52	C	N	0.20
19	38	C	Y	0.60	34	C	N	0.54
20	37	C	N	0.48	42	C	N	0.38
21	38	C	Y	0.60	31	AA	Y	0.55
	mean	%AA	%Y	mean	mean	%AA	%Y	mean
	38	10	52	0.51	40	10	38	0.46

To match two controls to each treated subject with fine balance, stack two copies of the distance matrix, one on top of the other, and then add extras as before to remove imbalances in the nominal covariate. In this way, the number of columns or controls does not change, but each treated subject is represented twice, once in each of two rows. In an optimal match, one control is paired with the first copy of a treated subject and a different control is paired with the second copy, yielding a 2-to-1 match. (Obviously, this cannot be done with the welder data, because at least 42 potential controls are needed, and only 26 are available.) The procedure for k-to-1 finely balanced matching is similar, but with k copies of the distance matrix rather than two. If needed, a precise algorithmic description is given in [6, §4.2]. It is easy to prove that a minimum distance match with this enlarged distance matrix is a minimum distance match subject to the fine balance constraint, and that the minimum distance match has infinite total distance if and only if the latter problem is infeasible [6, Proposition 1]. In particular, in the statistical package R, the `pairmatch` function in Hansen's [3] `optmatch` package may be used, whereas in SAS `proc assign` may be used. Alternatively, the minimum distance finely balanced match may be determined directly as a minimum-cost flow in a network [5, §3.2].

Table 10.3 Table of treatment groups by (race × smoking) for the welder data, with race as C=caucasian, AA=African American, and current smoker coded as Y=yes or N=no.

Smoker	Yes	Yes	No	No	
Race	AA	C	AA	C	Total
Welder	1	10	1	9	21
Control	3	6	2	15	26

10.3 Controlling Imbalance When Exact Balance Is Not Feasible

When exact balance is not possible, the fine balance procedure can be used to obtain any specified, possible imbalance. The procedure is essentially the same as in §10.2, except that one must choose an acceptable imbalance from those that are possible.

Table 10.3 displays the options for race and smoking in the welder data. The obstacle to balance on the joint distribution of these two variables is a deficit of smoking, Caucasian controls: there are six, where ten are needed. That is, from the four categories given by the columns of Table 10.3, we would like to select controls with frequencies $(1, 10, 1, 9)$ to agree with the welder group, but the second entry is limited to six, so for a pair match some other entries must increase to compensate. One attractive choice is $(3, 6, 0, 12)$, because this choice uses all nine smoking controls, where 11 would be needed for balance, and is only slightly imbalanced for race, with three African Americans where two would produce balance on race. This choice presumes that smoking is more important than race, as is likely in genetic toxicology, and it provides no information about a treatment-by-race-by-smoking interaction. In any event, for this purely illustrative example, the important point that is fine balance can produce any specified possible imbalance, such as $(3, 6, 0, 12)$.

To produce the imbalance $(3, 6, 0, 12)$, one notices that: no controls are to be removed from the first two columns of Table 10.3, two are to be removed from the third column, and three are to be removed from the fourth column. So five rows are added to Table 8.5, two at zero distance from nonsmoking African Americans, three at zero distance from nonsmoking Caucasians, with the other entries set to ∞; see Table 10.4. The first control in Table 8.3 is an African American nonsmoker, and so in column 1 of Table 10.4, he is at zero distance from both of the first two extra rows. The second control in Table 8.3 is a Caucasian nonsmoker, and so, in column 2 of Table 10.4, he is at zero distance from the last three extra rows. The third control in Table 8.3 is a Caucasian smoker, and cannot be eliminated, and so, in column 3 of Table 10.4, he is at infinite distance from all five rows. As in §10.2, one finds an optimal pair matching in the extended square 26 × 26 distance matrix and discards the extras and their matched controls.

Table 10.5 is the result. It achieves the prescribed imbalance, $(3, 6, 0, 12)$, and subject to that constraint it minimizes the distance among all pair matchings. Notable in Table 10.5 is the use of all nine smoking controls without a large imbalance in race. This was produced by matching welder #10, a nonsmoking African

Table 10.4 Mahalanobis distances within propensity score calipers with extra rows to finely balance race and smoking. Rows are the 21 welders plus five extras (E), and columns are for the first 6 of 26 potential controls. An ∞ signifies that the caliper is violated.

Welder	Control 1	Control 2	Control 3	Control 4	Control 5	Control 6
1	∞	∞	6.15	0.08	∞	∞
2	∞	∞	∞	0.33	∞	∞
3	∞	∞	∞	∞	∞	∞
4	∞	∞	12.29	∞	∞	∞
5	∞	∞	∞	∞	∞	∞
6	∞	∞	∞	∞	∞	∞
7	∞	∞	∞	∞	∞	∞
8	∞	∞	0.02	5.09	∞	∞
9	∞	∞	∞	∞	∞	∞
10	0.51	∞	∞	∞	10.20	11.17
11	∞	∞	0.18	∞	∞	∞
12	∞	∞	7.06	0.33	∞	∞
13	∞	∞	7.57	∞	∞	∞
14	∞	∞	6.58	0.18	∞	∞
15	∞	∞	∞	∞	∞	∞
16	∞	∞	8.13	∞	∞	∞
17	∞	∞	∞	∞	∞	∞
18	9.41	∞	∞	∞	0.18	0.02
19	∞	∞	∞	∞	∞	∞
20	∞	∞	6.58	0.18	∞	∞
21	∞	∞	∞	∞	∞	∞
E1	0	∞	∞	∞	∞	∞
E2	0	∞	∞	∞	∞	∞
E3	∞	0	∞	0	0	0
E4	∞	0	∞	0	0	0
E5	∞	0	∞	0	0	0

American, to a smoker, even though a nonsmoking African American control was available.

In Tables 8.5, 10.1 and 10.4, it was convenient for display to use ∞ to indicate the penalties. In computation, large numerical penalties are used. Table 10.6 shows the numerical penalties used in producing the match in Table 10.5. Recall that the caliper on the propensity score in Chapter 8 imposed no penalty if the caliper was respected, a small penalty for small violations of the caliper, and a dramatic penalty for large violations of the caliper; this is seen in the first 21 rows of Table 10.6. The caliper on the propensity score was 0.086, and control #2 in Table 8.3, with $\widehat{e}(\mathbf{x}) = 0.09$, violated the caliper for all 21 welders. For welders #10 and #18, with $\widehat{e}(\mathbf{x}) = 0.20$ and $\widehat{e}(\mathbf{x}) = 0.19$, respectively, the violation is not large, and the penalties in column 2 and rows 10 and 18 of Table 10.6 are not extremely large. To give greater emphasis to the fine balance constraint than to the calipers, the penalties for the extra rows in Table 10.6 are five times the largest distance in the first 21 rows and 26 columns.

10.4 Fine Balance and Exact Matching

Table 10.5 Optimal pair match using the Mahalanobis distance within propensity score calipers, with fine balance for race and smoking. The caliper is half the standard deviation of the propensity score, or $0.172/2 = 0.086$. Covariates are age, race (C=Caucasian, AA=African American), current smoker (Y=yes, N=no), and the estimated propensity score $\widehat{e}(\mathbf{x})$. This match is constrained to use all three AA smokers, all six C smokers, neither AA nonsmoker, and 12 C nonsmokers.

	Welders				Matched Controls			
Pair	Age	Race	Smoker	$\widehat{e}(\mathbf{x})$	Age	Race	Smoker	$\widehat{e}(\mathbf{x})$
1	38	C	N	0.46	44	C	N	0.34
2	44	C	N	0.34	47	C	N	0.28
3	39	C	Y	0.57	36	C	Y	0.64
4	33	AA	Y	0.51	41	AA	Y	0.35
5	35	C	Y	0.65	35	C	Y	0.65
6	39	C	Y	0.57	36	C	Y	0.64
7	27	C	N	0.68	30	C	N	0.63
8	43	C	Y	0.49	45	C	N	0.32
9	39	C	Y	0.57	35	C	N	0.52
10	43	AA	N	0.20	56	AA	Y	0.13
11	41	C	Y	0.53	44	C	Y	0.47
12	36	C	N	0.50	41	C	N	0.40
13	35	C	N	0.52	40	C	N	0.42
14	37	C	N	0.48	42	C	N	0.38
15	39	C	Y	0.57	39	C	Y	0.57
16	34	C	N	0.54	38	C	N	0.46
17	35	C	Y	0.65	34	C	Y	0.67
18	53	C	N	0.19	52	C	N	0.20
19	38	C	Y	0.60	31	AA	Y	0.55
20	37	C	N	0.48	42	C	N	0.38
21	38	C	Y	0.60	34	C	N	0.54
	mean	%AA	%Y	mean	mean	%AA	%Y	mean
	38	10	52	0.51	40	14	43	0.45

10.4 Fine Balance and Exact Matching

Fine balance can produce exact balance on a nominal covariate without matching exactly for that covariate. Nonetheless, the two techniques may be combined. In parallel with §9.3, this is one way to implement fine balance in extremely large matching problems: the problem is divided into a few smaller problems.

In fact, this was done in the study of ovarian cancer in Chapter 7 and [7]. Three separate matches were constructed, one for each of the three intervals 1991–1992, 1993–1996, and 1997–1999 of year of diagnosis, with fine balance for SEER sites. We did this not primarily for computational reasons but because we had originally envisioned an additional analysis, not reported in [7], that would have compared the speed with which GOs and MOs adopted new chemotherapies for ovarian cancer. The three time intervals 1991–1992, 1993–1996, and 1997–1999 correspond with innovations in chemotherapy for ovarian cancer, and for that analysis it would have been convenient to have pairs matched for the time interval. That is, in each pair, the same chemotherapies would have been available, so it would make sense to ask who used what. Although the matching procedure worked well, we never

Table 10.6 Mahalanobis distances within propensity score calipers with extra rows to finely balance race and smoking. Rows are the 21 welders plus five extras (E), and columns are for the first 6 of 26 potential controls. This table displays the numerical penalties rather than ∞'s.

Welder	Control 1	Control 2	Control 3	Control 4	Control 5	Control 6
1	237.66	293.68	6.15	0.08	141.69	171.85
2	115.36	166.78	50.06	0.33	17.98	47.65
3	358.16	408.89	20.23	75.94	259.61	289.35
4	287.20	361.09	12.29	18.96	206.61	237.16
5	439.02	492.83	101.71	156.10	341.41	371.48
6	358.16	408.89	20.23	75.94	259.61	289.35
7	467.74	532.22	141.84	184.10	374.36	405.42
8	273.83	321.48	0.02	5.09	174.33	203.74
9	358.16	408.89	20.23	75.94	259.61	289.35
10	0.51	45.85	193.46	134.45	10.20	11.17
11	316.15	365.34	0.18	34.28	217.12	246.70
12	280.67	338.23	7.06	0.33	185.17	215.50
13	302.27	360.61	7.57	20.03	207.01	237.42
14	259.11	315.90	6.58	0.18	163.37	193.61
15	358.16	408.89	20.23	75.94	259.61	289.35
16	323.84	382.94	8.13	41.42	228.81	259.30
17	439.02	492.83	101.71	156.10	341.41	371.48
18	9.41	15.61	196.03	142.97	0.18	0.02
19	378.88	430.37	41.10	96.48	280.55	310.38
20	259.11	315.90	6.58	0.18	163.37	193.61
21	378.88	430.37	41.10	96.48	280.55	310.38
E1	0	2661.08	2661.08	2661.08	2661.08	2661.08
E2	0	2661.08	2661.08	2661.08	2661.08	2661.08
E3	2661.08	0	2661.08	0	0	0
E4	2661.08	0	2661.08	0	0	0
E5	2661.08	0	2661.08	0	0	0

determined whether GOs or MOs were quicker in adopting new chemotherapies. Instead, we determined that GOs and MOs are both much quicker in adopting new chemotherapies than Medicare is in creating a code for new chemotherapies.

10.5 Further Reading

Matching with fine balance is discussed in detail in [6]. The network algorithm in [5, §3.2] will be of interest primarily to programmers because it can be more efficient in its use of space than the procedure described here and in [6].

References

1. Costa, M., Zhitkovich, A., Toniolo, P.: DNA-protein cross-links in welders: Molecular implications. Cancer Res **53**, 460–463 (1993)

2. Feller, W.: An Introduction to Probability Theory and Its Applications, Volume 1. New York: Wiley (1968)
3. Hansen, B.B.: Optmatch: Flexible, optimal matching for observational studies. R News **7**, 18–24 (2007)
4. Rosenbaum, P.R., Rubin, D.B.: The central role of the propensity score in observational studies for causal effects. Biometrika **70**, 41–55 (1983)
5. Rosenbaum, P.R.: Optimal matching in observational studies. J Am Statist Assoc **84**, 1024–1032 (1989)
6. Rosenbaum, P.R., Ross R.N., Silber, J.H.: Minimum distance matched sampling with fine balance in an observational study of treatment for ovarian cancer. J Am Statist Assoc **102**, 75–83 (2007)
7. Silber, J.H., Rosenbaum, P.R., Polsky, D., Ross, R.N., Even-Shoshan, O., Schwartz, S., Armstrong, K.A., Randall, T.C.: Does ovarian cancer treatment and survival differ by the specialty providing chemotherapy? J Clin Oncol **25**, 1169–1175 (2007)

Chapter 11
Matching Without Groups

Abstract Optimal matching without groups, or optimal nonbipartite matching, offers many additional options for matched designs in both observational studies and experiments. One starts with a square, symmetric distance matrix with one row and one column for each subject recording the distance between any two subjects. Then the subjects are divided into pairs to minimize the total distance within pairs. The method may be used to match with doses of treatment, or with multiple control groups, or as an aid to risk-set matching. An extended discussion of Card and Krueger's study of the minimum wage is used to illustrate.

11.1 Matching Without Groups: Nonbipartite Matching

What is nonbipartite matching?

In previous chapters, treated subjects were matched to controls. That is, there were two groups, treated and control, which existed before the matching began, and members of these two groups were placed in pairs or sets to minimize the distance between treated and control subjects in the same pair or set. The optimization algorithm solved the so-called assignment problem or bipartite matching problem, where 'bipartite' makes reference to 'two parts.' A different optimization problem begins with a single group and divides it into pairs to minimize the distance between paired units; this is called by the awkward name 'nonbipartite' matching. For textbook discussions of the contrast between these two matching problems, see [5, Chapter 5] or [21, Chapter 11]. Nonbipartite matching is highly flexible, and it substantially enlarges the collection of matched designs for observational studies.

In Chapter 8, the distance matrix had one row for each treated subject and one column for each potential control. In contrast, in nonbipartite matching, the distance matrix is square, with one row and one column for every subject. Table 11.1 is a small, artificial example with six subjects, so it is 6×6. The distance matrix is symmetric because the distance from k to ℓ in row k and column ℓ is the same as the

Table 11.1 A 6×6 distance matrix for nonbipartite matching for six units. Unlike treatment-control matching, every unit appears as both a row and a column of this distance matrix. The optimal nonbipartite matching (1,2), (3,6), (4,5) is shown in **bold** with a minimum total distance of 106+25+34 = 165.

ID	1	2	3	4	5	6
1	0	**106**	119	231	110	101
2	**106**	0	207	126	192	68
3	119	207	0	156	247	**25**
4	231	126	156	0	**34**	67
5	110	192	247	**34**	0	212
6	101	68	**25**	67	212	0

distance from ℓ to k in row ℓ and column k. Every subject is at zero distance from itself, but of course a subject cannot be paired with itself.

In Table 11.1, an optimal nonbipartite match would pair the six units into three pairs to minimize the total distance within pairs. The optimal match is shown in bold type: it is (1,2), (3,6), and (4,5) for a total distance of $106 + 25 + 34 = 165$. Notice that subject 1 is closer to subject 6 than to subject 2, but if 1 were matched to 6, then the match for 3 would be much worse. Optimal nonbipartite matching may be done in the statistical package R.[1]

Treatment-control matching using an algorithm for nonbipartite matching

Generally, one should not use a nonbipartite matching algorithm to perform bipartite matching because the former requires a larger distance matrix and is somewhat slower. Nonetheless, it is instructive to do it once. Imagine that the first three subjects in Table 11.1 were treated and the last three were controls. To force treated subjects to be matched to controls, infinite distances are placed among treated subjects and among controls, as in Table 11.2. An optimal nonbipartite match in Table 11.2 avoids the infinite distances and pairs (1,5), (2,4), and (3,6), so treated subjects

---

[1] Fortran code by Derigs [9] has been made available inside R by Bo Lu [19] through a function nonbimatch(n, d), where d is the vector form of an n×n symmetric matrix of nonnegative integer distances.

```
> dm
    1   2   3   4   5   6
1   0 106 119 231 110 101
2 106   0 207 126 192  68
3 119 207   0 156 247  25
4 231 126 156   0  34  67
5 110 192 247  34   0 212
6 101  68  25  67 212   0
> nonbimatch(6,as.vector(dm))
[1] 2 1 6 5 4 3
```

This says that unit 1 is paired with unit 2, unit 2 is paired with unit 1, unit 3 is paired with unit 6, unit 4 is paired with unit 5, unit 5 is paired with unit 4, and unit 6 is paired with unit 3.

11.1 Matching Without Groups: Nonbipartite Matching

Table 11.2 Using nonbipartite matching to perform treatment-versus-control matching (or bipartite matching). Subjects 1, 2, and 3 are treated, and infinite distances prevent them from being matched to each other. Subjects 4, 5, and 6 are controls, and infinite distances prevent them from being matched to each other. The optimal match is (1,5), (2,4), (3,6) with distance $110 + 126 + 25 = 261$. Use of a bipartite matching algorithm would have required only a 3×3 distance matrix and be expected to run somewhat more quickly.

ID	1	2	3	4	5	6
1	0	∞	∞	231	**110**	101
2	∞	0	∞	**126**	192	68
3	∞	∞	0	156	247	**25**
4	231	**126**	156	0	∞	∞
5	**110**	192	247	∞	0	∞
6	101	68	**25**	∞	∞	0

are paired with controls. In this sense, nonbipartite matching is a generalization of bipartite matching. Of course, the bipartite matching could have been performed with a 3×3 distance matrix rather than the 6×6 distance matrix in Table 11.2.

11.1.1 Matching with doses

Suppose that we do not have treated subjects and controls but rather individuals who received various doses of treatment. In this case, we might wish to form pairs of individuals who are similar in terms of covariates but quite different in terms of doses. As an illustration, imagine that the six individuals 1, 2, ..., 6 in Table 11.1 received the treatment at doses 1, 2, ..., 6, respectively. In this case, we might wish to match so that pairs differ in dose by at least 2; this would preclude matching 1 with 2, for instance.

In Table 11.3, an infinite distance appears whenever two subjects have doses that differ by less than 2. Unlike Table 11.1, the infinite distances constrain who may be matched to whom. Unlike Table 11.2, these constraints are not the sort that can be addressed with a bipartite matching algorithm; that is, the pattern of ∞'s in Table 11.3 does not divide the six units into two nonoverlapping groups. The optimal match in Table 11.3 pairs (1,5), (2,4), and (3,6): that is, this minimizes the distance subject to the constraint that pairs differ in doses by at least 2.

A practical example of matching with doses was given by Bo Lu, Elaine Zanutto, Robert Hornik, and me [17] in a study of a national media campaign against drug abuse. The dose was the degree of exposure to the media campaign. Another practical example is discussed in §11.3.

Table 11.3 A 6×6 distance matrix for matching with doses of treatment. For simplicity in this artificial illustration, the doses equal the ID numbers, and the requirement is that paired individuals have doses that differ by at least 2. Where the dose difference is less than 2, the distance is set to ∞. The optimal match is (1,5), (2,4), (3,6).

ID	1	2	3	4	5	6
1	0	∞	119	231	**110**	101
2	∞	0	∞	**126**	192	68
3	119	∞	0	∞	247	**25**
4	231	**126**	∞	0	∞	67
5	**110**	192	247	∞	0	∞
6	101	68	**25**	67	∞	0

Table 11.4 A 6×6 distance matrix for matching with three groups. Subjects 1 and 2 are in the same group, as are subjects 3 and 4 and subjects 5 and 6. Because subjects in the same group cannot be matched to one another, there is an infinite distance between them. The optimal match is (1,3), (2,6), (4,5) with a total distance of $119 + 68 + 34 = 221$. This is a very small balanced incomplete block design in the sense that every group is paired with every other group exactly once.

ID	1	2	3	4	5	6
1	0	∞	**119**	231	110	101
2	∞	0	207	126	192	**68**
3	**119**	207	0	∞	247	25
4	231	126	∞	0	**34**	67
5	110	192	247	**34**	0	∞
6	101	**68**	25	67	∞	0

11.1.2 Matching with several groups

Suppose that we do not have a treated and control group but rather several groups. In this case, we might wish to pair similar individuals from different groups. For instance, in Table 11.1, suppose that subjects 1 and 2 are in one group, 3 and 4 are in a second group, and 5 and 6 are in a third group. In Table 11.4, individuals in the same group are at infinite distance from one another. The optimal match is (1,3), (2,6), and (4,5). In the optimal match, each individual is paired with an individual from another group to minimize the distance within matched pairs.

The optimal match in Table 11.4 has an additional property. In a balanced incomplete block design, each pair of treatment groups appears together in the same block with the same frequency; see [4, Chapter 11] or [7, §4.2]. The optimal match in Table 11.4 is a very small balanced incomplete block design in the sense that each of the three groups is paired with each of the other groups exactly once. There is a certain equity to a balanced incomplete block design: the number of pairs offering a direct comparison of two groups is the same for every pair of two groups. Bo Lu and I [18] showed how, with three groups, a balanced incomplete block design may be constructed using optimal nonbipartite matching. See [18] for a practical example, and see §11.2 for a few specifics.

Table 11.5 Matching with an odd number of subjects. With five subjects, a sink is added at zero distance to all subjects. One subject is paired with the sink and discarded. In this case, the optimal match is (1,2), (4,5), (3,6) with a distance of $106 + 34 + 0 = 140$ and subject 3 is discarded.

ID	1	2	3	4	5	sink
1	0	**106**	119	231	110	0
2	**106**	0	207	126	192	0
3	119	207	0	156	247	**0**
4	231	126	156	0	**34**	0
5	110	192	247	**34**	0	0
sink	0	0	**0**	0	0	0

11.2 Some Practical Aspects of Matching Without Groups

An odd number of subjects

An odd number of subjects cannot be paired. Suppose that we had only the first five subjects in Table 11.1. Then only two pairs may be constructed, with one individual discarded. The best choice of a subject to discard is the one who will make the remaining pairs as close as possible. Beginning with the 5×5 distance matrix for the first five subjects, a 'sink' is added at zero distance from all five subjects, making a 6×6 distance matrix; see Table 11.5. When optimal nonbipartite matching is applied to this 6×6 distance matrix, one subject is paired to the sink and discarded. Because any one subject may be discarded at a cost of 0, the matching discards the subject who most improves the total distance for the remaining four subjects within two pairs.

Discarding some subjects

Closer matches may be obtained by discarding some subjects. This is done by introducing one 'sink' for each subject to be discarded. The sinks are at zero distance from each subject and at infinite distance from one another. The infinite distances prevent sinks from being matched to each other. Optimal nonbipartite matching discards and pairs subjects to minimize the distance within the pairs that remain.

The process is illustrated in Table 11.6. Two sinks remove two subjects to form two pairs. As seen in this small example, discarding the subjects who are most difficult to match can leave behind closely matched pairs.

When there are several groups, as in Table 11.4, one may wish to discard a specific number of subjects from each group, perhaps two subjects from group one, seven subjects from group two, etc. To do this, one introduces sinks in the specified numbers that are at zero distance from the specified groups and at infinite distance from other groups. To discard two subjects from group one, two sinks are introduced at zero distance from subjects in group one and at infinite distance from subjects in other groups. As always, sinks are at infinite distance from one another. In this

Table 11.6 Optimal choice of two pairs from six subjects discarding two subjects. The 8 × 8 distance matrix introduces two sinks at zero distance from each subject and at infinite distance from each other. Two subjects are matched to sinks and discarded. The optimal match is (1,8), (2,7), (3,6), (4,5) with a distance of $0 + 0 + 25 + 34 = 59$. Subjects 1 and 2 discarded.

ID	1	2	3	4	5	6	sink	sink
1	0	106	119	231	110	101	0	**0**
2	106	0	207	126	192	68	**0**	0
3	119	207	0	156	247	**25**	0	0
4	231	126	156	0	**34**	67	0	0
5	110	192	247	**34**	0	212	0	0
6	101	68	**25**	67	212	0	0	0
sink	**0**	**0**	0	0	0	0	0	∞
sink	**0**	0	0	0	0	0	∞	0

way, the required subjects in each group are paired with sinks and discarded, and the pairs that remain are as close as possible; see [18] for a proof and an example.

Balanced incomplete block designs with three groups

The procedure indicated in Table 11.4 and discussed in §11.1 suffices to form matched pairs from several groups. In general, the groups will be paired to minimize the distance within pairs, and the pairing may be far from balanced. For instance, group one might always be paired with group two, while group three is always paired with group four. Whether an unbalanced comparison is undesirable depends upon the context; it may be sensible given the distribution of covariates in the various groups.

In some contexts, one may prefer to force some degree of balance, or even to have a balanced incomplete block design, in which every group is paired equally often with every other group. With three groups, it is possible to control the number of subjects from one group who are paired to subjects from another group, within the limits of arithmetic. Indeed, with the three groups, by simple algebra, the number of subjects used from each group determines the number of subjects in one group paired with another group; see [18, Expression (2.1)]. For instance, in the example in [18, §3.2], 56 subjects were retained in each of three groups, which produced a balanced incomplete block design of minimum distance with 28 pairs of group one with group two, 28 pairs of group one with group three, and 28 pairs of group two with group three.

Propensity scores for several groups

There are several generalizations of the propensity score for use with doses or with several groups. These generalizations depend upon additional assumptions and yield a variety of properties. One generalization uses an ordinal logit model for ordered doses, and matching on the linear portion of this score tends to balance observed covariates when the model is correct; see [15, page 331]. Scores of this form have been used in nonbipartite matching [17, §2.3]. Another generalization uses several binary logit models as a device for weighting observations [14]. Multivariate models yielding multidimensional scores are also possible; see [15, page 331] and [13].

11.3 Matching with Doses and Two Control Groups

11.3.1 Does the minimum wage reduce employment?

The example uses data from David Card and Alan Krueger's [2, 3] study of the effects on employment of increasing the minimum wage; see also §4.5. It is often said by economists that minimum wage laws harm the people the laws are intended to benefit. For instance, in 1946, George Stigler [26] wrote:

> Each worker receives the value of his marginal product under competition. If a minimum wage is effective, it must therefore have one of two effects: first, workers whose services are worth less than the minimum wage are discharged ...; or, second the productivity of low-efficiency workers is increased.

After arguing that the second effect is not plausible, Stigler continues: "The higher the minimum wage, the greater will be the number of covered workers who are discharged." To my mind, Stigler is discussing a 'reason for an effect,' in the sense of §4.6, rather than presenting direct evidence of an effect. For a modern textbook discussion of the minimum wage, see Pierre Cahuc and André Zylberberg [1, §12.1].

In one of the many empirical studies of the effects of the minimum wage on employment, Card and Krueger [2] looked at changes in employment at fast food restaurants, such as Burger King and Kentucky Fried Chicken, in the adjacent U.S. states of New Jersey and Pennsylvania after New Jersey increased its state minimum wage from $4.25 to $5.05 per hour on 1 April 1992. Did the increase in the minimum wage in New Jersey reduce employment at fast food restaurants?

The increase in the minimum wage in New Jersey from $4.25 to $5.05 is expected to have its largest effect on New Jersey restaurants whose starting wage before the increase was at or near the old minimum, namely $4.25, with little or no effect on restaurants in Pennsylvania where the minimum wage did not increase, and with reduced effect on restaurants in New Jersey having a higher starting wage before the increase. For instance, a New Jersey restaurant paying $4.25 an hour before the increase would need to raise its starting wage by $0.80 whereas one paying $4.75 an hour would comply with the new law by raising its wage by $0.30; presumably, the latter restaurant is less affected by the law. The theory that the minimum wage depresses employment then yields two predictions, one about a comparison of New Jersey and Pennsylvania, the other about a comparison within New Jersey.

11.3.2 Optimal matching to form two independent comparisons

There are 351 restaurants with both starting wages and employment data both before and after the wage increase. They will be divided into 175 pairs of two restaurants, discarding one restaurant, where $351 = 2 \times 175 + 1$. In 65 pairs, there is one Pennsylvania restaurant and one New Jersey restaurant. In the remaining 110 pairs, there are two New Jersey restaurants, one with a low starting wage before the increase, the other with a high starting wage before the increase.

The matching used a 351×351 distance matrix comparing every restaurant to every other restaurant. The distances were a combination of distances on covariates and penalties to force an appropriate comparison. There were five covariates: three binary indicators to distinguish the four restaurant chains (BK = Burger King, KFC = Kentucky Fried Chicken, RR = Roy Rogers, W = Wendy's), a binary indicator of whether or not the restaurant was company owned, and the number of hours the restaurant was open each day.[2] The rank based Mahalanobis distance was calculated from these five covariates; see §8.3.

A 66×66 submatrix of the 351×351 distance matrix refers to distances among Pennsylvania restaurants. We do not wish to match Pennsylvania restaurants to one another, so the entries in this 66×66 submatrix were increased by adding a large number, specifically 20 times the largest of the covariate distances, $\mathtt{m} = 37.44$, or $20 \times 37.44 = 748.8$. Any sufficiently large penalty would suffice.

If we match a Pennsylvania restaurant to a New Jersey restaurant, we would like the two matched restaurants to have similar starting wages before the wage increase. A caliper of \$0.20 was used, and it was implemented with a penalty function; see §8.4. Specifically, if a New Jersey restaurant k and a Pennsylvania restaurant ℓ had starting wages before the increase of w_k and w_ℓ, respectively, then the distance between them was increased by adding the penalty $100 \times \mathtt{m} \times \max(0, |w_k - w_\ell| - 0.2)$. If $|w_k - w_\ell| \leq \$0.20$, the penalty is zero. If $|w_k - w_\ell| = \$0.21$, the penalty is $|w_k - w_\ell| = 100 \times 37.44 \times 0.01 = 37.4$, whereas if $|w_k - w_\ell| = \$0.25$, the penalty is $100 \times 37.44 \times 0.05 = 187.2$. As discussed in §8.4, the advantage of a graduated penalty function over a single large penalty is that slight violations of the penalty

---

[2] Card and Krueger's [2, 3] data are at http://www.irs.princeton.edu/links/. Restaurants were interviewed twice, once in February 1992, before New Jersey increased its minimum wage, and once in November 1992, after New Jersey increased its minimum wage. Card and Krueger define full-time equivalent employment as the number of managers plus the number of full-time employees plus half the number of part-time employees, which is NMGRS+EMPFT+EMPPT/2 in their first interview and NMGRS2+EMPFT2+EMPPT2/2 in the second, and the change in employment is the difference between these two quantities, November−minus−February. The price of a full meal refers to the sum of the price of a soda, fries and an entree, or PSODA+PFRY+PENTREE in the first interview and PSODA2+PFRY2+PENTREE2 in the second, and the change in price is the difference of these two quantities. The starting wage before the increase in the minimum wage is WAGE_ST from the first interview. Other variables used are the restaurant chain (CHAIN), whether the restaurant was company-owned (CO_OWNED), and hours open (HRSOPEN). There are 410 restaurants, but the analysis here uses the 351 restaurants with complete data on employment and starting wage. The variable SHEET is Card and Krueger's identification number for a restaurant.

11.3 Matching with Doses and Two Control Groups

Table 11.7 Distance matrix for 6 restaurants. Covariates are: (i) Chain = restaurant chain (BK = Burger King, KFC = Kentucky Fried Chicken, RR = Roy Rogers, W = Wendy's); CO = company owned (Yes, No); and HRS = number of hours open per day. Other variables are State (NJ = New Jersey, PA = Pennsylvania); Wage = starting wage before the wage increase in dollars per hour; and Sheet = Card and Krueger's identification number. The last six columns contain the distances between these six restaurants.

Chain	CO	HRS	State	Wage	Sheet	301	310	477	434	208	253
KFC	Yes	10.5	NJ	5.00	301	-	7.4	2067.6	2065.5	233.5	698.0
W	No	11.5	NJ	4.25	310	7.4	-	7.3	4.7	254.8	6.4
BK	No	16.5	PA	4.25	477	2067.6	7.3	-	749.5	641.2	1583.1
BK	No	16.0	PA	4.25	434	2065.5	4.7	749.5	-	642.5	1579.7
RR	Yes	17.0	NJ	4.62	208	233.5	254.8	641.2	642.5	-	475.7
RR	Yes	13.0	NJ	4.87	253	698.0	6.4	1583.1	1579.7	475.7	-

are preferred to large violations. When the matching was completed, among the 65 matched pairs of a restaurant in New Jersey with a restaurant in Pennsylvania, the median difference, $w_k - w_\ell$, NJ−minus−PA, was \$0.00, the median absolute difference, $|w_k - w_\ell|$ was \$0.00, the mean absolute difference was \$0.07, and the maximum absolute difference was \$0.25, so the \$0.20 caliper was violated a few times, but not by much.

If we match a New Jersey restaurant to another New Jersey restaurant, we would like the two matched restaurants to have had very different starting wages before the increase. This too was implemented using a penalty function, with a penalty for a difference less than \$0.50. For two New Jersey restaurants, k and ℓ, with starting wages before the increase of w_k and w_ℓ, the distance between them was increased by adding the penalty $50 \times \max(0, 0.5 - |w_k - w_\ell|)$. If $|w_k - w_\ell| \geq \$0.50$, the penalty is zero, but if $|w_k - w_\ell| = \$0.40$, the penalty is 187.2. When the matching was completed, among the 110 pairs of a high-wage and a low-wage New Jersey restaurant, the high−minus−low differences in starting wages before the increase ranged from \$0.25 to \$1.50, with a mean of \$0.58 and a median of \$0.50.

The left side of Table 11.7 shows the baseline information for six of the 351 restaurants, while the right side of Table 11.7 shows the 6×6 portion of the 351×351 distance matrix for these six restaurants. Restaurants with Sheet or identification numbers 301 and 310 are at a distance of 7.4, with no penalties applied, because they are two New Jersey restaurants with very different starting wages before the increase; however, they are not close on the covariates because one is a KFC, the other a Wendy's, one is company owned and the other is not. A large penalty prevents restaurants 301 and 477 from being matched because one is in New Jersey, the other in Pennsylvania, and their starting wages differ by \$0.80. And so on.

An odd number of restaurants cannot be paired. The 351×351 distance matrix is increased in size to 352×352 by adding a row and column of zeros. The added 'sink' is at a distance of zero from all actual restaurants. One restaurant — the one restaurant that is hardest to match — is matched to the sink for a cost of zero, and the remaining 350 restaurants are paired into 175 pairs. If 51 sinks were added instead of one sink, with zero distances between restaurants and sinks and infinite distances between one sink and another, then optimal nonbipartite matching would

Table 11.8 Balance on one covariate, hours open per day, in All 175 pairs of a more affected and a less affected restaurant, in 65 NJ-vs.-PA pairs, and in 110 NJ-vs.-NJ pairs. In addition to this one covariate, the pairs were also matched for restaurant chain, with 88% matched exactly, and for company owned, with 93% matched exactly. The NJ-vs.-PA pairs had similar starting wages before the increase in the NJ minimum wage. The NJ-vs.-NJ pairs had quite different starting wages before the increase, the less affected restaurant having a higher starting wage.

Group	Type	Minimum	Lower Quartile	Median	Upper Quartile	Maximum
More Affected	All	7.0	11.5	15.5	16.0	24.0
Less Affected	All	8.0	12.0	16.0	16.5	24.0
More Affected	NJ-vs.-PA	7.0	11.5	16.0	16.5	19.0
Less Affected	NJ-vs.-PA	8.0	12.0	16.0	16.5	24.0
More Affected	NJ-vs.-NJ	10.0	11.5	15.0	16.0	24.0
Less Affected	NJ-vs.-NJ	9.5	12.0	15.0	16.0	24.0

Table 11.9 Comparison of starting wages in New Jersey, before the increase in the minimum wage, for the 110 pairs of a low-wage (more affected) restaurant and a high-wage (less affected) restaurant. The minimum wage was increased to $5.05, so the less affected restaurants were, on average, forced to raise the starting wage by $0.14, while the more affected restaurants were, on average, forced to raise the starting wage by $0.72. If raising the minimum wage reduces employment, one would expect it to have a larger effect on the more affected restaurants.

Group	Type	Mean	Min	Lower Quartile	Median	Upper Quartile	Max
More Affected	NJ-vs.-NJ	4.33	4.25	4.25	4.25	4.5	4.50
Less Affected	NJ-vs.-NJ	4.91	4.50	4.75	4.83	5.0	5.75
Difference	NJ-vs.-NJ	0.58	0.24	0.50	0.58	0.50	1.25

pair 51 restaurants with sinks, leaving 300 restaurants to be paired with each other to form 150 pairs. Use of additional sinks would discard some restaurants, which would be unfortunate, but it would produce closer matches on the pairs that remain. In the current example, one sink was used and one restaurant was discarded.[3]

The optimal nonbipartite matching algorithm was applied to the 352 × 352 distance matrix to pair the restaurants to minimize the total distance within pairs. Implicitly, the algorithm decides: (i) which one restaurant to discard, (ii) which New Jersey restaurants should be paired with Pennsylvania restaurants and which should be paired with other New Jersey restaurants, (iii) which individual restaurants should be paired.

For the optimal match, Table 11.8 describes the covariates. In Table 11.8, 'more affected' refers to the New Jersey restaurant in a NJ-vs.-PA pair, and it refers to the restaurant with the lower starting wage before the increase in a NJ-vs.-NJ pair. Of course, 'less affected' refers to the other restaurant in the same pair. In 88% of the pairs, the restaurants were the same restaurant chain, and in 93% of the pairs, they were the same in terms of company ownership. In addition, the distribution of hours open was similar, as seen in Table 11.8.

For the optimal match, Figure 11.1 and Table 11.9 show the distribution of starting wages before the wage increase. As was intended, the distribution of starting

---

[3] As it turned out, the discarded restaurant was a company-owned Pennsylvania KFC open 10 hours per day and paying a starting wage before the increase of $5.25. It was Sheet #481.

11.3 Matching with Doses and Two Control Groups

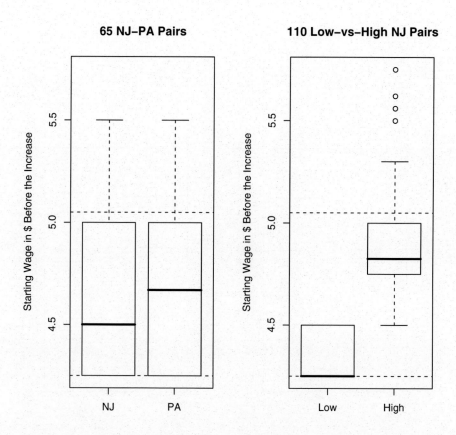

Fig. 11.1 Starting wages in dollars per hour in February 1992, before New Jersey increased its minimum wage on April 1, 1992. In the 65 pairs of a New Jersey (NJ) and a Pennsylvania (PA) restaurant, restaurants were matched to have similar starting wages, restaurant chains, company ownership, and hours open in February 1992. In the 110 pairs of two New Jersey restaurants, the matching selected one restaurant with a low starting wage and another with a high starting wage, so the low-wage restaurant was required to raise wages by a larger amount to comply with the new NJ minimum wage. The horizontal dashed lines are at the old minimum wage of $4.25 and the new minimum wage of $5.05.

wages was quite similar in the 65 NJ-vs.-PA pairs, and it was very different in the 110 NJ-vs.-NJ pairs. On average, the more affected restaurant in a NJ-vs.-NJ pair would have to raise the starting wage by $0.73 to comply with the increase in the minimum wage to $5.05, whereas the less affected restaurant would need an increase of just $0.14.

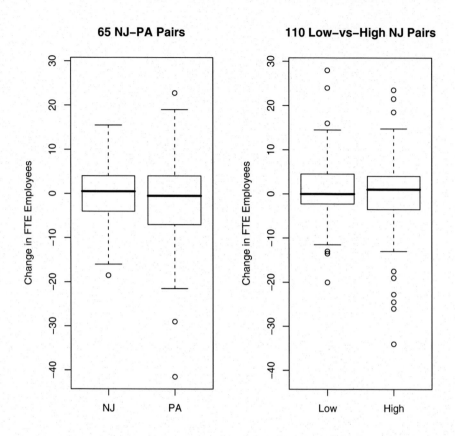

Fig. 11.2 Change in full-time-equivalent employment (FTE), after−minus−before, from before the increase in New Jersey's minimum wage to after the increase. If increasing the minimum wage tended to decrease employment at fast-food restaurants, one might reasonably expect to see larger declines in employment in the 65 New Jersey restaurants in the 65 NJ-vs.-PA pairs and in the 110 lower-wage New Jersey restaurants in the 110 NJ-vs.-NJ pairs; however, there is no sign of this.

11.3.3 Difference in change in employment with 2 control groups

Did the increase in New Jersey's minimum wage reduce employment in more affected restaurants in the 175 pairs? The outcome is the change in full-time-equivalent employment (FTE) following the wage increase, after−minus−before. In the discussion that follows, a 'worker' or an 'employee' refers to one FTE worker, even if that means two people each working half time. The typical restaurant had about 20 workers. A change of −1 means one less worker in a restaurant after the wage increase. Table 11.10 and Figure 11.2 display the distribution of changes. A few of the changes look implausibly large, whether positive or negative, perhaps

11.3 Matching with Doses and Two Control Groups

Table 11.10 Change in full-time-equivalent employment, after-minus-before, in All 175 pairs of a more affected and a less affected restaurant, in 65 NJ-vs.-PA pairs, and in 110 NJ-vs.-NJ pairs. The NJ-vs.-PA pairs had similar starting wages before the increase in the NJ minimum wage. The NJ-vs.-NJ pairs had quite different starting wages before the increase, the less affected restaurant having a higher starting wage.

Group	Type	Minimum	Lower Quartile	Median	Upper Quartile	Maximum
More Affected	All	−20.0	−3.0	0.0	4.2	28.0
Less Affected	All	−41.5	−5.0	0.0	4.0	23.5
More Affected	NJ-vs.-PA	−18.5	−4.0	0.5	4.0	15.5
Less Affected	NJ-vs.-PA	−41.5	−7.0	−0.5	4.0	22.8
More Affected	NJ-vs.-NJ	−20.0	−2.2	0.0	4.4	28.0
Less Affected	NJ-vs.-NJ	−34.0	−3.5	1.0	4.0	23.5

Table 11.11 Difference-in-differences estimates for FTE employment and the price of a full meal. The matched pair difference in after-minus-before changes is examined using Wilcoxon's signed rank test and the associated Hodges-Lehmann point estimate and confidence interval. The point estimates suggest an increase, not the predicted decrease, in employment in more affected restaurants, but none of the differences in employment is significantly different from zero at the 0.05 level. There is slight evidence for an increase in the price of a full meal, but it, too, is not significantly different from zero at the 0.05 level, and the magnitude of the increase in price is estimated to be less than 5 cents.

Outcome	Type	HL-estimate	95% CI	P-value
Employment	All	1.00	[−0.62, 2.63]	0.22
Employment	NJ-vs.-PA	1.25	[−1.38, 4.25]	0.40
Employment	NJ-vs.-NJ	0.88	[−1.00, 2.87]	0.38
Price	All	0.040	[−0.005, 0.085]	0.07
Price	NJ-vs.-PA	0.025	[−0.045, 0.115]	0.47
Price	NJ-vs.-NJ	0.045	[−0.005, 0.105]	0.08

due to some form of measurement error, but most values are plausible. The median change in both more and less affected restaurants was zero.

Table 11.11 compares the changes in more versus less affected restaurants. These are differences-in-differences: matched pair differences, more-minus-less affected restaurants, in the change, after-minus-before. The upper half of Table 11.11 refers to employment and the lower half to the price of a full meal, meaning soda, fries, and entree. Inferences use Wilcoxon's signed rank test, the associated confidence interval for an additive effect, and the Hodges-Lehmann point estimate of an additive effect; see Chapter 2. Although Stigler [26] predicted that increases in the minimum wage reduce employment, the point estimates for employment in Table 11.11 are positive, not negative, and none is significantly different from zero at the 0.05 level. In the 'All' pairs comparison, a 1-employee decline falls outside the 95% confidence interval. There is slight evidence of a small increase in prices in more affected restaurants, but the point estimate is less than five cents, and it is not significantly different from zero at the 0.05 level.

11.4 Further Reading

Statistical uses and applications of optimal nonbipartite matching are discussed in [11, 17, 18, 19, 23, 24, 25]. As these references indicate, Bo Lu has played an important role in advancing the use of nonbipartite matching in statistics; see [11, 17, 18, 19] and Note 1. The specific method in §11.3 illustrates both matching with doses [17] and matching with two control groups [18]. Optimal nonbipartite matching is also useful in 'risk-set' matching which is discussed in Chapter 12 and in [16, 19, 24, 25]. Nonbipartite matching permits matching before randomization in randomized experiments [11]. Tapered matching is discussed by Shoshana Daniel and colleagues [8]; it has been used for research into the causes of disparities.

Jack Edmonds [10] developed some of the results used in algorithms for optimal nonbipartite matching. See [21, §11.3] and [5, §5.3] for textbook discussions and [6] for a survey article. Ulrich Derigs [9] provides an implementation in Fortran that Bo Lu [19] has made accessible from within the statistical package R. An implementation in C is also available [6].

Section 11.3 uses data and ideas from the fine study by Card and Krueger [2].

References

1. Cahuc, P., Zylberberg, A.: Labor Economics. Cambridge, MA: MIT Press (2004)
2. Card, D., Krueger, A.B.: Minimum wages and employment: A case study of the fast-food industry in New Jersey and Pennsylvania. Am Econ Rev **84**, 772–793 (1994)
3. Card, D., Krueger, A.B.: Myth and Measurement: The New Economics of the Minimum Wage. Princeton, NJ: Princeton University Press (1995) Data: http://www.irs.princeton.edu/
4. Cochran, W.G., Cox, G.M.: Experimental Designs. New York: Wiley (1957)
5. Cook, W.J., Cunningham, W.H., Pulleyblank, W.R., Schrijver, A.: Combinatorial Optimization. New York: Wiley (1998)
6. Cook, W., Rohe, A.: Computing minimum-weight perfect matchings. INFORMS J Comput **11**, 138–148 (1999) Software: http://www2.isye.gatech.edu/~wcook/
7. Cox, D.R., Reid, N.: The Theory of the Design of Experiments. New York: Chapman and Hall/CRC (2000)
8. Daniel, S., Armstrong, K., Silber, J.H., Rosenbaum, P.R.: An algorithm for optimal tapered matching, with application to disparities in survival. J Comput Graph Statist **17**, 914–924 (2008)
9. Derigs, U.: Solving nonbipartite matching problems by shortest path techniques. Ann Operat Res **13**, 225–261 (1988)
10. Edmonds, J.: Matching and a polyhedron with 0-1 vertices. J Res Nat Bur Stand **65B**, 125–130 (1965)
11. Greevy, R., Lu, B., Silber, J.H., Rosenbaum, P.R.: Optimal matching before randomization. Biostatistics **5**, 263–275 (2004)
12. Hornik, R., Jacobsohn, L., Orwin, R., Piesse, A., Kalton, G.: Effects of the national youth anti-drug media campaign on youths. Am J Public Health **98**, 2229–2236 (2008)
13. Imai, K., van Dyk, D.A.: Causal inference with general treatment regimes: generalizing the propensity score. J Am Statist Assoc **99**, 854–866 (2004)
14. Imbens, G.W.: The role of the propensity score in estimating dose-response functions. Biometrika **87**, 706–710 (2000)

15. Joffe, M.M., Rosenbaum, P.R.: Propensity scores. Am J Epidemiol **150**, 327–333 (1999)
16. Li, Y.F.P., Propert, K.J., Rosenbaum, P.R.: Balanced risk set matching. J Am Statist Assoc **96**, 870–882 (2001)
17. Lu, B., Zanutto, E., Hornik, R., Rosenbaum, P.R.: .Matching with doses in an observational study of a media campaign against drug abuse. J Am Statist Assoc **96**, 1245–1253 (2001)
18. Lu, B., Rosenbaum, P.R.: Optimal matching with two control groups. J Comput Graph Statist **13**, 422–434 (2004)
19. Lu, B.: Propensity score matching with time-dependent covariates. Biometrics **61**, 721–728 (2005) http://cph.osu.edu/divisions/biostatistics/
20. Lu, B., Greevy, R., Xu, X., Beck, C.: Optimal nonbipartite matching and its statistical applications. Manuscript. (2009)
21. Papadimitriou, C.H., Steiglitz, K.: Combinatorial Optimization: Algorithms and Complexity. Englewood Cliffs, NJ: Prentice Hall (1982)
22. R Development Core Team.: R: A Language and Environment for Statistical Computing. Vienna: R Foundation, http://www.R-project.org (2007)
23. Rosenbaum, P.R.: An exact, distribution free test comparing two multivariate distributions based on adjacency. J Royal Statist Soc B **67**, 515–530 (2005)
24. Rosenbaum, P.R., Silber, J.H.: Sensitivity analysis for equivalence and difference in an observational study of neonatal intensive care units. J Am Statist Assoc **104**, 501–511 (2009)
25. Silber, J.H., Lorch, S.L., Rosenbaum, P.R., Medoff-Cooper, B., Bakewell-Sachs, S., Millman, A., Mi, L., Even-Shoshan, O., Escobar, G.E.: Additional maturity at discharge and subsequent health care costs. Health Serv Res **44**, 444–463 (2009)
26. Stigler, G.J.: The economics of minimum wage legislation. Am Econ Rev **36**, 358–365 (1946)

Chapter 12
Risk-Set Matching

Abstract When a treatment may be given at various times, it is important to form matched pairs or sets in which subjects are similar prior to treatment but avoid matching on events that were subsequent to treatment. This is done using risk-set matching, in which a newly treated subject at time t is matched to one or more controls who are not yet treated at time t based on covariate information describing subjects prior to time t.

12.1 Does Cardiac Transplantation Prolong Life?

In the late 1960s and early 1970s, heart transplantation was a new surgical procedure. In 1972, Mitchell Gail [5] published a brief but influential and insightful critique of two empirical studies from that time that evaluated cardiac transplantation. Hearts were not immediately available for transplantation; a candidate for a heart had to wait until one became available. In "Does cardiac transplantation prolong life? A reassessment," Gail wrote:

> Two recent reports of experience with cardiac transplantation conclude that this procedure appears to prolong life ... [however] the observed differences can be explained in terms of probable selection bias.... In both [studies] the patient is assigned to the nontransplant group by default. That is, a potential transplant recipient becomes a member of the nontransplant group because no suitable donor becomes available before the potential recipient dies ... [T]his assignment method biases the results in favor to the transplanted group. [5, page 815] ... [T]he survival time of the nontransplanted group is shorter than would have been observed with a random assignment method, because the method used assigns an unfair proportion of the sicker patients to this group ... [and] the survival time of the transplanted group is longer than would have been observed with a random assignment method for two reasons. First, an unfairly large number of good risk patients have been assigned to this group, introducing a bias. Second, the patients in this group are guaranteed (by definition) to have survived at least until a donor was available, and this grace period has been implicitly added into the survival time of the transplanted group. [5, page 816]

If you die early, before a heart becomes available, you are a 'control.' If you survive long enough for a heart to become available for you, then you become a treated subject. In an experiment, exclusion criteria are applied before random treatment assignment, ensuring that the same exclusion criteria are applied to treated and control groups (§1.2.5), but here, dying early causes you to be included in the control group and excluded from the treatment group. The term "immortal time bias" has sometimes been used to describe this situation [20]; see also [17].

The difficulty is created by the delay, waiting for a suitable heart. Gail then suggested a design for a randomized experiment that removes this difficulty. When a heart becomes available, the two living individuals most compatible with that heart are identified, paired, and then one is selected at random to receive it, the other becoming a control, with survival measured from that date of randomization [5, page 817]. This randomized experiment addresses the problem that hearts are not immediately available, yet creates an appropriate randomized control group, one that could be used in randomization inference.

A similar but slightly different randomized experiment would begin as Gail's hypothetical experiment begins, pairing individuals when a heart becomes available, assigning the heart at random to one member of the pair; however, unlike Gail's experiment, the control, who did not receive the newly available heart, would remain eligible to receive a suitable heart if one became available at a later time.

In general, this slightly different experiment would estimate the effect of delay, that is, the effect of treating now as opposed to waiting and possibly treating later. There are contexts, common in medicine, in which this is the practical choice: treat now or wait and see; there is the option of treating later if treatment is delayed. Is treating immediately advantageous? Does it improve outcomes? Or is waiting better? Would delay lead to a better decision about who to treat? Would delay eliminate unnecessary treatment without harming outcomes? Or is treatment inevitable? If treatment is inevitable, is delay pointless or harmful? A randomized experiment of this form could answer such questions. Risk set matching creates an observational study analogous to this randomized experiment.

12.2 Risk-Set Matching in a Study of Surgery for Interstitial Cystitis

Interstitial cystitis (IC) is a chronic urologic disorder characterized by bladder pain and irritative voiding, similar to the symptoms of urinary tract infection but without evidence of infection. In an effort to better understand the disorder and its treatment, the National Institute of Diabetes, Digestive and Kidney Diseases created the Interstitial Cystitis Data Base (ICDB) [15]. In [9], Paul Li, Kathleen Propert, and I proposed risk-set matching as a method for studying the effects of a surgical intervention, cystoscopy and hydrodistention, on symptoms of interstitial cystitis using data from the ICDB. Aspects of this study are described here.

12.2 Risk-Set Matching in a Study of Surgery for Interstitial Cystitis

To enter the ICDB, a patient must have symptoms of IC for at least the previous six months. Patients are evaluated upon entry into the ICDB and at approximately three month intervals thereafter. Three quantities were measured regularly: pain and urgency, both recorded on scale of 0 to 9, where higher numbers indicate greater intensity; and nocturnal frequency of voiding. Periodically, an additional patient was treated with the surgical procedure, cystoscopy and hydrodistention. For brevity in what follows, cystoscopy and hydrodistention is described as 'surgery' or 'the treatment.'

Patients were not selected at random for surgery. Presumably, a patient who finds current symptoms to be intolerable is more likely to opt for surgery. This presents a problem. One cannot reasonably compare all patients in the ICDB who received surgery with all those who did not, because to know that a patient never received surgery is to have reason to suspect the patient's symptoms never became intolerable. Implicitly, to know that a patient never received surgery is to know a bit about the patient's entire course of symptoms. We would like to create pairs of patients who were similar up to the moment that one of them received surgery, but without any knowledge of their subsequent symptoms. Matching should make pairs comparable prior to treatment; what happens later is an outcome. Therefore, a new surgical patient is paired with a control patient with similar symptoms up to the point of surgery for the surgical patient; however, this control may receive surgery at a later time. This paired comparison estimates the effect of surgery now versus delaying surgery into the indefinite future, with the possibility that surgery will never be performed. It estimates the effect of the choice that patients and surgeons keep facing.

Whenever a patient received surgery, that patient was paired with a patient who had not yet received surgery but who had similar symptoms at baseline, upon entry into the ICDB, and also at the time of surgery for the surgical patient. Based on their symptoms at baseline and just prior to surgery, it would have been difficult to guess which patient would receive surgery because their symptoms at that time were similar. By definition, this control patient did not receive surgery in the three month recording interval following surgery for the treated patient, but the control might have surgery at any time thereafter.[1]

Figure 12.1 depicts both the baseline symptoms used for matching and the same symptoms six months after the date of surgery for the surgical patient. In Figure 12.1, baseline refers to the time of entry into the ICDB, time 0 months refers to the time just prior to surgery for the surgical patient, and 3 months refers to three months after surgery. For instance, if a patient was treated surgically nine months after entry into the ICDB, then for the pair that includes this patient, time 0 is nine months after entry into the ICDB, and time 3 months is 12 months after entry into the ICDB for both patients. In Figure 12.1, the matching appears to have been successful in the specific sense that surgical patients and their matched risk-set controls

---

[1] The matching algorithm used a distance as in §8.3, but with an important change. The distance between a patient who received surgery at time t and a patient who had not yet received surgery at time t was computed from the covariates for these two patients up to time t, without reference to information obtained after time t.

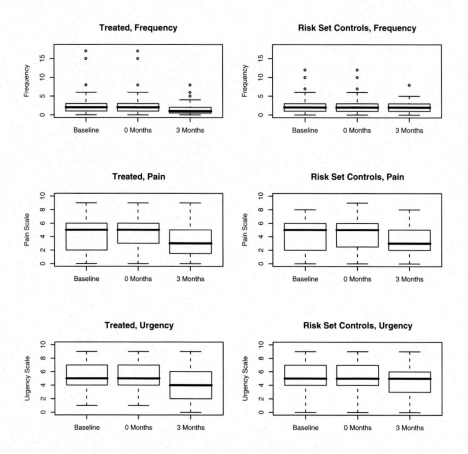

Fig. 12.1 Frequency, urgency, and pain at baseline, at treatment, and three months after treatment for treated patients and their matched not-yet-treated risk-set controls.

had similar distributions of symptoms at baseline and prior to the surgical patient's date of surgery. There is a very slight uptick in the distribution of pain prior to surgery, but matching has ensured that the uptick is present for both surgical patients and matched controls. Indeed, the symptoms of the two matched individuals were similar over time. Specifically, the correlations between the symptoms of the two matched individuals were high: at baseline, for frequency, pain, and urgency, the Spearman rank correlations were, respectively, 0.92, 0.94, and 0.87, whereas at time 0 they were 0.92, 0.91 and 0.92.

In Figure 12.1, pain and urgency scores improved three months after surgery for surgical patients, but they improved by just about the same amount for matched controls. In Figure 12.1, frequency improved somewhat for surgical patients but not for controls.

12.2 Risk-Set Matching in a Study of Surgery for Interstitial Cystitis

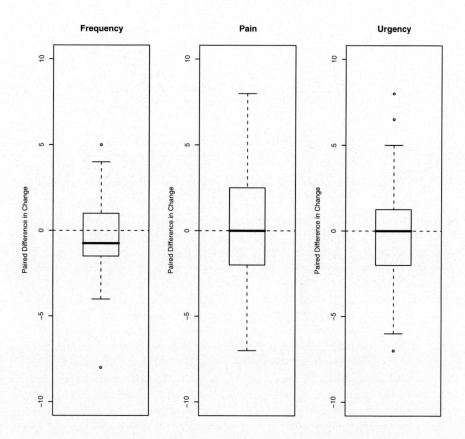

Fig. 12.2 Treated-minus-control difference in changes in frequency, urgency and pain.

Figure 12.2 sharpens this comparison. For each of the three symptoms, frequency, pain, and urgency, for each patient, the change is computed between the measurement at three months and the average of the measurements at baseline and at time zero. Figure 12.2 depicts the treated-minus-control matched pair difference in these differences. For pain and urgency, the median difference in the changes is zero, suggesting little effect of surgery. For frequency, the median difference in the changes is -0.75, a fairly small change. Table 12.1 applies Wilcoxon's signed rank test to the matched pair differences in changes, where the difference in frequency is significantly different from zero at the 0.05 level.

Table 12.1 Inferences about the matched pair differences in changes in IC symptoms. The two-sided P-value is from Wilcoxon's signed rank test and tests the null hypothesis that the differences are symmetric about zero. Associated with the signed rank test is the Hodges-Lehmann (HL) point estimate of the center of symmetry and the 95% confidence interval (CI) formed by inverting the test.

	Frequency	Pain	Urgency
P-value	0.004	0.688	0.249
HL-estimate	−0.50	0.00	−0.50
95% CI	[−1.00, 0.00)	[−0.50, 0.75]	[−1.0, 0.25]

12.3 Maturity at Discharge from a Neonatal Intensive Care Unit

A premature baby is kept in the neonatal intensive care unit (NICU) until it has matured sufficiently to go home [1]. Once babies have matured sufficiently to go home, they typically stay a few more days, just to be sure, and the number of days included in these 'few more days' can vary considerably from one baby to the next. Recently, Jeffrey Silber, Scott Lorch, Barbara Medoff-Cooper, Susan Bakewell-Sachs, Andrea Millman, Lanyu Mi, Orit Even-Shoshan, Gabriel Escobar, and I asked [18, 19]: Does a longer stay in the NICU after maturity benefit the babies who receive it?

We looked at 1402 premature infants born in five hospitals of the Northern California Kaiser-Permanente Medical Care Program between 1998 and 2002. The relevant age for a premature baby is not age from birth but rather postmenstrual age or PMA, and gestational age refers to the age at birth. All 1402 babies in this study were born with a gestational age of 34 weeks or less and were discharged from the hospital alive. Using risk-set matching [9, 10], we divided the 1402 babies into 701 pairs of two babies, an 'early baby' and a 'late baby,' in such a way that the two babies were similar on the day, the PMA, that the early baby was discharged, but the late baby stayed a few more days.[2] As anticipated, the late baby was not only older on its day of discharge but also heavier and had maintained various measures of maturity for a longer period of time. Did the late baby benefit from a few more days to grow older, heavier, and more mature inside the NICU? Or would the quite substantial cost of those extra days be better spent improving outpatient services for such babies?

The matching balanced the variables in Table 12.2. In Table 12.2, the first three groups of variables do not vary with time. In consequence, their values are the same in the last two columns of Table 12.2. The first group of covariates describes the babies at birth. On average, they were 31 weeks old at birth and weighed about 1.7 kilograms (3.75 pounds). The second group of variables describes the baby's history of significant health problems. The third group of covariates describes mom.

The fourth group of variables measures maturity, so they do vary with time; they differ in the last two columns of Table 12.2. Aspects of maturity include: (i) maintenance of body temperature, (ii) coordinated sucking, (iii) sustained weight gain, and (iv) maturity of cardiorespiratory function. Six dimensions of maturity were scored

---

[2] The matching was an optimal nonbipartite matching; see Chapter 11.

12.3 Maturity at Discharge from a Neonatal Intensive Care Unit

Table 12.2 Balance on fixed and time-dependent covariates after risk-set matching for 1402 premature babies in 701 matched pairs. Matching ensured that paired babies were similar on the day the early baby was discharged home from the neonatal intensive care unit, but the late baby was more mature (older, heavier) on the later day of discharge for the late baby. Of course, the fixed covariates are the same on both days; only the time-dependent covariates change.

Covariate Group	Covariate	Early Baby at Early Baby Discharge	Late Baby at Early Baby Discharge	Late Baby at Late Baby Discharge
	Number of Babies	701	701	701
Baby at Birth (fixed)	Gestational Age (weeks) at birth	31.1	31.1	31.1
	Weight at birth (grams)	1669	1686	1686
	SNAP-II 20 to 59	0.15	0.13	0.13
	SNAP-II 10 to 19	0.18	0.20	0.20
	SNAP-II 0 to 9	0.67	0.67	0.67
	Male Sex	0.51	0.52	0.52
Baby's Health History (fixed)	Bronchopulmonary Dysplasia	0.09	0.11	0.11
	Necrotizing Enterocolitis	0.01	0.01	0.01
	Retinopathy Stage ≥ 2	0.06	0.06	0.06
	Intraventricular Hemorrhage ≥ 3	0.02	0.01	0.01
Mom (fixed)	Maternal Age (years)	29.9	30.3	30.3
	Marital Status Single	0.24	0.24	0.24
	Other children = 0	0.40	0.37	0.37
	Other children = 1	0.34	0.37	0.37
	Other children ≥ 2	0.26	0.26	0.26
	Income $	59,517	59,460	59,460
	White Race	0.47	0.48	0.48
	Black	0.10	0.09	0.09
	Asian	0.20	0.23	0.23
	Hispanic	0.22	0.18	0.18
Baby's Time Dependent Variables	Postmenstrual Age (days)	247.4	247.4	250.9
	Propensity to discharge	0.67	0.64	1.33
	Apnea smoothed score	0.04	0.05	0.03
	Brady smoothed score	0.06	0.07	0.04
	Methyl smoothed score	0.04	0.03	0.02
	Oxygen smoothed score	0.11	0.11	0.07
	Gavage smoothed score	0.22	0.23	0.10
	Incubator smoothed score	0.15	0.15	0.08
	Combined maturity score	0.62	0.63	0.34
	Current weight	2153	2148	2231
	Current weight < 1700 grams	0.02	0.03	0.01
	$1700 \leq$ weight < 1800	0.06	0.06	0.02
	Current weight ≥ 1800 grams	0.92	0.91	0.97

daily as binary variables, 1 indicating that this dimension had not yet been achieved, and 0 indicating that it had been achieved. In almost all cases, babies had six zeros on their day of discharge, so the relevant question is how long the baby had a zero score before discharge. We measured this by applying exponential smoothing [2] to the binary variables, so that the smoothed scores were between 0 and 1, and a score near 0 indicated that the baby had achieved and maintained a zero score for quite a few days. In Table 12.2, these are the Apnea, Brady, Methyl, Oxygen, Gavage

Table 12.3 Absolute difference in covariate means in units of the standard deviation for 20 fixed covariates and 13 time-dependent covariates. For the time-dependent covariates, two values are reported, one comparing the babies on the day the early baby was discharged, the other comparing the babies on their own days of discharge. The babies are quite similar on the day the early baby was discharged, but quite different on the day of their own discharge. On their own day of discharge, three of 13 variables had absolute standardized differences above 0.6, namely the time-dependent propensity score, the combined maturity score, and the smoothed gavage score.

Quantile	min	25%	50%	75%	max
Fixed	0.00	0.01	0.04	0.06	0.09
Time-Dependent at Early Discharge	0.00	0.01	0.02	0.06	0.09
Time-Dependent at Own Discharge	0.09	0.16	0.19	0.34	0.75

and Incubator smoothed scores; notice that these are similar on the day the early baby went home, but somewhat closer to zero for the late baby on the day the late baby went home. Another time-dependent covariate is the baby's current weight, which is recorded in several forms in Table 12.2; the late baby went home weighing about 100 grams more. Also, there is the baby's current age or PMA; the late baby was about 3.5 days older at discharge. The time-dependent propensity score [9, 10] was based on Cox's [3] proportional hazards model using both the fixed and time-dependent variables; it is the linear portion or log-hazard of discharge from the model, and it varies from day to day.

Table 12.3 summarizes Table 12.2 in terms of standardized differences; see §9.1. For the covariates in Table 12.2, absolute standardized differences in means are displayed in Table 12.3. Although the babies were similar on the day the early baby was discharged, the late baby was substantially more mature on its own day of discharge. For three time-dependent variables, the difference is more than 0.6 standard deviations.

A detailed analysis of outcomes appears in [18, 19]; see also §18.2 and §19.6. Here, it suffices to say that the early and late babies had similar experiences after discharge, so there was little indication of benefit or harm from the extra days in the NICU. Of course, the extra days were quite expensive.

The matching was implemented as follows. As noted, the proportional hazards model contributed a time-dependent propensity score [9, 10]. Then a 1402 × 1402 distance matrix was computed comparing the babies to one another. If the baby in row i and the baby in column j were discharged on the same day (the same PMA), then the distance between them was infinite. Otherwise, one baby was discharged earlier than the other, and the distance in row i and column j describes both babies on the day the earlier baby was discharged. The distance used a penalty to implement a caliper on the time-dependent propensity score (§8.4) and a Mahalanobis distance using the current values of the time-dependent covariates, with small penalties to improve balance on recalcitrant variables. Optimal nonbipartite matching was applied to this distance matrix to divide the 1402 babies into 701 pairs so that the total distance within pairs was minimized; see Chapter 11. The computations used Fortran code developed by Derigs [4] that Lu [10] has made available inside R.

12.4 Joining a Gang at Age 14

Is joining a gang at age 14 a major turning point in a boy's life? Does it initiate a life of violence, a violent career? Amelia Haviland, Daniel Nagin, Richard Tremblay, and I [8] examined this question using data from the Montréal Longitudinal Study of Boys [21], which followed a cohort of boys in kindergarten in 1984 through 1995 when their average age was 17. The boys were from 53 schools in low socioeconomic areas of Montréal, Canada, and all were white, born in Canada to French-speaking parents. The data in the study are based on assessments by parents, teachers, peers, self-reports, and records from schools and the juvenile court. Violence was scored as a weighted count of violent incidents [8, page 425]. It should be mentioned at the outset that Montréal is not one of earth's most violent places, and the boys we studied were not the most violent boys in the Montréal Longitudinal Study; see [8] for a detailed description of the group under study.

The study looked at boys who had not joined a gang prior to age 14 and who were not part of a small group of exceptional boys with very high and consistent violence prior to age 14. Among these boys, 59 joined a gang at age 14. The 59 joiners were quite different from the typical boy in this group prior to age 14: they were more violent prior to age 14, less popular, more aggressive, more oppositional, more sexually active, and their mothers were younger. Each joiner (J) was matched to two controls (C) from the same group who did not join a gang at age 14. In the 59 matched sets, the 59 joiners did not differ greatly from the $2 \times 59 = 118$ matched controls in terms of the covariates. The matching used the techniques described in Chapter 8 together with Nagin's [24] latent trajectory groups fitted to violence prior to age 14. This is the simplest form of risk-set matching because the risk-set is defined at a single time, namely age 14.

Figure 12.3 depicts the results at various ages for joiners (J) and their matched controls (C). The upper portion of the figure depicts violence, while the lower portion depicts gang membership. By definition of the study group, none of these boys were in gangs at ages 12 and 13, and all of the joiners and none of the matched controls were in gangs at age 14. Because of the matching, the boys were similar in terms of violence at ages 12 and 13. For joiners, there is a step up in violence at age 14 that is smaller at age 15 and is no longer significantly different from zero at ages 16 and 17. Most of the joiners had quit the gang by age 15, and a few of the controls had joined. By ages 16 and 17, the difference in the percentage of gang membership between joiners and controls is quite small. Only four of 59 joiners were in gangs in all four years, ages 14–17.

In this cohort, joining a gang at age 14 does not appear to launch a violent career: most joiners soon quit without a permanent elevation in violence, and controls often subsequently joined and quit gangs. Appearances suggest that joining a gang has highly transient effects. A conclusion of this sort is possible only if individuals who were similar at a certain point in time are followed forward through time; that is, the groups are not defined by what happened to individuals subsequently. One could imagine an analysis that sliced boys into boy-years, counting a boy as a gang member in years of membership and as a control in other years, but such an analysis

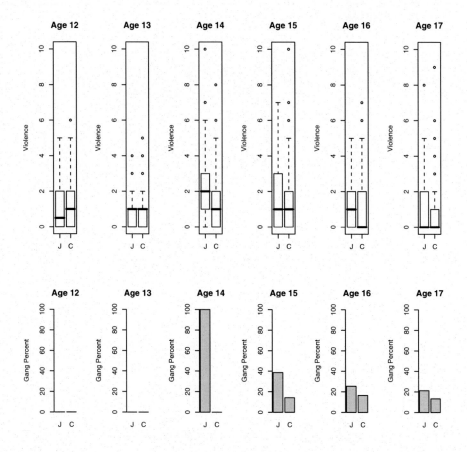

Fig. 12.3 Gang violence and gang membership for 59 boys who joined gangs for the first time at age 14 and for two risk-set controls who had not joined gangs through age 14.

could easily miss the highly transient nature of gang membership in the lives of intact boys.

12.5 Some Theory

The theory of risk-set matching is not difficult and is developed in about four pages in [9, §4]. A brief, informal sketch of this theory follows.

You are waiting for an event to occur, but you do not know when it will occur. In the most common medical applications, the event is death, but in risk-set matching, the event is the start of treatment, perhaps surgery, or discharge from a neonatal intensive care unit, or joining a gang. At any moment, given that the event has not yet occurred, there is a small chance it will occur in the next moment. Given that you are alive reading this sentence, there is a small chance you will not live to read its last word. These small chances, divided by the length of the moment, are called hazards.[3] If an event has not yet occurred, up to the current instant, for both of two people, and those two people have the same hazard of this event, and if the event occurs for exactly one of them, then each of the two people has an equal chance, $\frac{1}{2}$, of being the recipient of this event. This logic, expanded in various ways, underlies much of what is done with hazards, including much of Cox's proportional hazards model [3]. Suppose that two people are identical up to a certain time with regard to certain measured time-varying covariates, such as pain in §12.2, and suppose that the hazard of treatment depends only on these measured covariates, and suppose, finally, that exactly one of them is treated in the next moment; then they each have equal chance, $\frac{1}{2}$, of treatment, not unlike the 'slightly different experiment' discussed at the end of §12.1. That 'slightly different randomized experiment' is built in an unusual way, but the essential properties of its randomization inferences are about the same as those in Chapter 2. Under the suppositions just mentioned, exact risk-set matching for time-varying covariates recreates the randomization distribution of treatment assignments in the 'slightly different randomized experiment;' in essence, it recreates the naive model of Chapter 3. The supposition that we have measured all of the covariates that affect the hazard of treatment is no small supposition; indeed, it is not especially plausible. We might imagine that there is an unmeasured time varying-covariate that affects the hazard of treatment and that was not controlled by risk-set matching for the observed covariates. With only a little structure, it is possible to show that matching for the observed time-varying covariates then reproduces the sensitivity analysis model of Chapter 3. In other words, risk-set matching for measured covariates brings us back to the two simple models of Chapter 3. Again, the technical details are not difficult and are developed in a few pages in [9, §4].

12.6 Further Reading

Cardiac transplantation has advanced since the 1970s, but Mitchell Gail's [5] short paper is of continuing methodological interest. Risk set matching refers to matching from the set of people who are at risk of receiving the treatment. The theory of risk-set matching is related to hazard models, such as David Cox's [3] proportional hazards model, where a time-varying hazard of treatment acts in a manner similar to the propensity score; the theory is discussed in [9, §4]. Bo Lu [10] develops the balancing properties of time-dependent propensity scores, and studies their use in conjunction with nonbipartite matching. Sue Marcus, Juned Siddique,

---

[3] This is actually a calculus notion: the moments of time grow in number while becoming shorter and shorter in duration, so the chance that anything happens in such a short interval is shrinking, but the time interval is shrinking as well, so perhaps the ratio, the hazard, tends to a limit; if so, the limit is the instantaneous hazard function at a specific time.

Tom Ten Have, Robert Gibbons, Liz Stuart, and Sharon-Lise Normand [11] discuss balancing covariates in longitudinal studies. In risk-set matching, sensitivity of conclusions to bias from unmeasured time-varying covariates is discussed in [9]. The examples in §12.2–§12.4 are discussed in greater detail in [8, 9, 18, 19]. An alternative approach to matching in the Montréal Longitudinal Study of Boys is discussed in [7], where many more controls are used by matching a variable number of controls to each joiner; see also §8.5. Paul Nieuwbeerta, Daniel Nagin, and Arjan Blokland [14] used risk-set matching in a study of the possible effects of first imprisonment on the development of a criminal career. Wei Wu, Stephen West, and Jan Hughes [24] study the effects of retention in first grade by matching retained students to students who appeared to be similar at the end of first grade but were promoted to second grade, following them forward for several years. Jamie Robins and colleagues [16] have developed a general approach to treatments that evolve over time; see the book by Mark van der Laan and Jamie Robins [22] for detailed discussion of this approach.

References

1. American Academy of Pediatrics.: Hospital discharge of the high-risk neonate – proposed guidelines. Pediatrics 102, 411–17 (1998)
2. Cox, D.R.: Prediction by exponentially weighted moving averages and related methods. J R Statist Soc B **23**, 414–422 (1961)
3. Cox, D.R.: Regression models and life-tables. J Roy Statist Soc B **34**, 187–220.
4. Derigs, U.: Solving nonbipartite matching problems by shortest path techniques. Ann Operat Res **13**, 225–261 (1988)
5. Gail, M.H.: Does cardiac transplantation prolong life? A reassessment. Ann Intern Med **76**, 815–817 (1972)
6. Haviland, A.M., Nagin, D.S.: Causal inferences with group based trajectory models, Psychometrika **70**, 557–578 (2005)
7. Haviland, A.M., Nagin, D.S., Rosenbaum, P.R.: Combining propensity score matching and group-based trajectory analysis in an observational study. Psychol Methods **12**, 247–267 (2007)
8. Haviland, A.M., Nagin, D.S., Rosenbaum, P.R., Tremblay, R.E.: Combining group-based trajectory modeling and propensity score matching for causal inferences in nonexperimental longitudinal data. Dev Psychol **44**, 422–436 (2008)
9. Li, Y.F.P., Propert, K.J., Rosenbaum, P.R.: Balanced risk set matching. J Am Statist Assoc **96**, 870–882 (2001)
10. Lu, B.: Propensity score matching with time-dependent covariates. Biometrics **61**, 721–728 (2005)
11. Marcus, S.M., Siddique, J., Ten Have, T.R., Gibbons, R.D., Stuart, E., and Normand, S-L.: Balancing treatment comparisons in longitudinal studies. Psychiatric Ann **38**, 805–812 (2008)
12. Messmer, B.J., Nora, J.J., Leachman, R.D., et al: Survival-times after cardiac allografts. Lancet **1**, 954–956 (1969)
13. Nagin, D.S.: Group-Based Modeling of Development. Cambridge, MA: Harvard University Press (2005)
14. Nieuwbeerta, P., Nagin, D.S., Blokland, A.A.J.: The relationship between first imprisonment and criminal career development: A matched samples comparison. J Quant Criminol, to appear.

15. Propert. K.J., Schaeffer, A.J., Brensinger, C.M., Kusek, J.W., Nyberg, L.M., Landis, J.R.: A prospective study of interstitial cystitis: Results of longitudinal followup of the interstitial cystitis data base cohort. J Urol **163**, 1434–1439. (2000)
16. Robins, J.M., Blevins, D., Ritter, G., Wulfsohn, M.: G-estimation of the effect of prophylaxis therapy for pneumocystis carinii pneumonia on the survival of AIDS patients. Epidemiology **3**, 319–336.
17. Rosenbaum, P.R.: The consequences of adjustment for a concomitant variable that has been affected by the treatment. J Roy Statist Soc A **147**, 656–666 (1984)
18. Rosenbaum, P.R., Silber, J.H.: Sensitivity analysis for equivalence and difference in an observational study of neonatal intensive care units. J Am Statist Assoc **104**, 501–511 (2009)
19. Silber, J.H., Lorch, S.L., Rosenbaum, P.R., Medoff-Cooper, B., Bakewell-Sachs, S., Millman, A., Mi, L., Even-Shoshan, O., Escobar, G.E.: Additional maturity at discharge and subsequent health care costs. Health Serv Res **44**, 444–463 (2009)
20. Suissa, S.: Immortal time bias in pharmacoepidemiology. Am J Epidemiol **167**, 492–499 (2008)
21. Tremblay, R.E., Desmarais-Gervais, L., Gagnon, C., Charlebois, P.: The preschool behavior questionnaire: Stability of its factor structure between culture, sexes, ages, and socioeconomic classes. Int J Behav Devl **10**, 467–484 (1987)
22. van der Laan, M., Robins, J.: Unified Methods for Censored Longitudinal Data and Causality. New York: Springer (2003)
23. Wermink, H., Blokland, A., Nieubeerta, P., Nagin, D., Tollenaar, N.: Comparing the effects of community service and imprisonment on recidivism: A matched samples approach. Manuscript.
24. Wu, W., West, S.G., Hughes, J.N.: Effect of retention in first grade on children's achievement trajectories over 4 years: A piecewise growth analysis using propensity score matching. J Educ Psychol **100**, 727–740 (2008)

Chapter 13
Matching in R

Abstract The statistical package R is used to construct several matched samples from one data set. The focus is on the mechanics of using R, not on the design of observational studies. The process is made tangible by describing it in detail, closely inspecting intermediate results; however, essentially, three steps are involved, (i) creating a distance matrix, (ii) adding a propensity score caliper to the distance matrix, and (iii) finding an optimal match. One appendix contains a short introduction to R. A second appendix contains short R functions used to create distance matrices used in matching.

13.1 R

Multivariate matching is easy to do in the statistical package R, as is illustrated in the current chapter. It is convenient to use one data set to illustrate various types of matching. For that reason, and for that reason only, four matched samples will be constructed from Susan Dynarski's [9] fine study of the termination in 1982 of the Social Security Student Benefit Program, which provided tuition assistance to the children of deceased Social Security beneficiaries; this study was discussed in §1.5. The construction of one of the four matched samples is presented in step-by-step detail. Outcomes are briefly compared in §13.8, but the focus is on constructing matched control groups in R. The chapter is structured to illustrate various aspects of R, not to produce a recommended analysis for Dynarski's [9] study.

The favorite statistical package of research statisticians, R is distributed free at http://cran.r-project.org/ by the Comprehensive R Archive Network (CRAN). Although there is also free documentation at that webpage, beginners often find a book helpful, e.g., [15]. See the first appendix to this chapter for a brief introduction to R.

Matching is easy in R, largely due to the efforts of Ben Hansen [11], who created an R function, fullmatch, to do optimal matching from Fortan code created by Demetri Bertsekas [3, 4]. To use `fullmatch`, you will need to install the `optmatch` package; see the first appendix.

Statistical methodology changes slowly, but software changes rapidly. This chapter is intended to give a sense of matching in R, but R changes constantly, so you should look inside R for updates and new developments.

13.2 Data

The first match, the only one considered in detail, concerns seniors in the period 1979–1981, when the Social Security Student Benefit Program was still in operation; see the left half of Figure 1.2. During this period, a child of a deceased Social Security beneficiary might receive a substantial tuition benefit from the program.

Eight covariates are used in the match.[1] They are: family income (`faminc`), income missing (`incmiss`), black, hispanic, the Armed Forces Qualifications Test (`afqtpct`), mother's education (`edm`), mother's education missing (`edmissm`) and gender (`female`);[2] these are contained in Xb with 2820 rows and eight columns. The treatment indicator zb is a vector with a 1 for a senior whose father is deceased and 0 for other seniors. The first 20 of 2820 rows of the treatment indicator, zb and covariates Xb are displayed in Table 13.1. In R, a variable such as faminc in a data.frame such as Xb is referred to as Xb$faminc.

Mother's education but not father's education was used because one group was defined by a deceased father. The AFQT is the only educational measure among the covariates. The AFQT was missing for less than 2% of subjects, and these subjects are not used in the matching. With this restriction,[3] there are 131 high school seniors with deceased fathers and 2689 other high school seniors in the 1979–1981 cohort, before the Social Security Student Benefit Program ended in 1982; see Table 13.2. Also, $131 + 2689 = 2820$. In consequence, R finds:

---

[1] Dynarski [9, Table 2] presents two analyses, one with no covariates, the other with many more covariates, obtaining similar estimates of effect from both analyses. That is a reasonable approach in the context of her paper. In constructing a matched control group in this chapter, I have omitted a few of the covariates that Dynarski used, including 'single-parent household' and 'father attended college,' from a comparison involving a group defined by a deceased father. In general, if one wants to present parallel analyses with and without adjustment for a particular covariate, say 'single-parent household,' then one should not match on that covariate, but should control for it in one of the parallel analyses using analytical techniques; e.g., §18.2 or [10, 23, 28, 29] and [27, §3.6].

[2] In detail, the covariates used in the match are (i) faminc: family income in units of $10,000; (ii) incmiss: income missing (incmiss=1 if family income is missing, incmiss=0 otherwise); (iii) black (black=1 if black, black=0 otherwise), (iv) hispanic (hispanic=1 if hispanic, hispanic=0 otherwise), (v) afqtpct: Armed Forces Qualifcations Test (AFQT), (vi) edmissm: mother's education missing (edmissm=1 if missing, edmissm=0 otherwise), (vii) edm: mother's education (edm=1 for less than high school, edm=2 for high school, edm=3 for some college, edm=4 for BA degree or more), (viii) female (female=1 for female, female=0 for male).

[3] Because of this restriction, the counts of seniors in various groups are slightly different than in [9].

13.2 Data

Table 13.1 For the first 20 people in the 1979–1981 cohort, the treatment zb and the eight covariates Xb are listed. Here, zb is 1 if the father is deceased, and zb is 0 otherwise. In R, this table is displayed by cbind(zb,Xb)[1:20,], which instructs R to bind the vector zb and the matrix Xb as columns, and to display rows 1 to 20 and all columns. The id numbers (actually rownames in R) are not consecutive because zb and Xb contain only the 1979–1981 cohort, before the cancellation of the Social Security Student Benefit Program in 1982. As discussed in Chapter 9, when missing value indicators are used in matching (here incmiss and edmissm), the missing values in the variables themselves are arbitrary. Here, faminc is set to 2 when incmiss is 1 indicating that faminc is missing, and edm is set to 0 when edmissm is 1 indicating that edm is missing. Had different values been filled in, the matching would have been the same.

id	zb	faminc	incmiss	black	hisp	afqtpct	edmissm	edm	female
1	0	5.3	0	0	0	71.9	0	2	0
2	0	2.5	0	0	0	95.2	0	2	1
4	0	9.5	0	0	0	95.8	0	2	1
5	0	0.4	0	0	0	90.6	0	2	1
6	0	10.4	0	0	0	81.8	0	2	0
7	0	6.3	0	0	0	97.9	0	4	0
10	1	3.2	0	0	0	61.9	0	2	1
11	0	9.5	0	0	0	20.4	0	2	1
12	0	2.7	0	0	0	57.2	0	1	0
14	0	2.0	1	0	0	8.3	0	1	1
15	0	7.1	0	0	0	50.9	0	2	1
16	0	7.8	0	0	0	71.1	0	2	1
17	0	12.5	0	0	0	74.9	0	2	0
18	0	2.0	1	0	0	98.4	0	2	1
21	0	9.4	0	0	0	98.7	0	4	1
23	0	2.0	1	0	0	99.7	0	4	0
24	0	2.0	1	0	0	73.3	0	3	1
29	0	2.0	1	1	0	1.8	0	1	0
31	0	2.0	1	0	0	42.9	0	2	1
34	1	2.6	0	0	0	99.6	0	2	0

Table 13.2 Frequency table showing the number of seniors available and the number used in matching. In 1979–1981 and 1982–1983, seniors whose fathers were deceased are matched to ten seniors whose fathers were not deceased, yielding, respectively, 131 and 54 matched sets of 11 seniors each. Also, 1038 pairs of seniors whose fathers are not deceased are found, with one from 1979–1981 and the other from 1982–1983. Among seniors with deceased fathers, a variable match is also found, using all 54+131 seniors, with 54 matched sets containing one senior in 1982–1983 and between 1 and 4 seniors in 1979–1981.

	1979–1981	1982–1983	Matched 1979–1981
Father Deceased (FD)	131	54	131
Father Not Deceased (FND)	2689	1038	1038
Matched FND (10 controls)	1310	540	

```
> length(zb)
[1] 2820
> dim(Xb)
[1] 2820    8
> sum(zb)
[1] 131
> sum(1-zb)
[1] 2689
```

The 131 seniors with deceased fathers will each be matched to 10 controls whose fathers were not deceased.

13.3 Propensity Score

The propensity score is estimated using a logit model. In R, logit models are fitted using the metaphor of generalized linear models [16]; however, if logit models are familiar and generalized linear models are unfamiliar, then you may think of this as R's syntax for logit regression because it is the same logit regression. In R, the estimated propensity scores are the fitted probabilities of a deceased father given the eight covariates, and one obtains these as the $fitted.values in a generalized linear model glm with the family=binomial. Specifically, the vector p contains the 2820 estimated propensity scores, $\widehat{e}(\mathbf{x}_\ell)$, $\ell = 1, 2, \ldots, 2820$:

```
> p<-glm(zb~Xb$faminc+Xb$incmiss+Xb$black+Xb$hisp
  +Xb$afqtpct+Xb$edmissm+Xb$edm+Xb$female,
    family=binomial)$fitted.values
```

It is reasonable to work to improve this model, perhaps including interaction terms or transformations or polynomials or whatnot, but this chapter will take a rather minimal approach, seeking an acceptable match in a few simple steps.

Figure 13.1 displays the estimated propensity scores for the 131 seniors with deceased fathers and the other 2689 high school seniors. The difference between the two distributions is substantial: (i) the standardized difference in means (§9.1) of $\widehat{e}(\mathbf{x}_\ell)$ is 0.67, or $\frac{2}{3}$ of a standard deviation, and (ii) in Figure 13.1, the median $\widehat{e}(\mathbf{x}_\ell)$ in the treated group is about equal to the upper quartile among potential controls. Nonetheless, the distributions in Figure 13.1 overlap substantially, so matching appears to be possible.

13.4 Covariates with Missing Values

Missing covariate values are discussed in §9.4; please review that discussion. The situation with missing covariate values is a common source of misunderstanding.

13.4 Covariates with Missing Values

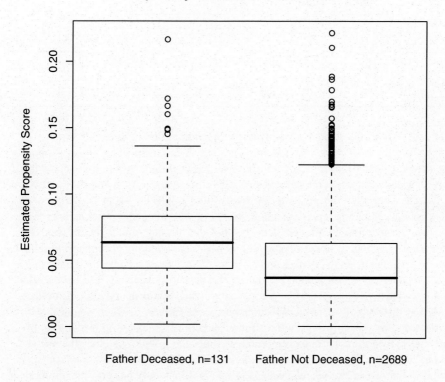

Fig. 13.1 Estimated propensity scores for 131 seniors with deceased fathers and 2689 other seniors in the 1979–1981 cohort, before the Social Security Student Benefit Program was eliminated.

As in §9.4, missing covariate values are filled-in with an arbitrary but fixed number; then, missing value indicators are appended as additional covariates; see Table 13.1 and the logit model for the propensity score in §13.3. If the missing values for family income, faminc, or mother's education, edm, had been filled in with different constants, the estimated propensity scores $\widehat{e}(\mathbf{x}_\ell)$ would have been the same because of the presence of the missing value indicators, incmiss and edmissm. The coefficients in the logit model would change if different values had been used, but they change to compensate for the change in the filled-in values, keeping $\widehat{e}(\mathbf{x}_\ell)$ unchanged.[4] To prevent round-off error in the computer's arithmetic work, it makes sense to fill in values that are not wild — a family income of −9999 is not recom-

---

[4] Technically, the fitted probabilities in logit regression are invariant under affine transformations of the predictors.

Table 13.3 First five rows and columns of the 131 × 2689 distance matrix using the rank-based Mahalanobis distance. Notice that the 131 rows and 2689 columns are labeled with distinct numbers $\ell = 1, 2, \ldots, 2820 = 131 + 2689$.

ℓ	Control 1	Control 2	Control 3	Control 4	Control 5
Treated 7	3.86	2.61	5.06	6.86	6.72
Treated 20	3.47	3.03	7.58	6.23	5.82
Treated 108	9.60	19.47	20.02	24.62	13.03
Treated 126	6.81	8.05	12.93	10.74	9.88
Treated 145	8.70	15.09	17.74	18.86	12.37

mended — but aside from round-off error, the value filled in does not matter. To emphasize, the missing values are still missing. The propensity score is estimating the probability that a senior will have a deceased father when that senior has certain specific measured covariates and a specific pattern of missing covariates; it is not guessing what the missing covariate values might be. That is, the propensity score is estimating a particular conditional probability about observable events, because the patterns of missing data (but not the missing values themselves) are observable. Matching or stratifying on this particular conditional probability relating observables tends to balance those observables, that is, the observed covariates and the pattern of missing data, but there is no basis for expecting it to balance the missing values [21, Appendix]. Specifically, a successful match should produce a similar distribution of observed family incomes (faminc, when incmiss=0) in treated and control groups and a similar percent of seniors with missing family incomes (incmiss=1); however, whether or not the missing incomes themselves (faminc, when incmiss=1) are similar is not known. With regard to missing covariates, the propensity score succeeds in its ambitions only because its ambitions are quite limited, namely balancing observable quantities.

13.5 Distance Matrix

The distance matrix is constructed in two steps. The first step computes the rank-based Mahalanobis distance in §8.3. The function smahal(·,·), which is given in the second appendix to this chapter, creates the 131 × 2689 distance matrix. The first five rows and columns of dmat are given in Table 13.3.

```
> dmat<-smahal(zb,Xb)
> dim(dmat)
  [1] 131 2689
```

The second step adds the caliper on the propensity score in §8.4. The function addcaliper(·,·,·), which is given in the second appendix to this chapter, takes dmat and adds a penalty when the propensity scores differ by more than 0.2 times the standard deviation of $\widehat{e}(\mathbf{x})$. In the welder data in Chapter 8, the caliper was set at 0.5 times the standard deviation of $\widehat{e}(\mathbf{x})$, but aside from very small problems, a value

13.6 Constructing the Match

Table 13.4 First five rows and columns of the 131 × 2689 distance matrix after adding the propensity score calipers. The caliper was at $0.2 \times \text{sd}(\hat{e}(\mathbf{x}))$. Only two of these 25 entries were unaffected by the caliper.

ℓ	Control 1	Control 2	Control 3	Control 4	Control 5
Treated 7	18.60	20.64	42.04	79.91	46.10
Treated 20	46.32	3.03	72.66	51.18	73.30
Treated 108	82.94	47.40	115.60	39.07	111.01
Treated 126	57.81	13.64	86.16	47.54	85.51
Treated 145	8.70	54.51	33.32	113.31	30.34

of 0.1 to 0.25 is more common. The result is a new 131 × 2689 distance matrix. The first five rows and columns of the revised dmat are given in Table 13.4. Among these 25 entries, only two respected the caliper and did not incur a penalty.

```
> dmat<-addcaliper(dmat,zb,p)
> dim(dmat)
  [1] 131 2689
```

13.6 Constructing the Match

From the distance matrix, dmat, the match is obtained using Hansen's [11] fullmatch function in his optmatch package. After optmatch is installed, you must load it, either using the packages menu or with the command

```
> library(optmatch)
```

We wish to match ten controls to each senior with a deceased father. The fullmatch function needs to know the distance matrix, here dmat, the minimum number of controls, here 10, the maximum number of controls, here 10, and the fraction of controls to omit, here $(2689 - 10 \times 131)/2689 = 0.51283$. That is, matching 10-to-1 means 10×131 will be used, omitting $2689 - 10 \times 131$, which is 51%.

```
> 2689-10*131
  [1] 1379
> 1379/2689
  [1] 0.51283
```

With these quantities determined, we do the match:

```
> m<-fullmatch(dmat,min.controls=10,max.controls=10,
   omit.fraction=1379/2689)
```

After a few seconds, R places the match in m. There is an entry in m for each of the 2820 seniors.

```
> length(m)
```

    [1] 2820

The first ten entries in m are

```
> m[1:10]
  m.34 m.10 m.01 m.87 m.02 m.26 m.1 m.03 m.17 m.117
```

This says that the first senior of the 2820 seniors is in matched set #34 and the second senior is in matched set #10. The third senior was one of the 1379 unmatched controls; this is the meaning of the zero in m.01. The fourth senior is in matched set #87, the fifth is unmatched, and so on. The function matched(·) indicates who is matched. The first ten entries of matched(m) are

```
> matched(m)[1:10]
TRUE TRUE FALSE TRUE FALSE TRUE TRUE FALSE TRUE TRUE
```

which says, again, that the first two seniors were matched but the third was not, and so on. There are 1441 matched seniors, where $1441 = 131 \times 11$, because the 131 seniors with deceased fathers were each matched to ten controls, making 131 matched sets each of size 11.

```
> sum(matched(m))
  [1] 1441
> 1441/11
  [1] 131
```

The first three matched sets are in Table 13.5. The first match consists of 11 female high school seniors, neither black nor hispanic, whose mothers had a high school education, with family incomes between $30,000 and $40,000, mostly with test scores between 59% and 77%. In the second matched set, incomes were lower but test scores were higher. And so on.

Having constructed the match, the remaining work is done with standard R functions.

13.7 Checking Covariate Balance

Three other matches were constructed: a 10-to-1 match for the period 1982–1983, after the Social Security Student Benefit Program had ended, a 1-to-1 match of seniors with fathers who were not deceased in 1982–1983 and in 1979–1981, and a variable match (§8.5) for all seniors whose fathers were deceased, with 54 matched sets containing one senior from 1982–1983 and between one and four seniors from 1979–1981; see Table 13.2. In producing the variable match, the call to fullmatch set min.controls=1, max.controls=4, omit.fraction=0, which says that all 131 potential controls are to be used. The small variable-match is included for illustration.

13.7 Checking Covariate Balance

Table 13.5 The first three of 131 matched sets, each set containing one treated subject and ten matched controls.

id	set	zb	faminc	incmiss	black	hisp	afqtpct	edmissm	edm	female
10	1380	1	3.22	0	0	0	61.92	0	2	1
350	1380	0	3.56	0	0	0	59.58	0	2	1
365	1380	0	3.60	0	0	0	61.24	0	2	1
465	1380	0	3.56	0	0	0	56.37	0	2	1
518	1380	0	3.79	0	0	0	67.01	0	2	1
550	1380	0	3.79	0	0	0	63.50	0	2	1
1294	1380	0	3.56	0	0	0	67.31	0	2	1
2072	1380	0	3.79	0	0	0	64.98	0	2	1
2082	1380	0	3.79	0	0	0	63.62	0	2	1
2183	1380	0	3.97	0	0	0	76.52	0	2	1
3965	1380	0	3.76	0	0	0	72.58	0	2	1
396	1381	1	2.37	0	0	0	88.51	0	2	1
2	1381	0	2.46	0	0	0	95.16	0	2	1
147	1381	0	2.27	0	0	0	77.09	0	2	1
537	1381	0	2.60	0	0	0	95.96	0	2	1
933	1381	0	2.85	0	0	0	96.11	0	2	1
974	1381	0	1.90	0	0	0	99.60	0	3	1
987	1381	0	2.13	0	0	0	81.18	0	2	1
1947	1381	0	2.05	0	0	0	91.45	0	3	1
2124	1381	0	2.30	0	0	0	72.40	0	2	1
2618	1381	0	2.21	0	0	0	68.92	0	2	1
3975	1381	0	2.37	0	0	0	90.74	0	2	1
3051	1382	1	3.41	0	0	1	62.87	0	1	0
606	1382	0	4.18	0	0	1	81.74	0	1	0
664	1382	0	4.39	0	0	1	91.57	0	1	0
884	1382	0	2.85	0	0	1	48.77	0	1	0
995	1382	0	3.13	0	0	1	55.12	0	1	0
1008	1382	0	3.32	0	0	1	51.61	0	1	0
1399	1382	0	3.44	0	0	1	90.02	0	2	0
2908	1382	0	3.80	0	0	1	80.28	0	1	0
3262	1382	0	3.79	0	0	1	57.28	0	1	0
3400	1382	0	2.93	0	0	1	80.88	0	2	0
3624	1382	0	3.32	0	0	1	74.08	0	1	1

Table 13.6 and Figure 13.1 display the imbalance in the eight covariates and the propensity score in these four comparisons, before and after matching. For the variable match, the means in the control group are weighted; see §8.5.

The four pairs of two groups were quite different before matching but were much closer after matching. The family income of seniors with deceased fathers was lower, they were more often black, and their mothers had less education. Between 1979–1981 and 1982–1983, AFQT test scores decline. The imbalances are much smaller after matching.

Table 13.6 Balance on covariates before and after matching for four matched comparisons. FD = father deceased and FND = father not deceased. For covariate k, $\bar{x}_{tk}$ is the mean in the first group in the comparison, $\bar{x}_{ck}$ is the mean in the second group in the comparison, and $\bar{x}_{mck}$ is the mean in the matched comparison group formed from the second group. As in Chapter 9, sd_{bk} is the standardized difference in means before matching and sd_{mk} is the standardized difference in means after matching; they have the same denominator but different numerators. The largest of 36 standardized differences before matching was 0.80, compared with 0.19 after matching.

Comparison	set	$\widehat{e}(\mathbf{x})$	faminc	incmiss	black	hisp	afqtpct	edmissm	edm	female
1979–1981	$\bar{x}_{tk}$	0.07	2.78	0.15	0.35	0.15	49.58	0.08	1.62	0.49
FD vs. FND	$\bar{x}_{cmk}$	0.06	2.77	0.15	0.34	0.15	49.10	0.04	1.61	0.50
	$\bar{x}_{ck}$	0.05	4.58	0.19	0.29	0.15	52.39	0.04	1.91	0.50
131 sets	sd_{bk}	0.67	0.71	0.11	0.13	0.02	0.10	0.19	0.33	0.03
10-to-1	sd_{mk}	0.09	0.00	0.00	0.02	0.03	0.02	0.19	0.02	0.02
1982–1983	$\bar{x}_{tk}$	0.08	2.31	0.30	0.46	0.15	37.41	0.04	1.56	0.46
FD vs. FND	$\bar{x}_{mck}$	0.07	2.53	0.33	0.41	0.13	39.13	0.03	1.62	0.49
	$\bar{x}_{ck}$	0.05	4.31	0.24	0.30	0.16	43.81	0.06	1.83	0.47
54 sets	sd_{bk}	0.76	0.80	0.12	0.35	0.04	0.22	0.11	0.34	0.02
10-to-1	sd_{mk}	0.18	0.09	0.08	0.11	0.06	0.06	0.03	0.07	0.05
FND	$\bar{x}_{tk}$	0.30	4.31	0.24	0.30	0.16	43.81	0.06	1.83	0.47
1982–1983	$\bar{x}_{cmk}$	0.30	4.33	0.24	0.29	0.15	44.06	0.05	1.81	0.47
vs 1979–1981	$\bar{x}_{ck}$	0.27	4.58	0.19	0.29	0.15	52.39	0.04	1.91	0.50
1038	sd_{bk}	0.37	0.09	0.12	0.01	0.03	0.30	0.10	0.09	0.06
pairs	sd_{mk}	0.02	0.01	0.00	0.01	0.02	0.01	0.05	0.03	0.00
FD	$\bar{x}_{tk}$	0.35	2.31	0.30	0.46	0.15	37.41	0.04	1.56	0.46
1982–1983	$\bar{x}_{cmk}$	0.33	2.49	0.25	0.43	0.15	38.55	0.06	1.65	0.43
vs 1979–1981	$\bar{x}_{ck}$	0.27	2.78	0.15	0.35	0.15	49.58	0.08	1.62	0.49
54 sets	sd_{bk}	0.67	0.24	0.35	0.23	0.01	0.43	0.20	0.09	0.05
variable	sd_{mk}	0.16	0.09	0.12	0.07	0.00	0.04	0.12	0.12	0.06

13.8 College Outcomes

A treatment effect should enhance college attendance in the one group that received an incentive to attend, namely seniors in 1979–1981 whose fathers were deceased. If the other groups differ substantially from each other after matching for observed covariates, then this cannot be an effect of the treatment, and must indicate that matching has failed to render the groups comparable [5, 20, 26] and [30, Chapter 5]. Table 13.7 compares three measures of college attendance in four matched comparisons. In Table 13.7, the Mantel-Haenszel test and odds ratio are reported [10, §13.3] and percentages in the variable match are weighted see §8.5.

In 1979–1981, when the Social Security Student Benefit Program provided tuition aid to students of a deceased Social Security beneficiary, seniors with deceased fathers were more likely to attend college and complete one year of college than were matched controls, with an odds ratio of about 1.65, but there is no sign of this in 1982–1983 after the program ended. Among seniors whose fathers were not deceased, college outcomes were slightly but not significantly better in 1982–1983 than among matched controls in 1979–1981. Seniors with deceased fathers in 1982–1983 ($n = 54$) and their variably matched seniors with deceased fathers in 1979–

13.9 Further Reading

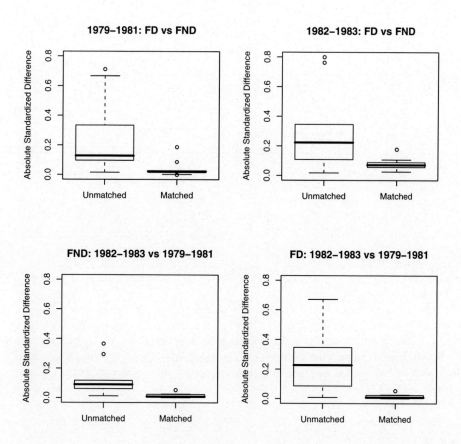

Fig. 13.2 Standardized differences in covariate means before and after matching in three matched comparisons. Each boxplot displays standardized differences in means, for the nine covariates and the propensity score.

1981 ($n = 131$) did not differ significantly on any outcome, but the sample size is small, and the point estimates are not incompatible with reduced college attendance in 1982–1983, when the tuition benefit was withdrawn. There was no indication of a difference in completion of four years of college for any comparison.

13.9 Further Reading

Susan Dynarski's [9] fine study is very much worth reading. The reanalysis presented here was intended just to illustrate many aspects of matching in R, not as a recommended analysis for a study of this form. Because of the several sample sizes

Table 13.7 College outcomes by age 23: attended, completed a year, and completed four years. In matched groups, the table gives percentages, the Mantel-Haenszel odds ratio, and the two-sided *P*-value from the Mantel-Haenszel test. Percents for the variable match are weighted.

Outcome	Comparison	Group 1 %	Group 2 %	MH–odds ratio	p-value
Attend College	1979–1981: FD vs. FND	53	43	1.67	0.019
	1982–1983: FD vs. FND	35	37	0.90	0.853
	FND: 1982–1983 vs. 1979–1981	43	39	1.21	0.070
	FD: 1982–1983 vs. 1979–1981	35	46	0.63	0.264
Complete Any College	1979–1981: FD vs. FND	50	40	1.65	0.022
	1982–1983: FD vs. FND	31	34	0.87	0.798
	FND: 1982–1983 vs. 1979–1981	40	37	1.20	0.100
	FD: 1982–1983 vs. 1979–1981	31	36	0.81	0.369
Complete 4 Years	1979–1981: FD vs. FND	15	14	1.03	0.915
	1982–1983: FD vs. FND	09	12	0.70	0.630
	FND: 1982–1983 vs. 1979–1981	14	13	1.13	0.460
	FD: 1982–1983 vs. 1979–1981	9	12	0.88	0.805

in Dynarski's [9] study, several different forms of matching could be used with one data set.

Ben Hansen's fullmatch function is well described in Hansen [11]. Documentation about R is available at http://cran.r-project.org/. See also [13]. Four good books about R, in order of increasing difficulty, are [8, 15, 1, 7].

If you want to know what fullmatch is doing, see [12, 24, 25], but if you *really* want to know what fullmatch is doing, see [3, 4]. If you plan to use proc assign in SAS, see [17].

13.10 Appendix: A Brief Introduction to R

This chapter is not intended to teach R, although I do teach it in the classroom. My impression is that R is easy to learn, and the only obstacle is that, in a world of menu-driven software, R is not menu-driven, so you need to know keywords. Once you know a few keywords, it is easy to learn more, but the first 20 minutes with R can cause heart palpitations in people who are easily frustrated. You install R by going to the website http://cran.r-project.org/ and following the instructions. Once R is installed, the following comments may help with those first twenty minutes.

You find keywords with apropos. For instance, if you were looking for Wilcoxon's signed rank test, you might type
> apropos("wilc")
and get back
[1] "dwilcox" "pairwise.wilcox.test" "pwilcox"
[4] "qwilcox" "rwilcox" "wilcox.test"
Here, wilcox.test looks promising, so you type
> help(wilcox.test)

13.10 Appendix: A Brief Introduction to R

and receive instructions about Wilcoxon's test.

If you'd like to mess around with any old data, type
> help(data)
and
> data()
where, yes, there is nothing inside the ().

Some keywords to learn early include data.frame, NA, ls, rm, summary, plot, and boxplot. For instance, type
> help(data.frame)
Speaking informally, a data.frame is an object that resides in your workspace and contains data with variables as columns and subjects as rows. Missing values are coded as NA. When you use an R function, you should check its help to see what it does with NA's; here, you will find the solution to many otherwise mysterious events. If you type ls(), then you will see a list of the objects in your current workspace. Keep a tidy workspace. Remove an unwanted object named unwanted with rm(unwanted). R is not Windows: it will not ask you "Are you absolutely sure you want to discard unwanted forever?" Save copies of anything that is important using SAVE WORKSPACE on the file menu. Save different research projects in different workspace's. The command summary(object) provides an appropriate description for many different types of object.

A key feature of R is that you can download packages from the Web. You can search the Web using help.search. For instance,
> help.search("match")
produces lots of things, some of them related to the current chapter, whereas
help.search("fullmatch")
guides you to Ben Hansen's [11] fullmatch function, which is used in the current chapter.

To use a package from the Web, you must install it, which is most easily done using the packages menu, clicking on install package(s). For the current chapter, you will want to install Hansen's [11] optmatch package, then type
> library(optmatch)
> help(fullmatch)
and you are good to go.

At some point, you will want to use R with your own data. A basic command is read.table, so type
> help(read.table)
When reading data from files created by other statistical packages, a helpful package is foreign. For instance, to read a stata file, install foreign, and type
> library(foreign)
> help(read.dta)
To read a SAS file, install foreign, and type
> library(foreign)
> help(read.xport)

Although read.table is the basic command and read.dta is an extension, reading a file that has already been formatted – say, reading a stata file with read.dta – often turns out to be easier than reading a numeric file with read.table, where you may have more details to attend to.

Most of the data sets that inhabit R that are listed by data() are not observational studies. The web page for Jeffrey Woodridge's nice book [31] includes stata files for several interesting observational studies published in economics journals. David Card and Alan Krueger have made the data from their study [6] of the minimum wage data available at http://www.irs.princeton.edu/. Several versions of the NSW experiment in Chapter 2 are available in downloadable packages in R, but you need to identify each version with a different paper from the large literature spawned by Robert LaLonde's [14] study. The data from Angrist and Lavy's study [2] are available from Joshua Angrist on his webpage at MIT.

13.11 Appendix: R Functions for Distance Matrices

This appendix contains several simple R functions that do the arithmetic involved in creating a distance matrix or adding a penalty. The functions perform a few steps of vector arithmetic. In all of these functions, z is a vector of length(z)=n with z[i]=1 for treated and z[i]=0 for control, X is an n×k matrix of covariates, dmat is a distance matrix created by mahal or smahal, p is a vector of length(p)=n typically containing a propensity score, and f is a vector of length(f)=n with a few values used in almost exact matching.

The function mahal creates a distance matrix using the Mahalanobis distance; see §8.3. It requires that the MASS package has been installed.

```
> mahal
function(z,X){
X<-as.matrix(X)
n<-dim(X)[1]
rownames(X)<-1:n
k<-dim(X)[2]
m<-sum(z)
cv<-cov(X)
out<-matrix(NA,m,n-m)
Xc<-X[z==0,]
Xt<-X[z==1,]
rownames(out)<-rownames(X)[z==1]
colnames(out)<-rownames(X)[z==0]
library(MASS)
icov<-ginv(cv)
for (i in 1:m) {
  out[i,]<-mahalanobis(Xc,Xt[i,],icov,inverted=T)}
out
}
```

13.11 Appendix: R Functions for Distance Matrices

The function `smahal` creates a distance matrix using the rank-based Mahalanobis distance; see §8.3. It requires the MASS package has been installed.

```
> smahal
function(z,X){
# Rank based Mahalanobis distance.
X<-as.matrix(X)
n<-dim(X)[1]
rownames(X)<-1:n
k<-dim(X)[2]
m<-sum(z)
for (j in 1:k) X[,j]<-rank(X[,j])
cv<-cov(X)
vuntied<-var(1:n)
rat<-sqrt(vuntied/diag(cv))
cv<-diag(rat)%*%cv%*%diag(rat)
out<-matrix(NA,m,n-m)
Xc<-X[z==0,]
Xt<-X[z==1,]
rownames(out)<-rownames(X)[z==1]
colnames(out)<-rownames(X)[z==0]
library(MASS)
icov<-ginv(cv)
for (i in 1:m) {
   out[i,]<-mahalanobis(Xc,Xt[i,],icov,inverted=T)}
out
}
```

The function `addcaliper` adds a penalty to a distance matrix `dmat` for violations of the caliper on p; see §8.4 for discussion of this penalty function. The default caliper width is $0.2*sd(p)$. The magnitude of the penalty is `penalty` multiplied by the magnitude of the violation, where `penalty` is set to 1000 by default.

```
> addcaliper
function(dmat,z,p,caliper=0.2,penalty=1000){
  sdp<-sd(p)
  adif<-abs(outer(p[z==1],p[z==0],"-"))
  adif<-(adif-(caliper*sdp))*(adif>(caliper*sdp))
  dmat<-dmat+adif*penalty
  dmat
}
```

The function `addalmostexact` adds a penalty to a distance matrix `dmat` for failure to match exactly on `f`; see §9.2. The magnitude of the penalty is `mult` multiplied by the largest value in `dmat`.

```
> addalmostexact
function(dmat,z,f,mult=10){
  penalty<-mult*max(dmat)
  mismatch<-outer(f[z==1],f[z==0],"!=")
  dmat<-dmat+mismatch*penalty
  dmat
}
```

References

1. Aitkin, M., Francis, B., Hinde, J., Darnell, R.: Statistical Modelling in R. New York: Oxford University Press (2009)
2. Angrist, J.D., Lavy, V.: Using Maimonides' rule to estimate the effect of class size on scholastic achievement. Quart J Econ **114**, 533–575 (1999)
3. Bertsekas, D.P.: A new algorithm for the assignment problem. Math Program **21**, 152–171 (1981)
4. Bertsekas, D.P.: Linear Network Optimization. Cambridge, MA: MIT Press (1991)
5. Campbell, D.T.: Factors relevant to the validity of experiments in social settings. Psychol Bull **54**, 297–312 (1957)
6. Card, D., Krueger, A.: Minimum wages and employment: A case study of the fast-food industry in New Jersey and Pennsylvania. Am Econ Rev **84**, 772–793 (1994) Data: http://www.irs.princeton.edu/
7. Chambers, J.: Software for Data Analysis: Programming with R. New York: Springer (2008)
8. Dalgaard, P.: Introductory Statistics with R. New York: Springer (2002)
9. Dynarski, S.M.: Does aid matter? Measuring the effect of student aid on college attendance and completion. Am Econ Rev **93**, 279–288 (2003)
10. Fleiss, J.L., Levin, B., Paik, M.C.: Statistical Methods for Rates and Proportions. New York: Wiley (2001)
11. Hansen, B.B.: Optmatch: Flexible, optimal matching for observational studies. R News **7**, 18–24 (2007)
12. Hansen, B.B., Klopfer, S.O.: Optimal full matching and related designs via network flows. J Comp Graph Statist **15**, 609–627 (2006)
13. Ho, D., Imai, K., King, G., Stuart, E.A.: Matching as nonparametric preprocessing for reducing model dependence in parametric causal inference. Polit Anal **15**, 199–236 (2007)
14. LaLonde, R.J.: Evaluating the econometric evaluations of training programs with experimental data. Am Econ Rev **76**, 604–620 (1986)
15. Maindonald, J., Braun, J.: Data Analysis and Graphics Using R. New York: Cambridge University Press (2001)
16. McCullagh, P., Nelder, J.A.: Generalized Linear Models. New York: Chapman and Hall/CRC (1989)
17. Ming, K., Rosenbaum, P.R.: A note on optimal matching with variable controls using the assignment algorithm. J Comput Graph Statist **10**, 455–463 (2001)
18. R Development Core Team.: R: A Language and Environment for Statistical Computing. Vienna: R Foundation, http://www.R-project.org (2007)
19. Rosenbaum, P.R., Rubin, D.B.: The central role of the propensity score in observational studies for causal effects. Biometrika **70**, 41–55 (1983)

20. Rosenbaum, P.R.: From association to causation in observational studies. J Am Statist Assoc **79**, 41–48 (1984)
21. Rosenbaum, P.R., Rubin, D.B.: Reducing bias in observational studies using subclassification on the propensity score. J Am Statist Assoc **79**, 516–524 (1984)
22. Rosenbaum, P.R., Rubin, D.B.: Constructing a control group by multivariate matched sampling methods that incorporate the propensity score. Am Statistician **39**, 33–38 (1985)
23. Rosenbaum, P.R.: Permutation tests for matched pairs with adjustments for covariates. Appl Statist **37**, 401–411 (1988) (Correction: [27, §3])
24. Rosenbaum, P.R.: Optimal matching in observational studies. J Am Statist Assoc **84**, 1024–32 (1989)
25. Rosenbaum, P.R.: A characterization of optimal designs for observational studies. J Roy Statist Soc B **53**, 597–610 (1991)
26. Rosenbaum, P.R.: Stability in the absence of treatment. J Am Statist Assoc **96**, 210–219 (2001)
27. Rosenbaum, P.R.: Observational Studies (2nd ed.). New York: Springer (2002)
28. Rosenbaum, P.R.: Covariance adjustment in randomized experiments and observational studies (with Discussion). Statist Sci **17**, 286–327 (2002)
29. Rubin, D.B.: Using multivariate matched sampling and regression adjustment to control bias in observational studies. J Am Statist Assoc **74**, 318–328 (1979)
30. Shadish, W. R., Cook, T. D., Campbell, D.T.: Experimental and Quasi-Experimental Designs for Generalized Causal Inference. Boston: Houghton-Mifflin (2002)
31. Wooldridge, J.M.: Econometric Analysis of Cross Section and Panel Data. Cambridge, MA: MIT Press. (2002)

Part III
Design Sensitivity

Chapter 14
The Power of a Sensitivity Analysis and Its Limit

Abstract In an experiment, power and sample size calculations anticipate the outcome of a statistical test that will be performed when the experimental data are available for analysis. In parallel, in an observational study, the power of a sensitivity analysis anticipates the outcome of a sensitivity analysis that will be performed when the observational data are available for analysis. In both cases, it is imagined that the data will be generated by a particular model or distribution, and the outcome of the test or sensitivity analysis is anticipated for data from that model. Calculations of this sort guide many of the decisions made in designing a randomized clinical trial, and similar calculations may usefully guide the design of an observational study. In experiments, the power in large samples is used to judge the relative efficiency of competing statistical procedures. In parallel, the power in large samples of a sensitivity analysis is used to judge the ability of design features, such as those in Chapter 5, to distinguish treatment effects from bias due to unmeasured covariates. As the sample size increases, the limit of the power of a sensitivity analysis is a step function with a single step down from power 1 to power 0, where the step occurs at a value $\widetilde{\Gamma}$ of Γ called the design sensitivity. The design sensitivity is a basic tool for comparing alternative designs for an observational study.

14.1 The Power of a Test in a Randomized Experiment

What is the power of a test?

The power of a statistical test anticipates the judgment that the test will issue when the test is put to use. Conceptually, the power of a test is the probability that the test will recognize that a false null hypothesis is indeed false and reject it. If the test will reject the null hypothesis when the P-value is less than or equal to, say, the conventional 0.05 level, then the power of the test is the probability that the P-value will be less than or equal to 0.05 when the null hypothesis is indeed false.

If the test is a valid test of its null hypothesis, and if the null hypothesis were true, the probability that the P-value would be less than or equal to 0.05 is itself less than or equal to 0.05. This is the defining property of a valid test or P-value. The P-value is unlikely to be less than or equal to 0.05 if the null hypothesis is true — it happens in only one experiment in 20. Having defined a test so that rejection of a true null hypothesis is improbable, we now want to make rejection of a false null hypothesis highly probable; that is, we would like to have a powerful test.

The power of a test depends upon many things. First and foremost, it depends upon what is true. If the null hypothesis is false, something else is true instead. If the null hypothesis is false but barely so, the power is likely to be little better than chance, 0.05. If the null hypothesis is far from the truth, the power is likely to be much higher. The power depends also upon the sample size, the duration of follow-up, the specifics of the experimental design, and the procedures used in statistical analysis.

Power is a basic tool in designing a randomized experiment. How many patients are needed? Well, calculate the power with 100, 200 and 300 patients. Is it better to study the survival of 200 patients for five years or 300 patients for three years? Calculate the power in each situation. The new treatment is expensive and difficult to apply. To reduce cost, the experimenter is considering randomizing one-third of the patients to treatment and two-thirds to control. Will this design be substantially inferior to randomizing half the patients to treatment and half to control? Calculate the power for both designs. There are two ways to measure the response, one precise but expensive, the other imprecise but inexpensive. If the inexpensive device is used, the money saved will permit a larger sample size. For fixed total cost, which is better, more precision with fewer patients, or less precision with more patients? Calculate the power in both situations. With 20 schools, it would be convenient to randomly assign ten schools to treatment, the rest to control, and to use the same treatment for the hundreds of students in each school. Is that a terrible idea? Would it be vastly better to randomize the many hundreds of students as individuals? Calculate the power for the two designs.

Power is also a basic tool in evaluating the statistical methods that are used to analyze the results of a randomized experiment. Faced with the same data from the same statistical models, two different tests will typically have different power. The power of two different tests in a variety of circumstances is often the basis for choosing which test to use. The t-test has slightly better power than the Wilcoxon test for data from a Normal distribution, but substantially inferior power for distributions with longer tails, which is a basic reason that the Wilcoxon test is preferred; its power is robust.

The remainder of this section briefly reviews the idea of power and its computation for Wilcoxon's signed rank statistic when used in a randomized experiment. The concept of power is important throughout Part III. The details of the computation of power are relevant to the details of the discussion in Part III, but the concepts of Part III should be accessible if the concept of power is clear.

14.1 The Power of a Test in a Randomized Experiment

A pep talk about statistical power

Design sensitivity and the power of a sensitivity analysis are simple but useful extensions of two other concepts, namely sensitivity analysis and the power of a statistical test. The concept of a sensitivity analysis was introduced in §3.4 and §3.5, and this might be a good moment to take a second look at those sections. The power of a statistical test will be very familiar to some readers and less so to others. If power is very familiar, please skip the next few paragraphs, continuing with the subsection "Computing power in a randomized experiment: The two steps."

For people to whom statistical power is less familiar, let me mention a common source of discomfort and misunderstanding. In testing a null hypothesis, we tentatively suppose that the null hypothesis is true, simply to push through the logic of hypothesis testing. That is, we say, 'suppose the null hypothesis were true.' If it were true, what is the chance of data as inconsistent with the null hypothesis as the data we observed? That chance is the P-value, and if the P-value is small, then we begin to doubt the supposition that the null hypothesis is true. That is the logic of hypothesis testing. In computing the power, we suppose that the null hypothesis isn't true after all, and we ask about the probability of a small P-value when the null hypothesis is false. This second question, the one about power, can seem perfectly reasonable at one moment and incoherent the next, not unlike certain unstable optical illusions. The reason is that a power calculation supposes two contradictory things — that the null hypothesis is true and that it isn't — and that can seem reasonable if the suppositions occur at different levels and incoherent if they collapse into a single contradictory supposition.

Rather than talk about power, let's talk about fishing. Suppose that I go fishing in a lake, but suppose there are no fish in the lake. Will I catch any fish? One perfectly reasonable answer is 'no': if there are no fish in the lake, I won't catch any fish. Another answer is: if I supposed there were no fish in the lake, I would not go fishing in that lake; perhaps I would fish in some other lake, and then yes, perhaps I would catch a fish. The first answer was willing to suppose a situation in which I supposed something false, namely it supposed that I supposed there were fish in the lake, hence went fishing in the lake, but my supposition was false. In the second answer, the two levels of supposing have collapsed into one, and the idea of going fishing in a lake with no fish seems incoherent. The fish-in-lake situation is clear enough, but add some technical concepts, a few Greek symbols, a little jargon, and suppositional collapse can seem quite puzzling. In a power calculation, I suppose the null hypothesis is false, but I also suppose that I am ignorant of the fact that it is false, so I test it, and in my ignorance, in the process of testing the null hypothesis, I suppose it to be true. To think about the possibility of error, I must make suppositions at two levels: I must suppose something is true, and I must suppose that I can be ignorant of that truth. If the concept of power ever feels incoherent, go back to fish in a lake. In our ignorance, we scientists spend a lot of time fishing in lakes with no fish, and you need to get the basic concepts down.

In an analogous way, when in §14.2 we compute the power of a sensitivity analysis, we suppose that there is a treatment effect and there is no bias from unmeasured covariates, but we are ignorant of these facts, so we perform a sensitivity analysis when an omniscient investigator would not.[1] Again, there is the danger of misunderstanding from suppositional collapse. The sensitivity analysis entertains the possibility of no treatment effect and a bias of magnitude Γ, for several values of Γ, but the power of the sensitivity analysis is computed under the supposition of a treatment effect and no bias. The power of a sensitivity analysis asks: If the treatment worked, and there was no bias from unobserved covariates, then how would the sensitivity analysis turn out? This is parallel to the power of a test of no effect in a randomized experiment that asks: If the treatment worked, then how would the test of no effect turn out?

Think about what you want a sensitivity analysis to do. In an observational study, if you are looking at a treatment effect without bias, then you will not know this from the data, but you want the sensitivity analysis to report that the ostensible effects of the treatment are highly insensitive to unobserved bias. Perhaps that will happen, perhaps not. The power of the sensitivity analysis is the probability that it will happen when, indeed, there is a treatment effect without bias. Again, it is important to avoid suppositional collapse. You are imagining that the world is a certain way, but you are ignorant of this, and you are performing a statistical analysis, in this case a sensitivity analysis, in a state of ignorance. You want to know: Under what circumstances will a sensitivity analysis do what you want it to do? Under what circumstances will the analysis sort things out correctly?

Formulas for power are needed to produce numerical results and to prove assertions. If you are okay with formulas, then that is fine, but if not, take the assertions and numerical results as true, focus on conceptual issues, and move on quickly to Chapter 15 rather than getting bogged down here with a formula. For instance, right now, look ahead to Figure 14.1 and realize that this figure is intelligible without reference to the technical material that precedes and produces it. Figure 14.1 says that, in a randomized experiment, the power is greater if the treatment effect is larger ($\tau = 1/2$ rather than $\tau = 1/4$), and the power increases to 1 as the sample size I increases. Figure 14.1 captures the way we think about randomized experiments. Figures 14.2 and 14.3 are also intelligible without reference to the technical material that precedes and produces them; look at them now. Figures 14.2 and 14.3 capture the key difference between power in a randomized experiment and power in an observational study. Figures 14.2 and 14.3 say that the power of a sensitivity analysis increases to 1 as the sample size I increases for small values of Γ and decreases to 0 for large values of Γ, the dividing point being a quantity, $\widetilde{\Gamma}$, called the design sensitivity (which is $\widetilde{\Gamma} = 3.171$ in Figure 14.2). Here, Γ is an aspect of your analysis and $\widetilde{\Gamma}$ is an aspect of the world; they cannot coexist at the same suppositional level.

---

[1] The omniscient investigator would not test hypotheses. Then again, the omniscient investigator would not run a laboratory or collect data. So, the omniscient investigator would be unable to publish, and would be denied tenure. The omniscient investigator would protest the denial of tenure on the grounds that the university routinely grants tenure to investigators who remain fallible despite their expensive laboratories. The omniscient investigator would end up in an asylum. Be content that you are not an omniscient investigator, and instead try to objectively and publicly control the probability of error.

14.1 The Power of a Test in a Randomized Experiment

If you understand Figures 14.1–14.3, then you understand most of the concepts of this chapter, even if you do not know how to produce these figures.

Computing power in a randomized experiment: The two steps

The power of Wilcoxon's signed rank statistic, T, in a randomized experiment is computed in two steps. First, for a particular α, conventionally $\alpha = 0.05$, one finds the critical value, ζ_α, such that the chance that $T \geq \zeta_\alpha$ if the null hypothesis is true is α.[2] The null hypothesis would be rejected in a one-sided 0.05 level test if $T \geq \zeta_{0.05}$ because the one-sided P-value would be less than or equal to 0.05. Second, one computes the power as the probability that $T \geq \zeta_\alpha$ when the null hypothesis is false and something else is true instead. The power is then the probability of rejecting the null hypothesis with a P-value less than or equal to 0.05 when the null hypothesis is, in fact, false.

Step 1: Determining the critical value assuming the null hypothesis is true

Consider the first step, the behavior of Wilcoxon's signed rank statistic, T, in a paired randomized experiment with I pairs when the null hypothesis of no treatment effect is true. It is convenient to assume there are no ties.[3] Recall from §2.3.3 that if the null hypothesis of no effect is true, then the expectation and variance of T are given by

$$E(T \mid \mathscr{F}, \mathscr{Z}) = \frac{I(I+1)}{4}, \qquad (14.1)$$

$$\mathrm{var}(T \mid \mathscr{F}, \mathscr{Z}) = \frac{I(I+1)(2I+1)}{24}, \qquad (14.2)$$

and, after subtracting the expectation and dividing by the standard deviation, the distribution of T is well approximated by the standard Normal distribution if the

---

[2] This statement is correct in concept, but it omits a small technical detail. The distribution of Wilcoxon's statistic, T, is discrete, so its distribution attaches dollops of probability to the possible integer values of T, namely $0, 1, \ldots, I(I+1)/2$. In consequence, there may be no value $\zeta_{0.05}$ such that $\Pr(T \geq \zeta_{0.05} \mid \mathscr{F}, \mathscr{Z}) = 0.05$. In consequence, the best feasible value, $\zeta_{0.05}$, is used, that is, the smallest value such that $\Pr(T \geq \zeta_{0.05} \mid \mathscr{F}, \mathscr{Z}) \leq 0.05$ when the null hypothesis is true. With this value, $\zeta_{0.05}$, the probability of rejection of a true hypothesis is less than or equal to 0.05 — that is, the test has level 0.05 — but the probability of false rejection may be slightly below 0.05 — that is, the test has size slightly less than 0.05. In moderately large samples, I, the dollops of probability at integer values of T are very small, so this technical detail may be ignored for conceptual purposes. When an appeal is made to the central limit theorem as $I \to \infty$ to justify an approximate Normal null distribution for T, the technical detail disappears.

[3] Power will be computed for various continuous distributions of the responses, such as a Normal distribution, and for continuous distributions ties never occur; that is, the probability of a tie is zero. As has been seen several times, ties are a minor inconvenience, easy to address in practice, but they tend to clutter theoretical arguments. As is typically done, ties are assumed absent until a circumstance arises in which it actually becomes important to address the issue.

number of pairs I is moderately large, that is, the standardized deviate converges in distribution to the standard Normal distribution,

$$\frac{T - E(T|\mathscr{F}, \mathscr{Z})}{\sqrt{\text{var}(T|\mathscr{F}, \mathscr{Z})}} \longrightarrow N(0,1) \quad \text{as } I \to \infty; \tag{14.3}$$

see Lehmann [3, §3.2]. Recall also that the standard Normal cumulative distribution function is denoted $\Phi(\cdot)$ and its inverse, the quantile function, is denoted $\Phi^{-1}(\cdot)$, so that, for instance, $\Phi(1.65) = 0.95$ and $\Phi^{-1}(0.95) = 1.65$.

With $I = 100$ pairs, compute $E(T|\mathscr{F},\mathscr{Z}) = I(I+1)/4$ as $100(100+1)/4 = 2525$, $\text{var}(T|\mathscr{F},\mathscr{Z}) = I(I+1)(2I+1)/24$ as $100(100+1)(2 \times 100+1)/24 = 84587.5$. Using (14.3) to approximate the probability that $T \geq 3005$, we calculate $(3005 - 2525)/\sqrt{84587.5} = 1.65$, and $\Pr(T \geq 3005|\mathscr{F},\mathscr{Z}) \doteq 1 - \Phi(1.65) = 1 - 0.95 = 0.05$. In §2.3.3, computations along these lines yielded approximate one-sided P-values for Wilcoxon's statistic. More generally, under the null hypothesis H_0, for any fixed ζ,

$$\Pr(T \geq \zeta|\mathscr{F},\mathscr{Z}) \doteq 1 - \Phi\left\{\frac{\zeta - E(T|\mathscr{F},\mathscr{Z})}{\sqrt{\text{var}(T|\mathscr{F},\mathscr{Z})}}\right\}. \tag{14.4}$$

The Normal approximation works increasingly well as the sample size increases, $I \to \infty$; however, $I = 100$ is quite adequate for most practical uses of (14.4).

For power calculations, we do not want the realized P-value computed in §2.3.3 from (14.4), but rather the quantile, ζ_α, of the distribution of T that corresponds with a P-value less than or equal to α, conventionally $\alpha = 0.05$. Rearranging (14.4) yields $\Pr(T \geq \zeta_\alpha|\mathscr{F},\mathscr{Z}) \doteq \alpha$ if

$$\zeta_\alpha \doteq E(T|\mathscr{F},\mathscr{Z}) + \Phi^{-1}(1-\alpha)\sqrt{\text{var}(T|\mathscr{F},\mathscr{Z})}. \tag{14.5}$$

For instance, with $I = 100$ pairs and $\alpha = 0.05$, $\Phi^{-1}(1 - 0.05) = 1.645$, so the critical value $\zeta_{0.05}$ is computed as approximately $2525 + 1.645\sqrt{84587.5} = 3005$. Of course, (14.4) and (14.5) agree that $\Pr(T \geq 3005|\mathscr{F},\mathscr{Z}) \doteq 0.05$ for $I = 100$ pairs, but (14.5) reverses the computation in (14.4) determining ζ_α from α. With $I = 100$ pairs in a randomized experiment, we would reject the null hypothesis of no treatment effect in a one-sided $\alpha = 0.05$ level test if T turned out to be at least $\zeta_{0.05} = 3005$.

Wilcoxon's signed rank statistic T has the convenient property that the quantiles ζ_α of its null distribution are determined by α and I without reference to $\mathscr{F}$. See the discussion of 'distribution free' statistics in §2.3.3. In particular, the approximation (14.5) is computed from α and I using the standard Normal distribution $\Phi(\cdot)$ and (14.1) and (14.2).

14.1 The Power of a Test in a Randomized Experiment

Step 2: Determining the power assuming the null hypothesis is false

Consider now the second step, determining the power as the probability that $T \geq \zeta_\alpha$ when the null hypothesis is false. If the null hypothesis is false, then something else is true, and the power depends upon what is, in fact, true. One way to proceed would be to specify in detail what is true, that is to specify $\mathscr{F}$, in particular to specify the responses, (r_{Tij}, r_{Cij}), each subject ij would exhibit under treatment and under control. This specification would permit the calculation of the conditional power given $\mathscr{F}$, that is, the conditional probability that $T \geq \zeta_\alpha$ given $\mathscr{F}, \mathscr{Z}$. Somewhat more precisely, the conditional power is $\Pr(T \geq \zeta_\alpha | \mathscr{F}, \mathscr{Z})$ for some specific $\mathscr{F}$ for which there is a nonzero treatment effect. Computing the conditional power for $I = 100$ pairs of two subjects would entail specifying (r_{Tij}, r_{Cij}) for two hundred individuals, which sounds like a thankless task. Instead, we imagine that $\mathscr{F}$ was generated by a stochastic model, and compute the unconditional power — invariably just called 'the power' — as the probability that $T \geq \zeta_\alpha$ for all random $\mathscr{F}$'s generated by this model. That is, the (unconditional) power for Wilcoxon's statistic is $\Pr(T \geq \zeta_\alpha | \mathscr{Z})$ which is the expectation of $\Pr(T \geq \zeta_\alpha | \mathscr{F}, \mathscr{Z})$ over the specified model generating $\mathscr{F}$'s.

In principle, with a sufficiently fast computer, the problem of computing the power is now 'solved.' That is, (i) simulate an $\mathscr{F}$ from the specified model, (ii) randomize treatments in the I pairs, (iii) compute Wilcoxon's T from the treated-minus-control differences in observed responses, Y_i, (iv) score a 1 if $T \geq \zeta_\alpha$, score a 0 if $T < \zeta_\alpha$, and (v) repeat steps (i)–(iv) many times, estimating the power as the proportion of 1's. In practice, it is often helpful to work a little harder and obtain an expression for the power, $\Pr(T \geq \zeta_\alpha | \mathscr{Z})$.

A simple case: constant effect with random errors

This definition of power is quite flexible, and its flexibility is put to use in later chapters, but to make things tangible, consider the simplest nontrivial case, though nonetheless an important case. In the simplest case, the treated-minus-control differences in observed responses, $Y_i = (Z_{i1} - Z_{i2})(R_{i1} - R_{i2})$, are independent and identically distributed (iid) observations drawn from a distribution $F(\cdot)$. For instance, a simple model to generate $\mathscr{F}$ has differences in responses to control, $r_{Ci1} - r_{Ci2}$, that are iid observations from a Normal distribution with expectation 0 and variance ω^2 — that is, $r_{Ci1} - r_{Ci2} \sim_{iid} N(0, \omega^2)$ — and an additive constant treatment effect, $r_{Tij} - r_{Cij} = \tau$ for all ij. Then, in a randomized experiment, the treated-minus-control difference in observed responses, $Y_i = (Z_{i1} - Z_{i2})(R_{i1} - R_{i2})$, is $Y_i = \tau + (Z_{i1} - Z_{i2})(r_{Ci1} - r_{Ci2})$, where given $\mathscr{F}$ and $\mathscr{Z}$, the quantity $Z_{i1} - Z_{i2}$ is ± 1 with equal probabilities, so $Y_i \sim_{iid} N(\tau, \omega^2)$. The power in this case is the chance that $T \geq \zeta_\alpha$ when Wilcoxon's signed rank statistic is computed from I independent observations drawn from $N(\tau, \omega^2)$.

When computing the power for differences Y_i that are sampled independently from a continuous distribution $F(\cdot)$, Lehmann [3, §4.2] gives an approximation to the power of Wilcoxon's signed rank statistic, T. The approximation appeals to the central limit theorem twice as the sample size I increases, $I \to \infty$, once to obtain the critical value ζ_α in step 1 as has already been done, and the second time to approximate the behavior of T when the null hypothesis is false. For $Y_i \sim_{iid} F(\cdot)$, define $p = \Pr(Y_i > 0)$, $p'_1 = \Pr(Y_i + Y_{i'} > 0)$ and $p'_2 = \Pr(Y_i + Y_{i'} > 0 \text{ and } Y_i + Y_{i''} > 0)$ with $i < i' < i''$. Then Lehmann shows that the expectation μ_F of Wilcoxon's T when the null hypothesis is false and the Y_i are sampled independently from $F(\cdot)$ is

$$\mu_F = \frac{I(I-1)p'_1}{2} + Ip, \tag{14.6}$$

the variance of T is

$$\sigma_F^2 = I(I-1)(I-2)\left(p'_2 - p'^{\,2}_1\right) \tag{14.7}$$

$$+ \frac{I(I-1)}{2}\left\{2\left(p - p'_1\right)^2 + 3p'_1\left(1 - p'_1\right)\right\} + Ip(1-p), \tag{14.8}$$

and that the central limit theorem yields approximate power as $\Pr(T \geq \zeta_\alpha | \mathscr{Z}) \doteq 1 - \Phi\{(\zeta_\alpha - \mu_F)/\sigma_F\}$.[4]

The required quantities p, p'_1, and p'_2 are determined by the distribution $F(\cdot)$. For the Normal distribution, p and p'_1 have simple expressions, while p'_2 may be obtained by numerical integration or simulation. For any $F(\cdot)$, it is easy to estimate p, p'_1, and p'_2 very precisely by simulating one very large sample of Y_i's drawn from $F(\cdot)$.[5]

For instance, if $Y_i \sim_{iid} N(\tau, \omega^2)$ with $\tau = \frac{1}{4}$ and $\omega = 1$ (or more generally with $\tau/\omega = \frac{1}{4}$), then $p = 0.599$, $p'_1 = 0.638$, $p'_2 = 0.482$. For $I = 100$, we found earlier that $\zeta_{0.05} \doteq 3005$. For $I = 100$ pairs, $\mu_F = 3217.16$, $\sigma_F^2 = 76400.5$, and $\Pr(T \geq \zeta_\alpha | \mathscr{Z}) \doteq 1 - \Phi\{(\zeta_\alpha - \mu_F)/\sigma_F\} = 0.78$. In other words, if the null hypothesis of no effect is false in a randomized experiment in such a way that the treated-minus-control differences in observed outcomes are Normal with expectation equal to one quarter of their standard deviation, then the probability that the null hypothesis will be rejected at the 0.05 level in a one-sided test is about 78% when there are 100 pairs. In the same way, with $\tau = \frac{1}{2}$ and $\omega = 1$ (or more generally with $\tau/\omega = \frac{1}{2}$) and $I = 100$ pairs, $\mu_F = 3834.67$, $\sigma_F^2 = 57041.2$, and $\Pr(T \geq \zeta_\alpha | \mathscr{Z}) \doteq 1 - \Phi\{(\zeta_\alpha - \mu_F)/\sigma_F\} = 0.999$.

Figure 14.1 plots the power of Wilcoxon's test for $Y_i \sim_{iid} N(\tau, 1)$ with $\tau = \frac{1}{4}$ or $\tau = \frac{1}{2}$ for $I = 20, \ldots 200$ pairs. With $I = 50$ pairs, the power is above 95% if the

---

[4] This is expression (4.28) in Lehmann [3, §4.2] except for the exclusion of the continuity correction.

[5] The probability $p'_1 = \Pr(Y_i + Y_{i'} > 0)$ will turn out to be the most important of the three probabilities. Notice that in (14.6), the quantity p'_1 dominates the expectation of T, because p'_1 is multiplied by $I(I+1)/2$ while p is multiplied by I, so in large samples, as $I \to \infty$, the dominant role is played by p'_1. Looking back to §2.5, we see that p'_1 is the expectation of $H_\mathscr{I}$ for $\mathscr{I} = \{i, i'\}$. As seen in §2.5, the quantity $H_\mathscr{I}$ is highly interpretable, and the interpretation carries over to p'_1.

effect is larger, $\tau = \frac{1}{2}$, but the power is a coin toss, about 50%, if the effect is smaller, $\tau = \frac{1}{4}$.

14.2 Power of a Sensitivity Analysis in an Observational Study

What is the power of a sensitivity analysis?

As seen in Chapter 2, if Wilcoxon's signed rank test is used to test the null hypothesis of no treatment effect in a randomized experiment, it yields a single P-value because we know the distribution of treatment assignments that was created by the random assignment of treatments. As seen in Chapter 3, in an observational study, randomization is not used to assign treatments, so this single P-value lacks a justification or warrant; it would be valid under the naïve model, but naïveté is not justification for an inference. Recall from §3.4 and §3.5 that for any given magnitude $\Gamma \geq 1$ of departure from random assignment in (3.13), an interval $[P_{\min}, P_{\max}]$ of possible P-values may be determined; see Table 3.2. The study is judged sensitive to unmeasured biases of magnitude Γ if the corresponding $P_{\min}$ is small and $P_{\max}$ is large, for instance $P_{\min} \leq 0.01$ and $P_{\max} > 0.05$. In Table 3.2, the study by Werfel et al. [9] was highly insensitive to unmeasured biases, specifically insensitive to biases of magnitude $\Gamma = 4$ but sensitive to $\Gamma = 5$.

If an observational study were actually free from unmeasured biases — if the naïve model (3.5)–(3.8) of §3.3 were true — then we would not know this from the observable data. We might see that the treated-minus-control differences in observed responses, Y_i, were typically positive, but we would be uncertain whether this reflected an effect caused by the treatment or some bias in the way treatments

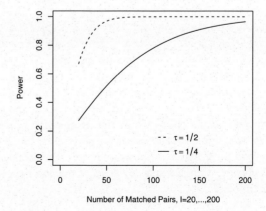

Fig. 14.1 Approximate power of Wilcoxon's signed rank test, T, in a paired randomized experiment when the treatment has an additive constant effect τ and the treated-minus-control differences in observed responses Y_i are independently sampled from a Normal distribution, $N(\tau, 1)$, with expectation τ and variance 1, for samples of size $I = 20, 21, \ldots, 200$ pairs.

were assigned or a combination of the two. The best we could hope to say in this favorable situation is that the treatment appears to work and that this appearance is highly insensitive to unmeasured biases, that is, only a very large value of Γ could produce this appearance from bias alone. Speaking informally, the power of a sensitivity analysis is the probability that our hopes will be realized.

Somewhat more precisely, define the 'favorable situation' as a treatment that actually causes meaningful effects in an observational study that is free of unmeasured biases. Specifically, in the favorable situation, the null hypothesis of no effect is false, and not merely false but the treatment effect is large enough to be interesting and substantively important. Also, in the favorable situation, the study is free of unmeasured biases, people who look comparable are comparable, in the sense that (3.5)–(3.8) are true. That is, the treated-minus-control differences in responses, Y_i, are not centered at zero because the treatment worked in the absence of bias. If we were in the favorable situation, we would not know it from the observable data. It is in this favorable situation that we would like to report a high degree of insensitivity to unmeasured biases.

In testing the null hypothesis of no treatment effect at level α, conventionally $\alpha = 0.05$, rejection of the null hypothesis is insensitive to a bias of magnitude Γ if $P_{\max} \leq \alpha$; see, again, the example in Table 3.2. In the favorable situation, the power of a sensitivity analysis conducted with a specified α and Γ is the probability that $P_{\max} \leq \alpha$. That is, if the treatment had an effect and there is no unmeasured bias, then the power of the sensitivity analysis for specified α and Γ is the probability that we will be able to say that a bias of magnitude Γ would not lead to acceptance of the null hypothesis of no effect when tested at level α. In Table 3.2, it was found that $P_{\max} \leq 0.05$ for $\Gamma = 4$; the power is the probability of such an event computed before the data are examined. For $\Gamma = 1$, this is the power of a test of no effect in a randomized experiment as computed in §14.1.

Computing the power of a sensitivity analysis: The two steps

As in §14.1, the computations leading to the power of a sensitivity analysis are done in two steps. As in §14.1, the first step is to compute a critical value, now $\zeta_{\Gamma,\alpha}$, so that $\Pr(T \geq \zeta_{\Gamma,\alpha} | \mathscr{F}, \mathscr{Z}) \leq \alpha$ for all treatment assignment probabilities that satisfy (3.16)–(3.18) when the null hypothesis of no treatment effect is true.[6] If we observed $T \geq \zeta_{\Gamma,\alpha}$, then we would report that the upper bound on the one-sided P-value for testing no effect is at most α for all biases of magnitude at most Γ; see again Table 3.2. The second step is to compute the probability that $T \geq \zeta_{\Gamma,\alpha}$ in the 'favorable situation' in which the treatment has an effect and there is actually no bias from unmeasured covariates. It is in this favorable situation that we hope to report a high degree of insensitivity to unmeasured bias, and if $T \geq \zeta_{\Gamma,\alpha}$ then we will be able to report that rejection of the null hypothesis is insensitive to a bias of magnitude Γ.

---

[6] To be precise, we seek the smallest $\zeta_{\Gamma,\alpha}$ such that $\Pr(T \geq \zeta_{\Gamma,\alpha} | \mathscr{F}, \mathscr{Z}) \leq \alpha$.

14.2 Power of a Sensitivity Analysis in an Observational Study

As in §14.1, the computation of the critical value is little more than a rearrangement of the computation of the P-value for a one-sided test. Specifically, rearranging the computation (3.23) in §3.5 yields

$$\zeta_{\Gamma,\alpha} \doteq E\left(\overline{\overline{T}}\middle|\mathscr{F},\mathscr{L}\right) + \Phi^{-1}(1-\alpha)\sqrt{\operatorname{var}\left(\overline{\overline{T}}\middle|\mathscr{F},\mathscr{L}\right)} \qquad (14.9)$$

so that using (3.19) and (3.20) yields

$$\zeta_{\Gamma,\alpha} \doteq \frac{\Gamma}{1+\Gamma} \cdot \frac{I(I+1)}{2} + \Phi^{-1}(1-\alpha)\sqrt{\frac{\Gamma}{(1+\Gamma)^2}\frac{I(I+1)(2I+1)}{6}}. \qquad (14.10)$$

For $\Gamma = 1$ in a randomized experiment, expressions (14.9) and (14.10) reduce to (14.5).

Continuing the numerical illustration from §14.1, with $I = 100$ pairs, $\alpha = 0.05$, and $\Gamma = 2$, the critical value $\zeta_{2,0.05}$ is approximately

$$\frac{2}{1+2} \cdot \frac{100(100+1)}{2} + 1.645\sqrt{\frac{2}{(1+2)^2}\frac{100(100+1)(2\cdot 100+1)}{6}} \qquad (14.11)$$

or $\zeta_{2,0.05} = 3817.7$, which is noticeably higher than $\zeta_{0.05} = \zeta_{1,0.05} \doteq 3005$ for a randomized experiment $\Gamma = 1$ in §14.1. In words, with $I = 100$ pairs, if $T \geq 3817.7$ then the maximum possible P-value is ≤ 0.05 for all treatment assignment probabilities that satisfy (3.16)–(3.18) with $\Gamma = 2$. This completes the first step.

Step 2: Determining the power when the null hypothesis is false and there is no unobserved bias

Again, we wish to compute the power of the sensitivity analysis — the chance that $T \geq \zeta_{\Gamma,\alpha}$ — in the favorable situation in which there is no bias from unmeasured covariates and the treatment has an effect. It is in this situation that we hope to report that the treatment appears to work and the appearance is insensitive to moderately large biases.

More precisely, the naïve model (3.5)–(3.8) is assumed to hold, or equivalently (3.16)–(3.18) holds with $\Gamma = 1$, and the data $\mathscr{F}$ were generated by a stochastic model with a treatment effect, and we wish to determine $\Pr(T \geq \zeta_{\Gamma,\alpha}|\mathscr{L})$ averaging $\Pr(T \geq \zeta_{\Gamma,\alpha}|\mathscr{F},\mathscr{L})$ over the model that generates $\mathscr{F}$. This is, of course, identical to the second step in §14.1 for a randomized experiment except the critical value, $\zeta_{\Gamma,\alpha}$, has changed. In consequence, the second step is performed in the same way as in §14.1 with the new critical value. As in §14.1, the second step may in general be performed by simulating many sets of data $\mathscr{F}$ from the model, determining the proportion with $T \geq \zeta_{\Gamma,\alpha}$, and in particular cases the power of the sensitivity analysis may be determined analytically.

Fig. 14.2 Approximate power of a sensitivity analysis when there is no bias from unmeasured covariates, the treatment has an additive constant effect $\tau = 0.5$ and the treated-minus-control differences in observed responses Y_i are independently sampled from a Normal distribution, $N(\tau, 1)$, with expectation τ and variance 1, for samples of size $I = 20, 21, \ldots, 2000$ pairs. The several curves give the power for several values of Γ. The power tends to 1 as I increases for $\Gamma < 3.171$ and it tends to 0 for $\Gamma > 3.171$.

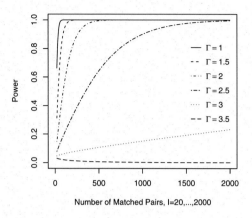

Continuing the numerical example, if $Y_i \sim_{iid} N(\tau, \omega^2)$, $i = 1, \ldots, I = 100$ with $\tau = \frac{1}{2}$ and $\omega = 1$ then as in §14.1, $\mu_F = 3834.67$, $\sigma_F^2 = 57041.2$. To compute $\Pr(T \geq \zeta_{\Gamma, \alpha} | \mathscr{Z})$ for $\Gamma = 2$ and $\alpha = 0.05$, we find in the first step that $\zeta_{2, 0.05} = 3817.7$. In the second step, we approximate $\Pr(T \geq \zeta_{\Gamma, \alpha} | \mathscr{Z})$ by $1 - \Phi\{(\zeta_{\Gamma, \alpha} - \mu_F)/\sigma_F\}$ which is $1 - \Phi\{(3817.7 - 3834.67)/\sqrt{57041.2}\} = 0.528$. If the treatment had an effect of $\tau = \frac{1}{2}$ and there is no bias from unobserved covariates, there is a 53% chance that a sensitivity analysis will yield an upper bound on the P-value that is at most 0.05 when performed with $\Gamma = 2$. This contrasts with 99.9% power in a randomized experiment, $\Gamma = 1$, as calculated in §14.1.

A first look at the power of a sensitivity analysis

Figure 14.2 shows the power in the 'favorable situation' with $Y_i \sim_{iid} N\left(\frac{1}{2}, 1\right)$, $i = 1, \ldots, I$ for $I = 20, 21, \ldots, 2000$ and for $\Gamma = 1, 1.5, 2, 2.5, 3, 3.5$. Again, the 'favorable situation' is characterized by a treatment effect without bias, so it is the situation where we hope to report insensitivity to small and moderate unmeasured biases. The rather special case in Figure 14.2 illustrates several general properties of the power of a sensitivity analysis. The power is higher in a randomized experiment, where Γ is known to equal 1, because there is no need to take account of bias from nonrandom treatment assignment. The power degrades as Γ increases. The most striking feature of Figure 14.2 is that the power is increasing in the sample size I for small Γ but is decreasing in I for $\Gamma = 3.5$. There is, in fact, a value, $\widetilde{\Gamma}$, such that the power $\Pr(T \geq \zeta_{\Gamma, \alpha} | \mathscr{Z})$ tends to 1 as $I \to \infty$ for all $\Gamma < \widetilde{\Gamma}$ and $\Pr(T \geq \zeta_{\Gamma, \alpha} | \mathscr{Z})$ tends to 0 as $I \to \infty$ for all $\Gamma > \widetilde{\Gamma}$. In the situation in Figure 14.2, the design sensitivity is $\widetilde{\Gamma} = 3.171$. Because $3 < \widetilde{\Gamma} = 3.171$, the power for $\Gamma = 3$ will eventually rise to 1 as $I \to \infty$ even though it has not yet reached 30% at $I = 2000$

in Figure 14.2. In a sufficiently large sample, I, the 'favorable situation' with $Y_i \sim_{iid} N\left(\frac{1}{2},1\right)$, $i = 1,\ldots,I$ can be distinguished from bias of magnitude $\Gamma < \widetilde{\Gamma} = 3.171$, but not from biases of magnitude $\Gamma > \widetilde{\Gamma} = 3.171$. Think of this statement as the quantitative replacement for the statement 'association does not imply causation.' In the 'favorable situation' with $Y_i \sim_{iid} N\left(\frac{1}{2},1\right)$, $i = 1,\ldots,I$, the association between treatment Z and response Y can be distinguished from biases less than $\widetilde{\Gamma}$ but not from biases greater than $\widetilde{\Gamma}$.

It is time to consider the design sensitivity, $\widetilde{\Gamma}$, in greater detail.

14.3 Design Sensitivity

A first look at design sensitivity

In many simple settings, the power of a sensitivity analysis tends to 1 as the sample size I increases if the analysis is performed with a sufficiently small value of Γ, but the power tends to 0 if the analysis is performed with a sufficiently large value of Γ. The switch occurs at a value, $\widetilde{\Gamma}$, called the design sensitivity [5]. In the situation in Figure 14.2, the design sensitivity was $\widetilde{\Gamma} = 3.171$. It would be natural to seek a design that results in a larger design sensitivity and to avoid designs that result in a smaller design sensitivity. It would be natural to appraise the performance of the methods in Chapter 5 in terms of their impact on design sensitivity.

Figure 14.2 plotted the power as a function of the sample size for several values of Γ. Although such a plot resembles the familiar plot of power for a randomized experiment (e.g., Figure 14.1), it is actually less instructive than plotting the power as a function of Γ for several sample sizes I. Figure 14.3 plots power against Γ for samples of size $I = 200, 2000, 20000$ in the 'favorable situation' in which there is a treatment effect but no unmeasured bias, with $Y_i \sim_{iid} N(\tau, 1)$, $i = 1,\ldots,I$, for $\tau = \frac{1}{2}$ and $\tau = 1$.

The first thing to notice about Figure 14.3 is that, as the sample size I increases, the power viewed as a function of Γ is tending to a step function with a single step down from power 1 to power 0, with the step located at the design sensitivity, $\widetilde{\Gamma}$. For $\tau = \frac{1}{2}$, the step is at $\widetilde{\Gamma} = 3.171$, whereas for $\tau = 1$ the step is at $\widetilde{\Gamma} = 11.715$. For all sample sizes, I, there is negligible power for $\Gamma > \widetilde{\Gamma}$.

The six power curves in Figure 14.3 refer to three sample sizes, $I = 200, 2000, 20000$, and two design sensitivities, $\widetilde{\Gamma} = 3.171$ (for $\tau = \frac{1}{2}$) and $\widetilde{\Gamma} = 11.715$ (for $\tau = 1$). In this specific 3×2 array of power curves, the difference in design sensitivities matters more than the difference in sample sizes, although sample size does have a substantial effect on power. For instance, at $\Gamma = 5$, the power is close to 1 for $I = 200$ pairs with $\tau = 1$ but the power is close to zero with $I = 20000$ pairs and $\tau = \frac{1}{2}$. A bias of magnitude $\Gamma = 5$ could explain the observed association between treatment and outcome if $\tau = \frac{1}{2}$ no matter how large the sample size becomes, whereas a mere $I = 200$ pairs could eliminate bias of magnitude $\Gamma = 5$ as a plausible explanation

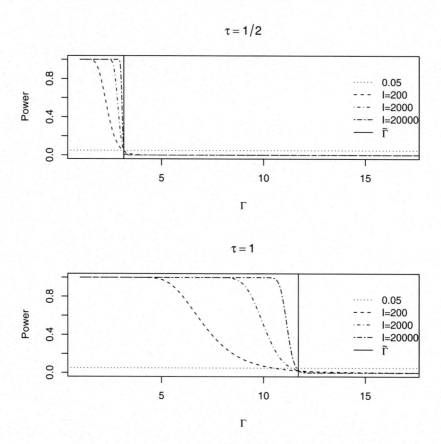

Fig. 14.3 Approximate power of a sensitivity analysis when there is no bias from unmeasured covariates, the treatment has an additive constant effect τ and the treated-minus-control differences in observed responses Y_i are independently sampled from a Normal distribution, $N(\tau, 1)$, with expectation τ and variance 1, for samples of size $I = 200$, 2000, and 20000 pairs. The solid vertical line is at the design sensitivity, $\tilde{\Gamma}$. The power tends to 1 as I increases for $\Gamma < \tilde{\Gamma}$ and tends to 0 for $\Gamma > \tilde{\Gamma}$. The power is computed for a 0.05 level test, and the horizontal dotted line is at 0.05.

if $\tau = 1$. A smaller study with a larger effect ($I = 200$, $\tau = 1$) is very likely to be less sensitive to unmeasured bias than a vastly larger study with a smaller effect ($I = 20000$, $\tau = \frac{1}{2}$). It is often asserted that, in observational studies, biases from unmeasured covariates constitute a greater source of uncertainty than does a limited sample size, and Figure 14.3 is one quantitative expression of this assertion.

14.3 Design Sensitivity

A formula for design sensitivity

Design sensitivity is a general concept applicable to many statistics, not just Wilcoxon's signed rank statistic, and to many sampling situations, not just matched pairs [5]. However, in the case of Wilcoxon's signed rank statistic for matched pairs, there is a simple explicit formula for the design sensitivity [1, 8], which is given in Proposition 14.1. Notice that this explicit form uses the quantity $p'_1 = \Pr(Y_i + Y_{i'} > 0)$, which was encountered in §14.1 in connection with Lehmann's formula for the approximate power of Wilcoxon's signed rank statistic when applied in a randomized experiment.[7]

Proposition 14.1. *In I pairs matched exactly for observed covariates, $\mathbf{x}_{ij}$, suppose that the treated-minus-control matched pair differences in outcomes, Y_i, $i = 1,\ldots,I$, are independently sampled from a distribution $F(\cdot)$ and there is no bias from unmeasured covariates in the sense that treatments are actually assigned by the naïve model (3.5)–(3.8). For a sensitivity analysis applied to Wilcoxon's signed rank statistic T in a one-tailed α-level test of no effect with $\alpha > 0$, the power of the sensitivity analysis $\Pr(T \geq \zeta_{\Gamma,\alpha} | \mathscr{Z})$ satisfies*

$$\Pr(T \geq \zeta_{\Gamma,\alpha} | \mathscr{Z}) \to 1 \text{ as } I \to \infty \text{ for } \Gamma < \widetilde{\Gamma} \tag{14.12}$$

and

$$\Pr(T \geq \zeta_{\Gamma,\alpha} | \mathscr{Z}) \to 0 \text{ as } I \to \infty \text{ for } \Gamma > \widetilde{\Gamma}, \tag{14.13}$$

where

$$\widetilde{\Gamma} = \frac{p'_1}{1 - p'_1} \text{ with } p'_1 = \Pr(Y_i + Y_{i'} > 0), \; i \neq i'. \tag{14.14}$$

The proof is given in the appendix to this chapter.

Computing design sensitivity with additive effects and iid errors

If the Y_i, $i = 1,\ldots,I$, are independently sampled from a distribution $F(\cdot)$, then it is always easy to determine $p'_1 = \Pr(Y_i + Y_{i'} > 0)$, $i \neq i'$, and hence also $\widetilde{\Gamma}$, by simulation. In the case of an additive effect τ with Normal or Cauchy errors, explicit formulas are available. Recall that $\Phi(\cdot)$ is the standard Normal cumulative distribution, $N(0,1)$, and write $\Upsilon(\cdot)$ for the cumulative Cauchy distribution. As in [3, pages 166–167], if $Y_i \sim_{iid} N(\tau, \omega^2)$ then $(Y_i - \tau)/\omega \sim_{iid} N(0,1)$ and $(Y_i + Y_{i'} - 2\tau)/\sqrt{2\omega^2} \sim N(0,1)$, so

$$\Pr(Y_i + Y_{i'} > 0) = \Pr\left(\frac{Y_i + Y_{i'} - 2\tau}{\omega\sqrt{2}} > \frac{-2\tau}{\omega\sqrt{2}}\right) = \Phi\left(\sqrt{2}\frac{\tau}{\omega}\right). \tag{14.15}$$

---

[7] See also Note 5 in §14.1.

For instance, if $Y_i \sim_{iid} N\left(\frac{1}{2}, 1\right)$, as in Figures 14.2 and 14.3, then $\Pr(Y_i + Y_{i'} > 0) = \Phi\left(\sqrt{2} \cdot \frac{1}{2}\right) = 0.76025$ and $\widetilde{\Gamma} = 0.76025/(1 - 0.76025) = 3.171$. If $(Y_i - \tau)/\omega \sim \Upsilon(\cdot)$, then $\{(Y_i - \tau)/\omega + (Y_{i'} - \tau)/\omega\}/2 \sim \Upsilon(\cdot)$, so

$$\Pr(Y_i + Y_{i'} > 0) = \Pr\left(\frac{Y_i + Y_{i'}}{2} > 0\right) = \Pr\left(\frac{Y_i + Y_{i'} - 2\tau}{2\omega} > \frac{-2\tau}{2\omega}\right) = \Upsilon\left(\frac{\tau}{\omega}\right). \tag{14.16}$$

For instance, if $Y_i - \frac{1}{2} \sim \Upsilon(\cdot)$, then $\Pr(Y_i + Y_{i'} > 0) = \Upsilon\left(\frac{1}{2}\right) = 0.64758$ and $\widetilde{\Gamma} = 0.64758/(1 - 0.64758) = 1.838$.

14.4 Summary

As seen in Figure 14.3 and Proposition 14.1, no matter how large the sample size I becomes, a study will be sensitive to bias of magnitude Γ if $\Gamma > \widetilde{\Gamma}$ where $\widetilde{\Gamma}$ is the design sensitivity. More precisely, when testing the null hypothesis of no effect, the power of a sensitivity analysis will tend to 1 as $I \to \infty$ for all $\Gamma < \widetilde{\Gamma}$ and to 0 for all $\Gamma > \widetilde{\Gamma}$. In this sense, $\widetilde{\Gamma}$ measures the limit of the ability of a particular study design to distinguish a treatment effect from a bias due to failure to control for an unmeasured covariate u_{ij}. For this reason, the design sensitivity $\widetilde{\Gamma}$ is a natural numerical measure for comparing design strategies that hope to distinguish treatment effects from unmeasured biases, such as the strategies discussed in Chapter 5. The remainder of Part III will make such comparisons.

14.5 Further Reading

The remainder of Part III is further reading for this chapter. Design sensitivity was introduced in [5] and is developed further in [1, 6, 7, 8].

Appendix: Technical Remarks and Proof of Proposition 14.1

This appendix contains a proof of Proposition 14.1 in §14.3. The formula and proof [1, 8] for Wilcoxon's signed rank statistic are a special case of a general result [5, §3] with the attractive feature that the design sensitivity $\widetilde{\Gamma}$ has an explicit closed form (14.14).

The approximate power is $\Pr(T \geq \zeta_{\Gamma,\alpha} | \mathscr{L}) \doteq 1 - \Phi\{(\zeta_{\Gamma,\alpha} - \mu_F)/\sigma_F\}$, where μ_F is the expectation of T in the 'favorable situation.' An intuitive, heuristic derivation [5, 8] uses the notion that if $\zeta_{\Gamma,\alpha} > \mu_F$, then in large samples, as $I \to \infty$, the signed rank statistic T eventually will be less than $\zeta_{\Gamma,\alpha}$, whereas if $\zeta_{\Gamma,\alpha} < \mu_F$ then T will eventually be greater than $\zeta_{\Gamma,\alpha}$. So the heuristic equates $\zeta_{\Gamma,\alpha}$ in (14.10) and

14.5 Further Reading

μ_F in (14.6) and solves for Γ. It is slightly better to equate $\zeta_{\Gamma,\alpha}/\mu_F$ to 1, let $I \to \infty$, ignore terms that are small compared with I^2, and solve for Γ,

$$\frac{\zeta_{\Gamma,\alpha}}{\mu_F} \doteq \frac{\frac{\Gamma I(I+1)}{2(1+\Gamma)} + \Phi^{-1}(1-\alpha)\sqrt{\frac{\Gamma I(I+1)(2I+1)}{6(1+\Gamma)^2}}}{I(I-1)p_1'/2 + Ip} \tag{14.17}$$

$$\doteq \frac{\Gamma}{p_1'(1+\Gamma)}, \tag{14.18}$$

where $1 = \Gamma / \{p_1'(1+\Gamma)\}$ has solution $\tilde{\Gamma} = p_1'/(1-p_1')$ as in (14.14). The heuristic is attractive because it uses only the expectation μ_F of T in the favorable situation, so the calculation is simple. Alas, the heuristic is 'heuristic' precisely because the power depends on $(\zeta_{\Gamma,\alpha} - \mu_F)/\sigma_F$, not on $\zeta_{\Gamma,\alpha} - \mu_F$, and the heuristic ignores σ_F, hoping that things will work out. Indeed, things do work out, as the proof that follows demonstrates [1]. The general result in [5, §3] shows that the heuristic calculation works quite generally, but the proof that follows is self-contained and does not depend upon the general result.

Proof. For large I, the power, $\Pr(T \geq \zeta_{\Gamma,\alpha} | \mathscr{Z})$, is approximately equal to $1 - \Phi\{(\zeta_{\Gamma,\alpha} - \mu_F)/\sigma_F\}$, so the power tends to 1 as $I \to \infty$ if $(\zeta_{\Gamma,\alpha} - \mu_F)/\sigma_F \to -\infty$ and the power tends to 0 as $I \to \infty$ if $(\zeta_{\Gamma,\alpha} - \mu_F)/\sigma_F \to \infty$. Write $\kappa = \Gamma/(1+\Gamma)$. Using the expressions for μ_F and σ_F^2 from (14.6) and (14.7) in §14.1 and the expression for $\zeta_{\Gamma,\alpha}$ from (14.10) in §14.2 yields

$$\frac{\zeta_{\Gamma,\alpha} - \mu_F}{\sigma_F} = \tag{14.19}$$

$$\frac{\kappa I(I+1)/2 + \Phi^{-1}(1-\alpha)\sqrt{\kappa(1-\kappa)I(I+1)(2I+1)/6} - I(I-1)p_1'/2 - Ip}{\sqrt{I(I-1)(I-2)\left(p_2' - {p_1'}^2\right) + \frac{I(I-1)}{2}\left\{2(p-p_1')^2 + 3p_1'(1-p_1')\right\} + Ip(1-p)}}. \tag{14.20}$$

Expression (14.20) looks untidy at first. Notice however that every term in the numerator and denominator of (14.20) involves I, and $I \to \infty$. So divide the numerator and denominator of (14.20) by $I\sqrt{I}$, and let $I \to \infty$. The last term in the numerator, $-Ip$, becomes $-Ip/(I\sqrt{I}) \to 0$ as $I \to \infty$, whereas the first term in the numerator, $\kappa I(I+1)/2$, becomes $\kappa I(I+1)/(2I\sqrt{I}) \approx \sqrt{I}\kappa/2$ in the sense that their ratio is tending to 1 as $I \to \infty$. Continuing in this way, we find that

$$\frac{\zeta_{\Gamma,\alpha} - \mu_F}{\sigma_F} \approx \frac{\sqrt{I}(\kappa - p_1')}{2\sqrt{p_2' - {p_1'}^2}} \text{ as } I \to \infty, \tag{14.21}$$

in the sense that the ratio of the left and right sides of (14.21) tends to 1 as $I \to \infty$ provided $\kappa \neq p_1'$. If $\kappa > p_1'$, then (14.21) tends to ∞ as $I \to \infty$, so the power tends to

0. If $\kappa < p'_1$ then (14.21) tends to $-\infty$ as $I \to \infty$, so the power tends to 1. Moreover, $\kappa = \Gamma/(1+\Gamma) > p'_1$ if and only if $\Gamma > p'_1/\left(1-p'_1\right)$.

References

1. Heller, R., Rosenbaum, P.R., Small, D.: Split samples and design sensitivity in observational studies. J Am Statist Assoc **104**, to appear (2009)
2. Hodges, J.L., Lehmann, E.L.: Estimates of location based on ranks, Ann Math Statist **34**, 598–611 (1963)
3. Lehmann, E.L.: Nonparametrics, San Francisco: Holden Day (1975). Reprinted New York: Springer (2006)
4. Rosenbaum, P. R.: Hodges-Lehmann point estimates of treatment effect in observational studies. J Am Statist Assoc **88** 1250–1253 (1993)
5. Rosenbaum, P. R.: Design sensitivity in observational studies. Biometrika **91**, 153–164 (2004)
6. Rosenbaum, P. R.: Heterogeneity and causality: Unit heterogeneity and design sensitivity in observational studies. Am Statistician **59**, 147–152 (2005)
7. Rosenbaum, P.R.: What aspects of the design of an observational study affect its sensitivity to bias from covariates that were not observed? Festshrift for Paul W. Holland. Princeton, NJ: ETS (2009)
8. Small, D., Rosenbaum, P. R.: War and wages: The strength of instrumental variables and their sensitivity to unobserved biases. J Am Statist Assoc **103**, 924–933 (2008)
9. Werfel, U., Langen, V., Eickhoff, I., Schoonbrood, J., Vahrenholz, C., Brauksiepe, A., Popp, W., Norpoth, K.: Elevated DNA single-strand breakage frequencies in lymphocytes of welders. Carcinogenesis **19**, 413–418 (1998)

Chapter 15
Heterogeneity and Causality

Abstract Before R.A. Fisher introduced randomized experimentation, the literature on causal inference emphasized reduction of heterogeneity of experimental units. To what extent is heterogeneity relevant to causal claims in observational studies when random assignment of treatments is unethical or infeasible?

15.1 J.S. Mill and R.A. Fisher: Reducing Heterogeneity or Introducing Random Assignment

In his *System of Logic: Principles of Evidence and Methods of Scientific Investigation*, John Stuart Mill [11] proposed "four methods of experimental inquiry," including the "method of difference":

> If an instance in which the phenomenon ... occurs and an instance in which it does not ... have every circumstance save one in common ... [then] the circumstance [in] which alone the two instances differ is the ... cause or a necessary part of the cause ... [Mill wanted] "two instances ... exactly similar in all circumstances except the one" [under study] [11, III, §8]

Notice Mill's emphasis on a complete absence of heterogeneity: "have every circumstance save one in common;" that is, on treated and control units that are identical but for the treatment. In the modern biology laboratory, nearly identical, genetically engineered mice are compared under treatment and control; this is a modern expression of Mill's 'method of difference.' It is clear from the quote that Mill believed, rightly or wrongly, that heterogeneity of experimental units is directly relevant to causal claims, and does not refer simply to reducing the standard error of an estimate.

Ronald Fisher [3, Chapter 2] took a starkly different view. In 1935, in Chapter 2 of his *Design of Experiments*, Fisher introduced randomized experimentation for the first time in book form, discussing his famous experiment of the lady tasting tea. Fisher was directly critical of the 'method of difference':

It is not sufficient remedy to insist that "all the cups must be exactly alike" in every respect except that to be tested. For this is a totally impossible requirement in our example, and equally in all other forms of experimentation... These are only examples of the differences probably present; it would be impossible to present an exhaustive list of such possible differences... because [they]... are always strictly innumerable. When any such cause is named, it is usually perceived that, by increased labor and expense, it could be largely eliminated. Too frequently it is assumed that such refinements constitute improvements to the experiment... [3, page 18]

In the first omission, "...," in this quote, Fisher discussed the many ways two cups of tea may differ.

Fisher is, of course, engaged in an enormously important task: he is introducing the logic of randomized experimentation to a broad audience for the first time. Moreover, it would be reasonable to say that Fisher was correct and Mill was wrong in certain critical respects. In a randomized clinical trial conducted in a hospital, the patients are heterogeneous and not much can be done about it. There is no opportunity to replace the hospital's patients by genetically engineered, nearly identical patients. And yet it is possible to randomly assign treatments to heterogeneous patients and draw valid causal inferences in just the manner that Fisher advocated.

Despite this, one senses that there is at least something that is sensible in Mill's method of difference, something that is sensible in the fanatical efforts of the basic science laboratory to eliminate heterogeneity, in the use of nearly identical mice. One senses that Fisher, in his understandable enthusiasm for his new method, has gone just a tad too far in his criticism of Mill's drive to eliminate heterogeneity. Indeed, the issue may be particularly relevant in observational studies where random assignment of treatments is either unethical or infeasible.

Some care is required in discussing heterogeneity. Heterogeneity is itself heterogeneous; there are several kinds of heterogeneity. In the biology laboratory, it is often wise to use several different strains or species of genetically engineered, nearly identical laboratory animals, making sure that each strain or species is equally represented in treated and control groups, to verify that any conclusion is not a peculiarity of a single strain. It is uncontrolled rather than controlled heterogeneity that is the target for reduction or elimination. Controlled heterogeneity has numerous uses.[1]

See Paul Holland's [8] essay "Statistics and causal inference" for further contrasts of the views of Mill and Fisher.

---

[1] Recall the related discussion of Bitterman's concept of "control by systematic variation" in §5.2.2, where some factor is demonstrated to be irrelevant by systematically varying that factor.

15.2 A Larger, More Heterogeneous Study Versus a Smaller, Less Heterogeneous Study

Large I or small σ: Which is better?

To explore the issue raised in §15.1, consider the following simple situation. There are I matched pairs in an observational study with treated-minus-control differences in outcomes Y_i, $i = 1,\ldots,I$. Because it is an observational study, not a randomized experiment, we cannot assume that matching for observed covariates has removed all bias from nonrandom assignment — we cannot assume the naïve model of Chapter 3, and will report a sensitivity analysis. Although we cannot know this from the observed data, the situation is, in fact, the 'favorable situation,' in which there is a treatment effect and the matching has succeeded in removing bias, so the naïve model is correct and treatment assignment is effectively randomized within matched pairs; see §14.2. In this favorable situation, the investigator hopes to report that the treatment appears to be effective and that appearance is insensitive to small and moderate biases. Indeed, the situation is simpler still: the treatment has an additive, constant effect, $\tau = r_{Tij} - r_{Cij}$, so that $Y_i = \tau + (2Z_{i1} - 1)(r_{Ci1} - r_{Ci2})$; see §2.4.1. Moreover, the $r_{Ci1} - r_{Ci2}$ are independent and identically distributed observations drawn from a continuous distribution symmetric about zero; see §14.1. Because this is the favorable situation, $2Z_{i1} - 1 = \pm 1$, each with probability $\frac{1}{2}$ independently of $r_{Ci1} - r_{Ci2}$, so $Y_i - \tau$ itself has this same continuous distribution symmetric about zero.

The investigator faces a choice between a larger study with more heterogeneous responses or a smaller study with less heterogeneous responses, both in the 'favorable situation.' In §15.1, Mill would have advocated the smaller, less heterogeneous study. Is there any merit to Mill's claim? The heterogeneity here refers to heterogeneity that remains after matching for observed covariates, that is, heterogeneity within pairs; heterogeneity between pairs is not at issue. Specifically, the investigator faces the following admittedly stylized choice: observe either $4I$ pairs with additive effect τ and $(r_{Ci1} - r_{Ci2})/\omega \sim F(\cdot)$, where $F(\cdot)$ is a continuous distribution symmetric about zero, or alternatively observe I pairs with additive effect τ and $(r_{Ci1} - r_{Ci2})/(\omega/2) \sim F(\cdot)$. In words, the choice is between $4I$ pairs with dispersion ω or I pairs with dispersion $\omega/2$. The choice is stylized in the following sense. If $F(\cdot)$ were the standard Normal distribution, then the sample mean difference, $\overline{Y} = (1/I)\sum_{i=1}^{I} Y_i$, would be Normally distributed $\overline{Y} \sim N\{\tau, \omega^2/(4I)\}$ with expectation τ and variance $\omega^2/(4I)$ in both the larger, more heterogeneous study and the smaller less heterogeneous study. If ω were known in a randomized experiment, the larger more heterogenous study and the smaller less heterogeneous study would barely be worth distinguishing, because the sufficient statistic, $\overline{Y}$, has the same distribution in both studies.

Of course, this is not a randomized experiment. Does that matter for this choice? If so, how does it matter?

A simulated example

Figure 15.1 depicts a simulated example of the choice between a smaller, less heterogeneous study (SL) with $Y_i \sim N\left\{\tau, (\omega/2)^2\right\}$ for $i = 1, \ldots, I = 100$, and a larger, more heterogeneous study (LM) with $Y_i \sim N\left\{\tau, \omega^2\right\}$ for $i = 1, \ldots, I = 400$. In Figure 15.1, $\tau = 1/2$, $\omega = 1$. The boxplots for SL and LM have 100 and 400 pairs, respectively.

If SL and LM were analyzed as if they were randomized experiments, the inferences would be very similar. In SL, the mean difference is $\overline{Y} = 0.487$ with estimated standard error 0.054, while in LM the mean difference is $\overline{Y} = 0.485$ with estimated standard error 0.049; however, the true standard error is 0.05 in both SL and LM. Using Wilcoxon's signed rank statistic to test the null hypothesis of no effect yields a very small P-value in both SL and LM, less than 10^{-10}. The Hodges-Lehmann point estimate $\widehat{\tau}$ of τ is $\widehat{\tau} = 0.485$ for SL and $\widehat{\tau} = 0.489$ for LM. The 95% confidence interval from the randomization distribution of Wilcoxon's statistic is $[0.374, 0.600]$ from SL and $[0.390, 0.587]$ from LM. If the choice were between two randomized experiments with the distributions SL and LM yielding Figure 15.1, there would be little reason to prefer one over the other.

Suppose, however, that SL and LM came from observational studies, so the behavior of the Y_i might reflect either a treatment effect or an unmeasured bias or a combination of the two. How sensitive are the conclusions from SL and LM to departures from the naïve model (3.5)-(3.8) that underlies the inferences in the previous paragraph?

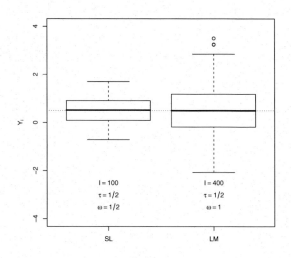

Fig. 15.1 A simulated example of the choice between a smaller, less heterogeneous study (SL) and a larger, more heterogeneous study (LM). In SL there are $I = 100$ independent matched pair differences, Y_i, that are Normal with expectation τ and standard deviation $\omega = 1/2$. In LM there are $I = 400$ independent matched pair differences, Y_i, that are Normal with expectation τ and standard deviation $\omega = 1$. The horizontal dotted line is at $1/2$.

15.2 A Larger, More Heterogeneous Study Versus a Smaller, Less Heterogeneous Study

Table 15.1 Sensitivity analysis for the larger, more heterogeneous study (LM) and the smaller, less heterogeneous study (SL). Upper bounds on the one-sided P-value from Wilcoxon's signed rank test when testing the null hypothesis of no treatment effect are given. Although the randomization inferences are similar ($\Gamma = 1$), the smaller, less heterogeneous study is much less sensitive to bias from unmeasured covariates ($\Gamma \geq 3$).

Γ	1	2	3	4	5
LM	$< 10^{-10}$	0.00046	0.39	0.97	1.00
SL	$< 10^{-10}$	0.000016	0.0021	0.022	0.083

Table 15.2 Sensitivity analysis for the larger more heterogeneous study (LM) and the smaller less heterogeneous study (SL). For $\Gamma = 1$, the table gives the value of the Hodges-Lehmann point estimate $\hat{\tau}$ of the treatment effect, τ. For $\Gamma = 2$, the table gives the interval of possible point estimates, $[\hat{\tau}_{min}, \hat{\tau}_{max}]$, and the length of that interval, $\hat{\tau}_{max} - \hat{\tau}_{min}$. For a bias of magnitude $\Gamma = 2$, the range of possible point estimates is much longer for the larger, more heterogeneous study than for the smaller, less heterogeneous study.

	$\Gamma = 1$ $\hat{\tau}$	$\Gamma = 2$ $[\hat{\tau}_{min}, \hat{\tau}_{max}]$	$\Gamma = 2$ $\hat{\tau}_{max} - \hat{\tau}_{min}$
LM	0.489	[0.19, 0.79]	0.60
SL	0.485	[0.32, 0.66]	0.34

Table 15.1 displays two sensitivity analyses, one for LM, the other for SL, giving the upper bound on the one-sided P-value for testing the hypothesis of no treatment effect using Wilcoxon's signed rank statistic; see §3.4. As noted above, the randomization inferences ($\Gamma = 1$) are quite similar for LM and SL. In sharp contrast, the smaller, less heterogeneous study, SL, is much less sensitive to bias from an unmeasured covariate. A bias of $\Gamma = 3$ could produce a boxplot similar to the LM boxplot in Figure 15.1 (the upper bound on the P-value is 0.39), but a bias of $\Gamma = 3$ is very unlikely to produce the boxplot for SL (the upper bound on the P-value is 0.0021). To put this in context, SL is just slightly more sensitive to unmeasured bias than Hammond's [5] study of heavy smoking as a cause of lung cancer (see [14, §4]), one of the least sensitive observational studies, whereas LM is much more sensitive.

In parallel with Table 3.4, Table 15.2 displays two sensitivity analyses, one for LM, the other for SL, giving the interval of possible point estimates, $[\hat{\tau}_{min}, \hat{\tau}_{max}]$, of the treatment effect τ. For $\Gamma = 1$, the interval is a point, namely the Hodges-Lehmann point estimate, $\hat{\tau}_{min} = \hat{\tau}_{max} = \hat{\tau}$, and it is about the same for LM and SL. For $\Gamma = 2$, the interval for LM, namely [0.19, 0.79], is considerably longer than the interval for SL, namely [0.32, 0.66].

In Tables 15.1 and 15.2, the smaller, less heterogeneous study is better than the larger, more heterogeneous study, better in the sense of being less sensitive to unmeasured biases. It is important to keep in mind that Figure 15.1 depicts two 'favorable situations,' that is, treatment effects without unmeasured biases. Because these are observational studies, the investigator does not know when she is in the 'favorable situation,' and so she cannot assert that Figure 15.1 depicts effects, not biases. The investigator can, however, assert that a bias of magnitude $\Gamma = 4$ is too small to explain away as unmeasured bias the ostensible effect in the smaller, less

Table 15.3 Power of a sensitivity analysis for a larger, more heterogeneous study (LM, $4I$ pairs, $\omega = 1$), and a smaller, less heterogeneous study, (SL, I pairs, $\omega = 1/2$), with the same additive, constant treatment effect, $\tau = 1/2$, in the favorable situation with $\{r_{Ci1} - r_{Ci2}\}/\omega \sim_{iid} F(\cdot)$. The power is similar when $\Gamma = 1$, but is higher for SL when Γ is larger.

Study	Distribution $F(\cdot)$	Number of Pairs	Dispersion ω	Power $\Gamma = 1$	Power $\Gamma = 1.5$	Power $\Gamma = 2$
LM	Normal	120	1	1.00	0.96	0.60
SL	Normal	30	$\frac{1}{2}$	1.00	1.00	0.96
LM	Logistic	120	1	0.93	0.31	0.04
SL	Logistic	30	$\frac{1}{2}$	0.93	0.61	0.32
LM	Cauchy	200	1	0.98	0.32	0.02
SL	Cauchy	50	$\frac{1}{2}$	0.95	0.60	0.28

heterogeneous study, but could not assert this about the larger, more heterogeneous study.

Power comparisons with Normal, logistic and Cauchy errors

Table 15.3 contrasts the larger, more heterogeneous study (LM) and the smaller, less heterogeneous study (SL) in terms of the power of the sensitivity analysis. Table 15.3 refers to a one-sided, 0.05 level test of the null hypothesis of no treatment effect. The power is the probability that the upper bound on the one-sided P-value is at most 0.05. The power is calculated as in §14.2.

The power of the randomization test ($\Gamma = 1$) is similar for LM and SL, but for larger Γ, particularly for $\Gamma = 2$, the power is higher for the smaller, less heterogeneous study (SL). The pattern seen in Table 15.1 is not a peculiarity of one simulation; rather, it is the anticipated pattern based on a comparison of the powers of LM and SL.

Design sensitivity

Proposition 14.1 and expressions (14.15) and (14.16) may be used to determine the design sensitivity $\widetilde{\Gamma}$ of the larger, more heterogeneous study (LM) and the smaller, less heterogeneous study (SL) as

$$\widetilde{\Gamma} = \frac{\Phi\left(\sqrt{2}\frac{\tau}{\omega}\right)}{1 - \Phi\left(\sqrt{2}\frac{\tau}{\omega}\right)} \qquad (15.1)$$

for Normal errors and as

$$\widetilde{\Gamma} = \frac{\Upsilon\left(\frac{\tau}{\omega}\right)}{1 - \Upsilon\left(\frac{\tau}{\omega}\right)} \qquad (15.2)$$

for Cauchy errors. For Normal errors with $\tau = 1/2$, the designs LM ($\omega = 1$) and SL ($\omega = 1/2$) have design sensitivities $\tilde{\Gamma}$ of 3.171 and 11.715, respectively. In light of this, the power of both designs with Normal errors tends to 1 as $I \to \infty$ for the values of Γ in Table 15.3, but for $\Gamma = 5$ the power of LM would tend to 0 while the power of SL would tend to 1. In parallel, for Cauchy errors with $\tau = 1/2$, the designs LM ($\omega = 1$) and SL ($\omega = 1/2$) have design sensitivities $\tilde{\Gamma}$ of 1.838 and 3, respectively. In light of this, with Cauchy errors, for $\Gamma = 2$ in Table 15.3, the power of SL tends to 1 as $I \to \infty$ while the power of LM tends to 0.

15.3 Heterogeneity and the Sensitivity of Point Estimates

In the current chapter, the treatment has an additive constant effect, $\tau = r_{Tij} - r_{Cij}$, and in a randomized experiment, the Hodges-Lehmann estimate $\hat{\tau}$ is a consistent estimate of τ. For a given deviation Γ from randomized treatment assignment, Tables 3.4 and 15.2 displayed the interval of possible Hodges-Lehmann point estimates $[\hat{\tau}_{\min}, \hat{\tau}_{\max}]$ of τ, where $\hat{\tau}_{\min} = \hat{\tau}_{\max} = \hat{\tau}$ for $\Gamma = 1$; see §3.5 and [13]. As the sample size increases, $I \to \infty$, the endpoints of this interval converge in probability to the endpoints of a fixed interval, $[\tau_{\min}, \tau_{\max}]$; this interval reflects the uncertainty about τ that is due to a potential bias of magnitude Γ when there is no longer any sampling uncertainty.

In the 'favorable situation,' with errors having Normal $\Phi(\cdot)$ or Cauchy $\Upsilon(\cdot)$ cumulative distributions, the following proposition gives the form of this limiting interval. See [16, Appendix] for proof of Proposition 15.1.[2]

Proposition 15.1. *If* $(D_i - \tau)/\omega \sim_{iid} \Phi(\cdot)$ *then* $[\tau_{\min}, \tau_{\max}]$ *is* $\tau \pm \omega \Phi^{-1}(\kappa)/\sqrt{2}$, *where* $\kappa = \Gamma/(1+\Gamma)$. *If* $(D_i - \tau)/\omega \sim_{iid} \Upsilon(\cdot)$ *then* $[\tau_{\min}, \tau_{\max}]$ *is* $\tau \pm \omega \Upsilon^{-1}(\kappa)$.

Proposition 15.1 is consistent with Mill's view that heterogeneity of experimental units, ω, is directly relevant to causal claims. In Proposition 15.1, there is no sampling variability, because Proposition 15.1 refers to the limit as $I \to \infty$. The uncertainty addressed in Proposition 15.1 is quantified by the length of the limiting interval, $\tau_{\max} - \tau_{\min}$, and despite the absence of sampling variability, the length of that interval is directly proportional to ω.

Proposition 15.1 does not contradict Fisher's view, but it does emphasize that this view is applicable only when biases from nonrandom treatment assignment have been avoided by randomization. Reducing heterogeneity ω and increasing sample

---

[2] Although the proof has a few details, it is simple in concept. The lower endpoint $\hat{\tau}_{\min}$ of the interval of possible point estimates is obtained by equating Wilcoxon's signed rank statistic T computed from $Y_i - \tau_0$, say T_{τ_0}, to the maximum null expectation of T, namely $E\left(\overline{T} \mid \mathscr{F}, \mathscr{Z}\right) = \Gamma I(I+1)/\{2(1+\Gamma)\}$ from (3.19) and solving the equation for $\hat{\tau}_{\min}$. Dividing the equation by $I(I+1)/2$ yields the equivalent equation $2T_{\tau_0}/\{I(I+1)\} = \Gamma/(1+\Gamma)$. If $(Y_i - \tau)/\omega \sim \Phi(\cdot)$ or $(Y_i - \tau)/\omega \sim \Upsilon(\cdot)$ then as $I \to \infty$, the left side of the equation, $2T_{\tau_0}/\{I(I+1)\}$, converges in probability to a function of $(\tau - \tau_0)/\omega$, and the rest of the proof is detail devoted to showing that equation can be solved to give the solutions in the statement of Proposition 15.1.

size I compete for resources in a randomized experiment because bias is known to have been avoided, so the analysis can be conducted with $\Gamma = 1$. More precisely, in Proposition 15.1, if it were known that $\Gamma = 1$, then $\kappa = 1/2$, so $\Phi^{-1}(\kappa)/\sqrt{2} = \Phi^{-1}(1/2)/\sqrt{2} = 0$ and $\Upsilon^{-1}(\kappa) = \Upsilon^{-1}(1/2) = 0$, and the length of the limiting interval is $\tau_{max} - \tau_{min} = 0$ for every value of ω.

The length of the interval is also affected by the magnitude of potential bias, Γ, though $\kappa = \Gamma/(1+\Gamma)$. The two components determine the length $\tau_{max} - \tau_{min}$ of the limiting interval in a multiplicative manner; for the Normal, $\tau_{max} - \tau_{min} = 2\omega \, \Phi^{-1}(\kappa)/\sqrt{2}$. A given magnitude Γ of deviation from a randomized experiment does more harm when the units are more heterogeneous, that is, when ω is larger. If you were deceptively trying to bias a randomized trial by covertly tilting the treatment assignment probabilities by a magnitude of Γ in (3.16)-(3.18), then you could do more harm if the units were the heterogeneous patients in a clinical trial than if they were the homogeneous genetically engineered mice in a laboratory experiment.

In the 'favorable situation' in an observational study, increasing the sample size I reduces the standard error, but it does not materially reduce sensitivity to unmeasured biases. In contrast, in this situation, reducing the heterogeneity of experimental units, ω, reduces both the standard error and sensitivity to unmeasured biases. In an observational study, LM and SL of Figure 15.1 are not at all the same: SL is much better.

15.4 Examples of Efforts to Reduce Heterogeneity

Twins

What are the economic returns to additional education? You cannot compare the mid-life earnings of surgeons and high school dropouts — they differed in the middle of high school before the dropout left school. Not in every case, but typically, the child who went on to become a surgeon was receiving better grades and standardized test scores in high school, was more strongly motivated for school studies, had better educated, wealthier parents, and not inconceivably differed in some relevant genes. You would like to compare two children of the same parents with different education growing up at the same time in the same home with the same genes. Ashenfelter and Rouse [1] compared the earnings of identical twins with differing education, estimating about a 9% increase in earnings per year of additional education.

The use of twins is the canonical example of trading sample size for reduced heterogeneity. Twin pairs are quite heterogeneous between pairs and in several important respects fairly homogeneous within pairs, so the use of twins reflects the type of heterogeneity discussed in this chapter.

15.4 Examples of Efforts to Reduce Heterogeneity

Road hazards

What permanent features of a road affect the risk of collisions with roadside objects? Road hazards are a fairly small part of accident risk. Also relevant are: the driver's sobriety, skill and risk tolerance; ambient light; the weather — ice, snow, rain and fog; safety equipment — use of seat belts, quality and condition of brakes, tires, air bags, traction and stability control devices. These factors are related. The risk-averse driver will drive near the legal speed limit, but will also invest in safety devices and wear seat belts. In the rain or snow, one drives on the highway to work but not on the dirt road to the picnic area, so weather and road hazards vary together. Sobriety is more common at noon than at midnight, so sobriety and ambient light vary together. You would like to compare different road hazards with the same driver, in the same car, in the same weather, with the same ambient light, in the same state of sobriety, with seat belts in the same state of use. Is this possible?

Using a simple, clever study design, Wright and Robertson [22] did just that. They compared 300 fatal roadside collisions in Georgia in 1974–1975 to 300 nonaccidents involving the same driver, in the same car, in the same ambient light, and so on. The nonaccidents occurred one mile back from the crash site, a location passed without incident by the driver just moments before the crash. At crash sites, Wright and Robertson found a substantial excess of roads that curved more than 6 degrees with downhill gradients of greater than 2%. (Technically, this is a 'case-crossover' study of the type proposed by Malcolm Maclure [10], except that it is defined by geography rather than time; see also the 'case-specular' design of Sander Greenland [4].)

The genetically engineered mice of microeconomics

Many businesses that provide products or services over large regions adopt a strategy known as 'replication' in which nearly identical outlets are reproduced at high speed in diverse locations [21]. Starbucks and Tesco are two of the many such businesses. This strategy confers various benefits to businesses that use it, but it also creates nearly identical copies of a business in locations that may have adopted different regulations, taxes or other policies. For instance, Card and Kreuger's [2] study of the minimum wage and employment compared Burger Kings in New Jersey to Burger Kings in Pennsylvania, KFCs in New Jersey to KFCs in Pennsylvania, etc., and in this way eliminated one of several sources of extraneous variation between the two states; see §4.5 and §11.3 for further discussion of this study.

Motorcycle helmets

Do helmets reduce the risk of death in motorcycle crashes? Crashes occur at different speeds with different forces, and neither speeds nor forces are likely to be measured. Motorcyclists hit different objects — pedestrians or Hummers — in dense or light traffic, with emergency services near or far away. One would like to compare two people, one with a helmet, the other without, on the same type of motorcycle, driving at the same speed, crashing into the same object, in the same traffic, with equal proximity to medical aid. Is that possible?

It is when two people ride one motorcycle, one with a helmet, the other without. Norvell and Cummings [12] looked at such crashes, finding about 40% lower risk associated with helmet use.

15.5 Summary

In a randomized experiment, an unbiased estimate of treatment effect is available, so increasing the sample size, I, or reducing the unit heterogeneity, ω, both serve to reduce the standard error of an unbiased estimate. The situation is strikingly different in an observational study. In the 'favorable situation' in an observational study, the treatment is effective and there are no unmeasured biases. If the favorable situation arose, the investigator would not know it, and at best would hope to report that the treatment appears to be effective and that appearance is insensitive to small and moderate biases. In this situation, reducing heterogeneity, even purely random heterogeneity, ω, confers benefits that cannot be obtained by increasing the sample size, I. Specifically, reducing heterogeneity reduces sensitivity to unmeasured biases. Several cleverly designed studies have illustrated efforts to reduce heterogeneity.

15.6 Further Reading

This chapter is based on [16], where additional discussion may be found.

References

1. Ashenfelter, O., Rouse, C.: Income, schooling and ability: Evidence from a new sample of identical twins. Q J Econ **113**, 253–284 (1998)
2. Card, D., Krueger, A.: Minimum wages and employment: A case study of the fast-food industry in New Jersey and Pennsylvania. Am Econ Rev **84**, 772–793 (1994)
3. Fisher, R.A.: Design of Experiments. Edinburgh: Oliver and Boyd (1935)
4. Greenland, S.: A unified approach to the analysis of case-distribution (case-only) studies. Statist Med **18**, 1–15 (1999)
5. Hammond, E.C.: Smoking in relation to mortality and morbidity. J Natl Cancer Inst **32**, 1161–1188 (1964)
6. Heller, R., Rosenbaum, P.R., Small, D.: Split samples and design sensitivity in observational studies. J Am Statist Assoc **104**, to appear (2009)
7. Hodges, J.L., Lehmann, E.L.: Estimates of location based on ranks, Ann Math Statist **34**, 598–611 (1963)

8. Holland, P.W.: Statistics and causal inference. J Am Statist Assoc **81**, 945–960 (1986)
9. Lehmann, E.L.: Nonparametrics, San Francisco: Holden Day (1975) Reprinted New York: Springer (2006)
10. Maclure, M.: The case-crossover design: A method for studying transient effects on the risk of acute events. Am J Epidemiol **133**, 144–152 (1991)
11. Mill, J.S.: A System of Logic: The Principles of Evidence and the Methods of Scientific Investigation. Indianapolis: Liberty Fund (1867)
12. Norvell, D.C., Cummings, P.: Association of helmet use with death in motorcycle crashes: A matched-pair cohort study. Am J Epidemiol **156**, 483–487 (2002)
13. Rosenbaum, P.R.: Hodges-Lehmann point estimates of treatment effect in observational studies. J Am Statist Assoc **88**, 1250–1253 (1993)
14. Rosenbaum, P.R.: Observational Studies (2nd ed.). New York: Springer (2002)
15. Rosenbaum, P. R.: Design sensitivity in observational studies. Biometrika **91**, 153–164 (2004)
16. Rosenbaum, P. R.: Heterogeneity and causality: Unit heterogeneity and design sensitivity in observational studies. Am Statistician **59**, 147–152 (2005)
17. Rosenbaum, P.R.: What aspects of the design of an observational study affect its sensitivity to bias from covariates that were not observed? In: Festshrift for Paul W. Holland. Princeton, NJ: ETS (2009)
18. Salsburg, D.: The Lady Tasting Tea. San Francisco: Freeman (2001)
19. Small, D., Rosenbaum, P.R.: War and wages: The strength of instrumental variables and their sensitivity to unobserved biases. J Am Statist Assoc **103**, 924–933 (2008)
20. Werfel, U., Langen, V., Eickhoff, I., Schoonbrood, J., Vahrenholz, C., Brauksiepe, A., Popp, W., Norpoth, K.: Elevated DNA single-strand breakage frequencies in lymphocytes. Carcinogenesis **19**, 413–418 (1998)
21. Winter, S.G., Szulanski, G.: Replication as strategy. Organizat Sci **12**, 730–743 (2001)
22. Wright, P.H., Robertson, L.S.: Priorities for roadside hazard modification. Traffic Eng **46**, 24–30 (1976)

Chapter 16
Uncommon but Dramatic Responses to Treatment

Abstract Large effects in moderate to large studies are typically insensitive to small and moderate unobserved biases, but the concept of a 'large effect' is vague. What if most subjects are not much affected by treatment, but a small fraction, perhaps 10% or 20% of subjects, are strongly affected? On average, such an effect may be small, but not at all small for the affected fraction. Is such an effect insensitive to small and moderate unobserved biases?

16.1 Large Effects, Now and Then

Are large but rare effects insensitive to unmeasured biases?

In §2.5, in the National Supported Work Experiment, depicted in Figure 2.1, many men appear to have received little or no benefit from the treatment, but the few men with high earnings were fairly consistently in the treated group. In §2.5, if pairs of men were examined two at a time (i.e., four men), then 61% of the time, the man with the highest aligned earnings was a treated man, where 50% is expected by chance, but if pairs were examined 20 at a time (i.e., 40 men), 86% of the time the man with the highest aligned earnings was a treated man, where again 50% is expected by chance. Big gains in earnings are consistently in the treated group.

A similar pattern is seen for toxicity in Chapter 7 in Figure 7.1 in connection with Jeffrey Silber et al.'s [14] study of intensity of chemotherapy in $I = 344$ pairs of women with ovarian cancer: the median level of toxicity is not very different in the MO and GO groups, but high levels are more common in the MO group. If pairs of women are examined two at a time (i.e., 4 women), then in 65% of pairs, the woman with the highest aligned toxicity was treated by an MO, but if we look at 20 pairs at a time (i.e., 40 women), then 90% of the time, the woman with the highest aligned toxicity was treated by an MO. Figure 16.1 depicts the distribution of the MO-minus-GO differences, Y_i, in toxicity. In Figure 7.1 and Figure 16.1, it appears

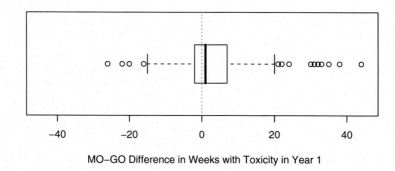

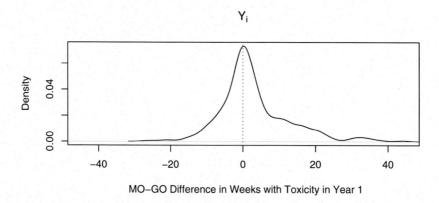

Fig. 16.1 Boxplot and density estimate of the MO-GO difference Y_i in weeks with toxicity in year 1 for $I = 344$ pairs of women with ovarian cancer; see Chapter 7 and [14]. The vertical dotted line is at zero difference. The density estimate uses the default settings in R.

that many MOs and GOs treat with similar intensity producing similar toxicity, but a fraction of MOs treat more intensively producing more toxicity.

In these cases, the hypothesis of a constant effect, $\tau = r_{Tij} - r_{Cij}$ for all i, j, does not look plausible. As discussed in §2.5, a more plausible hypothesis is that $r_{Tij} - r_{Cij}$ is zero or small for many subjects i, j, but $r_{Tij} - r_{Cij}$ is large for some subjects i, j. David Salsburg [11] argued that effects of this sort are fairly common, are often important, and that we tend to miss them because our methodology tends to focus on typical effects, but in this context large effects do occur but are not typical. The question addressed in the current chapter is whether an effect of this kind is highly sensitive to unmeasured biases. In both Figure 2.1 and Figure 7.1, the typical difference in outcomes is not large, yet the distributions of outcomes are quite different.

Review of §2.5: Measuring large but uncommon effects

Building upon certain technical results by Eric Lehmann [2, (6.1)], David Salsburg [11] and William Conover and Salsburg [1] made a strong argument that, in the comparison of unmatched treated and control groups, the pattern of large, uncommon effects is not properly associated with 'transformations' or 'outliers' or other similar concepts of data analysis, because the pattern may be seen in the ranks: the extremely high ranks are consistently in the treated group. In outline in §2.5 and in detail in [7], it was found that the highly interpretable rank scores, $\widetilde{q}_i$ in (2.8), proposed by W. Robert Stephenson [15] approximate the locally optimal rank scores of Conover and Salsburg [1]. For example, the statement about 20 pairs of men in the first paragraph of this Chapter was obtained in §2.5 by an interpretation of Stephenson's ranks.

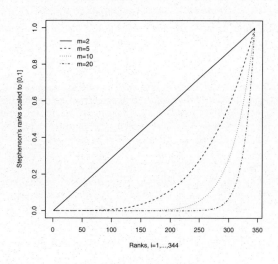

Fig. 16.2 Stephenson's ranks $\widetilde{q}_i$ from (2.8) for $m = 2, 5, 10, 20$ and for the $I = 344$ matched pairs in Silber et al.'s [14] study of intensity of chemotherapy for ovarian cancer; see Chapter 7. Wilcoxon's ranks are essentially the same as $m = 2$. For $m = 5$, the smallest 100 or so $|Y_i|$ are almost ignored. For $m = 20$, the smallest 300 or so $|Y_i|$ are almost ignored.

Recall that Stephenson's ranks score pairs m at a time, and are essentially the same as Wilcoxon's ranks for $m = 2$; see §2.5. For $I = 344$ matched pairs, as in Silber et al.'s [8, 14] study in Chapter 7, Figure 16.2 depicts Stephenson's ranks, $\widetilde{q}_i$ in (2.8), sorted into increasing order and scaled to be between 0 and 1 as $\widetilde{q}_i / \max \widetilde{q}_j$, for $m = 2, 5, 10, 20$. With $m = 20$, one looks at 20 pairs simultaneously, counting a 'success' if the largest of the 40 aligned responses in the 20 pairs is the response of a treated subject. For $m = 10$ in Figure 16.2, Stephenson's $\widetilde{q}_i$ largely ignore the smallest 200 of the 344 absolute differences in responses $|Y_i|$, judging the treatment effect by the remaining 144 pairs. Salsburg and Conover suggested the equivalent of $m = 5$ for the applications that they discussed.

16.2 Two Examples

Chemotherapy intensity and toxicity in ovarian cancer

Table 16.1 compares four sensitivity analyses for the ovarian cancer study [8, 14] in Chapter 7. The four sensitivity analyses are each conducted using Stephenson's [15] signed rank test, with four values of m, that is, with the absolute ranks in Figure 16.2. Wilcoxon's ranks are similar to Stephenson's ranks with $m = 2$. The ranks with $m = 20$ are concerned with the largest aligned response of 40 women in 20 pairs, so that they emphasize the fairly high levels of toxicity experienced by some women.

Table 16.1 reports upper bounds on the one-sided P-value testing the null hypothesis of no treatment effect. Wilcoxon's statistic ($m = 2$) becomes sensitive to a bias of magnitude $\Gamma = 1.5$, but Stephenson's statistic with $m = 20$ is insensitive to a bias of magnitude $\Gamma = 3.5$. In this one example, the degree of sensitivity to bias is strongly affected by whether the analysis emphasizes typical effects or less common but larger effects. Small biases could not produce the large but uncommon effects in Figure 16.1.

Table 16.1 Using Stephenson's statistic with several values of m, upper bounds are given on the one-sided P-value testing the null hypothesis of no treatment effect in the ovarian cancer data of Chapter 7 and [14]. For $m = 2$, Stephenson's ranks closely resemble Wilcoxon's ranks, but for larger values of m, Stephenson's ranks place greater emphasis on large $|Y_i|$'s. In this instance, the results are much less sensitive with $m = 20$ than with $m = 2$.

Γ	$m = 2$	$m = 5$	$m = 10$	$m = 20$
1	9.0×10^{-7}	5.2×10^{-9}	1.1×10^{-8}	1.1×10^{-6}
1.5	0.051	0.00017	0.000024	0.00019
2	0.71	0.016	0.0010	0.0020
2.5		0.14	0.0087	0.0081
3			0.035	0.021
3.5			0.088	0.041

The calculations to produce Table 16.1 closely parallel the calculations for Wilcoxon's signed rank statistic, except that Stephenson's ranks, $\tilde{q}_i$, are used in place of Wilcoxon's ranks, q_i. The null expectation and variance of Stephenson's statistic $\tilde{T} = \sum \text{sgn}(Y_i) \tilde{q}_i$ are bounded by expressions (3.25)-(3.26) with $\tilde{q}_i$ in place of q_i, and the standardized deviates are compared to the Normal distribution in parallel with (3.23).

DNA adducts among aluminum production workers

Bernadette Schoket, David Phillips, Alan Hewer, and István Vincze [13] compared DNA adducts in the lymphocytes of aluminum production workers and unexposed controls. Figure 16.3 shows 25 matched pairs of an aluminum production worker

16.2 Two Examples

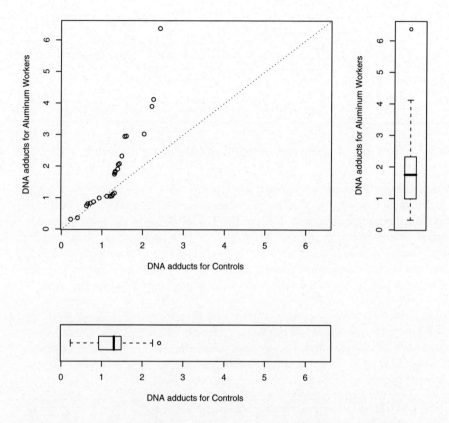

Fig. 16.3 DNA adducts per 10^8 nucleotides for 25 aluminum production workers and 25 controls matched for age and cigarettes-per-day [13]. The central plot is a quantile-quantile plot that ignores the pairing. The dotted line is the $x = y$ line. There appears to be little or no difference at lower quantiles, with a sharp rise at upper quantiles.

and a control matched for age and cigarettes smoked per day. The noticeable feature of Figure 16.3 is that there appears to be little or no difference in DNA adducts at lower quantiles, with a substantial divergence at upper quantiles.[1]

Table 16.2 displays two sensitivity analyses using Stephenson's signed rank statistic, one with $m = 2$ which is similar to Wilcoxon's statistic, and one with $m = 5$. (With just $I = 25$ pairs, it does not seem reasonable to compare $m = 20$ pairs at a time to determine the largest aligned response.) The table displays upper bounds on the one-sided P-value for testing the null hypothesis of no treatment effect. As Figure 16.3 might lead us to anticipate, the results are less sensitive for $m = 5$ than for $m = 2$.

---

[1] The paired data for the 25 pairs are listed in [7].

Table 16.2 Upper bounds on the one-sided P-values from Stephenson's test for $m = 2$ and $m = 5$ for the 25 pairs of an aluminum production worker and a matched control depicted in Figure 16.3. The results are less sensitive to bias for $m = 5$, presumably because the difference in the distribution of DNA adducts is apparent only at upper quantiles.

Γ	$m = 2$	$m = 5$
1	0.0076	0.0078
1.5	0.054	0.030
1.8	0.10	0.048
2	0.14	0.061

16.3 Properties of a Paired Version of Salsburg's Model

David Salsburg [11] proposed a model for treatment effects in unmatched groups in which a fraction $0 \leq \lambda \leq 1$ of the treated subjects experience a substantial treatment effect and the rest experience no effect at all. The following model is a slight modification for matched pairs of Salsburg's model for unmatched groups. Each pair i is characterized by a fixed pair parameter, η_i, which accounts for dependence within pairs, and is eliminated when matched pair differences Y_i are calculated. Then the $r_{Cij} - \eta_i$ are sampled from a continuous distribution $G(\cdot)$ with density $g(\cdot)$, and the $r_{Tij} - \eta_i$ are sampled from $(1 - \lambda) G(\cdot) + \lambda \{G(\cdot)\}^\nu$ where $0 \leq \lambda \leq 1$ and $\nu \geq 2$ is a positive integer. Responses from distinct subjects are mutually independent.[2]

The model has a simple interpretation. The model says, in effect, that a fraction, $1 - \lambda$, of treated subjects experience no treatment effect — their responses are sampled from $G(\cdot)$ which is the distribution of responses to control. A fraction λ of treated subjects do experience an effect. Instead of having a response drawn from $G(\cdot)$, the affected fraction have a response $r_{Tij} - \eta_i$ drawn from $\{G(\cdot)\}^\nu = G(\cdot) \times \cdots \times G(\cdot)$. If you sampled ν independent $r_{Cij} - \eta_i$'s from the distribution $G(\cdot)$ and took the maximum of these, that maximum would have the distribution $\{G(\cdot)\}^\nu$. As intuition suggests, $\{G(\cdot)\}^\nu$ is a 'larger' distribution[3] than $G(\cdot)$ in the sense that $\{G(r)\}^\nu \leq G(r)$ for all r, because $\nu \geq 2$; that is, the chance, namely $\{G(r)\}^\nu$, that the maximum of ν independent observations from $G(\cdot)$ is less than r is less than or equal to the chance, namely $G(r)$, that one observation from $G(\cdot)$ is less than r, and this is true for all r. Figure 16.4 depicts the two probability densities that are mixed with weights λ and $1 - \lambda$ to form the density of the

---

[2] The density of $r_{Tij} - \eta_i$ is $(1 - \lambda) g(r) + \lambda \nu g(r) \{G(r)\}^{\nu - 1}$. In the favorable situation, treatment assignment $Z_{i1} - Z_{i2}$ is ± 1 with probability $1/2$ independently of $(r_{Ci1}, r_{Ti1}, r_{Ci2}, r_{Ti2})$, and $Y_i = (Z_{i1} - Z_{i2})(R_{i1} - R_{i2})$ is distributed as the difference of two independent observations, one drawn from $(1 - \lambda) G(\cdot) + \lambda \{G(\cdot)\}^\nu$ and the other drawn from $G(\cdot)$. The matched pair difference Y_i has probability density

$$\int g(r) \left[(1 - \lambda) g(r + y) + \lambda \nu g(r + y) \{G(r + y)\}^{\nu - 1} \right] dr$$

which is easy to evaluate by numerical integration. In the calculations presented graphically, $G(\cdot)$ is the standard Normal cumulative distribution $\Phi(\cdot)$.

[3] It is stochastically larger; see Note 15 in Chapter 3.

16.3 Properties of a Paired Version of Salsburg's Model

treated response, $r_{Tij} - \eta_i$, for $v = 20$, when $G(\cdot)$ is the standard Normal cumulative distribution $G(\cdot) = \Phi(\cdot)$. In Figure 16.4, a fraction $1 - \lambda$ of treated subjects have a response $r_{Tij} - \eta_i$ drawn from the standard Normal density $\phi(r)$, the dashed curve, and a fraction λ of treated subjects have a response $r_{Tij} - \eta_i$ drawn from the the density $v \phi(r) \Phi(r)^{v-1}$ of the maximum of $v = 20$ independent observations from a standard Normal distribution, the solid curve. The treated-minus-control difference Y_i in pair i is the difference of a treated response $r_{Tij} - \eta_i$ formed from this mixture and a control response $r_{Cij} - \eta_i$ independently drawn from the standard Normal density $\phi(r)$.

In the unmatched two sample problem, Eric Lehmann [2, (6.1)] studied this model for $v = 2$ showing that Wilcoxon's ranks, with $m = 2$, perform well for small λ in large samples. William Conover and David Salsburg [1] showed that for general v, ranks similar to Stephenson's ranks perform well if $m = v$ for small λ in large samples.[4]

With $G(\cdot)$ equal to the standard Normal distribution, $G(\cdot) = \Phi(\cdot)$, Figure 16.5 depicts the distribution of matched pair differences, Y_i, under Salsburg's model with $\lambda = 0.2$ and $v = 50$. For comparison, Figure 16.5 also displays a Normal distribution with the same expectation and variance. Visually, the distinction between the two distributions is quite subtle: Salsburg's model has a slight skew to the right, with an expectation slightly to the right of its mode. Conceptually, the distributions are very different. In Salsburg's model, $1 - \lambda = 80\%$ of treated subjects experience no treatment effect: their treated responses, $r_{Tij} - \eta_i$, are sampled from the same Normal distribution as responses to control, $r_{Cij} - \eta_i$. In Salsburg's model, only $\lambda = 20\%$ of treated subjects experience a nonzero effect, and their responses are the maximum of $v = 20$ independent observations from a Normal distribution, rather than one observation from a Normal distribution. In Figure 16.4, the affected group had a much higher distribution of responses, but this is twice hidden in Figure 16.5, once because only a fraction λ of treated subjects experienced a treatment effect, and a second time by the need to compare treated subjects to controls to eliminate η_i by differencing to form the matched pair difference, Y_i. Figure 16.6 is similar to Figure 16.5, except $\lambda = 0.1$ and $v = 1000$; here, the difference is more noticeable, but still quite subtle. Compare Figure 16.6 to the estimated density in Figure 16.1.

Are the effects in Figures 16.5 and 16.6 highly sensitive to unmeasured biases? On the one hand, the affected group in Figure 16.4 is strongly affected. On the other hand, that strong effect is twice hidden in the observable distribution of Y_i in Figure 16.5. In Tables 16.1 and 16.2, effects were judged fairly sensitive by Wilcoxon's statistic ($m = 2$), which emphasizes typical effects, but were noticeably less sensitive

---

[4] Somewhat more precisely, Conover and Salsburg found the locally most powerful rank test for fixed v as $\lambda \to 0$ for the two-sample problem as the sample size increased. It is shown in [7] that these locally most powerful ranks are almost the same as Stephenson's [15] ranks in large samples, although the latter are easier to interpret in terms of the magnitude of the treatment effect. Stephenson and Ghosh [16] discuss the use of Stephenson's ranks in the unmatched two sample problem. The two sample model comparing $(1 - \lambda) G(\cdot) + \lambda \{G(\cdot)\}^v$ and $G(\cdot)$ is an instance of a 'Lehmann alternative' in the sense that the distribution of the ranks depends upon v and λ but not on $G(\cdot)$; see [2]. Reference [7] is not restricted to matched pairs, but in the current book, only the paired case is discussed.

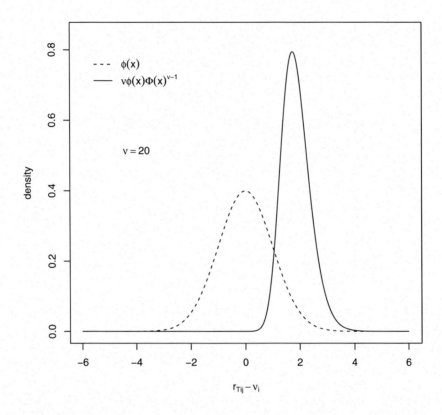

Fig. 16.4 The two densities that are mixed in Salsburg's model to form the distribution of $r_{Tij} - \eta_i$, with $\nu = 20$. The dashed density function is the standard Normal density. The solid density function is the density of the maximum of $\nu = 20$ independent observations from the standard Normal distribution.

to bias when Stephenson's statistic was used with larger values of m. Is the pattern in Tables 16.1 and 16.2 something that can be expected under Salsburg's model?

16.4 Design Sensitivity for Uncommon but Dramatic Effects

Design sensitivity of Stephenson's test

To answer this question, the design sensitivity $\widetilde{\Gamma}$ of Stephenson's [15] test is determined in general. Then $\widetilde{\Gamma}$ is calculated for Salsburg's model.

16.4 Design Sensitivity for Uncommon but Dramatic Effects

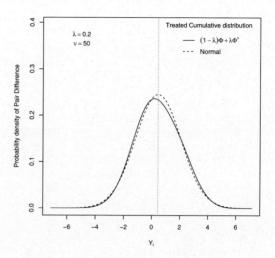

Fig. 16.5 Salsburg's model (solid line) for the matched pair difference Y_i with $\lambda = 0.2$ and $v = 50$ in the treated group. For comparison, a Normal density (dashed line) of Y_i with the same expectation and variance is plotted. The vertical dotted line is at their shared expectation.

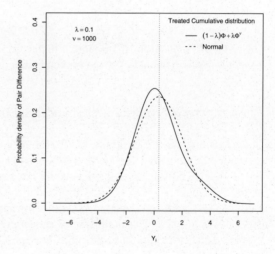

Fig. 16.6 Salsburg's model (solid line) for the matched pair difference Y_i with $\lambda = 0.1$ and $v = 1000$ in the treated group. For comparison, a Normal density (dashed line) of Y_i with the same expectation and variance is plotted. The vertical dotted line is at their shared expectation.

As in Chapter 14, the favorable situation is assumed, so there is a treatment effect without unmeasured biases, and therefore one hopes to report that rejection of the null hypothesis of no treatment effect is insensitive to small and moderate unmeasured biases. Therefore, suppose that the Y_i are independent observations from some continuous distribution $H(\cdot)$ and there are no unmeasured biases, in the sense that treatments were assigned by the naïve model (3.5)-(3.8).

Consider m independent observations Y_i drawn from $H(\cdot)$. One of these m observations, say Y_ℓ, has the largest value of $|Y_i|$; that is, $|Y_\ell| = \max_{1 \leq i \leq m} |Y_i|$. Define

$\widetilde{p}_m$ as the probability that the one Y_ℓ among these m with the largest value of $|Y_i|$ is positive, $Y_\ell > 0$.

The first thing to notice about $\widetilde{p}_m$ is that we have already made extensive use of $\widetilde{p}_2$ in Chapter 14 because $\widetilde{p}_2 = p_1' = \Pr(Y_i + Y_{i'} > 0)$; that is, for $i \neq i'$, the sign of $Y_i + Y_{i'}$ is determined by Y_i if $|Y_i| > |Y_{i'}|$ and by $Y_{i'}$ if $|Y_i| < |Y_{i'}|$. Indeed, by Proposition 14.1, the design sensitivity $\widetilde{\Gamma}$ of Wilcoxon's statistic ($m = 2$) is, $p_1'/\left(1 - p_1'\right) = \widetilde{p}_2/(1 - \widetilde{p}_2)$. The second thing to notice about $\widetilde{p}_m$ is that it can be easily calculated by simulation for any distribution $H(\cdot)$.

The following proposition generalizes Proposition 14.1.

Proposition 16.1. *In I pairs matched exactly for observed covariates, $\mathbf{x}_{ij}$, suppose that the treated-minus-control matched pair differences in outcomes, Y_i, $i = 1, \ldots, I$, are independently sampled from a distribution $H(\cdot)$ and there is no bias from unmeasured covariates in the sense that treatments are actually assigned by the naïve model (3.5)-(3.8). A sensitivity analysis is performed using Stephenson's signed rank statistic $\widetilde{T}$ with a specific value of m and Γ. The power of this sensitivity analysis for one-tailed α-level test of no effect with $\alpha > 0$ tends to 1 as $I \to \infty$ for $\Gamma < \widetilde{\Gamma}$ and the power tends to 0 for $\Gamma > \widetilde{\Gamma}$ where*

$$\widetilde{\Gamma} = \frac{\widetilde{p}_m}{1 - \widetilde{p}_m}. \qquad (16.1)$$

The proof of Proposition 16.1 is similar to the proof of Proposition 14.1 in Chapter 14 and is given in detail in [9] and is sketched in the appendix to this chapter.

Design sensitivity of Stephenson's test under Salsburg's model

Table 16.3 reports design sensitivities $\widetilde{\Gamma}$ for Stephenson's [15] signed rank statistic when applied to the paired version of Salsburg's [11] model. In Table 16.3, either $\lambda = 10\%$ or $\lambda = 20\%$ of treated subjects are affected by the treatment, receiving either the maximum of $\nu = 50$ or $\nu = 1000$ independent responses from the control distribution $G(\cdot)$, with $G(\cdot)$ equal to either the Normal or Cauchy distributions. The densities of two of these distributions were shown in Figures 16.5 and 16.6.

If only a small fraction of treated subjects respond to treatment, then the use of Wilcoxon's ranks ($m = 2$) is a mistake. In this case, much larger values of the design sensitivity $\widetilde{\Gamma}$ are obtained by using a test that focuses attention on large values of $|Y_i|$. This conclusion is specific to the situation in which many treated subjects are unaffected by treatment and a small fraction are strongly affected.

16.5 Summary

A treatment may affect some subjects and not others. In this case, the typical effect may be small, despite large effects for some subjects. Conventional statisti-

Table 16.3 Design sensitivity of Stephenson's test based on subsets of m pairs for Salsburg's model in which only a fraction λ of treated subjects are affected by the treatment, having a response that is the maximum of ν observations from the distribution of responses to control, $G(.)$, for $G(.)$ equal to the Normal or Cauchy distributions.

	Normal $\lambda=0.2, \nu=50$	Normal $\lambda=0.1, \nu=1000$	Cauchy $\lambda=0.2, \nu=50$	Cauchy $\lambda=0.1, \nu=1000$
$m=2$	1.8	1.4	2.0	1.5
$m=5$	2.6	2.0	3.3	2.3
$m=10$	3.4	3.0	4.7	4.4
$m=20$	4.4	4.9	5.7	11.5

cal methods look for typical treatment effects and may miss dramatic effects for a small subset of affected subjects. The issue arises in both randomized clinical trials [1, 11, 12] and observational studies [7, 9]. When the treatment has an effect of this form in an observational study, an analysis focused on detecting such an effect, for instance using Stephenson's [15] test, may be substanially less sensitive to unmeasured bias than an analysis focused on typical effects. The issue is relevant to planning statistical analyses as discussed in Part IV.

16.6 Further Reading

This chapter is built from components found in papers by Eric Lehmann [2, (6.1)], W. Robert Stephenson [15], David Salsburg [11, 12, 1], and William Conover [1]; see also [7, 9] for discussion of the use of these components in observational studies.

An alternative approach emphasizes the possibility that the magnitude of the treatment effect $r_{Tij} - r_{Cij}$ is not constant but rather increasing in r_{Cij} — a so-called dilated treatment effect — and estimates the effect at different quantiles of r_{Cij}; see [3, 5, §5]. With a dilated treatment effect, the effect at upper quantiles may be less sensitive to bias than the effect at lower quantiles.

16.7 Appendix: Sketch of the Proof of Proposition 16.1

As in the proof of Proposition 14.1 in Chapter 14, there are two versions of the proof. One version uses the heuristic and appeals to the general fact from [6, §3] that the heuristic works quite generally. The other version is complete as it stands and parallels the proof in Chapter 14. In either case, one starts with the observation from Stephenson [15] that in the favorable situation a scaled version of $\widetilde{T}$ converges in probability to $\widetilde{p}_m$, specifically

$$\frac{m!(I-m)!}{I!} \widetilde{T} \to \widetilde{p}_m.$$

One then checks that under the sensitivity analysis model, the maximum null expectation of $\widetilde{T}$ after scaling in the same way equals $\Gamma/(1+\Gamma)$. Equating $\widetilde{p}_m = \Gamma/(1+\Gamma)$ and solving completes the heuristic derivation. If no appeal is made to [6, §3], the detailed proof must again verify that if $\widetilde{p}_m \neq \Gamma/(1+\Gamma)$ then the difference in expectations overwhelms the variance terms as $I \to \infty$.

References

1. Conover, W.J., Salsburg, D.S.: Locally most powerful tests for detecting treatment effects when only a subset of patients can be expected to 'respond' to treatment. Biometrics, **44**, 189–196 (1988)
2. Lehmann, E.L.: The power of rank tests. Ann Math Statist **24**, 23–43 (1953)
3. Rosenbaum, P.R.: Reduced sensitivity to hidden bias at upper quantiles in observational studies with dilated treatment effects. Biometrics **55**, 560–564 (1999)
4. Rosenbaum, P.R.: Effects attributable to treatment: Inference in experiments and observational studies with a discrete pivot. Biometrika **88**, 219–231 (2001)
5. Rosenbaum, P.R.: Observational Studies (2nd ed.). New York: Springer (2002)
6. Rosenbaum, P.R.: Design sensitivity in observational studies. Biometrika **91**, 153–164 (2004)
7. Rosenbaum, P.R.: Confidence intervals for uncommon but dramatic responses to treatment. Biometrics **63**, 1164–1171 (2007)
8. Rosenbaum, P.R., Ross R.N., Silber, J.H.: Minimum distance matched sampling with fine balance in an observational study of treatment for ovarian cancer. J Am Statist Assoc **102**, 75–83 (2007)
9. Rosenbaum, P.R.: What aspects of the design of an observational study affect its sensitivity to bias from covariates that were not observed? Festshrift for Paul W. Holland. Princeton, NJ: ETS (2009)
10. Rosenbaum, P.R., Silber, J.H.: Amplification of sensitivity analysis in observational studies. J Am Statist Assoc, to appear (2009)
11. Salsburg, D.S.: Alternative hypotheses for the effects of drugs in small-scale clinical studies. Biometrics **42**, 671–674 (1986)
12. Salsburg, D.S.: The Use of Restricted Significance Tests in Clinical Trials. New York: Springer (1992)
13. Schoket, B., Phillips, D.H., Hewer, A., Vincze, I.: ^{32}P-Postlabelling detection of aromatic DNA adducts in peripheral blood lymphocytes from aluminum production plant workers. Mutation Res **260**, 89–98 (1991)
14. Silber, J.H., Rosenbaum, P.R., Polsky, D., Ross, R.N., Even-Shoshan, O., Schwartz, S., Armstrong, K.A., Randall, T.C.: Does ovarian cancer treatment and survival differ by the specialty providing chemotherapy? J Clin Oncol **25**, 1169–1175 (2007)
15. Stephenson, W.R.: A general class of one-sample nonparametric test statistics based on subsamples. J Am Statist Assoc **76**, 960–966 (1981)
16. Stephenson, W.R., Ghosh, M.: Two sample nonparametric tests based on subsamples. Commun Statist **14**, 1669–1684 (1985)
17. Wilcoxon, F.: Individual comparisons by ranking methods. Biometrics **1**, 80–83 (1945)

Chapter 17
Anticipated Patterns of Response

Abstract Design sensitivity is used to quantify the effectiveness of devices discussed in Chapter 5. Several of those devices anticipate a particular pattern of results, perhaps coherence among several outcomes, or a dose-response relationship. To what extent do these considerations reduce sensitivity to unmeasured biases?

17.1 Using Design Sensitivity to Evaluate Devices

In §5.2, several devices used to study unmeasured biases made reference to an anticipated pattern of responses. In §5.2.3, it was anticipated that several outcomes would be affected in a similar way, the pattern of coherence. In §5.2.5, it was anticipated that larger doses of treatment would yield larger responses, the pattern of dose-response. If such a pattern is anticipated, and if the anticipation is realized in the observed data, will the sensitivity to unmeasured biases be reduced? This did occur in particular examples in Tables 5.3 and 5.5. Under what circumstance should similar results be expected?

17.2 Coherence

Notation with several responses

In §5.2.3, the coherent signed rank statistic ([5] and [6, §9]) was simply the sum of the several signed rank statistics for several outcomes oriented in the anticipated direction. For instance, in Table 5.3, the signed rank statistics for the Math and Verbal test scores were added together, creating a statistic that would react strongly to improvements in both Math and Verbal scores, and weakly to, say, an increase in Math scores together with a decline in Verbal scores.

The coherent signed rank statistic so defined has advantages in analysis: the separate ranking of each outcome is one robust method for standardizing the outcomes, giving about equal weight to each of the outcomes. In studying design sensitivity for coherence, it is easiest if a slightly different statistic is used: Wilcoxon's signed rank statistic is applied to a weighted combination of the outcomes. In the first approach, weights are applied to the separate signed rank statistics, while in the second approach a single signed rank statistic is computed from a weighted combination of the outcomes. If the second approach is adopted, then in the multivariate Normal situation, there is an explicit formula for the design sensitivity for given weights, an explicit formula for the optimal weights, and so on. In analysis, data are not typically multivariate Normal, and the second approach leaves the investigator with the task of finding an appropriate, robust standardization of the outcomes. It is possible to compute the design sensitivity for both approaches, but the first approach requires some unenlightening Monte Carlo work; however, when this is done, similar results are found for both approaches. Ruth Heller, Dylan Small and I [2] looked at both approaches in quite a bit of detail; see also [8]. Here, a few of the simpler results from [2] are presented, and the reader may turn to [2] and [8] for additional approaches and numerical results.[1]

Instead of a single observed response R_{ij} from the jth person in pair i, that person now has M observed responses, $\mathbf{R}_{ij} = (R_{ij1}, \ldots, R_{ijM})^T$. In Table 5.3, $M = 2$, and R_{ij1} was the Math score, while R_{ij2} was the Verbal score. Wilcoxon's signed rank statistic will be applied to a weighted combination of the responses, $\sum_{m=1}^{M} \lambda_m R_{ijm}$, where the weights λ_m are selected before the data are collected. More precisely, Wilcoxon's signed rank statistic, T, is computed from the treated-minus-control difference in combined responses,

$$Y_i = (Z_{i1} - Z_{i2})\left(\sum_{m=1}^{M} \lambda_m R_{i1m} - \sum_{m=1}^{M} \lambda_m R_{i2m}\right) \tag{17.1}$$

$$= \sum_{m=1}^{M} \lambda_m Y_{im} \tag{17.2}$$

where

$$Y_{im} = (Z_{i1} - Z_{i2})(R_{i1m} - R_{i2m}). \tag{17.3}$$

---

[1] Why are simpler results available if the signed rank statistic is applied to a weighted combination of results, rather than applying weights to several signed rank statistics? As discussed in §2.3.3, the signed rank statistic is distribution-free; that is, its null distribution can be determined before any data are collected. In contrast, the coherent signed rank statistic is a randomization test in a randomized experiment, but, like the sample mean in Chapter 2, it is not distribution-free: its randomization distribution depends upon features of the observed data. When thinking about design, with no data on hand, it is convenient that the null distribution of the test statistic is known and does not depend upon the data you do not have. On the other hand, in principle, the design sensitivity is well defined in either case, but more Monte Carlo work may be needed to determine its value. See [2] for some comparisons.

17.2 Coherence

Again, this differs slightly from Table 5.3 where weights $\lambda_1 = 1$ and $\lambda_2 = 1$ were applied to two signed rank statistics, one computed from R_{ij1}, the other from R_{ij2}.

Responses with a multivariate Normal distribution

As in Chapter 14, design sensitivity is computed in the 'favorable situation' in which there is a treatment effect and no bias from unmeasured covariates. It is in the favorable situation that we hope to report that rejection of the hypothesis of no effect is insensitive to unmeasured biases.

The results are particularly simple for multivariate Normal $\mathbf{R}_{ij}$.[2] The multivariate Normal distribution may be defined in various equivalent ways, but one definition is that $\sum_{m=1}^{M} \lambda_m R_{ijm}$ has some Normal distribution for every choice of weights, $\lambda = (\lambda_1, \ldots, \lambda_M)^T$. In particular, it is assumed that each response R_{ijm} is affected by an additive treatment effect τ_m, so $R_{ijm} = r_{Cijm} + Z_{ij}\tau_m$, with the expectations, variances, covariances and correlations among the Y_{im}'s given by $E(Y_{im}) = \tau_m$, $\text{var}(Y_{im}) = \omega_m^2$, $\text{cov}(Y_{im}, Y_{im'}) = \omega_{mm'}$ and $\text{corr}(Y_{im}, Y_{im'}) = \rho_{mm'} = \omega_{mm'}/(\omega_m \omega_{m'})$, so $\omega_m^2 = \omega_{mm}$. The covariance matrix Ω is an $M \times M$ array with $\omega_{mm'}$ in row m and column m'. In this case, $Y_i = \sum_{m=1}^{M} \lambda_m Y_{im}$ is Normally distributed with expectation $\mu_\lambda = \sum_{m=1}^{M} \lambda_m \tau_m$ and variance $\omega_\lambda^2 = \sum_{m=1}^{M} \sum_{m'=1}^{M} \lambda_m \lambda_{m'} \omega_{mm'}$. Wilcoxon's signed rank statistic is calculated from the Y_i.

The situation is now the same as in Chapter 14, and the design sensitivity is

$$\widetilde{\Gamma} = \frac{p_1'}{1-p_1'} \tag{17.4}$$

where, for $i \neq i'$,

$$p_1' = \Pr(Y_i + Y_{i'} > 0) = \Pr\left(\frac{Y_i + Y_{i'} - 2\mu_\lambda}{\sqrt{2\omega_\lambda^2}} > \frac{-2\mu_\lambda}{\sqrt{2\omega_\lambda^2}}\right) = \Phi\left(\frac{\sqrt{2}\mu_\lambda}{\omega_\lambda}\right). \tag{17.5}$$

Moreover, the optimal value, say λ_{opt}, of λ with the largest design sensitivity $\widetilde{\Gamma}_{opt}$ is given by a simple formula involving the inverse of the covariance matrix, say Ω; specifically, λ_{opt} is any vector proportional to $\Omega^{-1}\tau$ where $\tau = (\tau_1, \ldots, \tau_M)^T$; see [2].[3]

---

[2] A good reference for basic facts about the multivariate Normal distribution and associated matrix manipulations is C.R. Rao's text [4].

[3] The proportionality constant cancels in (17.5). If $\lambda_{opt} = \Omega^{-1}\tau$ is replaced by $2\lambda_{opt}$, then both μ_λ and ω_λ are multiplied by 2, and (17.5) is unchanged. For this reason, it is often attractive to multiply λ_{opt} by a constant so that the values are easy to consider. Where possible, λ_{opt} is given as a vector of integers.

Table 17.1 Design sensitivity with bivariate Normal outcomes and an additive treatment effect. Four design sensitivities are given: for optimal weights, for outcome 1 alone, for outcome 2 alone, and for equal weights. The optimal weights are also given.

Case	Standard Deviation (ω_1, ω_2)	Correlation ρ_{12}	Effects (τ_1, τ_2)	Optimal Weights λ_{opt}	Optimal $\widetilde{\Gamma}_{opt}$	Outcome 1 $\widetilde{\Gamma}_1$	Outcome 2 $\widetilde{\Gamma}_2$	Equal Weights $\widetilde{\Gamma}_=$
A	$(1, 1)$	0	$(0.5, 0.5)$	$(1, 1)$	5.3	3.2	3.2	5.3
B	$(1, 1)$	0.25	$(0.5, 0.5)$	$(1, 1)$	4.4	3.2	3.2	4.4
C	$(1, 1)$	0.5	$(0.5, 0.5)$	$(1, 1)$	3.8	3.2	3.2	3.8
D	$(1, 0.5)$	0	$(0.5, 0.5)$	$(1, 4)$	16.6	3.2	11.7	8.7
E	$(1, 0.5)$	0.25	$(0.5, 0.5)$	$(1, 7)$	12.9	3.2	11.7	7.1
F	$(1, 0.5)$	0.5	$(0.5, 0.5)$	$(0, 1)$	11.7	3.2	11.7	6.0
G	$(1, 1)$	0	$(0.75, 0.25)$	$(3, 1)$	6.6	5.9	1.8	5.3
H	$(1, 1)$	0.25	$(0.75, 0.25)$	$(11, 1)$	6.0	5.9	1.8	4.4
I	$(1, 1)$	0.5	$(0.75, 0.25)$	$(5, -1)$	6.1	5.9	1.8	3.8

Numerical results for bivariate Normal responses

Table 17.1 computes the design sensitivities for bivariate Normal responses ($M = 2$), with an additive treatment effect (τ_1, τ_2), standard deviations (ω_1, ω_2) and correlation ρ_{12}; see also [2, 8] where $M > 2$ and other cases are considered. The table calculates four design sensitivities: (i) $\widetilde{\Gamma}_1$ and $\widetilde{\Gamma}_2$ are the design sensitivities for the two outcomes Y_{i1} and Y_{i2} used alone; (ii) $\widetilde{\Gamma}_=$ is the design sensitivity with equal weights, $\lambda = (1, 1)$, analogous to the coherent signed rank statistic used in Table 5.3; (iii) $\widetilde{\Gamma}_{opt}$ is the optimal design sensitivity and λ_{opt} are the weights that produce the optimal design sensitivity. It should be kept in mind that the investigator does not know λ_{opt}, and so the investigator can hope to get closer to $\widetilde{\Gamma}_{opt}$ in one way or another, but is unlikely to reach $\widetilde{\Gamma}_{opt}$.

In cases A–C, the effects (τ_1, τ_2) and standard deviations (ω_1, ω_2) are equal for the two outcomes and the optimal weights are equal, so $\widetilde{\Gamma}_1 = \widetilde{\Gamma}_2 \leq \widetilde{\Gamma}_{opt} = \widetilde{\Gamma}_=$. The advantage of using both outcomes with equal weights is substantial for uncorrelated outcomes, $\rho_{12} = 0$, and the advantage disappears as $\rho_{12} \to 1$.

Cases D–F resemble the situation in Chapter 15 in that the effects are equal, $\tau_1 = \tau_2$, but the standard deviation of Y_{i2} is half that of Y_{i1}; here, however, the investigator does not have to choose one outcome with a change in sample size, but can use both outcomes with the full sample size. In these cases, $\widetilde{\Gamma}_1 \leq \widetilde{\Gamma}_= \leq \widetilde{\Gamma}_2 \leq \widetilde{\Gamma}_{opt}$ and $\widetilde{\Gamma}_2 = \widetilde{\Gamma}_{opt}$ for $\rho_{12} = \frac{1}{2}$. The optimal procedure does not use the first outcome when $\rho_{12} = \frac{1}{2}$.

Cases G–I have equal standard deviations but unequal effects, $\tau_1 = \frac{3}{4}$, $\tau_2 = \frac{1}{4}$, so $\tau_1 + \tau_2$ is the same as in cases A–C, but τ_1 is larger. In these cases, $\widetilde{\Gamma}_2 \leq \widetilde{\Gamma}_= \leq \widetilde{\Gamma}_1 \leq \widetilde{\Gamma}_{opt}$, and the optimal procedure attaches negative weight to Y_{i2} for $\rho_{12} = \frac{1}{2}$.

By definition, $\widetilde{\Gamma}_{opt}$ is always the largest design sensitivity. In all the cases considered in A–I, equal weights were better than the inferior of the two outcomes, $\min\left(\widetilde{\Gamma}_1, \widetilde{\Gamma}_2\right) \leq \widetilde{\Gamma}_=$. Equal weights, $\widetilde{\Gamma}_=$, were sometimes better and sometimes worse than the better of the two outcomes, that is, $\max\left(\widetilde{\Gamma}_1, \widetilde{\Gamma}_2\right)$. The optimal weights, $\widetilde{\Gamma}_{opt}$, were sometimes much better than the second best alternative, $\max\left(\widetilde{\Gamma}_1, \widetilde{\Gamma}_2, \widetilde{\Gamma}_=\right)$, and sometimes no better. The design sensitivities vary considerably in cases A–I, from 1.8 to 16.6, so knowing which circumstance will arise and how to adapt the analysis to that circumstance will materially affect the design sensitivity.

Cases A–I in Table 17.1 have various practical implications. In favorable circumstances, a coherent statistic with equal weights can yield materially less sensitivity to unmeasured covariates than an analysis that focuses on a single outcome; however, not all circumstances are favorable in this sense. Similar results in [2] show that, in favorable circumstances, the gains in $\widetilde{\Gamma}$ from using an equally weighted coherent statistic can increase substantially with additional outcomes, $M > 2$, and that the sturdier, more practical coherent signed rank statistic (§5.2.3, [5] and [6, §9]) exhibits performance similar to the simpler but less practical statistic evaluated in Table 17.1. At the same time, in Table 17.1, the optimal weights diminish the importance of an outcome that is either unstable (i.e., larger ω) or only slightly affected (i.e., smaller τ), perhaps ignoring such an outcome with zero weight, or perhaps even attaching negative weight to such an outcome; see Chapter 15 for a different but related phenomenon. Finally, the substantial variability in design sensitivities in cases A–I, from 1.8 to 16.6, suggests that a correct plan for analysis may materially affect the sensitivity of results to unmeasured biases, so time and resources should be allocated to planning the analysis; see §18.1 and [2] for one practical approach based on split samples.

17.3 Doses

Recall from §5.2.5 that there are conflicting claims in the literature about whether or not dose-response relationships are important in appraising evidence about effects caused by a treatment. Also, recall the example in Table 5.5 in which doses did seem to reduce sensitivity to unmeasured biases. Calculations of design sensitivity in the current section will shed light on the various conflicting claims. In these calculations, the Y_i will be drawn from a continuous distribution, so ties do not occur.

Another way to write a signed rank statistic

Wilcoxon's signed rank statistic is $T = \sum_{i=1}^{I} \mathrm{sgn}(Y_i)\, q_i$, where q_i is the rank of $|Y_i|$, and the dose-weighted signed rank statistic [5, 7, 13] is $T_{\mathrm{dose}} = \sum_{i=1}^{I} \mathrm{sgn}(Y_i)\, q_i d_i$, where $d_i > 0$ is the dose of treatment applied to the treated subject in pair i.[4] In calculating the design sensitivity, $\widetilde{\Gamma}$, it is helpful to express T and T_{dose} in an equivalent

---

[4] Recall Note 8 of §5.2.5.

but slightly different form. The reason for this is that the expectation of T_{dose} will be needed in the 'favorable situation,' and the expectation has a simple form when T_{dose} is written in a different form.

Write
$$W_{ik} = \begin{cases} 1 & \text{if } |Y_i| \geq |Y_k| \text{ and } Y_i > 0 \\ 0 & \text{otherwise.} \end{cases} \quad (17.6)$$

It takes only a moment to realize that the dose-weighted signed rank statistic, $T_{\text{dose}} = \sum_{i=1}^{I} \text{sgn}(Y_i) q_i d_i$, is equal to $\sum_{i=1}^{I} \sum_{k=1}^{I} d_i W_{ik}$.[5] When all of the doses, d_i, equal 1, this simplifies to say that Wilcoxon's signed rank statistic $T = \sum_{i=1}^{I} \text{sgn}(Y_i) q_i$ equals $\sum_{i=1}^{I} \sum_{k=1}^{I} W_{ik}$. The advantage of writing T_{dose} in this way is that it yields a simple formula for $E(T_{\text{dose}})$, namely $E(T_{\text{dose}}) = \sum_{i=1}^{I} \sum_{k=1}^{I} E(d_i W_{ik})$.

Another similar quantity, V_{ik}, is also needed. Write
$$V_{ik} = \begin{cases} 1 & \text{if } |Y_i| \geq |Y_k| \\ 0 & \text{otherwise.} \end{cases} \quad (17.7)$$

Unlike W_{ik} in (17.6), the quantity V_{ik} ignores the sign of Y_i. In takes only another moment to realize that $q_i = \sum_{k=1}^{I} V_{ik}$ and $d_i q_i = \sum_{k=1}^{I} d_i V_{ik}$.[6] Obviously, $W_{ik} \leq V_{ik}$.

The favorable situation with doses

As in Chapter 14, we ask: Will we be in a position to report that the conclusions are insensitive to unmeasured bias when, indeed, the treatment did have an effect and there was no bias from unmeasured covariates? It is in this situation that we hope to report that the results are insensitive to small and moderate biases. In other words, it is assumed that subjects who look comparable in terms of observed covariates, x_{ij}, are indeed comparable, so the naïve model (3.5)-(3.8) is correct. In addition, a model is needed in which there is a treatment effect. Two such models will be considered, one a special case of the other more general case.

In the general case of the 'favorable situation,' there are $L \geq 1$ possible positive doses, $\delta_1, \ldots, \delta_L$. Of course, $L = 1$ represents the situation 'without doses' in the sense that every treated subject receives the same dose. The I doses d_i for the I matched pairs are independently sampled from the L possible doses, $\delta_1, \ldots, \delta_L$ with probabilities $\Pr(d_i = \delta_\ell) = \varphi_\ell$ for $\ell = 1, \ldots, L$; in effect, the doses are determined by sampling a multinomial distribution with L categories. Conditionally given that $d_i = \delta_\ell$, the treated-minus-control difference in observed responses is $Y_i = \tau_\ell + \varepsilon_i$ where ε_i are independently sampled from a continuous distribution $F_\ell(\cdot)$ that is symmetric

---

[5] To see this, compare $d_i \text{sgn}(Y_i) q_i$ to $\sum_{k=1}^{I} d_i W_{ik}$. If $Y_i \leq 0$, then $\text{sgn}(Y_i) = 0$ so $d_i \text{sgn}(Y_i) q_i = 0$, but also $d_i W_{ik} = 0$ for every k, so $0 = \sum_{k=1}^{I} d_i W_{ik}$. If $Y_i > 0$, then $\text{sgn}(Y_i) = 1$, so $d_i \text{sgn}(Y_i) q_i = d_i q_i$ is d_i times the rank q_i of $|Y_i|$. Also $d_i \sum_{k=1}^{I} W_{ik}$ is d_i times the number of $|Y_k|$ such that $|Y_i| \geq |Y_k|$, which is, of course, just d_i times the rank q_i of $|Y_i|$. So for every i, $d_i \text{sgn}(Y_i) q_i = \sum_{k=1}^{I} d_i W_{ik}$. This shows that $T_{\text{dose}} = \sum_{i=1}^{I} \sum_{k=1}^{I} d_i W_{ik}$. If you set $d_i = 1$ for all i, then this becomes $T = \sum_{i=1}^{I} \sum_{k=1}^{I} W_{ik}$.

[6] As in the previous footnote, $\sum_{k=1}^{I} V_{ik}$ is the number of $|Y_k|$ that are less than or equal to $|Y_i|$, which is just the rank q_i of $|Y_i|$.

17.3 Doses

about zero. That is, for each fixed dose, the 'favorable situation' is the same as in Chapter 14, namely an additive constant treatment effect τ_ℓ with independent, identically distributed symmetric, continuous errors, ε_i; in particular, with $L = 1$ dose, the situation is identical to that in Chapter 14. The new element is that the size of the treatment effect, τ_ℓ, may vary with the dose, $d_i = \delta_\ell$, and so may the error distribution, $F_\ell(\cdot)$. The model as just described first samples doses, $d_i = \delta_\ell$, then samples outcome differences, Y_i, conditionally given doses, but this implies a joint sampling scheme in which the dose outcome pairs (d_i, Y_i) are independent, identically distributed bivariate measurements (iid).

In this general model, two quantities, Ψ and Λ, will now be defined that are important for the design sensitivity, $\widetilde{\Gamma}$. Pick an i and a k with $i \neq k$. Because the bivariate (d_i, Y_i) are iid, it makes no difference which i and k are used, provided they are different. Define $\Psi = E(d_i W_{ik})$ and $\Lambda = E(d_i V_{ik})$, where W_{ik} is defined in (17.6) and V_{ik} is defined in (17.7). For specified $\delta_1, \ldots, \delta_L, \tau_1, \ldots, \tau_L$, and $F_1(\cdot), \ldots, F_L(\cdot)$, the quantities Ψ and Λ are well defined; for instance, they could be determined by Monte Carlo.[7] Because $d_i W_{ik} \leq d_i V_{ik}$, it follows that $\Psi \leq \Lambda$; moreover, if $\Pr(Y_i > 0) < 1$ then $\Psi < \Lambda$.

In addition to this general model, it is helpful to have a more tangible, more specific model to permit tabulation of a few numerical results. In this model, the doses are $\delta_1 = 1, \delta_2 = 2, \ldots, \delta_L = L$, and they have equal probabilities, $\varphi_\ell = 1/L$ for $\ell = 1, \ldots, L$. Then the expected and median dose are both $(L+1)/2$. Also, the effect is linear in the dose, $\tau_\ell = \beta \delta_\ell$ and the error distribution $F_\ell(\cdot)$ is the same $F(\cdot)$ for all ℓ. The median treatment effect is then $\beta(L+1)/2$, and this is also the expected treatment effect if $F(\cdot)$ has finite expectation. If the median effect, $\beta(L+1)/2$, is held fixed, we may ask: Is it better to have more dose categories, L? If you could choose between a somewhat larger median effect, $\beta(L+1)/2$, with $L = 1$ dose category, or a somewhat smaller median effect, $\beta(L+1)/2$, with several dose categories (larger L), which would be the better choice? How, if at all, does the sample size, I, affect these choices? Before looking at numerical results, in speculating about these questions, one might look again at the example in Table 5.5.

A formula for the design sensitivity with doses

The following proposition gives the design sensitivity of the dose weighted signed rank statistic T_{dose}. The proof is in the appendix to this chapter. Recall that $\Lambda \geq \Psi$ with equality if and only if $\Pr(Y_i > 0) = 1$.

Proposition 17.1. *Suppose that (i) doses d_i in the I pairs are independently sampled from a discrete distribution with $L \geq 1$ possible positive doses, $\delta_1, \ldots, \delta_L$ with*

---

[7] You can write Ψ or Λ as a weighted sum over the L^2 possible dose combinations in pairs i and k, and in simple situations the individual terms in this sum either have closed forms or can be evaluated numerically. For some details, see [10]. These forumlas are an aid to numerical computation, but they provide no insights, so they are not presented here.

$\Pr(d_i = \delta_\ell) = \varphi_\ell$, $\ell = 1, \ldots, L$; (ii) there is no bias from unmeasured covariates in the sense that conditions (3.5)-(3.8) are true, (iii) conditionally given that $d_i = \delta_\ell$, the treated-minus-control difference in observed responses is $Y_i = \tau_\ell + \varepsilon_i$ where ε_i are independently sampled from a continuous distribution $F_\ell(\cdot)$ which is symmetric about zero. Then, as $I \to \infty$, the power of a sensitivity analysis using the dose weighted signed rank statistic, T_{dose}, tends to 1 for $\Gamma < \widetilde{\Gamma}$ and to 0 for $\Gamma > \widetilde{\Gamma}$ where

$$\widetilde{\Gamma} = \frac{\Psi}{\Lambda - \Psi}, \tag{17.8}$$

$\Psi = E(d_i W_{ik})$, and $\Lambda = E(d_i V_{ik})$ with $i \neq k$.

Proposition 17.1 generalizes Proposition 14.1 of Chapter 14 in the following sense. Consider the case of a single dose, $L = 1$, with $d_i = 1$ for all i. With such a single dose, the dose-weighted signed rank statistic T_{dose} equals Wilcoxon's signed rank statistic T. In this case, $\Lambda = 1/2$ and $\Psi = p'_1/2 = \Pr(Y_i + Y_k > 0)/2$; so Proposition 17.1 implies Proposition 14.1 when there is a single dose.[8]

Recall that $\Lambda \geq \Psi$ with equality if and only if $\Pr(Y_i > 0) = 1$; in this case, correctly, $\widetilde{\Gamma} = \infty$ in (17.8).

Numerical evaluation of the design sensitivity

Table 17.2 provides numerical values of the design sensitivity $\widetilde{\Gamma}$ from (17.8) in the case of L equally probable, equally spaced integer doses, $\delta_1 = 1, \ldots, \delta_L = L$, effect $\tau_\ell = \beta \delta_\ell$ at dose δ_ℓ with errors that are Normal, logistic, or Cauchy. The median treatment effect is $\beta(L+1)/2$.

In cases A through D in Table 17.2, the median treatment effect is $\beta(L+1)/2 = 1/2$, but the number of dose levels varies from 1 in case A to 9 in case D. That is, the dose and the effect are constant at $1/2$ in case A, but the doses range from 1 to 9 in case D, with a median effect of $1/2$, and effects that vary from $1/10$ at dose 1 to $9/10$ at dose 9. The dose-weighted signed rank statistic, T_{dose}, gives greater weight to pairs with larger doses, but in case A it equals Wilcoxon's signed rank statistic because there is only one dose. With the median dose fixed, there is some advantage to having several dose levels, but the advantage is not extremely large.

In cases E through G, there is a single dose, with a slightly larger median effect. A single dose with a median effect of 0.60 is comparable in terms of design sensitivity to $L = 5$ doses with a median effect of 0.50. A single dose with a median effect of 0.75 or 0.83 in cases F and G produces larger design sensitivities $\widetilde{\Gamma}$ than are found in cases A–D. Faced with a choice between a single larger dose or a con-

---

[8] To see this, assume that $d_i = 1$ for every i. Observe that with continuously distributed Y_i, there is zero probability that $|Y_i| = |Y_k|$ with $i \neq k$. Because Y_i and Y_k are two independent observations from the same continuous distribution, $\Pr(|Y_i| > |Y_k|) = \Pr(|Y_i| < |Y_k|) = 1/2$. Therefore, because $d_i = 1$, $\Lambda = E(d_i V_{ik}) = E(V_{ik}) = 1/2$. Also, $Y_i + Y_k > 0$ can occur in two mutually exclusive ways, either $|Y_i| > |Y_k|$ and $Y_i > 0$ or $|Y_k| > |Y_i|$ and $Y_k > 0$, and each of these possibilities has the same probability, namely $E(W_{ik}) = E(W_{ki})$ which equals Ψ if $d_i = d_k = 1$. This shows $\Psi = p'_1/2$.

17.3 Doses

Table 17.2 Design sensitivity with doses. There are L equally probable, equally spaced, integer doses, $\delta_1 = 1, \ldots, \delta_L = L$ and the effect is proportional to dose with slope β, so the median effect is $\beta(L+1)/2$. The table gives the design sensitivity for the signed rank statistic with doses, which reduces to Wilcoxon's signed rank statistic when there is only one dose, $L = 1$.

Case	L	β	β(L+1)/2	Normal	Logistic	Cauchy
A	1	1/2	0.50	3.2	2.0	1.8
B	3	1/4	0.50	3.8	2.2	2.0
C	5	1/6	0.50	4.0	2.2	2.1
D	9	1/10	0.50	4.2	2.3	2.1
E	1	3/5	0.60	4.0	2.2	2.0
F	1	3/4	0.75	5.9	2.7	2.4
G	1	5/6	0.83	7.4	3.1	2.6

Table 17.3 Design sensitivity with doses, discarding pairs with doses below the median dose. Initially, there are L equally probable, equally spaced, integer doses, with L odd, $\delta_1 = 1, \ldots, \delta_L = L$ and the effect is proportional to dose with slope β. Unlike Table 17.2, pairs with doses strictly below $(L+1)/2$ are discarded, so the median dose for the remaining pairs is $(3L+1)/4$ and the median effect in these pairs is $\beta(3L+1)/4$. The table gives the design sensitivity for the signed rank statistic with doses applied to the pairs that are not discarded.

Case	L	β	β(3L+1)/4	Normal	Logistic	Cauchy
H	3	1/4	0.625	4.5	2.4	2.2
I	5	1/6	0.667	5.0	2.5	2.3
J	9	1/10	0.700	5.5	2.7	2.3

tinuum of doses with a smaller median dose, the former would be preferred. This addresses one of the controversies in §5.2.5.

It is important to realize that cases E through G are entirely practical. An investigator in case B could, in effect, switch to case F by using only the high-dose pairs, those with dose $\delta_3 = 3$, because in that dose category, the median effect is $\beta\delta_3 = (1/4) \times 3 = 0.75$. In parallel, an investigator in case C could, in effect, switch to case G by using only the high-dose pairs, those with $\delta_3 = 5$, because in that dose category, the median effect is $\beta\delta_5 = (1/6) \times 5 = 0.83$. Discarding the low dose pairs would decrease the sample size, but an increase in design sensitivity is often more important in a sensitivity analysis, as has been seen in various contexts, including Table 5.5 and Chapter 15.

Table 17.3 considers the design sensitivity that would result in cases B, C, and D of Table 17.2 if pairs with doses below the median were discarded, and T_{dose} were applied to the remaining pairs. The increases in design sensitivity, $\widetilde{\Gamma}$, are moderate for Normal errors, somewhat noticeable for logistic errors, and negligible for Cauchy errors.

Tables 17.2 and 17.3 refer to a dose that affects the magnitude of the effect. Not all doses do this. In one context, years of exposure may be unimportant, and yesterday's dose may be all-important. In another context, the dose a decade ago may matter, but yesterday's dose may be inconsequential for today's response. Doses may be measured with substantial errors, so that an indirect use of doses is supe-

Table 17.4 High, lower and mixed dose pairs in Angrist and Lavy's study of class size manipulated by Maimonides' rule. The table describes matched pair differences, larger-cohort-minus-smaller-cohort. The tabled values are trimeans, a robust location estimate that is twice the median plus the quartiles divided by four.

	86 High-Dose Pairs	49 Lower-Dose Pairs	135 Mixed-Dose Pairs
Class Size	-10.88	-5.33	-7.38
Math Test	3.97	3.47	3.66
Verbal Test	2.98	1.45	2.55

rior to their direct incorporation in a test statistic [9, 12]. The wrong dose may be measured or the right one may be measured poorly.

Arguably, Tables 17.2 and 17.3 speak not to the value of doses taken in isolation, but rather to the value of treated subjects who experienced only small effects when treatment effects are heterogeneous. When doses are actually indicative of the magnitude of the effect, they serve to identify treated subjects who will experience small effects. Here and in Chapter 16, design sensitivity is greater if little weight is given to subjects experiencing negligible effects.

17.4 Example: Maimonides' Rule

Recall from §1.3 and §5.2.3 the study by Angrist and Lavy [1] of class size manipulated by Maimonides' rule. So far, the discussion has considered only the 86 pairs in Figure 1.1 in which a school with between 41 and 50 students in the fifth grade was paired with a school with between 31 and 40 students in fifth grade. These are the 'high-dose' pairs, in that adherence to Maimonides' rule would cut a class of 40 students into two classes of average size 20.5 if one additional student enrolled. In Figure 1.1, adherence to Maimonides' rule is imperfect, so the typical reduction in class size is smaller than 20 students. Table 17.4 considers the addition of 49 more pairs, for schools with between 70 and 90 students in the fifth grade. These are the 'lower-dose' pairs, in that adherence to Maimonides' rule would cut two classes of 40 students into three classes of average size 27 if one additional student enrolled. Table 17.4 gives the typical difference in class size and test scores in matched pairs, larger-cohort-minus-smaller-cohort, for the 86 'high-dose' pairs, the 49 'lower-dose' pairs, and the 135 'mixed-dose' pairs formed by pooling the pairs, $135 = 86 + 49$. In Table 17.4, the typical difference in class size is larger for the 'high-dose' pairs, -10.9 students, than for the 'lower-dose' pairs, -5.3 students, and the difference in verbal test scores is also larger.

Table 17.5 compares the sensitivity analysis in Table 5.3 with two additional sensitivity analyses. One analysis uses Wilcoxon's signed rank statistic and the coherent signed rank statistic applied to the 135 'mixed-dose' pairs. The other analysis uses the dose-weighted signed rank statistic and the coherent version of that statistic [5] applied to the 135 'mixed-dose' pairs with doses of 2 and 1 for 'high-dose' and 'lower-dose' pairs, respectively. In this one example, the coherent statistic using the

17.6 Appendix: Proof of Proposition 17.1

Table 17.5 Sensitivity analysis and coherence in Angrist and Lavy's study of academic test performance and class size manipulated by Maimonides' rule. The table gives upper bounds on the one-sided P-values. The sensitivity analysis from Table 5.3 in §5.2.3 is the analysis for the 86 'high-dose' pairs. The remaining analyses add 49 'lower-dose' pairs to make 135 'mixed-dose' pairs. One analysis uses the signed rank statistic giving all pairs equal weights. The other analysis gives the 'high-dose' pairs twice the weight of the 'lower-dose' pairs. At $\Gamma = 1.65$, the only upper bound that is < 0.05 is for coherence with the 86 'high-dose' pairs.

	86 High-Dose Pairs			135 Mixed Dose Pairs			135 Dose-Weighted Pairs		
Γ	Math	Verbal	Coherent	Math	Verbal	Coherent	Math	Verbal	Coherent
1.00	0.0012	0.00037	0.00018	0.00034	0.00023	0.000081	0.00034	0.00014	0.000050
1.45	0.057	0.027	0.015	0.058	0.046	0.015	0.049	0.029	0.013
1.65	0.138	0.075	0.043	0.170	0.142	0.078	0.145	0.096	0.051

86 'high-dose' pairs is marginally less sensitive than the other analyses to departures from random treatment assignment. One should not make too much of individual examples, such as Tables 5.5 and 17.5; however, the results in those tables are not incompatible with the theoretical results about design sensitivity in this chapter.

17.5 Further Reading

Coherence and doses were discussed in Chapter 5, where general references may be found. The relationship between design sensitivity and coherence is discussed in [2, 8, 10]. The relationship between design sensitivity and doses is discussed in [8, 10]. Closely related issues are discussed in [5, 7] and [6, §9]. An alternative formulation of coherence is given by K. Joreskog and Arthur Goldberger [3]. Dose errors are discussed in [9, 12]. Christopher Wild [14] surveys the many types of doses that arise in modern cancer epidemiology. The proof in the appendix is from [10].

17.6 Appendix: Proof of Proposition 17.1

Proof. Define $\Psi_1 = E(d_i W_{ii})$ and $\Lambda_1 = E(d_i V_{ii}) = E(d_i)$. In the favorable situation in the statement of Proposition 17.1, the expectation of T_{dose} is

$$E(T_{\text{dose}}) = E\left(\sum_{i=1}^{I}\sum_{k=1}^{I} d_i W_{ik}\right) = \sum_{i=1}^{I} E(d_i W_{ii}) + \sum_{i=1}^{I}\sum_{k \neq i} E(d_i W_{ik})$$
$$= I\Psi_1 + I(I-1)\Psi.$$

In the sensitivity analysis for T_{dose}, the upper bound on the one-sided P-value is obtained by comparing T_{dose} to the distribution of the sum, $\overline{T}_{\Gamma, \text{dose}}$, of I independent random variables, $i = 1, \ldots, I$, taking the value $d_i q_i$ with probability $\theta = \Gamma/(1+\Gamma)$

and the value 0 with probability $1-\theta$, so $E\left(\overline{\overline{T}}_{\Gamma,\text{dose}} \mid \mathscr{F}\right) = \theta \sum_{i=1}^{I} d_i q_i$; see [5, 7]. Because $d_i q_i = \sum_{k=1}^{I} d_i V_{ik}$,

$$E\left(\overline{\overline{T}}_{\Gamma,\text{dose}} \mid \mathscr{F}, \mathscr{L}\right) = \theta \sum_{i=1}^{I} d_i q_i = \theta \sum_{i=1}^{I} \sum_{k=1}^{I} d_i V_{ik},$$

so that

$$E\left(\overline{\overline{T}}_{\Gamma,\text{dose}}\right) = E\left\{E\left(\overline{\overline{T}}_{\Gamma} \mid \mathscr{F}, \mathscr{L}\right)\right\} = \theta \sum_{i=1}^{I} \sum_{k=1}^{I} E\left(d_i V_{ik}\right)$$
$$= \theta \left\{I\Lambda_1 + I(I-1)\Lambda\right\}.$$

Using the result in [8, §3], the design sensitivity, $\widetilde{\Gamma}$, is the limit, as $I \to \infty$, of the solutions to $E(T_{\text{dose}}) = E\left(\overline{\overline{T}}_{\Gamma,\text{dose}}\right)$, or equivalently to

$$\frac{\Gamma}{1+\Gamma} = \frac{I\Psi_1 + I(I-1)\Psi}{I\Lambda_1 + I(I-1)\Lambda}, \qquad (17.9)$$

where the right side of (17.9) tends to Ψ/Λ as $I \to \infty$, yielding (17.8).

References

1. Angrist, J.D., Lavy, V.: Using Maimonides' rule to estimate the effect of class size on scholastic achievement. Q J Econ **114**, 533–575 (1999)
2. Heller, R., Rosenbaum, P.R., Small, D.: Split samples and design sensitivity in observational studies. J Am Statist Assoc **104**, to appear (2009)
3. Joreskog, K.G., Goldberger, A.S.: Estimation of a model with multiple indicators and multiple causes of a single latent variable. J Am Statist Assoc **70**, 631–639 (1975)
4. Rao, C.R.: Linear Statistical Inference and Applications. New York: Wiley (1973)
5. Rosenbaum, P.R.: Signed rank statistics for coherent predictions. Biometrics **53**, 556–566 (1997)
6. Rosenbaum, P.R.: Observational Studies (2nd edition). NY: Springer (2002)
7. Rosenbaum, P.R.: Does a dose-response relationship reduce sensitivity to hidden bias? Biostatistics **4**, 1–10 (2003)
8. Rosenbaum, P.R.: Design sensitivity in observational studies. Biometrika **91**, 153–164 (2004)
9. Rosenbaum, P.R.: Exact, nonparametric inference when doses are measured with random errors. J Am Statist Assoc **100**, 511–518 (2005)
10. Rosenbaum, P.R.: What aspects of the design of an observational study affect its sensitivity to bias from covariates that were not observed? In: Festshrift for Paul W. Holland. Princeton, NJ: ETS (2009)
11. Small, D., Rosenbaum, P.R.: War and wages: The strength of instrumental variables and their sensitivity to unobserved biases. J Am Statist Assoc **103**, 924–933 (2008)
12. Small, D.S., Rosenbaum, P.R.: Error-free milestones in error-prone measurements. Ann App Statist, to appear (2009)
13. van Eeden, C.: An analogue, for signed rank statistics, of Jureckova's asymptotic linearity theorem for rank statistics. Ann Math Statist **43**, 791–802 (1972)

14. Wild, C.P.: Environmental exposure measurement in cancer epidemiology. Mutagenesis **24**, 117–125 (2009)

Part IV
Planning Analysis

Chapter 18
After Matching, Before Analysis

Abstract Three design tasks may usefully follow matching and precede planning of the analysis. Splitting the sample of I pairs into a small planning sample and a large analysis sample may aid in planning the analysis in a manner that increases the design sensitivity. If there will be analytic adjustments for some unmatched variables, it is prudent to check that the matched samples exhibit sufficient overlap on unmatched variables to permit analytic adjustments. Quantitative analysis of matched samples may usefully be combined with qualitative examination and narrative description of a few closely matched pairs.

18.1 Split Samples and Design Sensitivity

In Chapters 14–17, the sensitivity of the design of an observational study to unmeasured biases, $\widetilde{\Gamma}$, was found to depend upon many things: unit heterogeneity, coherence among outcomes, doses, uncommon but dramatic responses to treatment, the strength of instrumental variables, among others. In some contexts, these considerations may be difficult to evaluate during design, prior to the examination of outcomes. An interesting option with good properties entails splitting the sample at random into a small planning sample and a large analysis sample, making some design decisions based on the planning sample, applying those decisions in the analysis sample. This section discusses the matter in brief outline based upon a paper by Ruth Heller, Dylan Small, and me [14].

Sample splitting is most familiar from cross-validation [31]. In cross-validation, a sample is split, usually at random, and a prediction is developed in one portion and tested in the other. Often the process is repeated with many random splits in an effort to develop an honest appraisal of the performance of some method of prediction.

Less familiar is the use of sample splitting to decide what question to ask, which hypothesis to test. David Cox [6] examined the use of split samples as an alternative to corrections for multiple testing in randomized experiments, such as corrections based on the Bonferroni inequality. Instead of testing many hypotheses, the sample is split, the first portion used to select a promising hypothesis, the second portion used to test that one hypothesis. In randomized experiments, Cox found that splitting was slightly inferior in terms of power to corrections for multiple testing in simple stylized settings. Cox noted the flexibility of splitting, which permits confirmation of unanticipated exploratory findings.

Sample splitting has much better properties in observational studies than in randomized experiments, outperforming, for example, the Bonferroni inequality [14]. In an idealized randomized experiment, there is no bias from unmeasured covariates, and the power of a conventional test is properly the focus of concern. In contrast, in an observational study, bias from failure to adjust for an unmeasured covariate is an ever-present worry, and the power of a sensitivity analysis is the relevant concern. As seen in Chapter 14, in favorable situations in which there is an effect but no unmeasured bias, as the sample size I increases, $I \to \infty$, the power of a sensitivity analysis tends to 1 for all $\Gamma < \widetilde{\Gamma}$ and to 0 for all $\Gamma > \widetilde{\Gamma}$, where $\widetilde{\Gamma}$ is the design sensitivity [21]. Again, the design sensitivity $\widetilde{\Gamma}$ depends upon many aspects of the design; see Chapters 14–17 and [21, 22, 23, 30]. Suppose that one could exchange a portion of the sample, say $(1-\rho)I$ observations with $0 < \rho < 1$, for a larger design sensitivity, say $\widetilde{\Gamma}^* > \widetilde{\Gamma}$, in the remaining portion. That is, suppose that a 'planning sample' of, say, $10\% = (1-\rho)$ of the observations led to better decisions about design, which in turn raised the design sensitivity, from $\widetilde{\Gamma}$ to $\widetilde{\Gamma}^*$, in the remaining 90%, the 'analysis sample.' If the sample size, I, is large, this is a wonderful swap. After all, as $I \to \infty$, without the split, the power is tending to zero for Γ in the interval $\left(\widetilde{\Gamma}, \widetilde{\Gamma}^*\right)$, whereas with the split the power is tending to one on the same interval. Admittedly, this is a limiting argument about a large sample, but as intuition might suggest and detailed calculations in [14] confirm, it does not take a very large sample size I for a larger design sensitivity, $\widetilde{\Gamma}^*$ instead of $\widetilde{\Gamma}$, to overwhelm a small reduction in sample size, to ρI from I.

The discussion in the previous paragraph presumed that a planning sample of size $(1-\rho)I$ could raise the design sensitivity in the analysis sample of size ρI. This presumption has two aspects, both more important than the loss of $(1-\rho)I$ observations. The first aspect is that there must be design decisions that would increase $\widetilde{\Gamma}$ if those decisions were correctly made. Concerning the first aspect, Chapters 14–17 offer some hope but no guarantees. The second aspect is that the planning sample must be large enough to lead to correct decisions about design when there are correct decisions to be made. Numerical results relevant to the second aspect are given in [14]. In several cases, $(1-\rho)I = 100$ planning observations with $\rho I = 900$ analysis observations worked extremely well; see [14] for specifics and detailed recommendations. Arguably, one needs a planning sample of size $(1-\rho)I$ adequate to make correct design decisions if there are substantial gains in design sensitivity to be had, with little to be gained by increasing the size of the planning sample beyond that.

The practical example in [14] seemed too small to hope for much. There were only $I = 36$ pairs, with $(1-\rho)I = 6$ used for planning and $\rho I = 30$ used for analysis. As an illustration, that analysis was repeated with 30 independent splits. In every one of the 30 splits, the six planning observations correctly guided the one important design decision that materially affected the sensitivity to unmeasured biases. That is too much to expect in general from six observations, but in larger studies, sample splitting has much to offer.

18.2 Are Analytic Adjustments Feasible?

In an observational study, matching sets up certain comparisons, sets up a certain analogous experiment: it builds adjustments for observed covariates into the design of an observational study. In Mervyn Susser's phrase, matching "simplifies the conditions of observation," aiding transparency (see Chapter 6).

Adjustments built into the design cannot be easily removed later. Under certain circumstances, one may plan two analyses, one with an adjustment for a certain variable, the other without that adjustment. In this case, the variable is not controlled by matching but is controlled in the statistical analysis. An aspect of the design of an observational study entails verifying that both analyses are feasible. This means checking that an unmatched variable exhibits sufficient overlap in treated and control groups that analytical adjustments are an interpolation, not an extrapolation or a fabrication.

Typically, one leaves unmatched a variable of ambiguous status. In that way, different analyses may assign the variable different roles [4]. If a concomitant variable is not truly a covariate, if it may have been affected by the treatment, then an adjustment for it may introduce a bias that would not otherwise have been present [15]. That said, a variable may have an ambiguous status, perhaps slightly affected by the treatment, but plausibly standing in as a surrogate for an important covariate that was not measured. For instance, two studies in 1982, by James Coleman, Thomas Hoffer, and Sally Kilgore [5] and Arthur Goldberger and Glen Cain [13], asked whether Catholic and other private high schools did more than public schools to increase the cognitive test performance of their students. Important covariates in this context are cognitive test scores before high school, but these were not available. Both studies make various comparisons of sophomore and senior test scores, with the view that sophomore test scores, though possibly affected by public-vs.-private schooling, might serve in one way or another as a proxy for the unmeasured covariates of test scores prior to high school. An important variable with an ambiguous status does not necessarily imply an ambiguous conclusion: perhaps the conclusion is unchanged by changing the status of the variable. If a variable of ambiguous status is left unmatched, analyses may examine issues of this sort.

Another type of variable of ambiguous status exhibits 'seemingly innocuous confounding.' That is, the covariate predicts a substantial fraction of the variation in treatment assignment, but seems fairly innocuous, without obvious importance to the outcomes under study. At the same time that the treatment is observed to be confounded with something innocuous, it may also be confounded with unmeasured covariates that are not innocuous. In certain circumstances, the innocuous confounding — the devil you know, in the words of the proverb — may seem less worrisome than confounding with unmeasured covariates — the devil you don't.[1] Should 'seemingly innocuous confounding' be removed? Perhaps. Perhaps it is not innocuous, and only seems so. And yet, if all the innocuous confounding is removed, then perhaps the haphazard aspect of treatment assignment is removed as well, and the remaining aspects of treatment assignment are governed by biases from relevant unmeasured covariates. Again, when a variable has an ambiguous status, conclusions may be firmer if shown to be untouched by this ambiguity. Consider an example.

Recall from §12.3 the study by Jeffrey Silber and colleagues [24, 29] of slightly earlier or slightly later discharges of 1402 premature infants from five neonatal intensive care units (NICUs) in the Northern California Kaiser-Permanente Medical Care Program between 1998 and 2002. The babies were paired into 701 pairs so that the two babies looked similar in terms of measured (time-dependent) covariates on the day the first baby, the early baby, was discharged, while the other baby, the late baby, stayed in the NICU for a few more days, growing older and more mature. Some of the variation in discharge day seems to have been governed by the day of the week: reach maturity on Thursday and you go home on Friday; reach maturity on Saturday and you go home on Tuesday, or so it appeared. There were five hospitals in the Northern California Kaiser-Permanente Medical Care Program that provided data for this study, and hence five NICUs, one at each hospital. Some of the variation in discharge day seems to reflect styles of practice at the different NICUs: some are faster than others. You could make a case, of course, that these sources of variation are not innocuous. You could make a case that the baby discharged Friday morning was judged healthier by the attending neonatologist than the baby held until Tuesday, observed covariates to the contrary not withstanding. Although these were all Kaiser-Permanente moms in hospitals run by Kaiser-Permanente, you could make a case that the moms at certain hospitals or the hospitals themselves were not really the same. Counter to this, you could make a case that the variation in discharge day predicted by the identity of the NICU or the day of the week is less worrisome than unexplained variation. Should we speculate and argue, or should we look and see? Let's look and see.

Table 18.1 shows the distribution of the 701 pairs of two babies at the five NICUs. The table counts pairs, not babies, so the total count is 701 pairs, not 1402 babies. Of the 701 pairs, 233 fall on the diagonal: the early and late baby are in the same NICU. Some NICUs discharge earlier than others: NICU C contributed more late babies than early babies, while NICU D did the opposite. On the other hand, despite the trend, there are many early babies from NICU C, specifically 95, and many late babies from NICU D, specifically 89. The variation between NICUs in discharge rate accounts for part but not all of the variation in discharge day; therefore, the variation due to the NICU can be removed analytically. In this way, we will see the situation both with innocuous confounding and without.

---

[1] Erasmus: "Nota res mala optima (an evil thing known is best)" [8, #85, page 123].

18.2 Are Analytic Adjustments Feasible? 319

Table 18.1 Pairs classified by the neonatal intensive care unit (NICU, A–E) of the early baby and the late baby for 701 matched pairs. The table counts pairs, so the total count is 701. NICU C had 161 late babies and 95 early babies, for a Late/Early ratio of $161/95 = 1.69$, while NICU D had 89 late babies and 165 early babies, for a Late/Early ratio of $89/165 = 0.54$. Percentages are rounded.

| | Late Baby | | | | | | | Late/Early |
Early Baby	A	B	C	D	E	Total	Percent	Ratio
A	18	23	31	13	29	114	16	0.79
B	20	65	24	17	36	162	23	0.93
C	10	19	30	12	24	95	14	1.69
D	26	24	41	36	38	165	24	0.54
E	16	19	35	11	84	165	24	1.28
Total	90	150	161	89	211	701		
Percent %	13	21	23	13	30		100	

Table 18.1 is part of the design. There is no information about outcomes in Table 18.1. The pattern in Table 18.1 reveals that it is safe to leave NICU unmatched; within these 701 pairs matched for many characteristics of baby and mom, there is enough overlap in the NICUs to permit adjustments. We may perform two analyses, one that ignores the NICU, viewing the NICU as an innocent source of variation in the day of discharge, the other analysis removing entirely the variation in discharge day that can be predicted from the NICU. Deciding whether to match for the NICU is part of the design, and Table 18.1 is part of that decision.

As will be discussed in detail in §19.6, one of the outcomes in Silber et al.'s [24, 29] study concerned sick-baby services in an interval approximately six months long following the discharge of the late baby, when both babies were home. Did early discharge prompt a greater need for sick-baby services after discharge? Well-baby services, such as checkups, were not included. Sick-baby costs include both minor events with minor costs, and major events, such as readmission to the NICU, with enormous costs. Services were converted to dollar amounts according to a fixed schedule, so that babies who received the same services had the same dollar score for those services. Here, postdischarge sick-baby cost is serving as a measure of the severity of the health problem, for which it is not a bad proxy. See §19.6 and [24, 29] for specifics and additional outcomes.

Table 18.2 is a letter value display [32] of the 701 matched pair differences, late-minus-early, in postdischarge sick baby costs. Five deaths are recorded as infinite costs, three among the early babies, two among the late babies; see §19.6 and [29] for more about the deaths. The median difference in cost is $5 and the average of the quartiles is $32. The distribution is extremely long-tailed relative to the Normal distribution, as is seen from the pseudo-spreads in Table 18.2, which would be approximately constant for a very large sample from a Normal distribution. There is no sign of increased postdischarge costs for early babies.

One analysis ignores the NICU. It allows the variation in practice style at different NICUs to influence the day of discharge. A baby who happened to be born in NICU C might stay a few extra days, and one born in NICU D may be discharged a few days early. For two babies with the same matched covariates in §12.3, this anal-

Table 18.2 Letter value display of difference in postdischarge sick-baby costs in dollars in 701 pairs of a late and an early baby. The difference is late-minus-early. Various quantiles are displayed, together with their averages (the mids), differences (the spreads), and the ratio of the spreads to the corresponding spreads of the standard Normal distribution. Five deaths are recorded as infinite costs, three among the early babies, two among the late babies; see §19.6 for detailed discussion of the deaths.

Quantile %	'Letter'	Lower	Upper	Mid	Spread	Pseudo-spread
50	1/2	5	5	5	0	
25	1/4	−223	287	32	510	378
12.5	1/8	−1063	2313	625	3376	1467
6.25	1/16	−4089	4956	433	9045	2948
3.125	1/32	−9893	12944	1526	22836	6130

ysis views the NICU as irrelevant to outcomes except to the extent that it accelerates or retards discharge. Ignoring the NICU, procedures associated with Wilcoxon's signed rank statistic (§2.4) give a point estimate of an additive effect τ of $17 and a 95% confidence interval of $[-\$20, \$56]$; see §19.6 for additional analyses ignoring the NICU, including a sensitivity analysis.

The analysis using Wilcoxon's statistic viewed all 2^{701} possible treatment assignments in the 701 pairs as equally probable. Table 18.1 suggests they are not equally probable; rather, they are predictable using the NICU. For instance, look at row D and column C of Table 18.1 in comparison with row C and column D. In those two cells of Table 18.1 are $41 + 12 = 53$ pairs of two babies, one born at NICU 'C,' the other at NICU 'D.' In those 53 pairs, the odds that the early baby is at NICU D are estimated at $41/12 = 3.4$ to 1. The alternative analysis ([17] and [20, §3]) does not consider all 2^{701} possible treatment assignments, but rather a subset of treatment assignments that exhibits the same imbalance as observed in Table 18.1. For instance, in the 53 pairs with one baby at NICU C and another at NICU D, Wilcoxon's statistic considers all $2^{53} = 9 \times 10^{15}$ possible treatment assignments as equally probable and uses them all, but the alternative procedure considers only the $53!/(41!\,12!) = 2.7 \times 10^{11}$ treatment assignments that keep 12 early babies in NICU C and 41 late babies in NICU 'C.' For the 233 pairs on the diagonal in Table 18.1, all possible treatment assignments are considered, but for the off-diagonal pairs, the observed imbalance is preserved. The distribution of Wilcoxon's statistic, or any other statistic, is compared to this restricted or conditional permutation distribution.[2]

The alternative analysis that adjusts for the NICU yields a point estimate of an additive effect τ of $-\$6$ and a 95% confidence interval of $[-\$36, \$53]$. Some of the variation in day of discharge can be predicted from the NICU, a seemingly

---

[2] When is this analysis appropriate? Suppose the NICUs are included among the observed covariates and after including the NICUs, (3.5)–(3.8) is then true. Suppose further that the NICUs enter the propensity score as additive constants on the logit scale. Then the conditioning just described eliminates the additive NICU parameters, leaving behind a known distribution of treatment assignments that gives the remaining assigments equal probabilities. The distribution of the Wilcoxon statistic is then approximated using its easily computed moments. See [17] and [20, §3] for the paired case, as here, and see [16, 19] and [20, §3] for the general case.

18.2 Are Analytic Adjustments Feasible?

innocuous form of confounding. Two analyses were performed, one that ignored the predictable variation, the other that removed the predictable variation. The two analyses produce similar inferences about τ. Although the status of the NICU is ambiguous, the conclusion is less so. Indeed the pair of analyses provides some reassurance: different sources of variation in the day of discharge produce similar estimates of the effect of slightly delayed discharge.

Is it safe to leave a covariate unmatched? Will analytic adjustments be feasible for that covariate? Is there sufficient overlap on an unmatched covariate? Will analytic adjustments require an interpolation, not an extrapolation or a fabrication? In the design of an observational study, a table such as Table 18.1 recording treatment assignment and unmatched covariates provides answers to these questions.

Tapered matching

An alternative to leaving an ambiguous covariate unmatched is to construct matched triples, one treated subject and two controls, so that one control is matched for the covariate and the other is not. This ensures that both analyses are feasible, and provides a formal test as to whether their answers differ, as well as a confidence statement about the magnitude of the difference. The matching is implemented using 'optimal tapered matching', as discussed by Shoshana Daniel, Katrina Armstrong, Jeffrey Silber, and me [7]. In optimal tapered matching, a single control group is optimally partitioned into two types of controls and optimally matched to treated subjects. The procedure is easy to implement. A distance matrix is computed between each treated subject and each control, ignoring the ambiguous covariate, with one row for each treated subject and one column for each control, as in Chapter 8. A second distance matrix is computed between each treated subject and each control including the ambiguous covariate in the distances. These two distance matrices are stacked, one on top of the other, so the number of rows is doubled but the number of columns is unchanged. An optimal pair match is determined for this stacked distance matrix. Each control is then used at most once, but each treated subject is used twice. In this way, each treated subject is paired with two controls, one close in terms of the first distance, the other close in terms of the second distance, and the two control groups do not overlap, so they may be compared using conventional methods, such as Wilcoxon's signed rank test. The procedure is not restricted to matched triples, and may use more than one control from either group, or may have more than two control groups. Tapered matching was originally developed in connection with research on disparities, where the concept of a covariate is inherently ambiguous. The paper [7] discusses an example in detail, demonstrates a certain optimal property of the procedure, and discusses some of the subtleties involved in the choice of distance.

18.3 Matching and Thick Description

Matching and thick description

Matching compares groups of treated and control subjects, often pairs of treated and control subjects, who look comparable in terms measured covariates $\mathbf{x}_{ij}$. A constant worry is that subjects who look comparable may not be comparable, that they look comparable only because important distinctions are omitted in the observed covariates $\mathbf{x}_{ij}$, distinctions that are to be found only in unobserved covariates u_{ij} for which subjects are not matched.

In matching, people remain intact as people. In contrast, in model-based adjustments, the people themselves disappear or recede into the background to be replaced by features of a model. Where there were once people, there are now parameters in a model. A relatively neglected but potentially important aspect of matching is the possibility of taking a close look at a few well-matched pairs of people, providing a narrative account of these pairs.[3] That is, one might provide a 'thick description' of a small subset of pairs. Jeffrey Silber and I tried this in a study of mortality after surgery in the Medicare population in Pennsylvania, and the current section is a brief summary of the resulting paper [18].

It is a staple of professional education — in medicine, in business, in law — that a narrative account of a few well-chosen cases is often enlightening and instructive. For instance, in this spirit, the *New England Journal of Medicine* often publishes its "Case Reports from the Massachusetts General Hospital"; see also the essay by Jan Vandenbroucke [33]. The professions do not believe that broad generalizations can be warranted by examination of a few cases; rather, they believe that to become acquainted with generalizations without engaging particular cases to which those generalizations are supposed to apply is to run the risk of developing a fairly scholastic perspective.

In certain social sciences, there is a division of labor, with some investigators, sometimes called 'qualitative researchers,' creating narrative accounts from a few cases, and other investigators, sometimes called 'quantitative researchers,' creating analyses from data describing many people. Matching provides a framework within which qualitative and quantitative research can usefully interact within a single investigation. A narrative account of a few pairs could go hand in hand with a statistical analysis of many pairs.

---

[3] As has been discussed several times, modern matching methods try to balance high-dimensional observed covariates in treated and control groups, viewing close individual matched pairs as a secondary concern. In light of this, treated and control groups may be balanced with respect to observed covariates using many pairs that are not individually close on observed covariates. The methodology discussed in the current section is most appropriate for the subset of matched pairs in which the individual pairs are closely matched. Typically, there will be many such pairs, even if many other pairs contribute to covariate balance without being closely matched as pairs with respect to covariates. If a distance, such as the Mahalanobis distance, is used in matching (see Chapter 8), then a closely matched pair will be recognizable by a small distance within the pair.

18.3 Matching and Thick Description

What is thick description?

The term 'thick description' was introduced by a philosopher, Gilbert Ryle [25, page 479], and was made popular by an anthropologist, Clifford Geertz [11]; see [18, pages 221-222] for quotes from Ryle and Geertz that give a brief sense of the nature of their concerns. The aim of thick description according to Geertz [12, page 152] is "to render obscure matters intelligibly by providing them with an informing context" [12, p. 152]. The importance of an informing context is conveyed by the following passage by Howard Becker [2, page 58]:

> [I]f we don't find out from people what meanings they are actually giving to things, we will still talk about those meanings. In that case, we will, of necessity, invent them, reasoning that the people we are writing about must have meant this or that, or they would not have done the things they did. But it is inevitably epistemologically dangerous to guess at what could be observed directly. The danger is that we will guess wrong, that what looks reasonable to us will not be what looked reasonable to them. This happens all the time, largely because we are not those people and do not live in their circumstances.

It is quite easy to misinterpret data from a large electronic data set [10]. The attempt to create a narrative account of a few cases may serve to test interpretations of data and the concepts in which those interpretations are cast. An example follows.

Example: Mortality after surgery

Silber and colleagues' [26, 27, 28] study of mortality after surgery used (inexpensive) administrative data to create a matched sample from the Medicare population in Pennsylvania, which was followed by (expensive) chart abstraction for the matched sample. In this study, deaths shortly following surgery were matched to ostensibly similar survivors based on administrative data from Medicare. Although the best data would come from chart abstraction for the matched sample, it was important to have a reasonable match based on administrative data, so that charts for suitable patients would be abstracted.

In [18], a preliminary matched sample was constructed, and Medicare records for a handful of well-matched pairs were contrasted with far more detailed information in the hospital chart. The preliminary match looked reasonable in terms of the measured variables from Medicare's administrative records. However, even though only a few cases were examined, it was immediately clear from chart data that the initial match was not adequate and that many definitions that we created to interpret the administrative records required revision. Many of the mistakes we unearthed were obvious mistakes when viewed from the step-by-step narrative perspective of a hospital chart, but were not obvious from the perspective of computerized administrative records, or at least not obvious to us.

For instance, we generally wanted to avoid matching on events that were subsequent to surgery, because these might be outcomes of surgery or of subsequent care. Although that principle is sound, our initial use of it was mechanical and thoughtless. For instance, if surgery reveals that a patient has metastatic cancer, that cancer existed prior to the surgery that discovered it. Metastatic cancer is, of course, important to the patient's prognosis and to proper interpretation of the care the patient subsequently received. In light of this, we changed our definitions related to cancer and several other medical conditions that could not have begun during a hospital stay. No one would make this mistake faced with a narrative account describing an individual patient, but it is less obvious in computerized administrative records, and it might not be easily discerned from the coefficients of a model. If some mistakes are difficult to make when a situation is viewed from a narrative perspective, then that constitutes a good reason for expending the time and effort required to view the situation from a narrative perspective.

We also redefined "died shortly after surgery." One of the examined charts described a patient who had 'survived after surgery' by our original definition, but after a long stay had died without leaving the hospital. This case prompted us to reexamine the administrative data, only to discover that the fate of this one patient was sufficiently common to merit a different definition of "died shortly after surgery."

For additional examples and further discussion, see [18]. The example in [18] is a fairly limited use of matching in conjunction with narrative description, but extensive use is possible. For three compelling examples of extended narrative accounts (without matching), see [1, 3, 9].

18.4 Further Reading

Section 18.1 is a brief outline of a paper by Ruth Heller, Dylan Small and me [14]. The method of adjustment in §18.2 is from [17]. Although the method is described correctly and in detail in [17], there is a misstatement of one of the assumptions, which is corrected in [20, §3] and in Note 2 of the current chapter. The method of §18.2 is the simplest case of a general method discussed in [16, 19]. Section 18.3 is derived from a paper with Jeffrey Silber [18] that contains further discussion and additional references. The literature on thick description, ethnography, and qualitative research is vast. The books by Lonnie Athens [1], Charles Bosk [3] and Sue Estroff [9] are fine examples of ethnography touching upon criminology, surgery, and psychiatry, and the volume containing Becker's essay [2] contains many other interesting essays as well.

References

1. Athens, L.: Violent Criminal Acts and Actors Revisited. Urbana, Illinois: University of Illinois Press (1997)
2. Becker, H.S.: The epistemology of qualitative research. In: R. Jessor, A. Colby, R. Shweder (eds.), Ethnography and Human Development, pp. 53–72. Chicago: University of Chicago Press (1996)

3. Bosk, C.L.: Forgive and Remember: Managing Medical Failure. Chicago: University of Chicago Press (1981)
4. Cochran, W.G.: Analysis of covariance: Its nature and uses. Biometrics **13**, 261–281 (1957)
5. Coleman, J., Hoffer, T., Kilgore, S.: Cognitive outcomes in public and private schools. Soc Educ **55**, 65–76 (1982)
6. Cox, D.R.: A note on data-splitting for the evaluation of significance levels. Biometrika **62**, 441–444 (1975)
7. Daniel, S., Armstrong, K., Silber, J.H., Rosenbaum, P.R.: An algorithm for optimal tapered matching, with application to disparities in survival. J Comput Graph Statist **17**, 914–924 (2008)
8. Erasmus: The Collected Works of Erasmus, Volume 34, Adages IIvii1 to IIIiii100. Toronto: University of Toronto Press (1992)
9. Estroff, S.E.: Making It Crazy: An Ethnography of Psychiatric Clients in an American Community. Berkeley: University of California Press (1985)
10. Bittner, E., Garfinkel, H.: 'Good' organizational reasons for 'bad' organizational records. In: H. Garfinkel, Studies in Ethnomethodology, pp. 186–207. Englewood Cliffs, NJ: Prentice Hall (1967).
11. Geertz, C.: Thick description: toward an interpretative theory of culture. In: C. Geertz, The Interpretation of Cultures, pp. 3–30. New York: Basic Books (1973)
12. Geertz, C.: Local Knowledge. New York: Basic Books (1983)
13. Goldberger, A.S., Cain, G.S.: The causal analysis of cognitive outcomes in the Coleman, Hoffer and Kilgore report. Soc Educ **55**, 103–122 (1982)
14. Heller, R., Rosenbaum, P. R., Small, D.: Split samples and design sensitivity in observational studies. J Am Statist Assoc **104**, to appear (2009)
15. Rosenbaum, P.R.: The consequences of adjustment for a concomitant variable that has been affected by the treatment. J Roy Statist Soc A **147**, 656–666 (1984)
16. Rosenbaum, P.R.: Conditional permutation tests and the propensity score in observational studies. J Am Statist Assoc **79**, 565–574 (1984)
17. Rosenbaum, P.R.: Permutation tests for matched pairs with adjustments for covariates. Appl Statist **37**, 401–411 (1988) (Correction: [20], §3])
18. Rosenbaum, P.R., Silber, J.H.: Matching and thick description in an observational study of mortality after surgery. Biostatistics **2**, 217–232 (2001)
19. Rosenbaum, P.R.: Covariance adjustment in randomized experiments and observational studies (with Discussion). Statist Sci **17**, 286–327 (2002)
20. Rosenbaum, P.R.: Observational Studies (2nd ed). New York: Springer (2002)
21. Rosenbaum, P.R.: Design sensitivity in observational studies. Biometrika **91**, 153–164 (2004)
22. Rosenbaum, P.R.: Heterogeneity and causality: Unit heterogeneity and design sensitivity in observational studies. Am Statistician **59**, 147–152 (2005)
23. Rosenbaum, P.R.: What aspects of the design of an observational study affect its sensitivity to bias from covariates that were not observed? In: Festshrift for Paul W. Holland. Princeton, NJ: ETS (2009)
24. Rosenbaum, P.R., Silber, J.H.: Sensitivity analysis for equivalence and difference in an observational study of neonatal intensive care units. J Am Statist Assoc **104**, 501–511 (2009)
25. Ryle, G.: Collected Papers, Volume 2. London: Hutchinson (1971)
26. Silber, J.H., Rosenbaum, P.R., Trudeau, M.E., Even-Shoshan, O., Chen, W., Zhang, X., Mosher, R.E.: Multivariate matching and bias reduction in the surgical outcomes study. Med Care **39**, 1048–1064 (2001)
27. Silber, J.H., Rosenbaum, P. R., Trudeau, M.E., Chen, W., Zhang, X., Lorch, S.L., Rapaport-Kelz, R., Mosher, R.E, Even-Shoshan, O.: Preoperative antibiotics and mortality in the elderly. Ann Surg **242**, 107–114 (2005)
28. Silber, J.H., Rosenbaum, P.R., Zhang, X., Even-Shoshan, O.: Estimating anesthesia and surgical time from medicare anesthesia claims. Anesthesiology **106**, 346–355 (2007)
29. Silber, J.H., Lorch, S.L., Rosenbaum, P.R., Medoff-Cooper, B., Bakewell-Sachs, S., Millman, A., Mi, L., Even-Shoshan, O., Escobar, G.E.: Additional maturity at discharge and subsequent health care costs. Health Serv Res **44**, 444–463 (2009)

30. Small, D.S., Rosenbaum, P.R.: War and wages: The strength of instrumental variables and their sensitivity to unobserved biases. J Am Statist Assoc **103**, 924–933 (2008)
31. Stone, M.: Cross-validatory choice and assessment of statistical predictions. J Roy Statist Soc B **36**, 111–147 (1974)
32. Tukey, J.W.: Exploratory Data Analysis. Reading, MA: Addison-Wesley (1977)
33. Vandenbroucke, J.: In defence of case reports and case series. Ann Intern Med **134**, 330–334 (2001)

Chapter 19
Planning the Analysis

Abstract "Make your theories elaborate" in observational studies, argued R.A. Fisher, so that the many predictions of such a theory may disambiguate the association between treatment and outcome. How should one plan the analysis of an observational study to check the predictions of an elaborate theory?

19.1 Plans Enable

A randomized clinical trial follows a protocol that describes the design of the trial and a plan for its analysis. The plan for analysis will identify a primary endpoint or outcome, possibly secondary endpoints, describe the comparisons that will be made, and so forth. The plan for a clinical trial is reviewed by a funding agency and, for large important trials, the plan may be published before the trial begins. In this sense, the plan is a public plan. There is no barrier to planning the analysis of an observational study, and there is much to gain from planning.

Plans enable. In all aspects of life, much can be done with a plan that cannot be done without one.

A study that follows a public plan is more convincing than a study that emerges ambiguously from a fog of data. A public plan is subject to public scrutiny before the study begins, so criticism that would otherwise surface after the fact arises before the fact, perhaps leading to a better study more resistant to legitimate criticism. If there is a public plan, and if the critic raises no objection to the plan, then criticism that could have been raised to the plan rings hollow when raised for the first time after a particular conclusion is announced.

A planned analysis is more transparent than an unplanned analysis. As discussed in Chapter 6, transparency means making evidence evident. If there is a public plan for analysis, then the investigators either followed their plan or they deviated from it. If they deviated from the plan, they will be obligated to explain and justify the deviation. Perhaps the deviation will be judged appropriate, perhaps more appropriate in light of events than adherence to the plan. No matter. What was done and why it was done are open to view.

Without a plan, an analysis will first use the methods that are most familiar, most ready to hand, then suffer a disorganized retreat to appropriate methods when unforeseen but foreseeable dilemmas arise. Even if appropriate methods are ultimately used, such an analysis looks like what it is: a dubious circular meander around data, methods, and conclusions. Here is a common example. We are all more familiar with tests and confidence intervals for differences than with tests and intervals for near equivalence, although there is sound, straightforward, standard methodology available for both tasks [6]. Meandering through a data set, we therefore apply tests for difference indiscriminately, then remember that tests for difference make a muddle of attempts to demonstrate near equivalence, then retreat to appropriate methods in a manner that rarely convinces us, much less anyone else. And yet a moment's thought before the fact would have revealed that to provide evidence for a certain conclusion means finding a difference here and near equivalence there. A moment's thought before the fact is readily available; it just takes a moment's thought. This dilemma is not in the data but in the absence of a plan.[1]

A plan does not preclude unplanned analyses. A plan distinguishes planned and unplanned analyses.[2]

Admittedly, you cannot plan an ornate and delicate analysis. You can plan the analysis of a randomized trial because randomization has done the heavy lifting of ensuring that comparable people are compared; ornate analysis is not needed. You can plan the analysis of an observational study if matching in the design has done the heavy lifting of ensuring that ostensibly comparable people are compared; ornate adjustments of outcomes for observed covariates are not essential.[3] Properly performed, matching is done without examining outcomes; it is, therefore, part of design. To say that you cannot plan an ornate and delicate analysis is to offer a telling argument against ornate and delicate analyses.

Novelty and originality may be acceptable in a plan, but soundness, not novelty or originality, is the mark of excellence in a planned analysis. The analysis of a standard design may follow a standard plan.

---

[1] See, for instance, Note 2 in Chapter 5, where I deliberately did the meandering analysis I intend to correct in the current chapter.

[2] John Tukey: "We do not dare either give up exploratory data analysis or make it our sole interest" [36, page 72]. "We often forget how science and engineering function. Ideas come from previous exploration more often than from lightning strokes. Important questions can demand the most careful planning for confirmatory analysis" [35, page 23].

[3] By 'ostensibly comparable,' I mean comparable in terms of measured covariates, x, leaving open the question of whether people who look comparable in terms of x are actually comparable in terms of unmeasured covariates u. 'Ostensibly comparable' is an abbreviation.

19.2 Elaborate Theories

R.A. Fisher's Advice

In his paper, "The planning of observational studies of human populations," William Cochran [9, §5] wrote:

> About 20 years ago, when asked in a meeting what can be done in observational studies to clarify the step from association to causation, Sir Ronald Fisher replied: 'Make your theories elaborate.' The reply puzzled me at first, since by Occam's razor, the advice usually given is to make theories as simple as is consistent with known data. What Sir Ronald meant, as subsequent discussion showed, was that when constructing a causal hypothesis one should envisage as many different consequences of its truth as possible, and plan observational studies to discover whether each of these consequences is found to hold.... [T]his multiphasic attack is one of the most potent weapons in observational studies.

Fisher's advice fits closely with the devices in §5.2 that disambiguate causal associations [26], and with the reduction in sensitivity to unmeasured biases that may follow from an anticipated pattern of treatment effects in Chapter 17. In Fisher's advice, the temporal order is important: one envisions consequences of a causal hypothesis and then plans the study to discover whether the consequences hold. There is no value in an elaborate theory built after the fact to fit unanticipated patterns in the data.

A simple example, discussed in §5.2.2 and §11.3, is the use of two control groups. If the treatment is the cause of its ostensible effects, the anticipated pattern with two control groups is clear: the treated group differs substantially from the two control groups, which do not differ substantially from each other. The prediction here is for two substantial differences and one near equivalence. What plan is appropriate for the analysis?

A second simple example discussed in § 5.2.3 and §17.2 concerns coherence of two outcomes, for instance an increase in both math and verbal test scores associated with smaller class sizes prompted by Maimonides' rule. There is the hypothesis of a coherent association and hypotheses about two outcomes. What plan is appropriate for the analysis?

What should a planned analysis accomplish?

An elaborate theory makes an elaborate prediction, and in Chapter 17 it was seen that certain forms of compatibility with an elaborate theory may make the study less sensitive to unmeasured biases. The investigator hopes that the use of two control groups rather than one, or the search for coherence among two outcomes, will enlarge what can legitimately be said, not diminish it, but there are pitfalls for the unwary.

In the first instance, the data may confirm part of the elaborate theory, refute another part, with a third part left in an ambiguous state. A binary 'confirm' or 'otherwise' is not appropriate. The evaluation of an elaborate theory will, therefore, involve more than one comparison. If several comparisons are made using statistical procedures, then each comparison runs a risk of error, with the resulting possibility that a series of small risks of error accumulate to a risk of error that is no longer small. If a correction for multiple testing were made to control this accumulation of error, say using the Bonferroni inequality, then the investigator might discover, to her shock and dismay, that if she had used the first control group alone, it would have had outcomes significantly different from the treated group, and if she had used the second control group alone, it too would have had outcomes significantly different from the treated group, but because she used both control groups and did two tests, neither difference is significant after correcting for multiple testing. This, too, is not appropriate.

Moreover, if the elaborate theory makes predictions of near equivalence, and if tests for difference are performed where tests for near equivalence are needed, then successes may be misclassified as failures and failures as successes. In a large study, two control groups might have nearly equivalent outcomes, yet differ significantly; this is a success that might be mistaken for a failure. Conversely, two control groups may have outcomes that do not differ significantly, and yet because of low power may provide no evidence that the outcomes are similar in the two groups; this is a failure that might be mistaken for a success. A test for near equivalence is one that can reject substantial inequivalence.

So a plan for analysis must do several things. To be a legitimate statistical analysis at all, it must control, in some appropriate sense, the risk of incorrect inferences. It must do this while making several comparisons, resulting in a graduated assessment of compatibility with the elaborate theory, rather than a simple 'yes' or 'no'. If some comparisons have priority over others, then the analysis should respect that priority: the most important comparisons should come first, in some sense, and not be hindered by added interest in other comparisons, as seemed to be the case above in connection with the Bonferroni inequality.

To be credible in their respective roles, an elaborate theory and priorities for its assessment must appear in a plan that precedes examination of outcomes. The elaborate theory is intended to make predictions, but it isn't making predictions if it is built to fit the data that will be used to assess those predictions. Priorities can aid in controlling the risk of incorrect inferences only if they shape the analysis rather than being shaped by it.

19.3 Three Simple Plans with Two Control Groups

The simplest plan for analysis with two control groups

To examine the logic of the simplest plan in a practical case, consider again Table 5.1 from the study by Bilban and Jakopin [7] of genetic damage among lead-zinc miners compared with two control groups. This simplest plan is, in fact, too simple

19.3 Three Simple Plans with Two Control Groups

even for this most basic situation. Better plans will be considered soon. The focus, for the moment, is on the logic of the plan, not yet on finding a good plan.

Let us imagine that Bilban and Jakopin originally planned to use one control group, say the Slovene residents far from the mine. Let us imagine further that they added the second control group of local residents near the mine who were not miners in response to concerns that pollution in the area of the mine, not the mine itself, might be responsible for greater genetic damage among miners. This second control group is a reasonable response to that specific concern. Finally, let us suppose that they wanted to give priority to the first of these two control groups.

Consider the following analytic plan. Test the hypothesis of equality of micronuclei frequencies among miners and the first control group, say at level 0.05. If that hypothesis is not rejected, stop. If that hypothesis is rejected, declare it rejected and test the second hypothesis of equality of micronuclei frequencies among miners and the second control group, also at level 0.05, declaring it rejected or not as appropriate.

Admittedly, as analytic plans go, this is not much of a plan. It does nothing to investigate the magnitude of the effect, which is surely important, and it does nothing to appraise the comparability of the two control groups, which is also important. The issue of sensitivity to unmeasured biases is not mentioned. This plan, such as it is, does not plan much. Perhaps what it plans is less important than what it ignores. No matter. The plan exemplifies a certain principle in the simplest possible setting.

In Table 5.1 from [7], the first test yields a t-statistic of 16.13, the second a t-statistic of 13.87. If the distributions of micronuclei were independent and Normal with expectations μ_t, μ_{c1} and μ_{c2} and standard deviations ω_t, ω_{c1}, and ω_{c2} in the treated group and the two control groups, with $c1$ referring to Slovene residents and $c2$ referring to local residents, then, at the 0.05 level, the plan would reject the first hypothesis, then test and reject the second.[4]

Notice first that, under this plan, the addition of a second control group costs the investigators nothing. If the hypothesis of no difference between miners and the first control group would have been rejected without the second control group, it would also be rejected by this plan in the presence of the second control group. So the situation is different from correcting for multiple testing using the Bonferroni inequality as described above. The second control group may allow more to be said, but less will not be said.

The second thing to notice is that the plan controls the probability of at least one false rejection in two tests provided the two separate tests do their separate jobs of controlling the probability of a false rejection. It is important to understand what is being claimed, then why the claim is true. The claim is that if each test has level

---

[4] In Table 5.1, it appears that ω_t, ω_{c1}, and ω_{c2} are not equal, so a pooled standard deviation is not used, and the 't-statistics' are not actually distributed with a t-distribution. Because the degrees of freedom are fairly large, the separate standard errors are used, and the 't-statistics' are compared with the standard Normal distribution, both here and in all calculations from Table 5.1. Although this is an imperfect approach, the largest imperfection in my analysis is in assuming the data are Normal merely because only means and standard deviations were reported. This example is being used to illustrate certain concepts while minimizing incidentals unrelated to those concepts.

0.05, falsely rejecting one time in 20, then the two-step plan tests two hypotheses but makes at least one false rejection with probability at most 0.05. It is like buying two lottery tickets rather than one without increasing your chance of winning, though not quite like that. Something has prevented an accumulation of risk of false rejection. What is that something? Consider the logic.

There are two null hypotheses, H_0 and $\widetilde{H}_0$. To be specific, the one sided null hypotheses are $H_0 : \mu_t \leq \mu_{c1}$ and $\widetilde{H}_0 : \mu_t \leq \mu_{c2}$. Here, H_0 refers to the first control group and $\widetilde{H}_0$ refers to the second, but it doesn't really matter what they refer to; they could be any two hypotheses about anything. The plan is to test H_0, and only if H_0 is rejected to test $\widetilde{H}_0$. Consider the cases, one at a time. If H_0 and $\widetilde{H}_0$ are both false, then you cannot falsely reject anything, so there is no risk of false rejection; any rejection is a correct rejection. If H_0 is false but $\widetilde{H}_0$ is true, then you cannot falsely reject H_0, so to falsely reject anything you have to, among other things, falsely reject $\widetilde{H}_0$, but the chance of this is at most 0.05. If H_0 and $\widetilde{H}_0$ are both true, then you commit a false rejection if and only if you reject H_0, and the chance of this is at most 0.05. So no matter which case is true, the chance of falsely rejecting at least one true hypothesis is at most 0.05.[5]

This first bit of logic ramifies in many useful directions [3, 15, 17, 29]. This logic is discussed explicitly in the context of clinical trials by Gary Koch and S.A. Gansky [17] and Peter Bauer [3], but appears to have been understood 20 years earlier by Ruth Marcus, Eric Peritz, and K. R. Gabriel [22], who were doing something related but subtler. The logic also has links to multiparameter tests [5, 18, 20] and has been combined in an interesting way with the Bonferroni inequality [14, §3].

There is a second piece to the logic.

A symmetric plan for analysis with two control groups

An awkward feature of the first plan is that the investigators had to give priority to one of the control groups. Can the two control groups be viewed symmetrically?

Consider a second plan. Test at level 0.05 the hypothesis that $\overline{H}_0 : \mu_t \leq (\mu_{c1} + \mu_{c2})/2$, which says that the expected level of micronuclei in the treated group is no higher than in the average of the two control groups. If $\overline{H}_0$ is not rejected, stop. If $\overline{H}_0$ is rejected, declare it rejected and test at the 0.05 level both $H_0 : \mu_t \leq \mu_{c1}$ and $\widetilde{H}_0 : \mu_t \leq \mu_{c2}$, rejecting neither, either, or both as appropriate.

---

[5] In his PhD thesis, Frank Yoon [38] improves upon this plan in a sensible way. Suppose that we can specify a 'meaningful difference' (sometimes called a 'clinically significant difference') of $\tau_{mf} \geq 0$. The notion is that a difference smaller than τ_{mf} may exist but be too small to merit attention. Yoon suggests testing $H_0^{(\tau)} : \mu_t - \tau \leq \mu_{c1}$ in order [29] at level α for $\tau \in (-\infty, \tau_{mf}]$, stopping with the smallest $H_0^{(\tau)}$ that is not rejected, say τ_{nr}, declaring with $1 - \alpha$ confidence that $\mu_t - \mu_{c1} \geq \tau_{nr}$; however, if all $\tau \in (-\infty, \tau_{mf}]$ are rejected, declaring with $1 - \alpha$ confidence that $\mu_t - \mu_{c1} \geq \tau_{mf}$ and continuing to test $\widetilde{H}_0^{(\tau)} : \mu_t - \tau \leq \mu_{c2}$, in order at level α for $\tau \in (-\infty, \tau_{mf}]$, with a parallel interpretation. His method tries to establish a meaningful difference for the first control group and, if that is successful, tries to establish a meaningful difference for the second control group. See Yoon [38] for detailed discussion with an example.

19.3 Three Simple Plans with Two Control Groups

This second plan is little better than the first, but it does exemplify a second piece of the logic in the simplest case. Here, the two control groups are handled symmetrically, and if the control groups are similar, there is likely to be more power in testing $\overline{H}_0$ than in testing either H_0 or $\widetilde{H}_0$, because in testing $\overline{H}_0$ all the controls are used at once.

If each of the tests is done with level 0.05, then the chance of at least one false rejection is at most 0.05, even though three tests may be performed. To see this, consider the possible cases. If $\overline{H}_0$ is true, then you falsely reject at least one true hypothesis if and only if you reject $\overline{H}_0$, and the chance of this is at most 0.05. Now suppose $\overline{H}_0$ is false. Then you cannot falsely reject $\overline{H}_0$; any rejection of $\overline{H}_0$ is a correct rejection. Moreover, if $\overline{H}_0$ is false, then $\mu_t > (\mu_{c1} + \mu_{c2})/2$, so $H_0: \mu_t \leq \mu_{c1}$ and $\widetilde{H}_0: \mu_t \leq \mu_{c2}$ cannot both be true; either H_0 or $\widetilde{H}_0$, or both, are false. If both H_0 and $\widetilde{H}_0$ are false, then a false rejection is impossible. If exactly one of H_0 and $\widetilde{H}_0$ is true, then to falsely reject at least one hypothesis you must falsely reject that one true hypothesis, and the chance of this is at most 0.05.

In Table 5.1 from the study by Bilban and Jakopin [7], the hypothesis $\overline{H}_0$ is tested by estimating the contrast $1 \times \mu_t - \left(\frac{1}{2}\right) \times \mu_{c1} - \left(\frac{1}{2}\right) \times \mu_{c2}$ as 8.25 using the sample means in place of the population means, with estimated standard error

$$0.519 = \sqrt{(1^2) \cdot 0.479 + \left(-\frac{1}{2}\right)^2 \cdot 0.143 + \left(-\frac{1}{2}\right)^2 \cdot 0.377}, \quad (19.1)$$

yielding a 't-statistic' of $8.25/0.519 = 15.9$, so $\overline{H}_0$ is rejected, and both H_0 and $\widetilde{H}_0$ are tested, with t-statistics of 16.1 and 13.9 as in Table 5.1, so all three hypotheses are rejected.

Where the first plan added a second control group at no cost, the second plan may confer a gain, because the sample size is increased in the first step by using both control groups at once. Perhaps $\overline{H}_0$ will be rejected when neither H_0 nor $\widetilde{H}_0$ is rejected. If the treated group is judged to produce responses that, on average, are higher than the two control groups, then the investigation continues by testing the two control groups one at a time.

The first bit of logic concerned testing hypotheses in an order of priority. The second bit of logic, which was just illustrated, concerned the possibility that some hypotheses preclude one another, so in testing several hypotheses you may know that they cannot all be true simultaneously, even though you do not know which one is true. The second bit of logic is used in building confidence intervals: then you test infinitely many hypotheses, but only one hypothesis is true, so no correction for multiple testing is needed. The second bit of logic shows up in a certain form in two methods for testing several hypotheses, specifically the method of Ruth Marcus, Eric Peritz, and K.R. Gabriel [22] and the method of Julliet Popper Shaffer [32], together with a large literature developing from these two papers. In the specific form used above to compare two control groups, H_0 and $\widetilde{H}_0$ do not preclude one another, but they do preclude one another if $\overline{H}_0$ is false, and in this case $\left\langle \overline{H}_0, \left\{ H_0, \widetilde{H}_0 \right\} \right\rangle$ is called a 'sequentially exclusive partition' of the three hypotheses, $\overline{H}_0, H_0, \widetilde{H}_0$, in the sense that a set of hypotheses in the sequence, here $\left\{H_0, \widetilde{H}_0\right\}$, contains at most one true hypothesis if all previous hypotheses in the sequence, here $\overline{H}_0$, are false; see [29].

Before developing a serious plan for an analysis with two control groups, one more bit of logic is needed, this time about equivalence tests.

Are the two control groups nearly equivalent?

The current section is concerned with demonstrating that the two control groups are nearly equivalent in their responses, that is, with an equivalence test [6]. To assert that the two control groups are close in their responses is to assert that $|\mu_{c1} - \mu_{c2}|$ is small, say $|\mu_{c1} - \mu_{c2}| < \delta$, for some specified $\delta > 0$. To reject the hypothesis that $H_{\neq}^{(\delta)} : |\mu_{c1} - \mu_{c2}| \geq \delta$ is to provide a basis for asserting $|\mu_{c1} - \mu_{c2}| < \delta$; that is, to reject the hypothesis $H_{\neq}^{(\delta)}$ of inequivalence is to be in a position to assert equivalence, $|\mu_{c1} - \mu_{c2}| < \delta$. Moreover, $H_{\neq}^{(\delta)} : |\mu_{c1} - \mu_{c2}| \geq \delta$ is true if either $\overleftarrow{H}_0^{(\delta)} : \mu_{c1} - \mu_{c2} \leq -\delta$ or $\overrightarrow{H}_0^{(\delta)} : \mu_{c1} - \mu_{c2} \geq \delta$ is true; that is, $H_{\neq}^{(\delta)}$ is the union of two hypotheses [5, 20], and the set comprised of these two hypotheses, $\left\{\overleftarrow{H}_0^{(\delta)}, \overrightarrow{H}_0^{(\delta)}\right\}$, is exclusive in the sense that at most one of the two hypotheses is true [29].

Suppose that we adopt the plan of testing both $\overleftarrow{H}_0^{(\delta)}$ and $\overrightarrow{H}_0^{(\delta)}$ in one-sided, 0.05 level tests, rejecting each hypothesis if the corresponding P-value is less than 0.05, rejecting $H_{\neq}^{(\delta)}$ if both $\overleftarrow{H}_0^{(\delta)}$ and $\overrightarrow{H}_0^{(\delta)}$ are rejected. In brief, two tests are performed, each at level 0.05, with four possible outcomes: (i) no rejections, (ii) reject $\overleftarrow{H}_0^{(\delta)}$ but not $\overrightarrow{H}_0^{(\delta)}$, (iii) reject $\overrightarrow{H}_0^{(\delta)}$ but not $\overleftarrow{H}_0^{(\delta)}$, (iv) reject $\overleftarrow{H}_0^{(\delta)}, \overrightarrow{H}_0^{(\delta)}$, and $H_{\neq}^{(\delta)}$. What is the chance of falsely rejecting at least one true hypothesis? The chance is at most 0.05. To see this, recall that $\left\{\overleftarrow{H}_0^{(\delta)}, \overrightarrow{H}_0^{(\delta)}\right\}$ is exclusive: at most one of $\overleftarrow{H}_0^{(\delta)}$ and $\overrightarrow{H}_0^{(\delta)}$ is true. If neither $\overleftarrow{H}_0^{(\delta)}$ nor $\overrightarrow{H}_0^{(\delta)}$ is true, then there is no true hypothesis to falsely reject, and the chance of a false rejection is zero. If $\overleftarrow{H}_0^{(\delta)}$ is true, then $\overrightarrow{H}_0^{(\delta)}$ is false, and a false rejection occurs if and only if $\overleftarrow{H}_0^{(\delta)}$ is rejected, which happens with probability at most 0.05. If $\overrightarrow{H}_0^{(\delta)}$ is true, then $\overleftarrow{H}_0^{(\delta)}$ is false, and a false rejection occurs if and only if $\overrightarrow{H}_0^{(\delta)}$ is rejected, which happens with probability at most 0.05.

With $\delta = 2$ in Table 5.1 from [7], two 't-statistics' are computed as

$$\overleftarrow{t} = \frac{(6.400 - 6.005) + 2}{\sqrt{0.143^2 + 0.377^2}} = 5.9 \text{ and } \overrightarrow{t} = \frac{(6.400 - 6.005) - 2}{\sqrt{0.143^2 + 0.377^2}} = -4.0, \quad (19.2)$$

where $\overleftarrow{H}_0^{(\delta)} : \mu_{c1} - \mu_{c2} \leq -\delta$ is rejected for large values of $\overleftarrow{t}$ and $\overrightarrow{H}_0^{(\delta)} : \mu_{c1} - \mu_{c2} \geq \delta$ is rejected for small values of $\overrightarrow{t}$. Here, both $\overleftarrow{t} \geq 1.65$ and $\overrightarrow{t} \leq -1.65$, where $\Phi(-1.65) = 0.05$ and $1 - \Phi(1.65) = 0.05$ give the approximate one-sided

19.3 Three Simple Plans with Two Control Groups

critical values from the Normal distribution. So we are in a position to assert that $|\mu_{c1} - \mu_{c2}| < \delta = 2$ having rejected the two alternatives to this assertion.

How should one define equivalence? That is, how should one pick a δ? In fact, there is no need to pick a single δ. Start with $\delta = \infty$, which corresponds with $\overleftarrow{T} = \infty$ and $\overrightarrow{T} = -\infty$, continuing to test smaller values of δ until a value δ^* is encountered such that either $\overleftarrow{H}_0^{(\delta)}$ or $\overrightarrow{H}_0^{(\delta)}$ is not rejected; then assert with 95% confidence that $|\mu_{c1} - \mu_{c2}| \leq \delta^*$. In Table 5.1, $\delta^* = 1.06$, so with 95% confidence, the frequency of micronuclei in the two control groups differs by at most 1.06.

The reasoning here closely parallels the reasoning behind the TOST, or two-one-sided tests procedure for equivalence testing [6, 31, 37] and the associated confidence interval [2, 16].[6] It is also closely related to the intersection-union principle; see [5, 6].

An elementary plan for analysis with two control groups

With two control groups, an elementary plan for analysis follows. The plan is stated in terms of rejection at the conventional level $\alpha = 0.05$, but any level α may be used instead.

Step 1: Test $\overline{H}_0 : \mu_t \leq (\mu_{c1} + \mu_{c2})/2$. If the P-value is greater than 0.05, stop; otherwise, reject $\overline{H}_0$ and perform Step 2.

Step 2: Test both $H_0 : \mu_t \leq \mu_{c1}$ and $\tilde{H}_0 : \mu_t \leq \mu_{c2}$, rejecting H_0 if its P-value is at most 0.05, rejecting $\tilde{H}_0$ if its P-value is at most 0.05. If either P-value is above 0.05, stop. If both P-values are at most 0.05, perform Step 3.

Step 3: Starting with $\delta = \infty$ and proceeding to smaller values of δ, test both $\overleftarrow{H}_0^{(\delta)} : \mu_{c1} - \mu_{c2} \leq -\delta$ and $\overrightarrow{H}_0^{(\delta)} : \mu_{c1} - \mu_{c2} \geq \delta$. If the P-value for $\overleftarrow{H}_0^{(\delta)}$ is at most 0.05, reject $\overleftarrow{H}_0^{(\delta)}$, and if the P-value for $\overrightarrow{H}_0^{(\delta)}$ is at most 0.05, reject $\overrightarrow{H}_0^{(\delta)}$. If either P-value is greater than 0.05, stop testing and assert with 95% confidence that $|\mu_{c1} - \mu_{c2}| \leq \delta$. If both P-values are at most 0.05, then continue testing smaller values of δ.

This three step plan may reach a variety of conclusions. It may decide that there is insufficient evidence to reject $\overline{H}_0$, so that even when combining the sample sizes from the two control groups, the treated group cannot be confidently asserted to have higher responses than the average of the two control groups. If $\overline{H}_0$ is rejected in Step

---

[6] Alternatively, if somewhat unconventionally, one may view this as testing hypotheses in order [29] with an infinite sequentially exclusive partition of hypotheses, $\left\{\overleftarrow{H}_0^{(\delta)}, \overrightarrow{H}_0^{(\delta)}\right\}$, $\delta \in (0, \infty)$, where $\left\{\overleftarrow{H}_0^{(\delta)}, \overrightarrow{H}_0^{(\delta)}\right\}$ is 'before' $\left\{\overleftarrow{H}_0^{(\delta')}, \overrightarrow{H}_0^{(\delta')}\right\}$ if $\delta > \delta'$. Notice this ordering: extreme hypotheses are tested first. This slightly unconventional view has two advantages in the current discussion. First, it allows equivalence tests to be embedded in other forms of sequentially exclusive partitions without requiring a separate formalism. This will be done in a moment. Second, for a single δ, it permits rejection of either $\overleftarrow{H}_0^{(\delta)}$ or $\overrightarrow{H}_0^{(\delta)}$ when both hypotheses cannot be rejected, something that is not always explicitly stated in discussions of TOST.

1, then with 95% confidence one may assert $\mu_t > (\mu_{c1} + \mu_{c2})/2$. In Step 2, neither, either, or both of H_0 and $\widetilde{H}_0$ may be rejected; that is, one may be confident that the treated group has higher responses than neither control group, than one control group but not the other, or than both control groups. If H_0 and $\widetilde{H}_0$ are both rejected in Step 2, then Step 3 provides a confidence statement for the degree of equivalence of the outcomes in the two control groups.

As we have already seen, when applied to Table 5.1 from [7], the three step procedure rejects $\overline{H}_0$, rejects both H_0 and $\widetilde{H}_0$, and rejects both $\overleftarrow{H}_0^{(\delta)}$ and $\overrightarrow{H}_0^{(\delta)}$ for all $\delta > 1.06$. With 95% confidence, the procedure asserts that the treated responses are above the responses in both control groups and the two control group means differ by $|\mu_{c1} - \mu_{c2}| \leq 1.06$.

The chance that this three-step procedure tests and falsely rejects at least one true hypothesis is at most 0.05. Perhaps this is intuitive now, having previously considered the steps in isolation, one at a time. A proof of a general kind is given in the appendix to this chapter. It uses the simple fact that

$$\left\langle \overline{H}_0, \left\{ H_0, \widetilde{H}_0 \right\}, \left\{ \overleftarrow{H}_0^{(\delta)}, \overrightarrow{H}_0^{(\delta)} \right\}, \delta \in (0, \infty) \right\rangle \tag{19.3}$$

is a sequentially exclusive partition of hypotheses, with $\left\{ \overleftarrow{H}_0^{(\delta)}, \overrightarrow{H}_0^{(\delta)} \right\}$ placed before $\left\{ \overleftarrow{H}_0^{(\delta')}, \overrightarrow{H}_0^{(\delta')} \right\}$ whenever $\delta > \delta'$; that is, as one proceeds forward in this sequence of sets of hypotheses, at most one hypothesis in a set is true when all the hypotheses in earlier sets are false. This is true because if $\overline{H}_0$ is false, then at most one hypothesis in $\left\{ H_0, \widetilde{H}_0 \right\}$ is true, and for all δ, at most one hypothesis in $\left\{ \overleftarrow{H}_0^{(\delta)}, \overrightarrow{H}_0^{(\delta)} \right\}$ is true. Intuitively, in the three-step procedure, whenever two hypotheses are tested at once, at most one of them is true, so you are taking only one 0.05 chance, not two 0.05 chances, of a false rejection.

An alternative plan for analysis with two control groups

A weakness of the analysis plan just described is that it might reject $H_0 : \mu_t \leq \mu_{c1}$ and $\widetilde{H}_0 : \mu_t \leq \mu_{c2}$ in Step 2, even though the treated group is close to one of the control groups, in the sense that $\mu_t - \max(\mu_{c1}, \mu_{c2})$ is small, and the two control groups are far apart, in the sense that $\max(\mu_{c1}, \mu_{c2}) - \min(\mu_{c1}, \mu_{c2})$ is large. One might reasonably argue that a difference between the control groups is problematic unless it is a small difference compared with the difference between the treated group and both control groups. We are not so interested in whether the difference between the two control groups is significantly different from zero, but rather whether the magnitude of bias needed to explain any difference between the control groups is far smaller than the magnitude of bias needed to explain the difference between the treated group and both control groups. More precisely, for some fixed number, τ, one might wish to assert

19.3 Three Simple Plans with Two Control Groups

$$(\mu_t - \tau) - \max(\mu_{c1}, \mu_{c2}) > \max(\mu_{c1}, \mu_{c2}) - \min(\mu_{c1}, \mu_{c2}) \qquad (19.4)$$

and one would have grounds for asserting this if one had rejected the hypothesis

$$H_\diamond^{(\tau)} : (\mu_t - \tau) - \max(\mu_{c1}, \mu_{c2}) \leq \max(\mu_{c1}, \mu_{c2}) - \min(\mu_{c1}, \mu_{c2}). \qquad (19.5)$$

For instance, rejection at level 0.05 of $H_\diamond^{(\tau)}$ with $\tau = 0$ asserts with 95% confidence that the treated mean, μ_t, exceeds the means, μ_{c1} and μ_{c2}, in both control groups by more than the control groups differ from each other.

The following procedure from [29] tests $H_\diamond^{(\tau)}$.

Step 1a: Test $\overline{H}_0^{(\tau)} : (\mu_t - \tau) \leq (\mu_{c1} + \mu_{c2})/2$. If the P-value is greater than 0.05, stop; otherwise, reject $\overline{H}_0^{(\tau)}$ and perform Step 2.

Step 2a: Test both $H_0^{(\tau)} : (\mu_t - \tau) \leq \mu_{c1}$ and $\widetilde{H}_0^{(\tau)} : \mu_t - \tau \leq \mu_{c2}$, rejecting $H_0^{(\tau)}$ if its P-value is at most 0.05, rejecting $\widetilde{H}_0^{(\tau)}$ if its P-value is at most 0.05. If either P-value is above 0.05, stop. If both P-values are at most 0.05, perform Step 3.

Step 3a: Test both $H_\triangleright^{(\tau)} : (\mu_t - \tau) - \mu_{c1} \leq \mu_{c1} - \mu_{c2}$ and $H_\triangleleft^{(\tau)} : (\mu_t - \tau) - \mu_{c2} \leq \mu_{c2} - \mu_{c1}$, rejecting $H_\triangleright^{(\tau)}$ if its P-value is at most 0.05 and rejecting $H_\triangleleft^{(\tau)}$ if its P-value is at most 0.05. If either P-value is above 0.05, stop. Otherwise, if both P-values are at most 0.05, reject $H_\diamond^{(\tau)}$.

For $\tau = 0$, Steps 1a and 2a are the same as Steps 1 and 2, so the conclusions about $\overline{H}_0$, H_0, and $\widetilde{H}_0$ are the same. In Step 3a, two cases need to be distinguished, namely $\max(\mu_{c1}, \mu_{c2}) = \mu_{c1}$ and $\max(\mu_{c1}, \mu_{c2}) = \mu_{c2}$. If $\max(\mu_{c1}, \mu_{c2}) = \mu_{c1}$, then $H_\diamond^{(\tau)}$ and $H_\triangleright^{(\tau)}$ are the same. Conversely, if $\max(\mu_{c1}, \mu_{c2}) = \mu_{c2}$, then $H_\diamond^{(\tau)}$ and $H_\triangleleft^{(\tau)}$ are the same. Therefore, $H_\diamond^{(\tau)}$ is false if both $H_\triangleright^{(\tau)}$ and $H_\triangleleft^{(\tau)}$ are false; so, in Step 3a, $H_\diamond^{(\tau)}$ is rejected if both $H_\triangleright^{(\tau)}$ and $H_\triangleleft^{(\tau)}$ are rejected.

In Table 5.1 from the study by Bilban and Jakopin [7], the five hypotheses in Steps 1a–3a are each tested by a contrast in group means. For instance, in Step 1a, $\overline{H}_0^{(\tau)}$ is the hypothesis that $1 \times \mu_t - \left(\frac{1}{2}\right) \times \mu_{c1} - \left(\frac{1}{2}\right) \times \mu_{c2} \leq \tau$, and in Step 3a, $H_\triangleright^{(\tau)}$ is the hypothesis that $1 \times \mu_t - 2 \times \mu_{c1} + 1 \times \mu_{c2} \leq \tau$. As before, for $\tau = 0$, in Step 1a, the 't-statistic' for $\overline{H}_0^{(\tau)}$ is 15.9, so $\overline{H}_0^{(\tau)}$ is rejected at the 0.05 level, and the procedure continues to Step 2a. Had $\overline{H}_0^{(\tau)}$ not been rejected, the procedure would have stopped in Step 1a. In Step 2a, the 't-statistics' for $H_0^{(\tau)}$ and $\widetilde{H}_0^{(\tau)}$ are, respectively, 16.1 and 13.8, so $H_0^{(\tau)}$ and $\widetilde{H}_0^{(\tau)}$ are both rejected at the 0.05 level and the procedure continues to Step 3a. Had either $H_0^{(\tau)}$ or $\widetilde{H}_0^{(\tau)}$ not been rejected, the procedure would have stopped in Step 2a. In Step 3a, the 't-statistics' for $H_\triangleright^{(\tau)}$ and $H_\triangleleft^{(\tau)}$ are, respectively 11.4 and 9.8, so both $H_\triangleright^{(\tau)}$ and $H_\triangleleft^{(\tau)}$ are rejected, and hence $H_\diamond^{(\tau)}$ is rejected. Therefore, we may assert with 95% confidence that the treated group mean exceeds both of the control group means by more than the control group means differ from each other. After considering the properties of this procedure, other values of τ will be evaluated.

In Steps 1a, 2a, and 3a, the chance of testing and rejecting at least one true hypothesis is at most 0.05. The reasoning here closely parallels what has been done before, except there are more steps of the same kind. The details of the reasoning constitute the remainder of this somewhat long, slightly technical paragraph, which you may skip if that is your preference. The general argument is given more concisely in the appendix to this chapter. First, we check that the sequence of hypotheses has the needed structure. The sequence of sets of hypotheses

$$\left\langle \overline{H}_0^{(\tau)}, \left\{ H_0^{(\tau)}, \widetilde{H}_0^{(\tau)} \right\}, \left\{ H_\triangleright^{(\tau)}, H_\triangleleft^{(\tau)} \right\} \right\rangle \tag{19.6}$$

is a sequentially exclusive partition in the sense that: (i) if $\overline{H}_0^{(\tau)}$ is false, then at most one hypothesis in $\left\{ H_0^{(\tau)}, \widetilde{H}_0^{(\tau)} \right\}$ is true, and (ii) if $\overline{H}_0^{(\tau)}, H_0^{(\tau)}, \widetilde{H}_0^{(\tau)}$ are all false, then at most one hypothesis in $\left\{ H_\triangleright^{(\tau)}, H_\triangleleft^{(\tau)} \right\}$ is true. To see this, note first that part (i) follows from the same reasoning used for $\overline{H}_0, H_0$, and $\widetilde{H}_0$. For part (ii), if $H_0^{(\tau)} : (\mu_t - \tau) \leq \mu_{c1}$ and $\widetilde{H}_0^{(\tau)} : \mu_t - \tau \leq \mu_{c2}$ are both false, then $(\mu_t - \tau) - \mu_{c1} > 0$ and $(\mu_t - \tau) - \mu_{c2} > 0$, but either $\mu_{c1} - \mu_{c2} \leq 0$ or $\mu_{c2} - \mu_{c1} \leq 0$, so at most one of the following two hypotheses is true $H_\triangleright^{(\tau)} : (\mu_t - \tau) - \mu_{c1} \leq \mu_{c1} - \mu_{c2}$ and $H_\triangleleft^{(\tau)} : (\mu_t - \tau) - \mu_{c2} \leq \mu_{c2} - \mu_{c1}$. Next, we verify the claim about the probability of at least one false rejection. There is, of course, a general pattern here, in which a sequentially exclusive partition of hypotheses, tested in an order such as Steps 1a–3a, will falsely reject a true hypothesis with probability at most 0.05. This is demonstrated in general in [29] and in the appendix to this chapter, but let us do one last particular case, namely Steps 1a-3a. If $\overline{H}_0^{(\tau)}$ is true, then a false rejection occurs if and only if $\overline{H}_0^{(\tau)}$ is rejected, which happens with probability at most 0.05. So suppose $\overline{H}_0^{(\tau)}$ is false; then either $H_0^{(\tau)}$ or $\widetilde{H}_0^{(\tau)}$, or both, are false. If exactly one of $H_0^{(\tau)}$ or $\widetilde{H}_0^{(\tau)}$ is true, then at least one false rejection occurs if and only if that one true hypothesis is rejected, which happens with probability at most 0.05. If both $H_0^{(\tau)}$ and $\widetilde{H}_0^{(\tau)}$ are false, then a false rejection cannot occur at Step 2a. So suppose $\overline{H}_0^{(\tau)}, H_0^{(\tau)}$ and $\widetilde{H}_0^{(\tau)}$ are false; then either $H_\triangleright^{(\tau)}$ or $H_\triangleleft^{(\tau)}$, or both, are false. If both are false, a false rejection is not possible. If exactly one of $H_\triangleright^{(\tau)}$ or $H_\triangleleft^{(\tau)}$ is true, then a false rejection occurs if and only if that one true hypothesis is rejected, which happens with probability at most 0.05.

There is no need to select a value of τ. Rather, apply Steps 1a–3a repeatedly, beginning with $\tau = -\infty$, considering larger and larger values of τ, until for the first time at τ^* the procedure stops because one of the hypotheses has not been rejected. Then declare with 95% confidence that

$$(\mu_t - \tau^*) - \max(\mu_{c1}, \mu_{c2}) > \max(\mu_{c1}, \mu_{c2}) - \min(\mu_{c1}, \mu_{c2}). \tag{19.7}$$

In Table 5.1, the first acceptance occurs at $\tau^* = 6.56$ when $H_\triangleright^{(\tau)}$ is just barely not rejected in Step 3a with a 't-statistic' of 1.64. In words, with 95% confidence, the mean level of micronuclei in the treated group exceeds both control group means by at least 6.56 more than the two control groups differ from each other.

Another example is discussed in [29]. There, unlike Table 5.1, the two control groups differ significantly from each other, but with 95% confidence, the difference between the treated group and both control groups is substantially larger than the difference between the two control groups.

Summary

In this section, several plans have been considered for the analysis of a study with a treated group and two control groups. Some of the plans were too primitive for practical use and were simply illustrations of general principles. Other plans are suitable for a primary analysis. With two control groups, the 'elaborate theory' predicts that the treated group will differ substantially from the control groups, which do not differ substantially from each other. The 'elaborate theory' predicts a difference here and equivalence there, and the plan uses both tests for difference and for equivalence at appropriate steps. The plan tests the most important elements of the predictions first, without a correction for multiple testing; then, it continues to try to establish the less important elements of the prediction. The planned analysis may yield confirmation of some predictions and not others, and may provide quantitative statements that have some resemblance to confidence intervals.

The current section has considered the simplest of study designs without a sensitivity analysis. Sensitivity analyses with multiple control groups are possible, but they would take us a little far afield; see [27, §8]. The next section will plan a sensitivity analysis for coherence between two outcomes.

19.4 Sensitivity Analysis for Two Outcomes and Coherence

In §5.2.3 and §17.2, it was found that coherence among several outcomes has the potential to reduce sensitivity to departures from random treatment assignment. With two outcomes, what is a suitable plan for analysis?

In Table 5.3, Angrist and Lavy's study of the effects of class size on test performance was found to be somewhat less sensitive for the coherent combination of math and verbal test scores than for either score alone. The general results in §17.2 suggested that this is to be expected with two outcomes that are similarly affected by treatment but imperfectly correlated.

In our initial examination of Table 5.3, there was no plan for analysis. Three tests were done for two outcomes. Are there problems of multiple testing, of increased risk of false positives from performing several tests?

Suppose that there are two outcomes, say (r_{Tij}, r_{Cij}) and $(\widetilde{r}_{Tij}, \widetilde{r}_{Cij})$, with corresponding treated-minus-control matched pair differences Y_i and $\widetilde{Y}_i$, say the math and verbal test scores in §5.2.3. The null hypothesis of no treatment effect for the first outcome says $H_0: r_{Tij} = r_{Cij}$, for all i, j, and the null hypothesis of no treatment effect for the second outcome is parallel, $\widetilde{H}_0: \widetilde{r}_{Tij} = \widetilde{r}_{Cij}$, for all i, j. Write $\overline{H}_0$ for the hypothesis that is the conjunction of these two hypotheses, H_0 and $\widetilde{H}_0$, or in logical notation, $H_0 \wedge \widetilde{H}_0$, which says neither outcome was affected. Consider the following plan.

Step 1b: Test $\overline{H}_0$ using the coherent signed rank test. If the P-value is above 0.05, stop and do not reject $\overline{H}_0$. If the P-value is at most 0.05, reject $\overline{H}_0$ and perform Step 2b.

Step 2b: Test both H_0 and $\widetilde{H}_0$ using Wilcoxon's signed rank test, rejecting H_0 if its P-value is at most 0.05, rejecting $\widetilde{H}_0$ if its P-value is at most 0.05.

If the three separate P-values do their separate jobs of falsely rejecting a true hypothesis with probability at most 0.05, then the chance that the three-step procedure falsely rejects at least one true hypothesis is at most 0.05. The reasoning is parallel to that in §19.3. If $\overline{H}_0$ is false, it must be the case that either H_0 or $\widetilde{H}_0$, or both, are false. That is, $\left\langle \overline{H}_0, \left\{ H_0, \widetilde{H}_0 \right\} \right\rangle$ is a sequentially exclusive partition of hypotheses; if $\overline{H}_0$ is false, there is at most one true hypothesis in $\left\{ H_0, \widetilde{H}_0 \right\}$. If $\overline{H}_0$ is true, then at least one true hypothesis is rejected if and only if $\overline{H}_0$ is rejected in Step 1b, and this happens with probability at most 0.05. If $\overline{H}_0$ is false, then either H_0 or $\widetilde{H}_0$ is false, so at most one true hypothesis is tested in Steps 1b and 2b, and the chance of falsely rejecting that one true hypothesis is at most 0.05. See [19] for a closely related procedure.

The sensitivity analysis uses the upper bounds on the P-values in Steps 1b and 2b. The appendix to this chapter and [30] discuss why this procedure works.

In Table 5.3, the two-step procedure rejects $\overline{H}_0$, H_0, and $\widetilde{H}_0$ for $\Gamma \leq 1.4$. For $\Gamma = 1.55$, it rejects $\overline{H}_0$ and $\widetilde{H}_0$, but not H_0. For $\Gamma = 1.65$, it rejects $\overline{H}_0$ but not H_0 nor $\widetilde{H}_0$. In other words, the coherent association using both outcomes is marginally less sensitive to deviations from random assignment than either outcome alone. In this instance, the plan consisting of Steps 1b and 2b would have supported our previous interpretation of Table 5.3, while addressing any concern about testing more than one hypothesis. The general results in §17.2 provide some guidance about when $\overline{H}_0$ will be less sensitive to bias than H_0 or $\widetilde{H}_0$.

If instead a Bonferroni adjustment had been made for three tests, none of the hypotheses would have been rejected at $\Gamma = 1.55$. If the coherent test were not used, and Wilcoxon's statistic were applied twice, once to H_0 and once to $\widetilde{H}_0$, with a Bonferroni adjustment, then neither H_0 nor $\widetilde{H}_0$ would be rejected for $\Gamma \geq 1.45$.[7] For discussion of sensitivity analysis adjusted by the Bonferroni inequality, see [12] and [30, §4.5].

---

[7] With two tests at $\Gamma = 1.4$, $\widetilde{H}_0$ is rejected but H_0 is not using the Bonferroni adjustment, but both $\widetilde{H}_0$ and H_0 are rejected using Holm's [13] procedure.

19.5 Sensitivity Analysis for Tests of Equivalence

Table 19.1 Pain scores for surgical patients and controls in $I = 100$ matched pairs. The change is pain at three months minus the average at entry and baseline, and the difference in changes refers to the treated-minus-control difference in these changes.

Label	Group	Time	Minimum	Lower Quartile	Median	Upper Quartile	Maximum	Mean
a	Treated	Entry	0	2.0	5	6.0	9	4.28
b	Treated	Baseline	0	3.0	5	6.0	9	4.38
c	Treated	3 Months After Surgery	0	1.8	3	5.0	9	3.55
d	Control	Entry	0	2.0	5	6.0	8	4.31
e	Control	Baseline	0	2.8	5	6.0	9	4.34
f	Control	3 Months Later	0	2.0	3	5.0	8	3.39
		Difference in Changes	−7	−2.0	0	2.5	8	0.16

19.5 Sensitivity Analysis for Tests of Equivalence

The absence of a treatment effect may be an important finding. See, for instance, the book *Costs, Risks and Benefits of Surgery* edited by John Bunker, Benjamin Barnes, and Frederic Mosteller [8] and the developments that flowed from it [24].

The absence of a treatment effect may be an important finding, but its importance may be missed if it is described as a 'null result' with a P-value testing for difference above 0.05. Lacking evidence of an effect is not at all the same as possessing evidence that the effect is not large, yet a P-value above 0.05 in a test for difference is consistent with both of these situations. It is not possible to demonstrate a total absence of effect, but it is possible to demonstrate that the effect is not large using an equivalence test in a large randomized trial. In an observational study, there are added uncertainties because treatments were not randomly assigned. Nonetheless, in an observational study, one may find that the evidence that the effect is not large is insensitive to small or moderate biases from nonrandom treatment assignment.

To illustrate sensitivity analysis for an equivalence test, consider again the example in §12.2 and Table 12.1 concerning the effects of a surgical intervention, cystoscopy and hydrodistention, on symptoms of interstitial cystitis (IC), a chronic urologic disorder characterized by bladder pain and irritative voiding [21, 25]. In Table 12.1, the surgical treatment was estimated to have no effect on the nine-point pain score. How sensitive is this ostensible absence of effect to unmeasured biases from nonrandom treatment assignment?

Recall that pairs were formed using risk-set matching, matching at entry into the database and at baseline just prior to surgery for the surgical patients. Table 19.1 presents the pain scores in detail.[8] The surgically treated and control patients look similar in terms of pain at entry and prior to surgery for the surgical patient, and both groups look slightly and equally improved three months later. For the $I = 100$ pairs of two patients in Table 19.1, there are 400 'pretreatment' pain scores, merging 'entry' and 'baseline' scores; these have a mean of 4.3, a median of 5, a standard deviation of 2.2, and a median absolute deviation from the median of 2. Is

---

[8] In [21], there is a sensitivity analysis for equivalence using all three outcomes at once in a coherent test. Here, only pain is considered.

Table 19.2 Sensitivity analysis for an equivalence test for the effect of surgery on pain scores in the IC data. The table gives upper bounds on P-values. For an effect of 2 or more in either direction to be plausible, the magnitude of bias would need to exceed $\Gamma = 2.4$, and that magnitude of bias would only be able to mask an increase of 2 units in pain, not a decrease of 2 units in pain, caused by surgery. To mask a decrease of 2 units in pain caused by surgery, the bias would need to exceed $\Gamma = 3$.

Γ	$\overleftarrow{H}_0^{(\varsigma)}$	$\overrightarrow{H}_0^{(\varsigma)}$	$H_{\neq}^{(2)}$
1	4.9×10^{-9}	1.8×10^{-7}	1.8×10^{-7}
2	0.0012	0.0098	0.0098
2.4	0.0081	0.048	0.048
2.7	0.022	0.11	0.11
3	0.049	0.19	0.19

it plausible that the effect of surgery is 2 units or more, but this is hidden from view in Table 12.1 because of bias in the assignment of patients to surgery or control?

The equivalence test applies Wilcoxon's signed rank statistic to test hypotheses about an additive treatment effect, $r_{Tij} - r_{Cij} = \tau$, with the null hypotheses asserting an effect that is not small, $H_{\neq}^{(\varsigma)} : |\tau| \geq \varsigma$, where $\varsigma > 0$ is set at 2 in this example; then, rejecting $H_{\neq}^{(\varsigma)}$ provides a basis for asserting with confidence that $|\tau| < \varsigma$. The hypothesis of inequivalence, $H_{\neq}^{(\varsigma)}$, is the union of two exclusive hypotheses, $\overleftarrow{H}_0^{(\varsigma)}$: $\tau \leq -\varsigma$ or $\overrightarrow{H}_0^{(\varsigma)} : \tau \geq \varsigma$. As a consequence, as in §19.3 and [2, 5, 6, 16, 29, 31, 37], $\overleftarrow{H}_0^{(\varsigma)}$ and $\overrightarrow{H}_0^{(\varsigma)}$ are each tested without a correction for multiple testing, and $H_{\neq}^{(\varsigma)}$ is rejected if both $\overleftarrow{H}_0^{(\varsigma)}$ and $\overrightarrow{H}_0^{(\varsigma)}$ are rejected. The sensitivity analysis for $H_{\neq}^{(\varsigma)}$ is then assembled from two standard sensitivity analyses, one for $\overleftarrow{H}_0^{(\varsigma)} : \tau \leq -\varsigma$, the other for $\overrightarrow{H}_0^{(\varsigma)} : \tau \geq \varsigma$; see [30] for a proof.

Table 19.2 is the sensitivity analysis for the equivalence test with $\varsigma = 2$. How much bias would need to be present for a moderately large treatment effect to appear as if it were no effect? More precisely: How much bias, Γ, would need to be present to produce the ostensible absence of effect in Table 12.1 if the treatment effect were 2 or more, $|\tau| \geq \varsigma = 2$? Table 19.2 gives upper bounds on P-values for the given values of Γ.[9] To mask an effect that is at least $|\tau| \geq 2$ units on the pain scale, the

---

[9] The upper bound on the P-value for $\overleftarrow{H}_0^{(\varsigma)} : \tau \leq -2$ is obtained by calculating Wilcoxon's signed rank statistic T from $Y_i + 2$ to test $\overrightarrow{H}_0^{(\varsigma)} : \tau = -2$ against $\tau > -2$ as in §3.5, rejecting if T is large. The upper bound on the P-value for $\overrightarrow{H}_0^{(2)} : \tau \geq 2$ is obtained by calculating Wilcoxon's signed rank statistic T from $Y_i - 2$ to test $\overrightarrow{H}_0^{(\varsigma)} : \tau = 2$ against $\tau < 2$ as in §3.5, rejecting if T is small. The P-value for testing $H_{\neq}^{(2)} : |\tau| \geq 2$ is the maximum of these two upper bounds; see [30] for a proof. The proof shows that the upper bound on the P-value for testing $H_{\neq}^{(2)} : |\tau| \geq 2$ is exactly equal to the maximum of the two separate P-values. As in Note 6 of this chapter, because $\overleftarrow{H}_0^{(\varsigma)}$ and $\overrightarrow{H}_0^{(\varsigma)}$ are exclusive, we can test both hypotheses at level α and yet take only an α risk of at least one false rejection. In consequence, we may reject either $\overleftarrow{H}_0^{(\varsigma)}$ or $\overrightarrow{H}_0^{(\varsigma)}$ even when we cannot reject both; see [29] and the appendix to this chapter.

bias would need to be $\Gamma > 2.4$, and to mask a reduction in pain of at least $\tau \leq -2$, the bias would need to be $\Gamma > 3$. If the surgical procedure actually reduces pain by 2 units or more, the bias masking this effect would have to be fairly large.

19.6 Sensitivity Analysis for Equivalence and Difference

A planned analysis may seek to demonstrate a predicted positive difference for one outcome and predicted near equivalence for another, that is, an alternative hypothesis of superiority for one outcome and near equivalence for another [2, 34]. The demonstration would occur if a hypothesis of equality or inequivalence is rejected.[10]

A prediction of this sort arose in §12.3. In the study [30, 33] in §12.3 and §18.2 of early or late discharge of premature babies from the neonatal intensive care unit (NICU), there were 701 pairs of an early baby and a late baby who were similar on the day the early baby was discharged, although the late baby stayed in the hospital a few more days. Were the extra days in the NICU of benefit to the late babies?

The six months after the discharge of the early baby were divided into two intervals. During the first interval, the early baby was discharged home and the late baby was in the NICU. During the second interval, both babies had been discharged home. In [30, 33], health services provided to healthy babies (e.g., check-ups) were set aside, and two outcomes in [30, 33] converted emergency and sick-baby health services into a dollar amount, one for the first interval, the other for the second interval. Because services were converted by schedule into costs, two babies who received the same services were coded as having the same costs; there is no variation in costs for the same service.

Let us devote a moment's thought to what patterns of outcomes might be seen and what they might mean. During the first interval, the late baby is in the NICU receiving hospital (or sick-baby) services, but the baby may be doing fine, growing older, chubbier, more mature, better able to face the world. During the first interval, the early baby is discharged home, so any emergency or sick-baby services are a sign that something has gone wrong, and if the dollar amount is very high, then something has gone very wrong, say a readmission through the emergency room. During the second interval, both babies are home, so sick-baby costs mean something has gone wrong. It is, therefore, important to distinguish the first and second intervals: for the late baby in the first interval, cost could be money and nothing more; elsewhere, cost is a sign that something has gone wrong. If substantial costs were at all common for the early babies in the first interval, this would suggest that

---

[10] This is a form of Roger Berger's [5] intersection-union test, in that one tests the disjunction of two hypotheses, $H_1 \vee H_2$, in an effort to demonstrate a conjunction, $(\sim H_1) \wedge (\sim H_2)$. Here, H_1 is equality or inferiority for one outcome and H_2 is inequivalence for another outcome, so superiority and equivalence is $(\sim H_1) \wedge (\sim H_2)$. In an intersection-union test, $H_1 \vee H_2$ is rejected at level α if H_1 and H_2 are rejected at level α; see [5, 20]. Intersection-union tests may be done in order, possibly rejecting H_1 when H_2 is not rejected, and a sensitivity analysis may be constructed from sensitivity analyses for the separate tests of H_1 and H_2; see the appendix to this chapter.

the early discharges were premature. If early babies had higher second-period costs, this too might suggest the early discharges were premature. If the later baby had consistently higher first period costs and second-period costs were nearly equivalent, then the cost of retaining the late baby might plausibly have been better spent in some other way, such as enhanced outpatient services for the babies. The analysis uses late-baby-minus-early-baby differences in costs for the first and second period.

The study concerned babies discharged alive. There were five deaths in the six months after discharge among these 1402 babies, or a rate of 3.6 per thousand babies per six months. In the United States as a whole in 2004, the first year mortality rate was 6.8 per thousand. These numbers are not comparable in several respects; one refers to six months, the other to a year; one refers to premature babies, the other to all babies; one refers to postdischarge mortality, the other includes deaths prior to discharge. For purposes of analysis, deaths were coded as infinite costs, that is, as the worst outcome [28]. There were three deaths (0.4%) among the early babies and two (0.3%) among the late babies. See [30, §2.1] for detailed discussion.

The Hodges-Lehmann point estimates (§2.4.3) of the typical difference in costs, late-baby minus early-baby, were \$4,940 in the first period and \$17 in the second period, with 95% confidence intervals for an additive effect under the naïve model for treatment assignment (§12.5) of [\$4,485, \$5,103] in the first period and [−\$20, \$56] in the second period. Similar results were found using Stephenson's test (§2.5, §16.1) which emphasizes consistent results in the extreme tails; see [30, Table 2]. All but 8/701 (1.1%) of early babies had less than \$300 of costs in the first period, and those 8 had costs in the range \$1,422 to \$9,574. About 6% of late babies had first period costs above \$9,556, but again the meaning of first period cost is different for early and late babies. The second period costs were typically low in both groups, with some extremely high costs, and no sign that late babies had lower second period costs; see [30, Table 1 and Figure 1].

Is the pattern of difference in the first period and equivalence in the second sensitive to unmeasured bias? The hypothesis is formulated in terms of additive effects τ_1 in period 1 and τ_2 in period 2. Inequivalence was defined as $|\tau_2| \geq \$500$, largely based on the thought that if you can be cured for less than \$500 in the United States, then you can't be very sick. Recall that the Hodges-Lehmann estimate of the typical cost of a delayed discharge is about \$5000, and \$500 is 1/10 of that. The plan is to test for a difference in the first period, and if one is found, to test for equivalence in the second. Formally, the testing plan uses Wilcoxon's signed rank statistic to test $H_0 : \tau_1 \leq 0$, and if this hypothesis is rejected, the statistic is used to test $H_0 : |\tau_2| \geq \$500$. For each test separately, there is a sensitivity analysis, as in §3.5 and §19.5, and the upper bound on the P-value for testing $\tau_1 \leq 0$ or $|\tau_2| \geq \$500$ — that is, for testing $H_0 : \tau_1 \leq 0 \vee |\tau_2| \geq \$500$ — is the maximum of the upper bounds of the P-values for testing $H_0 : \tau_1 \leq 0$ and $H_0 : |\tau_2| \geq \$500$ separately, and, as in §19.5, the hypothesis $H_0 : |\tau_2| \geq \$500$ unpacks as either $H_0 : \tau_2 \leq -\$500$ or $H_0 : \tau_2 \geq \$500$.[11]

---

[11] This is true because the sequence of hypotheses $\langle H_1, \{H_2, H_2\} \rangle$ is a sequentially exclusive partition with H_1 asserting $\tau_1 \leq 0$, H_2 asserting $\tau_2 \leq -500$, and H_3 asserting $\tau_2 \geq 500$ because $\{H_2, H_2\}$

Table 19.3 Sensitivity analysis for difference in first period costs (τ_1) and equivalence of second period costs (τ_2).

Γ	$H_0: \tau_1 \leq 0$	$H_0: \tau_2 \leq -\$500$	$H_0: \tau_2 \geq \$500$	$H_0: \tau_1 \leq 0 \vee \|\tau_2\| \geq \$500$
1	0.00001	0.00001	0.00001	0.00001
2	0.00001	0.00001	0.0020	0.0020
2.25	0.00001	0.00001	0.049	0.049
2.5	0.00001	0.00042	0.28	0.28
3	0.00001	0.056	0.90	0.90

Table 19.3 is the sensitivity analysis [30, Table 3]. Because $H_0 : \tau_1 \leq 0$ is rejected at the 0.05 level for each Γ in Table 19.3, the test plan continues and tests both $H_0 : \tau_2 \leq -\$500$ and $H_0 : \tau_2 \geq \$500$. For $\Gamma \leq 2.25$, all three hypotheses are rejected at the 0.05 level, so $H_0 : \tau_1 \leq 0 \vee |\tau_2| \geq \$500$ is rejected; that is, late discharge raised first period costs, but did not much affect second period costs. At $\Gamma = 2.5$, it is plausible that late discharge increased second period costs by \$500, but not plausible it reduced them by \$500. At $\Gamma = 3$, either an increase or a decrease of \$500 is plausible. It appears that late discharge raised first period costs, without a compensating reduction in second period costs, and a small bias from nonrandom treatment assignment could not have created this appearance.

19.7 Summary

To disambiguate an association between treatment and response, Fisher advised "make your theories elaborate;" see §19.2 and [9, 26]. How should one plan the analysis of an observational study to evaluate the predictions of an elaborate theory? The plan should enlarge what can be said, not diminish it, yet control the frequency of false inferences. This is done by giving priority in the analysis to the evaluation of certain predictions, placing other predictions in a secondary role. The plan should distinguish predictions of a difference from predictions of near equivalence. This is implemented with appropriate use of equivalence tests and similar methods. The plan should permit a sensitivity analysis, because in an observational study it is never possible to be certain that adjustments for observed covariates have eliminated bias from covariates that were not measured.

19.8 Further Reading

Fisher's striking phrase "make your theories elaborate" has often been noted [9, 10, 11, 26].

---

is exclusive. In fact, the three hypotheses may be tested in order, so H_1 may be rejected when H_2 and H_3 are not. See the appendix to the chapter for specifics.

There is a large literature about planning the analysis of clinical trials, and it provides some useful guidance for observational studies, where some of the issues are the same and others are quite different. In particular, the confirmation, or partial confirmation, of an elaborate theory plays no role in a randomized experiment, where treatment assignment is known to be randomized.

The paper by Gary Koch and Stuart Gansky [17] is a gentle introduction to the clinical trials literature, whereas the review paper by Roger Berger and Jason Hsu [6] is a somewhat more technical discussion emphasizing tests of equivalence; see also the fine papers by Peter Bauer and colleagues [1, 2, 3]. The clinical trials literature quickly becomes technical and terse, so the serious reader will want to become acquainted with some of the early papers that shaped what came later, particularly the papers on 'closed testing' by Ruth Marcus, Eric Peritz, and K.R. Gabriel [22], the intersection-union principle of Roger Berger [5], the device introduced by Julliet Popper Shaffer [32], and the false discovery rate of Yoav Benjamini and Yosef Hochberg [4]. Erich Lehmann's early paper [20] illuminates several issues and is of continuing interest. Two potentially important ideas that have received less attention are found in papers by Jason Hsu and Roger Berger [15] and Gerhard Hommel and Siegfried Kropf [14, §3]. Specifically, in [15], a multiple test procedure concludes with a confidence statement about a parameter, while in [14, §3] an ordered testing procedure continues past a few acceptances with the aid of a mild application of the Bonferroni inequality.

Less has been written about planning the analysis of an observational study. Issues in the current chapter are discussed in greater detail in [29, 30].

19.9 Appendix: Testing Hypotheses in Order

This chapter has considered several plans for testing one of Fisher's 'elaborate theories' in a sequence of steps. These steps permit partial or total confirmation of the 'elaborate theory.' The investigator may select the first step in this sequence on the basis of importance, priority, conjecture, or whim. The first step in the sequence is an ordinary test done at level α as it might have been done without a plan for analysis, so in this rather specific sense, the subsequent steps cost the investigator nothing because the conclusion of the first step is the same. In brief, ambition in testing an 'elaborate theory' is not penalized; at worst, the result is the partial confirmation that would have been obtained anyway in the single first test.

In addition, these plans permitted a sensitivity analysis to be performed using the upper bounds on the P-values for the component tests; one applies the steps to the upper bounds. In other words, if you know how to do a sensitivity analysis for each step, then you know how to do a sensitivity analysis for the entire plan.

This appendix states and proves a result from [29], which then permits the statement of another result from [30]. The first result concerns conventional testing of a plan, while the second result concerns sensitivity analysis of the testing of a plan. Although the first result is tailored for use with an elaborate theory in an observational study, it has many analogues and precedents from the literature on multiple testing [2, 14, 15, 17, 22, 32], equivalence testing [2, 6, 16, 31, 37], and multiparameter testing [5, 18, 20] in experiments. The second result takes more space to prove [30], so the proof is not reproduced here, but some discussion of it here is needed.

What is a sequentially exclusive partition of a sequence of hypotheses?

We have a collection of hypotheses, H_t for $t \in \mathscr{T}$, where $\mathscr{T}$ is a set of indices (or names) for the hypotheses. The set $\mathscr{T}$ is totally ordered (like a dictionary or telephone directory), and we write $t \prec t'$ if t is before t' in $\mathscr{T}$. To say that $\mathscr{T}$ is totally ordered is to say that it is always clear who is before whom: that is, if $t \in \mathscr{T}$ and $t' \in \mathscr{T}$ with $t \neq t'$ then either $t \prec t'$ or $t' \prec t$ (and never both $t \prec t'$ and $t' \prec t$). The order $\prec$ will partially determine which hypotheses are tested first, although other considerations will enter as well. Using numerical inequality, $<$, in place of $\prec$, five possible sets $\mathscr{T}$ are: $\mathscr{T} = \{1, 2, \ldots, k\}$, $\mathscr{T} = \{1, 2, \ldots\}$, $\mathscr{T} = \{\ldots, -1, 0, 1, 2, \ldots\}$, $\mathscr{T} = [0, \infty)$, and the positive rational numbers $\mathscr{T} = \{a/b : a, b \in \{1, 2, \ldots\}\}$. Other important totally ordered sets use the lexical order on a product of sets, as in the dictionary or telephone directory, where names are organized by first letter, then within first letter by second letter, and so on. The lexical order on $\mathscr{T} = \{1, 2, \ldots, k\} \times [0, \infty)$ says that infinitely many hypotheses $(1, a) \in \mathscr{T}$ with $a \in [0, \infty)$ come before $(2, 0)$. The lexical order on $\mathscr{T} = [0, \infty) \times \{1, 2, \ldots, k\}$ is very different: the k hypotheses $(0, a) \in \mathscr{T}$ come before all the rest. In §19.3, there were several orders that combined discrete and continuous aspects, testing one or two hypotheses, then building an interval estimate, or testing several hypotheses for infinitely many values of a parameter.

One assumption is made about the hypotheses, H_t for $t \in \mathscr{T}$, namely the 'structure assumption.' In practical terms, it is an innocuous assumption, but interestingly it seems to be needed nonetheless. A person disinclined to attend to innocuous assumptions might skip this paragraph. (While walking down a city street, a person disinclined to attend to innocuous assumptions might fall through a manhole whose cover is missing.) The structure assumption says: either all hypotheses H_t for $t \in \mathscr{T}$ are false or else there is a first true hypothesis. Stated precisely, either H_t is false for all $t \in \mathscr{T}$ or else there exists a $t' \in \mathscr{T}$ such that $H_{t'}$ is true and H_t is false for all $t \prec t'$. It takes a moment's thought to realize that the structure assumption is indeed an assumption: it might be false. For instance, let $\mathscr{T} = \{a/b : a, b \in \{1, 2, \ldots\}\}$ be the positive rational numbers ordered by inequality, $<$. Let H_t be the 'hypothesis' that the circumference of a circle is less than or equal to t times its diameter. Then H_t is false for $t < \pi$ and is true for $t > \pi$, but there is not a first rational number $t > \pi$, so the structure assumption is false in this case. In this case, the structure assumption would be true if the positive rational numbers, $\mathscr{T} = \{a/b : a, b \in \{1, 2, \ldots\}\}$, were replaced by the positive real numbers, $\mathscr{T} = (0, \infty)$. The structure assumption seems to be a reminder that statistical inference needs the real numbers, and should avoid topological games involving things like the rational numbers; however, as far as I can see, the structure assumption is innocuous in practice because it is not an assumption about the world itself. Without further mention, the structure assumption is assumed to be true.

We are headed for a definition of a sequentially exclusive partition of hypotheses, a term that has been used informally at several points in this chapter. First we need to define an interval in $\mathscr{T}$ and then a partition of $\mathscr{T}$ into disjoint intervals.

It is natural to speak of an interval in $\mathscr{T}$. A nonempty subset $\mathscr{I} \subseteq \mathscr{T}$ is an interval if, for any $t, t', t'' \in \mathscr{T}$ with $t \prec t' \prec t''$, it follows that $t, t'' \in \mathscr{I}$ implies $t' \in \mathscr{I}$. For example, in $\mathscr{T} = \{1, 2, \ldots, 20\}$, the sets $\mathscr{I} = \{2\}$ and $\mathscr{I} = \{17, 18, 19\}$ are intervals, but $\{2, 4, 6\}$ is not an interval. In $\mathscr{T} = \{1, 2, \ldots, k\} \times [0, \infty)$ with the lexical order, the set $\{2\} \times [7, \infty) \cup \{3\} \times [0, 1]$ is an interval.

Disjoint intervals are ordered by their contents. If $\mathscr{I}_1 \subseteq \mathscr{T}$ and $\mathscr{I}_2 \subseteq \mathscr{T}$ are disjoint intervals, so $\mathscr{I}_1 \cap \mathscr{I}_2 = \emptyset$, then write $\mathscr{I}_1 \prec \mathscr{I}_2$ if $t \prec t'$ for all $t \in \mathscr{I}_1$ and $t' \in \mathscr{I}_2$. For instance, in $\mathscr{T} = \{1, 2, \ldots, 20\}$, we have $\{2, 3\} \prec \{6, 7, 8\}$. Write $t \prec \mathscr{I}$ if $\{t\} \prec \mathscr{I}$ and $\mathscr{I} \prec t$ if $\mathscr{I} \prec \{t\}$. For instance, in $\mathscr{T} = \{1, 2, \ldots, 20\}$, we have $2 \prec \{6, 7, 8\}$.

A partition of $\mathscr{T}$ into disjoint intervals is a representation of $\mathscr{T}$ as the union of disjoint intervals $\mathscr{I}_\lambda$, indexed by a set Λ, that is, $\mathscr{T} = \bigcup_{\lambda \in \Lambda} \mathscr{I}_\lambda$ with $\mathscr{I}_\lambda \cap \mathscr{I}_{\lambda'} = \emptyset$ for $\lambda, \lambda' \in \Lambda$ with $\lambda \neq \lambda'$. In $\mathscr{T} = \{1, 2, \ldots, 20\}$, with $\Lambda = \{a, b, c\}$, one partition of $\mathscr{T}$ into disjoint intervals is $\mathscr{T} = \bigcup_{\lambda \in \Lambda} \mathscr{I}_\lambda$ where $\mathscr{I}_a = \{1, 2, \ldots, 9\}$, $\mathscr{I}_b = \{10\}$, $\mathscr{I}_c = \{11, 12, \ldots, 20\}$. Because the intervals, $\mathscr{I}_\lambda$, $\lambda \in \Lambda$, are disjoint, they are ordered; so write $\lambda \prec \lambda'$ if $\mathscr{I}_\lambda \prec \mathscr{I}_{\lambda'}$, and for $t \in \mathscr{T}$, write $t \prec \lambda$ if $t \prec \mathscr{I}_\lambda$ or $\lambda \prec t$ if $\mathscr{I}_\lambda \prec t$. For instance, in the partition of $\mathscr{T} = \{1, 2, \ldots, 20\}$ just mentioned, $\mathscr{I}_a \prec \mathscr{I}_c$ and $7 \prec \mathscr{I}_c$.

A set $\mathscr{S} \subseteq \mathscr{T}$ of hypotheses is exclusive if at most one of the hypotheses H_t, $t \in \mathscr{S}$, is true. For instance, if θ is a real parameter, then the set of hypotheses $\{H_t : t \in (-\infty, \infty)\}$ would be exclusive if H_t were $H_t : \theta = t$, but the same set would not be exclusive if the H_t were $H_t : \theta \leq t$; in the first case, distinct hypotheses contradict each other but in the second case they do not. The words 'at most one' are important: the set of hypotheses $\{H_t : t \in (12, 17)\}$ with $H_t : \theta = t$ is exclusive even if there is no reason to think θ is between 12 and 17.

Now, we can define a sequentially exclusive partition of hypotheses. A partition of $\mathscr{T}$ into disjoint intervals, $\mathscr{T} = \bigcup_{\lambda \in \Lambda} \mathscr{I}_\lambda$, is a sequentially exclusive partition if each interval $\mathscr{I}_\lambda$, $\lambda \in \Lambda$, would be exclusive if all hypotheses in earlier intervals were false; that is, at most one hypothesis H_t, $t \in \mathscr{I}_\lambda$, is true whenever all hypotheses, $H_{t'}$, $t' \prec \mathscr{I}_\lambda$, are false. For instance, with $\mathscr{T} = \{1, 2, 3\}$, the partition $\mathscr{T} = \{1\} \cup \{2, 3\}$ is a sequentially exclusive partition if at most one of H_2 and H_3 is true whenever H_1 is false; see §19.3 for several examples. Notice that $\{2, 3\}$ need not be exclusive if H_1 is true.

A sequentially exclusive partition always exists, namely the partition $\mathscr{T} = \bigcup_{t \in \mathscr{T}} \{t\}$. This is so because a set containing a single hypothesis is always exclusive; it contains at most one true hypothesis.

19.9 Appendix: Testing Hypotheses in Order

Testing hypotheses in order

In the following testing plan, $\mathscr{T} = \bigcup_{\lambda \in \Lambda} \mathscr{I}_\lambda$ is a sequentially exclusive partition, and for each hypothesis, H_t, $t \in \mathscr{T}$, there is a valid P-value, p_t, such that $\Pr(p_t \leq \alpha) \leq \alpha$ for all $\alpha \in [0,1]$ if H_t is true. Fix an α, conventionally $\alpha = 0.05$.

Testing plan. If $p_s \leq \alpha$ for all $s \in \mathscr{I}_\lambda$ for all $\lambda \prec \omega$, then reject all hypotheses H_t, $t \in \mathscr{I}_\omega$, with $p_t \leq \alpha$.

This testing plan is, in fact, the general form of the plan used throughout this chapter. For the partition $\mathscr{T} = \bigcup_{t \in \mathscr{T}} \{t\}$, the testing plan says to reject all hypotheses prior to the first acceptance and stop. In general, with $\mathscr{T} = \bigcup_{\lambda \in \Lambda} \mathscr{I}_\lambda$, the procedure stops with the first interval $\mathscr{I}_\omega$ that contains at least one hypothesis $t' \in \mathscr{I}_\omega$ with $p_{t'} > \alpha$; however, all hypotheses $t \in \mathscr{I}_\omega$ are tested, and all of those with $p_t \leq \alpha$ are rejected. Particular cases of Proposition 19.1 have been discussed throughout this chapter.

Proposition 19.1. *[29] If the testing plan is applied to a sequentially exclusive partition of hypotheses, H_t, $t \in \mathscr{T} = \bigcup_{\lambda \in \Lambda} \mathscr{I}_\lambda$, where $\Pr(p_t \leq \alpha) \leq \alpha$ if H_t is true, then the probability that the plan tests and rejects at least one true hypothesis is at most α.*

Proof. Let $\mathscr{W}$ be the event that the testing plan tests and rejects at least one true hypothesis. If all hypotheses, H_t, $t \in \mathscr{T}$, are false, then there is nothing to prove; in particular, $\Pr(\mathscr{W}) = 0$. Otherwise, let H_v be the first true hypothesis, let $\mathscr{R}_v$ be the event that the testing plan tests and rejects H_v, and let $\mathscr{I}_\omega$ be the unique interval that contains v. Then H_t is false for all $t \prec \mathscr{I}_\omega$. This implies that H_v is the only true hypothesis in $\mathscr{I}_\omega$, because $\mathscr{T} = \bigcup_{\lambda \in \Lambda} \mathscr{I}_\lambda$ is sequentially exclusive. It follows that the plan tests and rejects at least one true hypothesis if and only if it tests and rejects H_v, so $\mathscr{W} = \mathscr{R}_v$, and $\Pr(\mathscr{W}) = \Pr(\mathscr{R}_v) \leq \alpha$.

Sensitivity analysis for testing in order

In §3.5, a sensitivity analysis was performed in which an upper bound, say $P_{t,\max}$, was found for the one-sided P-value when testing a single hypothesis, say H_t; see Table 3.2. This upper bound, $P_{t,\max}$, was the largest P-value that could be produced for a given magnitude Γ of deviation from random treatment assignment in (3.13). That bound was sharp, in the sense that there was some set of treatment assignment probabilities π_{ij} satisfying (3.16)–(3.18) for which the bound $P_{t,\max}$ was the correct P-value.

Suppose that we have a sequence of such bounds, $P_{t,\max}$, for a sequence of hypotheses, H_t, $t \in \mathscr{T}$, where $\mathscr{T} = \bigcup_{\lambda \in \Lambda} \mathscr{I}_\lambda$ is a sequentially exclusive partition of hypotheses. Suppose that the testing plan were applied using the bounds in place of the unknown, actual P-values. What would happen? In fact, this was done several times in this chapter.

It is possible to show the following [30, Proposition 1]. First, if (3.13) is true for a given value of $\Gamma \geq 1$, the chance that the testing plan tests and rejects at least one true hypothesis is at most α. Second, this bound is sharp: there are treatment assignment probabilities π_{ij} satisfying (3.16)–(3.18) for which the testing plan would terminate with acceptance of the same hypothesis, H_v, in the proof of Proposition 19.1. In other words, if the testing plan were used once for every set of treatment assignment probabilities π_{ij} satisfying (3.16)–(3.18), then a hypothesis H_t would be rejected in all of these analyses if and only if it is rejected in the one analysis that uses the individual bounds in place of the actual P-values.

The sharpness of the bound is actually somewhat surprising. The reason it is surprising is that the treatment assignment probabilities π_{ij} that produce $P_{t,\max}$ are not generally the same as the π_{ij}'s that produce $P_{t',\max}$ for $t \neq t'$, so there may be no treatment assignment probabilities satisfying (3.16)–(3.18) that produce both $P_{t,\max}$ and $P_{t',\max}$. You can have $P_{t,\max}$ or you can have $P_{t',\max}$, but there is no assurance that you can have both $P_{t,\max}$ and $P_{t',\max}$, even though the proposal is to plug them both into the testing plan as if they could coexist. If the Bonferroni inequality is used with $P_{t,\max}$ and $P_{t',\max}$, then the result is conservative, not sharp [30, §4.5]: it can happen that $\min \left(P_{t,\max}, P_{t',\max} \right) > \alpha/2$, while the minimum of the two P-values for testing H_t and $H_{t'}$ is less than $\alpha/2$ for all treatment assignment probabilities π_{ij} satisfying (3.16)–(3.18). Speaking informally, the bound is sharp for the testing plan because the plan stops with a single P-value; see [30, Proposition 1] for a precise statement.

The results in [29, 30] are general; they are not restricted to matched pairs.

References

1. Bauer P.: Multiple testing in clinical trials. Statist Med **10**, 871–890 (1991)
2. Bauer, P., Kieser, M.: A unifying approach for confidence intervals and testing of equivalence and difference. Biometrika **83**, 934–937 (1996)
3. Bauer, P.: A note on multiple testing procedures in dose finding. Biometrics **53**, 1125–1128 (1997)
4. Benjamini, Y., Hochberg, Y.: Controlling the false discovery rate. J Roy Statist Soc B **57**, 289–300 (1995)
5. Berger, R.L.: Multiparameter hypothesis testing and acceptance sampling. Technometrics **24**, 295–300 (1982)
6. Berger, R.L., Hsu, J.C.: Bioequivalence trials, intersection-union tests and equivalence confidence sets. Statist Sci **11**, 283–319 (1996)
7. Bilban, M., Jakopin, C.B.: Incidence of cytogenetic damage in lead-zinc mine workers exposed to radon. Mutagenesis **20**, 187–191 (2005)
8. Bunker, J.P., Barnes, B.A., Mosteller, F.: Costs, Risks and Benefits of Surgery. Oxford: Oxford University Press (1977)
9. Cochran, W.G.: The planning of observational studies of human populations (with Discussion). J Roy Statist Soc A **128**, 234–265 (1965)
10. Cox, D.R.: Causality: Some statistical aspects. J Roy Statist Soc A **155**, 291–301 (1992)
11. Gail, M.: Statistics in action. J Am Statist Assoc **91**, 1–13 (1996)
12. Heller, R., Rosenbaum, P.R., Small, D.: Split samples and design sensitivity in observational studies. J Am Statist Assoc **104**, to appear (2009)

13. Holm, S.: A simple sequentially rejective multiple test procedure. Scand J Statist **6**, 65–70 (1979)
14. Hommel, G., Kropf, S.: Tests for differentiation in gene expression using a data-driven order or weights for hypotheses. Biomet J **47**, 554–562 (2005)
15. Hsu, J.C., Berger, R.L.: Stepwise confidence intervals without multiplicity adjustment for dose-response and toxicity studies. J Am Statist Assoc **94**, 468–475 (1999)
16. Hsu, J.C., Hwang, J.T.G., Liu, H-K., Ruberg, S.J.: Confidence intervals associated with tests for bioequivalence. Biometrika **81**, 103–114 (1994)
17. Koch, G.G., Gansky, S.A.: Statistical considerations for multiplicity in confirmatory protocols. Drug Inform J **30**, 523–533 (1996)
18. Laska, E.M., Meisner, M.J.: Testing whether an identified treatment is best. Biometrics **45**, 1139–1151 (1989)
19. Lehmacher, W., Wassmer, G., Reitmeir, P.: Procedures for two-sample comparisons with multiple endpoints controlling the experimentwise error rate. Biometrics **47**, 511–521 (1991)
20. Lehmann, E.L.: Testing multiparameter hypotheses. Ann Math Statist **23**, 541–552 (1952)
21. Li, Y.F.P., Propert, K.J., Rosenbaum, P.R.: Balanced risk set matching. J Am Statist Assoc **96**, 870–882 (2001)
22. Marcus, R., Peritz, E., Gabriel, K.R.: On closed testing procedures with special reference to ordered analysis of variance. Biometrika **63**, 655–60 (1976)
23. Masjedi, M.R., Heidary, A., Mohammadi, F., Velayati, A.A., and Dokouhaki, P.: Chromosomal aberrations and micronuclei in lymphocytes of patients before and after exposure to anti-tuberculosis drugs. Mutagenesis **15**, 489–494 (2000)
24. McPherson, K., Bunker, J.P.: Costs, Risks and Benefits of Surgery: A milestone in the development of health services research. J Roy Soc Med **100**, 387–390 (2007)
25. Propert, K.J., Schaeffer, A.J., Brensinger, C.M., Kusek, J.W., Nyberg, L.M., Landis, J.R.: A prospective study of interstitial cystitis: Results of longitudinal followup of the interstitial cystitis data base cohort. J Urol **163**, 1434–1439. (2000)
26. Rosenbaum, P.R.: From association to causation in observational studies. J Am Statist Assoc **79**, 41–48 (1984)
27. Rosenbaum, P.R.: Observational Studies (2nd ed). New York: Springer (2002)
28. Rosenbaum, P.R.: Comment on a paper by Donald B. Rubin: The place of death in the quality of life. Statist Sci **21**, 313–316 (2006)
29. Rosenbaum, P.R.: Testing hypotheses in order. Biometrika **95**, 248–252 (2008)
30. Rosenbaum, P.R., Silber, J.H.: Sensitivity analysis for equivalence and difference in an observational study of neonatal intensive care units. J Am Statist Assoc **104**, 501–511 (2009)
31. Schuirmann, D.L.: On hypothesis testing to determine if the mean of a normal distribution is contained in a known interval. Biometrics **37**, 617 (1981)
32. Shaffer, J.P.: Modified sequentially rejective multiple test procedures. J Am Statist Assoc **81**, 826–831 (1986)
33. Silber, J.H., Lorch, S.L., Rosenbaum, P.R., Medoff-Cooper, B., Bakewell-Sachs, S., Millman, A., Mi, L., Even-Shoshan, O., Escobar, G.E.: Additional maturity at discharge and subsequent health care costs. Health Serv Res **44**, 444–463 (2009)
34. Tamhane, A., Logan, B.: A superiority-equivalence approach to one-sided tests on multiple endpoints in clinical trials. Biometrika **91**, 715–727 (2004)
35. Tukey, J.W.: We need both exploratory and confirmatory. Am Statistician **34**, 23–25 (1980)
36. Tukey, J.W.: Sunset salvo. Am Statistician **40**, 72–76 (1986)
37. Westlake, W.J.: Response to Kirkwood. Biometrics **37**, 591–593 (1981)
38. Yoon, F.: New Methods for the Design and Analysis of Observational Studies. Doctoral Thesis, Department of Statistics, University of Pennsylvania.

Summary: Key Elements of Design

In an observational study, competing theories should make conflicting predictions. Many studies dissipate before they begin from a simple lack of focus. The study is intended to settle something, or at least take a step towards settling something, and for that, there needs to be something definite to settle and some prospect of and means for settling it; see Chapter 4. Some pairs of theories rarely yield conflicting predictions, so one may search for unusual opportunities to contrast them; see §5.1.

An observational study should be structured to resemble a simple experiment. A typical structure is the comparison of a treated and a control group that looked comparable prior to treatment in terms of observed covariates. An experiment is an unusual situation. In a carefully controlled experiment, one of the rarest of things happens: the effects caused by treatments are seen with clarity. Each step away from the experimental template is a step closer to the edge of an abyss; see §1.2 and §12.1.

Adjustments for observed covariates should be simple, transparent, and convincing. The major source of uncertainty about the conclusions of an observational study comes from the possible failure to control for covariates that were not measured. This possibility is raised in virtually every observational study. If you think this possibility will not be raised in evaluating your study, then you are kidding yourself. There is little hope of addressing this major source of uncertainty if the study becomes bogged down in unnecessarily complex, obscure, or unconvincing adjustments for observed covariates. One simple, transparent, and convincing way to adjust for observed covariates is to compare a treated and a control group with similar distributions of the observed covariates. Such a control group may often be constructed with the aid of multivariate matching; see Part II.

The most plausible alternatives to a treatment effect should be anticipated and addressed. In an observational study, it is not possible to address every conceivable alternative to a treatment effect. It is often possible to anticipate several plausible objections to a claim that a comparison of matched treated and control groups estimates the effects of the treatment. Such an objection claims that the comparison is ambiguous, that it could estimate a treatment effect or it could be distorted by some specific form of bias. With a specific form of bias in mind, design elements can often be added, such as two control groups or unaffected outcomes, that resolve specific ambiguities; see §5.2.

The analysis should address possible biases from unmeasured covariates. Typically, the analysis should include a sensitivity analysis of one form or another; see Chapter 3. A sensitivity analysis asks: how much bias from unmeasured covariates – what magnitude of deviation from random assignment – would need to be present to qualitatively alter the conclusions suggested by the naïve, straightforward comparison of matched treated and control groups? The degree of sensitivity to unmeasured bias is a fact of the matter, something that is determined without ambiguity from the data at hand. Whether or not biases of this magnitude are present remains a matter of reasoned conjecture and responsible debate, but that debate is now informed and constrained by the facts of the matter. A P-value does not rule out the possibility that bad luck produced the observed results; rather, it objectively measures how much bad luck would be required to produce the observed results. A sensitivity analysis does not rule out the possibility that unmeasured bias produced the observed results; rather, it objectively measures how much unmeasured bias would be required to produce the observed results.

To the extent possible, observational studies should be designed to be insensitive to biases from unmeasured covariates. To do this, one must know what makes some designs (or data generating processes) sensitive to unmeasured biases and others insensitive. With this knowledge, when faced with a choice, an insensitive design may be chosen. Many factors strongly affect design sensitivity; see Part III.

There should be a plan for a primary analysis. A randomized controlled trial invariably has a protocol detailing a plan for a primary analysis. An observational study also needs such a plan. With design elements intended to resolve otherwise ambiguous comparisons, the study design anticipates one of a few patterns of results, what R.A. Fisher called an 'elaborate theory'; see §19.2. The typical elaborate theory predicts a difference here, near equivalence there, the absence of a trend here, a discontinuity there. For instance, an elaborate theory might predict much higher responses in the treated group than in two control groups, with the two control groups differing negligibly from each other. A planned primary analysis will attempt to examine, possibly confirm, the predictions of an elaborate theory; see Chapter 19. The predictions of an elaborate theory are predictions only if they precede examination of the outcomes; there is no value in an elaborate theory constructed after the fact to accommodate a particular set of data. Therefore, the analysis of an elaborate theory must be a planned analysis. Planning may be aided by sample splitting; see §18.1. A plan for a primary analysis does not preclude unplanned exploratory analyses; rather, it distinguishes planned and unplanned analyses.

Solutions to Common Problems

The matching problem is too large. If a matching problem is too large, divide it into several smaller problems by exact matching for one or more important covariates; see §9.3.

Treated and control groups are too far apart to match. Before trying to solve this problem, make sure the problem is real. Compare boxplots of the propensity scores for treated and control groups. If the boxplots cover much the same range, but some covariates or the propensity score are poorly balanced, consider: (i) tightening the caliper on the propensity score so that, for example, the penalty engages at 10% of the standard deviation rather than 20% (see §8.4), (ii) use penalties to improve balance on one or two stubborn covariates (see §9.2), or (iii) try full matching (see §8.6). Otherwise, if the boxplots exhibit large regions with little or no overlap, consider redefining the study population using a few key covariates (see §3.6). For an example of redefining the study population, see [2].

Treated and control groups overlap, but at some values of x there are too few controls even for 1-to-1 matching. Try full matching; see §8.6.

People are treated at different times. How do I match? Consider risk-set matching; see Chapter 12.

I want to match for a variable with many categories, but there are not enough people in the categories to permit a close match. Try matching with fine balance; see Chapter 10.

I have two control groups. How do I match? There are several options. One approach is to form matched triples by matching one treated group twice, once to one control group, then to the other. Another approach forms matched pairs in an 'incomplete block' design. See Chapter 11. A different problem is to split one control group to form two for a specific purpose; see the discussion of tapered matching in §18.2.

My propensity score model does a poor job of predicting treatment assignment. Not a problem. In a large, completely randomized experiment, a propensity score model would have great difficulty predicting treatment assignments from covariates precisely because treatment assignment is random and does not depend upon the covariates. The propensity score is intended to fix a specific problem, namely imbalances in observed covariates. If your study does not have that problem, then that is just fine.

How do I judge whether my model for the propensity score is a good model? The propensity score has various uses, and the answer to this question depends upon how the propensity score will be used. In this book, propensity scores are used for matching. When used for matching, propensity scores are a means to an end, namely matched pairs or sets that balance observed covariates. When you have matched pairs or sets that balance observed covariates, the matching for observed covariates is done, and attention shifts to potential bias from unmeasured covariates. In light of this, judging the propensity score model when used in matching is essentially the same task as judging whether the matching has balanced observed covariates; see §9.1.

How do I select covariates to use in the propensity score? This question inverts the means and the end. The proper question is: Which covariates do you wish to balance by matching on the propensity score?

There is a covariate that strongly predicts treatment assignment Z but seems unlikely to matter much for the response R. What should I do? Read about 'seemingly innocuous confounding,' analytical adjustments, and tapered matching in §18.2.

There is a variable that is subsequent to treatment assignment and may have been affected by the treatment, so it is not a covariate, but I feel I should adjust for it anyway. Although this is sometimes reasonable, think long and hard before you do this. If you adjust for a concomitant variable that has been affected by the treatment, you may introduce a bias that would not otherwise have been present; see [5]. If in doubt, it may be best to leave such a variable unmatched, so that analyses with and without analytical adjustments for it are possible; see §18.2. One alternative is to both match and not match for such a variable; see the discussion of tapered matching in §18.2.

I have matched with a variable ratio of controls to treated subjects, and now I want a boxplot. It is not difficult to do, but at the time I write this, current software does not do it, so a few steps are needed. Essentially, you need to compute a weighted empirical distribution function, compute the quantiles from it, and make a boxplot using those quantiles. The R function 'bxp' will help: it will draw boxplots from quantiles you give it. See Note 3 in Chapter 8. The resulting boxplots look like ordinary boxplots, but they are weighted to reflect the variable numbers of controls. Examples are found in [2].

Should I match with a fixed or a variable ratio of controls to treated subjects? The choice is discussed in §8.5. The previous solution, about the boxplot, illustrates the biggest disadvantage of matching with variable controls; simple, straightforward tasks require special programming effort. Opposed to this, theory strongly suggests that matching with variable controls is an efficient way to produce closer matches; see [4].

People say that a specific unobserved covariate u is strongly related to the outcome, r_C, but I doubt it. To show that a specific unobserved covariate u is not strongly related to the outcome r_C, find two control groups that differ markedly in terms of the unobserved u, and show that outcomes do not greatly differ in these two control groups. See §5.2.2 and §19.3.

People say that a specific unobserved covariate u is strongly related to treatment assignment Z, but I doubt it. To show that a specific unobserved covariate u is not strongly related to treatment assignment, Z, find an outcome known to be unaffected by the treatment that is highly correlated with u, and show that this unaffected outcome has a similar distribution in treated ($Z = 1$) and control ($Z = 0$) groups. See §5.2.4.

I would like to perform a sensitivity analysis, but I do not have matched pairs. It's not difficult. See [6, Chapter 4]. A quick and easy approach for matching with multiple controls uses the stratified Wilcoxon rank sum statistic and Table 1 in [1]. For matched sets with one to four controls (n_i equal to two to five in that table), and for Γ equal to one to four, the Table gives the calculations needed for one matched set; these contributions are summed to produce the sensitivity analysis. For any rank statistic, say the aligned rank statistic, the general calculations in [1, §3] are easy to perform in R or with a spreadsheet. An alternative approach with multiple controls uses an m-test [7]. Also, [7, §5] discusses sensitivity analysis with covariance adjustment in matching with multiple controls.

I would like to perform a sensitivity analysis, but my outcomes are binary. Again, it is not difficult. See [6, Chapter 4]. That chapter also discusses outcomes that are censored survival times.

How do I interpret the parameter Γ? The parameter Γ is convenient in that it is a single parameter that can refer to a wide variety of situations. The resulting sensitivity analysis is one-dimensional: just one parameter varies. The sensitivity bounds implicitly refer to an unobserved covariate u that is strongly related to the response, r_C, and that has a controlled relationship with the treatment, Z, the control being provided by the value of Γ. Sometimes, that situation is not the one under discussion, because a very strong relationship between u and r_C isn't plausible. It is possible to reexpress the one-parameter analysis in terms of two parameters, where one parameter controls the relationship between the unobserved covariate u and the response r_C and the other controls the relationship between u and the treatment, Z. It is the same one-dimensional sensitivity analysis, with no new computations, but now with a two-dimensional interpretation; see [8].

I am worried that my findings will be sensitive to small unmeasured biases. Review Chapters 14–17 to see what issues affect the sensitivity of conclusions to unmeasured biases. Consider sample splitting to guide design decisions that affect design sensitivity; see §18.1 and [3].

I have doses of treatment, but I am not sure whether they are any good. Split the sample and find out; see §18.1 and [3]. Similar advice applies to many other design decisions.

The treatment that interests me is known to be assigned in a very biased fashion. Instead of studying the effect of that treatment, consider studying its differential effect compared with other treatments affected by the same biases; see §5.2.6. This may help in some circumstances.

There have been several studies of the treatment that interests me, but none is convincing. Is it possible to study the treatment again, this time removing one of the several problems that made previous studies unconvincing? See §4.5.

The same problems inevitably occur whenever efforts are made to study the treatment that interests me. Why is the treatment thought to work? What reasons are given? Why is there doubt about the effects of the treatment? What reasons are given? Can an empirical study shed light on whether these reasons are valid? Perhaps the reasons can be supported or refuted, even though direct investigation of the effects is difficult. See §4.6.

I am disappointed by my 'null result.' Do you possess evidence for the absence of a large effect? That might be an important finding; see §19.5. Or do you lack evidence about the magnitude of the effect? That is disappointing, but it is a common occurrence, one that happens now and then to everyone.

References

1. Gastwirth, J.L., Krieger, A.M., Rosenbaum, P.R.: Asymptotic separability in sensitivity analysis. J Roy Statist Soc B **62**, 545–555 (2000)
2. Haviland, A.M., Nagin, D.S., Rosenbaum, P.R.: Combining propensity score matching and group-based trajectory analysis in an observational study. Psychol Methods **12**, 247–267 (2007)
3. Heller, R., Rosenbaum, P.R., Small, D.: Split samples and design sensitivity in observational studies. J Am Statist Assoc **104**, to appear (2009)
4. Ming, K., Rosenbaum, P.R.: Substantial gains in bias reduction from matching with a variable number of controls. Biometrics **56**, 118–124 (2000)
5. Rosenbaum, P.R.: The consequences of adjustment for a concomitant variable that has been affected by the treatment. J Roy Statist Soc A **147**, 656–666 (1984)
6. Rosenbaum, P.R.: Observational Studies (2nd ed.). New York: Springer (2002)
7. Rosenbaum, P.R.: Sensitivity analysis for m-estimates, tests, and confidence intervals in matched observational studies. Biometrics **63**, 456–464 (2007)
8. Rosenbaum, P.R., Silber, J.H.: Amplification of sensitivity analysis in observational studies. J Am Statist Assoc, to appear

Symbols

$\|A\|$	If A is a finite set, $\|A\|$ is the number of elements in A.
$\mathbf{x}$	Observed covariates for any one person. See §2.1.2.
$\mathbf{x}_\ell$	Observed covariates for person ℓ, $\ell = 1,\ldots,L$. The ℓ subscript is used to denote people before they are matched. The ℓ subscript is used in the same way with Z, R, r_T, r_C, and u. See §3.1.
$\mathbf{x}_{ij}$	Observed covariates for person j, in pair i, $i = 1,2,\ldots,I$, $j = 1,2$. The ij subscripts are used to denote people after they are arranged into matched pairs. The ij subscripts are used in the same way with Z, R, r_T, r_C, and u. See §2.1.2.
Z	Treatment indicator, $Z = 1$ if treated, $Z = 0$ if control. See §2.1.2.
$\mathbf{Z}$	Treatment indicators for the $2I$ subjects in the I matched pairs, $\mathbf{Z} = (Z_{11},Z_{12},Z_{21},\ldots,Z_{I2})^T$. See §2.1.2.
$e(\mathbf{x})$	Propensity score, $e(\mathbf{x}) = \Pr(Z = 1 \mid \mathbf{x})$. See §8.2.
(r_T, r_C)	For any one person, r_T is the response the person would exhibit under treatment, and r_C is the response this same person would exhibit under control. See §2.2.1.
$(\mathbf{r}_T, \mathbf{r}_C)$	Vectors $\mathbf{r}_T = (r_{T11}, r_{T12},\ldots, r_{TI2})^T$ and $\mathbf{r}_C = (r_{C11}, r_{C12},\ldots, r_{CI2})^T$ of potential responses for all $2I$ subjects in I pairs. See §2.2.1.
R	Observed response, that is, $R = r_T$ if $Z = 1$ and $R = r_C$ if $Z = 0$, so $R = Zr_T + (1-Z)r_C$. See §2.1.2.
$\mathbf{R}$	Observed responses for the $2I$ subjects in the I matched pairs, $\mathbf{R} = (R_{11}, R_{12},\ldots, R_{I2})^T$. See §2.1.2.
sgn	$\text{sgn}(a) = 1$ if $a > 0$ and $\text{sgn}(a) = 0$ otherwise. Not to be confused with its sibling sign. See §2.3.3.
sign	$\text{sign}(a) = 1$ if $a > 0$, $\text{sign}(a) = 0$ if $a = 0$, $\text{sign}(a) = -1$ if $a < 0$. Not to be confused with its sibling sgn. See §2.9.
s_i	$s_i = 1$ if $\|Y_i\| > 0$ and $s_i = 0$ if $\|Y_i\| = 0$. See §2.3.3.
u	An unobserved covariate. See §2.1.2.

$\mathbf{u}$	Unobserved covariates for the $2I$ subjects in the I matched pairs, $\mathbf{u} = (u_{11}, u_{12}, \ldots, u_{I2})^T$. See §2.1.2.
Y_i	Treated-minus-control difference in observed responses within pair i, that is, $Y_i = (Z_{i1} - Z_{i2})(R_{i1} - R_{i2})$. See §2.1.2.
$\mathbf{Y}$	Treated-minus-control differences in observed responses for I pairs, $\mathbf{Y} = (Y_1, Y_2, \ldots, Y_I)^T$. See §2.1.2.
$\mathscr{F}$	The values of $(r_T, r_C, \mathbf{x}, u)$ for all subjects. Notice that these quantities do not change when Z changes, whereas R does change when Z changes. In this specific sense, $\mathscr{F}$ is fixed in Fisher's theory of randomization inference. See §2.2.1. In simple experiments, the dose of treatment is the same for every treated subject, and so doses are not explicitly mentioned when $\mathscr{F}$ is first defined in §2.2.1. However, if there are doses d_i that vary by pair i, as in §5.2.5, or that vary by individual, (d_{Tij}, d_{Cij}), as in §5.3, then these doses are also part of $\mathscr{F}$. In a strict sense, the doses are always present in $\mathscr{F}$, but when the doses are constant there is no need to mention them.
$\mathscr{Z}$	The set containing the 2^I possible values of $\mathbf{z} = (z_{11}, z_{12}, \ldots, z_{I2})^T$ of $\mathbf{Z}$. So $\mathbf{z} \in \mathscr{Z}$ if $z_{ij} = 0$ or $z_{ij} = 1$ with $z_{i1} + z_{i2} = 1$. In conditional probabilities, conditioning on $\mathscr{Z}$ means conditioning on the event $\mathbf{Z} \in \mathscr{Z}$. See §2.2.3.
d_i	The dose at which the treatment is given in pair i. See the discussion of dose-response in §5.2.5.
(d_{Tij}, d_{Cij})	For person j in pair i, d_{Tij} is the dose of treatment this person would accept if assigned to treatment, $Z_{ij} = 1$, and d_{Cij} is the dose of treatment this person would accept if assigned to control, $Z_{ij} = 0$. See the discussion of instrumental variables in §5.3.
D_{ij}	For person j in pair i, D_{ij} is the dose actually received, $D_{ij} = Z_{ij}d_{Tij} + (1 - Z_{ij})d_{Cij}$. See the discussion of instrumental variables in §5.3.
Γ	Sensitivity parameter in a sensitivity analysis. Two subjects with the same observed covariates may differ in their odds of treatment by at most a factor of Γ. The parameter Γ is a key element in the 'second model' for an observational study in Chapter 3.
$P_{\min}$ and $P_{\max}$	Interval of possible P-values in a sensitivity analysis. See §3.5, specifically Table 3.2.
$\widetilde{\Gamma}$	Design sensitivity. In the favorable situation, the power of a sensitivity analysis tends to 1 as the sample size increases if $\Gamma < \widetilde{\Gamma}$ and the power tends to 0 if $\Gamma > \widetilde{\Gamma}$. See Chapter 14.

Acronyms

AAA	ACE-inhibitor after anthracycline randomized trial. See §5.3.
AFDC	Aid to Families with Dependent Children, a Federally funded public assistance program in the United States
CI	Confidence interval.
FARS	Fatal Accident Reporting System. FARS records information about fatal road accidents in the United States
FTE	Full time equivalent employment.
GO	Gynecologic oncologist.
HL	Hodges-Lehmann point estimate.
IC	Interstitial cystitis, a urologic disorder.
iid	Independent and identically distributed. See §14.1.
IV	Instrumental variable.
ICDB	Interstitial Cystitis Data Base.
Medicare	U.S. government program providing health care to people over the age of 65.
MN	Micronuclei, a common measure in genetic toxicology.
MO	Medical oncologist.
NSAID	Nonsteroidal anti-inflammatory drug.
NICU	Neonatal intensive care unit.
NJ	New Jersey, a state in the United States.
NSW	National Supported Work Demonstration, a randomized experiment evaluating a program intended to aid the transition to employment.
PA	Pennsylvania, a state in the United States.
PMA	Postmenstrual age, a proxy for age from conception.
R	A statistical software package. http://cran.r-project.org/. R is distributed free.
SAS	A statistical software package. http://www.sas.com/.
SEER	Surveillance, Epidemiology and End Results Program of the U.S. National Cancer Institute. http://seer.cancer.gov/.
sib-TDT	Sibship transmission disequilibrium test.
stata	A statistical software package. http://www.stata.com/.

TDT   Transmission disequilibrium test.
TOST   Two-one-sided tests procedure for equivalence testing.

Glossary of Statistical Terms

Bonferroni inequality If A and B are two events, then the probability of the event 'A or B' is at most the sum of the probabilities of A and of B; that is, $\Pr(A \cup B) \leq \Pr(A) + \Pr(B)$. The inequality applies also with more than two events. The Bonferroni inequality is used in many different ways, but it shows up in this book in connection with testing more than one hypothesis. If there are two hypotheses, where A is the event of falsely rejecting the first hypothesis, and B is the event of falsely rejecting the second hypothesis, then the chance of falsely rejecting at least one hypothesis is at most $\Pr(A) + \Pr(B)$. In particular, if each test is done at level 0.025, then the chance of at least one false rejection in two tests is at most $0.025 + 0.025 = 0.05$. This use of the Bonferroni inequality is attractive because it is simple and very general; however, there are methods that are only slightly more complex and equally general but better; see Holm, S.: A simple sequentially rejective multiple test procedure, Scand J Statist **6**, 65–70 (1979).

boxplot A boxplot is a graph that looks like a transistor. Figure 1.1 contains several boxplots. The box contains half the observations. The box has a middle line at the median and it ends at the upper and lower quartiles. One-quarter of the observations are inside the box above the median, and one-quarter of the observations are inside the box below the median. One-quarter of the observations are above the box, and one-quarter of the observations are below the box. Extreme observations appear as individual points. Boxplots were invented by John Tukey. See Cleveland, W.W.: Elements of Graphing Data, Summit, NJ: Hobart Press, pp. 139–143.

censored If a patient was diagnosed two years ago and is alive today, then this patient's survival time from diagnosis is censored at two years. This means that we know the patient's total survival after diagnosis is at least two years, but we do not know the actual survival time. Typically, some survival times are censored and others are known or uncensored. Given a choice, you'd rather be censored.

close match A pair is closely matched for a covariate if the two individuals in the pair have very similar values of that covariate. It is generally very difficult to obtain a close match on many covariates at the same time. See also 'covariate balance' for a very different notion.

consistent estimate A consistent estimate is one that gets it right if the sample size is large enough. An estimator $\hat{\theta}$ is a consistent estimate of a parameter θ if $\hat{\theta}$ converges in probability to θ as the sample size increases. See 'convergence in probability' in this glossary.

consistent test A consistent test is one that gets it right if the sample size is large enough. A test of a null hypothesis H_0 is consistent against an alternative hypothesis H_A if the power of the test tends to one as the sample size increases when H_A is true.

convergence in distribution Convergence in distribution formalizes the notion that a random variable, say ξ_I, almost has a certain distribution. Typically in this book, ξ_I is a statistic computed from a sample of size I — for instance, a standardized version of Wilcoxon's signed rank statistic, T — and the convergence occurs as the sample size I increases. In this book, the approximating distribution is always the standard Normal distribution, $\Phi(.)$ and, to be tangible, this glossary will define convergence in distribution to the standard Normal distribution. Slightly more detail is needed to define convergence in distribution in general. A sequence of random variables, ξ_I, $I = 1, 2, \ldots$, converges in distribution to the standard Normal distribution, $\Phi(.)$, if $\lim \Pr(\xi_I \leq c) = \Phi(c)$ for all c, the limit being taken as the sample size I increases.

convergence in probability Convergence in probability formalizes the notion that a random variable, say ξ_I, is almost a certain constant. Typically in this book, ξ_I is a statistic computed from a sample of size I — for instance, Wilcoxon's signed rank statistic, T, divided by $I(I+1)/2$ — and the convergence occurs as the sample size I increases. A sequence of random variables, ξ_I, $I = 1, 2, \ldots$, converges in probability to the real number a if $\Pr(|\xi_I - a| > \varepsilon)$ tends to 0 as I increases for all $\varepsilon > 0$.

covariate A covariate is a variable measured prior to treatment assignment, and hence prior to treatment. Because it is measured prior to treatment assignment and prior to treatment, it cannot be affected by treatment. Therefore, a covariate exists in a single version, say x, because an individual has the same value of a covariate whether assigned to treatment or to control. In contrast, an outcome may be affected by the treatment, so an outcome exists in two versions, one seen under treatment, r_T, the other seen under control, r_C. See §2.1.2.

covariate balance If the distributions of covariates are similar in treated and control groups, then there is covariate balance. For instance, if age is one covariate, then age is balanced if the distribution of ages for treated subjects is similar to the distribution of ages for controls. Covariate balance is a property of the treated and control groups taken as two whole groups, not a property of an individual matched pair or set. See also the entry for 'close match' for a very different notion. Randomization tends to balance both observed and unobserved covariates. Matching on the propensity score tends to balance observed covariates but not unobserved covariates.

efficiency Speaking informally, a statistical procedure is efficient if it makes the best possible use of the available sample size, producing the shortest confidence intervals, the smallest standard errors, the most powerful test, and so on. One procedure, say A, is more efficient than another procedure, say B, if A can accomplish with, say, 50 observations what B needs 52 observations to accomplish. Understandably, efficiency is a major concern in randomized controlled clinical trials, in part because each observation is expensive and time-consuming, and in part because additional observations entail treating additional patients, perhaps with a treatment that the trial itself will find to be an inferior treatment. Efficiency is not the central concern in observational studies because a limited sample size is not the principal source of uncertainty. Bias of fixed size, not the standard error, is the principal source of uncertainty in nonrandomized studies. Many observational studies would not become more compelling if the sample size were increased. In addition, in some observational studies, large samples are available from extant data.

exclusion restriction The exclusion restriction arises in the definition of an instrument in §5.3. Randomized encouragement to accept a treatment is an instrument for the treatment only if encouragement affects the outcome only to the extent that it affects acceptance of the treatment; this is the exclusion restriction.

favorable situation In the favorable situation, an observational study is free of bias from unmeasured covariates, but the investigator does not know this, and there is actually a treatment effect, but the investigator does not know this either. The investigator sees that the treated-minus-control matched pair differences in outcomes, Y_i, tend to be positive, but does not know that they tend to be positive because of a treatment effect, not because of some uncontrolled bias. It is in this favorable situation that the investigator would like to report that the treatment appears to have an effect and that appearance is highly insensitive to unmeasured biases. See §14.2.

full matching In a full matching, a matched set may contain one treated subject and one or more controls, or one control and one or more treated subjects. It is the 'optimal form' of stratification in that it makes subjects in the same stratum as similar as possible. See §8.6.

instrument, instrumental variable, IV An instrument (or instrumental variable or IV) is a randomized nudge or encouragement to accept the treatment that affects the outcome only to the extent that it affects acceptance of the treatment. See §5.3.

level α test A statistical test has level α if, in the worst case, it rejects a true hypothesis with probability at most α, that is, if the supremum over true hypotheses of the probability of rejection is less than or equal to α. Compare the level of a test with the size α of a test as defined later in the glossary. In typical use, the level of the test is a promise about the test's performance and the size is a fact about its performance, where the achieved fact may be better than the promised performance. Suppose that a randomization test, such as Wilcoxon's signed rank test, is applied in a paired randomized experiment, as in Chapter 2, and the null hypothesis is rejected if the P-value is less than or equal to 0.05. Then the test will have level $\alpha = 0.05$.

However, if $|\mathscr{L}|$ is not divisible by 20, then the size of the test will be a little smaller than 0.05. In this case, one is advertising false rejections of true hypotheses at a 5% rate, but the rate is a tad better, a bit lower, than 5%. See E.L. Lehmann's book Testing Statistical Hypotheses, New York: John Wiley (1959).

logit model, logistic regression If p is the probability of an event, then $p/(1-p)$ is the odds of the event, and $\log(p/(1-p))$ is the logit or log-odds. If you were to build a linear model for p in terms of predictors, then the model might produce fitted probabilities below 0 or above 1. If the logit is linear in predictors, the probabilities stay between 0 and 1. Linear logit or logistic models assert that logits of probabilities are linear in predictors. The models are typically fitted by the method of maximum likelihood. In this book, logit models are used to estimate propensity scores, but propensity scores can be estimated in other ways, and logit models have other uses. Logit models have many attractive technical properties; see D. Cox and E. Snell, Analysis of Binary Data, New York: Chapman and Hall/CRC (1989).

lowess Locally weighted scatterplot smoother. Lowess draws a curve through the points of a scatter plot. Figure 7.2 displays a lowess smooth. The curve is estimated locally, in the sense that, at a given value of x, the closest points (x,y) to x are given the greatest weight. Lowess was invented by William Cleveland. See Cleveland, W.W.: Elements of Graphing Data, Summit, NJ: Hobart Press, pp. 168–180.

marginal distribution The distribution of one random variable is its marginal distribution. The term marginal distribution is typically used when there are several random variables in the current discussion and the speaker wishes to emphasize that the distribution is for one random variable, ignoring the others. The situation is similar to that of the fellow who was delighted to learn that he had been speaking prose all his life.

Monte Carlo By tradition, algorithms that use random numbers are named for cities where people gamble. For example, there is Monte Carlo integration and there are Las Vegas algorithms. Monte Carlo integration means finding the value of an integral using random numbers. Typically, in statistical work, the integral is the expectation of a random variable. In this case, the Monte Carlo algorithm samples many independent copies of the random variable and takes their average. People sometimes use the terms Monte Carlo and simulation as if they meant the same thing, but that is not correct. Monte Carlo evaluates integrals, whereas simulation may be used for various purposes. Monte Carlo is a substitute for numerical integration, and as such it should be carried through to be accurate to several significant figures, so the value of the integral is known for practical purposes. When someone says that such and such can be determined by Monte Carlo, the implication is that it may not be worth the time to go into further mathematical details because a child could do it. Perhaps a child with a network of supercomputers.

ostensibly comparable 'Ostensibly comparable' means comparable in terms of measured covariates, x, leaving open the question of whether people who look comparable in terms of x are actually comparable in terms of unmeasured covariates u.

To say that treated and control groups are ostensibly comparable after matching is to say that matching has done its job, has controlled for x, without presuming that this matching suffices for causal inference; it is to say that step one is complete and step two is pending.

outcome An outcome is a variable measured after treatment. Because it is measured after treatment, it can be affected by treatment. Therefore, an outcome exists in two versions, one seen under treatment, r_T, the other seen under control, r_C. In contrast, a covariate measured prior to treatment assignment exists in a single version, say x, because an individual has the same value of a covariate whether assigned to treatment or to control. See §2.2.1.

power (of a test) The power of a test is the probability that the test will reject the null hypothesis when the null hypothesis is false. You would like the power to be high. If the test rejects when the P-value is less than 0.05, then the power of the test is the chance that the P-value is less than 0.05 when the null hypothesis is false. The power depends upon many things. In particular, the power depends upon what is true. If the null hypothesis is false, then something else is true. For instance, consider the power of a test of the null hypothesis that the treatment has no effect, where the test rejects when the P-value is less than 0.05. If the actual treatment effect is very small, the power may be only 0.06, but if the actual treatment effect is very large, the power may be 0.99. In other words, there may be only a 6% chance of rejecting the null hypothesis if the treatment effect is very small, but a 99% chance of rejection if the effect is very large. (Remember, if you test a true null hypothesis and reject when the P-value is less than 0.05, then you typically run a 5% chance of rejecting the true null hypothesis; so, 6% power is no reason for pride.) The power also depends upon the sample size. Typically, the power is larger when the sample size is larger.

propensity score The propensity score is the conditional probability of assignment to treatment given the observed covariates. See §3.3.

quantile-quantile plot, or qq-plot A plot of the quantiles of one distribution against the quantiles of another. Figure 9.1 includes a quantile-quantile plot. For instance, the lower quartile of one distribution is plotted against the lower quartile of the other, and similarly for the medians, upper quartiles, and other quantiles. Quantile-quantile plots, or qq-plots, were invented by Martin Wilk and Ram Gnanadesikan. See Cleveland, W.W.: Elements of Graphing Data, Summit, NJ: Hobart Press, pp. 143–149.

sensitivity analysis In an observational study, the naïve model says that people who look comparable are comparable; that is, people who look the same in terms of observed covariates do not differ in consequential ways in terms of covariates that were not measured; see §3.3. A sensitivity analysis in an observational study asks how the conclusions of the study might change if people who looked comparable were actually somewhat different; see §3.4. More generally, a sensitivity analysis asks how a conclusion that depends upon assumptions might change if those assumptions are relaxed. Performing several analyses of the same data set is not a sensitivity analysis; see Note 11 of Chapter 3.

size α test A statistical test has size α if, in the worst case, it rejects a true hypothesis with probability α, that is, if the supremum over true hypotheses of the probability of rejection is equal to α. Compare this with a level α test. See E.L. Lehmann's book Testing Statistical Hypotheses, New York: John Wiley (1959).

unbiased estimate An unbiased estimate is one that is right on average. An estimator $\hat{\theta}$ is an unbiased estimate of a parameter θ if the expectation of $\hat{\theta}$ equals θ.

unbiased test Speaking informally, a test is unbiased if it is more likely to reject false hypotheses than to reject true hypotheses. An α level test is unbiased against a set of alternative hypotheses if the power of the test against these alternatives is at least α. See 'level α test' in this glossary.

uncensored See censored.

Some Books

Abstract The books listed here discuss aspects of causal inference in observational studies or experiments.

References

1. Ashenfelter, O. (ed.): Labor Economics. New York: Worth (1999)
2. Bickman, L. (ed.): Research Design. Thousand Oaks, CA: Sage (2000)
3. Boruch, R.: Randomized Experiments for Planning and Evaluation. Thousand Oaks, CA: Sage (1997)
4. Breslow, N.E., Day, N.E.: Statistical Methods in Cancer Research: The Analysis of Case-Control Studies. Lyon: IARC (1980)
5. Campbell, D.T., Stanley, J.C.: Experimental and Quasi-Experimental Designs for Research. Chicago: Rand McNally (1963)
6. Campbell, D.T.: Methodology and Epistemology for Social Science: Selected Papers. Chicago: University of Chicago Press (1988)
7. Cox, D.R., Reid, N.: The Theory of the Design of Experiments. New York: Chapman and Hall/CRC (2000)
8. Cox, D.R., Wermuth, N.: Multivariate Dependencies: Models, Analysis and Interpretation. New York: Chapman and Hall/CRC (1996)
9. Diggle, P.J., Heagerty, P., Liang, K-Y., Zeger, S.L.: Analysis of Longitudinal Data (2nd ed). New York: Oxford University Press (2002)
10. Evans, A.S.: Causation and Disease: A Chronological Journey. New York: Plenum (1993)
11. Evans, L.: Traffic Safety. Bloomfield Hills, MI: Science Serving Society (2004)
12. Fisher, R.A.: Design of Experiments. Edinburgh: Oliver and Boyd (1935)
13. Freedman, D.A.: Statistical Models. New York: Cambridge University Press (2005)
14. Friedman, M. : Essays in Positive Economics. Chicago: University of Chicago Press (1953)
15. Gelman, A., Hill, J.L.: Data Analysis Using Regression and Multilevel/Hierarchical Models. New York: Cambridge (2007)
16. Khoury, M.J., Little, J., Burke, W.: Human Genome Epidemiology. New York: Oxford University Press (2004)
17. Koepsell, T.D., Weiss, N.S.: Epidemiologic Methods. New York: Oxford University Press (2003)
18. Lauritzen, S.L.: Graphical Models. New York: Oxford University Press (1996)
19. Lilienfeld, A.M., Lilienfeld, D.E.: Foundations of Epidemiology. New York: Oxford University Press (1980)

20. MacMahon, B., Trichopoulos, D.: Epidemiology. Boston: Little Brown (1996)
21. Manski, C.F.: Identification Problems in the Social Sciences. Cambridge, MA: Harvard University Press (1995)
22. Manski, C.F.: Identification for Prediction and Decision. Cambridge, MA: Harvard University Press (2008)
23. Morgan, S.L., Winship, C.: Counterfactuals and Causal Inference: Methods and Principles for Social Science. New York: Cambridge University Press (2007)
24. Nagin, D.S.: Group-Based Modeling of Development. Cambridge, MA: Harvard University Press (2005)
25. Pearl, J.: Causality. New York: Cambridge University Press (2000)
26. Roberstson, L.S.: Injury Epidemiology. New York: Oxford University Press (1998)
27. Rosenbaum, P.R.: Observational Studies (2nd ed). New York: Springer (2002)
28. Rosenthal, R., Rosnow, R.L. (eds.): Artifact in Behavioral Research. New York: Academic (1969)
29. Rosnow, R.L., Rosenthal, R.: People Studying People: Artifacts and Ethics in Behavioral Research. New York: W.H. Freeman (1997)
30. Rossi, P.H., Freeman, H.E.: Evaluation. Newbury Park, CA: Sage (1993)
31. Rothman, K.J., Greenland, S.: Modern Epidemiology. Philadelphia: Lippincott-Raven (1998)
32. Rubin, D.B.: Matched Sampling for Causal Effects. New York: Cambridge University Press (2006)
33. Rutter, M.: Identifying the Environmental Causes of Disease: How Do We Decide What to Believe and When to Take Action? London: Academy of Medical Sciences (2007)
34. Shadish, W.R., Cook, T.D., Campbell, D.T.: Experimental and Quasi-Experimental Designs for Generalized Causal Inference. Boston: Houghton-Mifflin (2002)
35. Susser, E., Schwartz, S., Morabia, A., Bromet, E.J.: Psychiatric Epidemiology. New York: Oxford University Press (2006)
36. Susser, M.: Causal Thinking in the Health Sciences: Concepts and Strategies in Epidemiology. New York: Oxford University Press (1973)
37. Susser, M.: Epidemiology, Health and Society: Selected Papers. New York: Oxford University Press (1987)
38. Tufte, E.: The Quantitative Analysis of Social Problems. Reading, MA: Addison-Wesley (1970)
39. van der Laan, M., Robins, J.: Unified Methods for Censored Longitudinal Data and Causality. New York: Springer (2003)
40. Willett, W.: Nutritional Epidemiology. New York: Oxford University Press (1998)
41. Wooldridge, J.M.: Econometric Analysis of Cross Section and Panel Data. Cambridge, MA: MIT Press (2002)

Suggested Readings for a Course

Abstract The articles listed here are suggested supplementary readings for a course on the design of observational studies.

References

1. Angrist, J.D., Krueger, A.B.: Empirical strategies in labor economics. In: Ashenfelter, O., Card, D. (eds.), Handbook of Labor Economics, Volume 3, pp. 1277–1366. New York: Elsevier (1999)
2. Bross, I.D.J.: Statistical criticism. Cancer **13**, 394–400 (1961)
3. Campbell, D.T.: Factors relevant to the validity of experiments in social settings. Psych Bull **54**, 297–312 (1957)
4. Cochran, W.G.: The planning of observational studies of human populations (with Discussion). J Roy Statist Soc A **128**, 234–265 (1965)
5. Cornfield, J., Haenszel, W., Hammond, E., Lilienfeld, A., Shimkin, M., Wynder, E.: Smoking and lung cancer: Recent evidence and a discussion of some questions. J Nat Cancer Instit **22**, 173–203 (1959)
6. Cox, D.R.: Causality: some statistical aspects. J Roy Statist Soc A **155**, 291–301 (1992)
7. Fisher, R.A.: Design of Experiments, Chapter 2. Edinburgh: Oliver and Boyd (1935)
8. Frangakis, C.E., Rubin, D.B.: Principal stratification in causal inference. Biometrics **58**, 21–29 (2002)
9. Freedman, D.A.: As others see us: A case study in path analysis. J Educ Statist **12**, 101–128 (1987)
10. Freedman, D.A.: On the so-called 'Huber sandwich estimator' and 'robust standard errors'. Am Statistician **60**, 299–302 (2006)
11. Friedman, M.: The Methodology of Positive Economics. In: Essays in Positive Economics. Chicago: University of Chicago Press (1953)
12. Gelman, A.: Multilevel (hierarchical) modelling: What it can and cannot do. Technometrics **48**, 432–435 (2006)
13. Greenhouse, S.W.: Jerome Cornfield's contributions to epidemiology. Biometrics, Supplement, **28**, 33–45 (1982)
14. Hansen, B.B.: Optmatch: Flexible, optimal matching for observational studies. R News **7**, 18–24 (2007)
15. Hill, A.B.: The environment and disease: Association or causation? Proc Roy Soc Med **58**, 295–300 (1965)

16. Meyer, B.D.: Natural and quasi-experiments in economics. J Business Econ Statist **13**, 151–161 (1995)
17. Neyman, J.: On the application of probability theory to agricultural experiments: Essay on principles, Section 9. In Polish, but reprinted in English with Discussion by T. Speed and D. B. Rubin in Statist Sci **5**, 463–480 (1923, reprinted 1990)
18. Peto, R., Pike, M., Armitage, P., Breslow, N., Cox, D., Howard, S., Mantel, N., McPherson, K., Peto, J., Smith, P.: Design and analysis of randomised clinical trials requiring prolonged observation of each patient, I. Brit J Cancer **34**, 585–612 (1976)
19. Rosenzweig, M.R., Wolpin, K.I.: Natural 'natural experiments' in economics. J Econ Lit **38**, 827–874 (2000)
20. Rubin, D.B.: Estimating causal effects of treatments in randomized and nonrandomized studies. J Educ Psychol **66**, 688–701 (1974)
21. Rubin, D.B.: Direct and indirect causal effects via potential outcomes. Scand J Statist **31**, 161–170 (2004)
22. Shadish, W.R., Cook, T.D.: The renaissance of field experimentation in evaluating interventions. Annu Rev Psychol **60**, 607–629 (2009)
23. Sobel, M.E.: An introduction to causal inference. Sociol Methods Res **24**, 353–379 (1996)
24. Vandenbroucke, J.P.: When are observational studies as credible as randomized trials? Lancet **363**, 1728–1731 (2004)
25. West, S.G., Duan, N., Pequegnat, W., Gaist, P., Des Jarlais, D.C., Holtgrave, D., Szapocznik, J., Fishbein, M., Rapkin, B., Clatts, M., Mullen, P.D.: Alternatives to the randomized controlled trial. Am J Public Health **98**, 1359–1366 (2008)

Index

Abadie, A., 69, 90, 141
Achinstein, P., 97, 110
ACM, 173
acronyms, 361
additive treatment effect, *see* treatment effect
Ahuja, R.K., 185
Ainslie, G., 105, 110
Aitkin, M., 248, 252
aligned rank statistic
    generalization of Wilcoxon's signed rank statistic, 49
    sensitivity analysis, 357
aligned response, **49**
Almond, D., 101, 110
almost exact matching, 190
    in R, 252
Alonso, A., 7, 18, 19
analytic adjustments
    Are they feasible?, 317
Anderson, I., 16, 18
Andrews, D.F., 35, 61
Angrist, J.D., 7, 18, 67, 90, 118, 121, 125, 131, 133, 135, 137, 139, 141, 250, 252, 308–310, 339, 371
Anthony, J.C., 130, 141
Armitage, P., 125, 143, 372
Armstrong, K.A., 153, 154, 161, 186, 195, 197, 205, 220, 298, 321, 325
Ashenfelter, O., 282, 284, 369
assignment algorithm, 172
assignment problem, *see* assignment algorithm
Athens, L., 105, 110, 324
Athey, S., 69, 90
attributable effects, **49**
Avriel, M., 185, 194

Baker, R.M., 134, 141

Bakewell-Sachs, S., 186, 221, 228, 235, 318, 325, 343–345, 351
balanced incomplete block design, 210, 212
Barnard, J., 141
Barnes, B.A., 341, 350
Battistin, E., 141
Bauer, P., 332, 335, 342, 343, 346, 347, 350
Beaumont, J.J., 94, 100, 112, 122, 145
Beck, C., 221
Becker, H.S., 323, 324
Behrman, J.R., 141
Benjamini, Y., 346, 350
Bennett, D.M., 61
Berger, R.L., 332, 335, 342, 343, 346, 347, 350, 351
Bergstralh, E.J., 185
Berk, R.A., 115, 141
Bernanke, B.S., 123, 141
Bernardini, G., 98, 110
Bertrand, M., 69, 90
Bertsekas, D.P., 173, 183, 185, 238, 252
Besley, T., 18, 67, 90
bias
    known direction, 123
    test for, 118
Bickel, P., 35, 61
Bickman, L., 90, 369
Bilban, M., 116, 117, 141, 330, 331, 333, 334, 336, 337, 350
biologic plausibility, *see* reasons for effects
biological gradient, *see* doses
Bitterman, M., 118
Bittner, E., 325
Black, S., 141
Blacker, D., 13, 19
Blauw, G.J., 101, 111
Blevins, D., 235

373

Blokland, A.A.J., 234, 235
Boehnke, M., 12, 18
bold entry in index
　meaning of, **17**
Bollobás, B., 16, 18
Bonferroni inequality, 330, 340, 363
　versus split samples, 315
Boruch, R., 57, 61, 369
Bosk, C.L., 324, 325
Bound, J., 124, 134, 141
Bowers, J., 194
boxplot, 363
Braitman, L.E., 185
Braun, J., 185, 248, 252
Brensinger, C.M., 235, 351
Breslow, N., 125, 143, 369, 372
Brooks-Gunn, J., 91
Bross, I.D.J., 76, 90, 371
Brown, A., 101, 112
Brown, J.R., 97, 110
Bruckdorfer, K.R., 3, 19
Bunker, J.P., 341, 350, 351
Burke, W., 369

Cahuc, P., 102, 110, 213, 220
Cain, G.S., 317, 325
caliper
　adjusting, 169, 355
　for a time-dependent covariate, 230
　how to set, 169, 174, 242, 355
　in R, 242
　using a penalty function, 174
caliper matching, *see* matching, caliper
Campbell, D.T., 18, 20, 56, 63, 90, 93, 115,
　116, 118, 119, 142, 145, 252, 253,
　369–371
Card, D., 102, 110, 140, 142, 213, 220, 250,
　252, 283, 284
Carpaneto, G., 185
Case reports from the Massachusetts General
　Hospital, 322
Case, A., 18, 67, 90
case-crossover study, 283
Ceballos, J.M., 125, 143
censored, 363
censored survival times
　example, 157
Chamberlin, T.C., 97, 110
Chambers, J., 248, 252
characteristic function, 88
Chen, W., 93, 186, 325
Cheng, Y., 141
Choudhri, E.U., 123, 142
Cioran, E.M., 113, 142

Clark, B.J., 144
Clark, C.R., 104, 110
Clatts, M., 20, 145, 372
Cleveland, W.S., 158, 161, 194
Clinton, S.K., 108–110
close match, 363
Cnaan, A., 7, 19, 133–135, 139, 142, 144
Cochran, W.G., 4, 18, 61, 125, 142, 185, 194,
　210, 220, 317, 325, 329, 345, 350, 371
coherence, 69, 118–120, 299, 300, 339
　and equivalence tests, 341
　impact on design sensitivity, 299–303
coherent signed rank statistic, 120, 300, 308
Colditz, G.A., 143, 148, 149
Coleman, J., 317, 325
Coley, N., 110
comparability
　checking, 5, 74, 154, 156, 160, 187, 225,
　230, 231, 244
conditional independence, $\perp\!\!\!\perp$, 71
confidence interval
　for a multiplicative effect, 46
　for a Tobit effect, 47
　for an additive constant effect, 41
　for attributable effects, 54, 55
Conover, W.J., 55, 61, 289, 293, 297, 298
consistent estimate, 364
consistent test, 364
constant treatment effect, *see* treatment effect
control by systematic variation, *see* systematic
　variation
control construct, 121
control groups
　tapered, 321
　two, 69, 116–118, 218, 330–334, 339, 357
control outcome, 121
convergence
　in distribution, 364
　in probability, 364
Cook, T.D., 18, 20, 56, 57, 62, 63, 93, 115,
　142, 144, 253, 370, 372
Cook, W.J., 185, 194, 220
Copas, J.B., 90
Cornfield, J., 18, 76, 90, 371
Costa, M., 163, 164, 185, 198, 204
Couch, K.A., 22, 62
Coulibaly, D., 110
covariate, 5, 364
covariate balance, 154, 364
　compared to balance produced by
　　randomization, 190
　example, 156
　in R, 244
　in randomized experiments, 163

standardized difference, 187
  with time-dependent covariates, 225, 230
covariates
  distinguished from outcomes, 26
  missing values, 193, 240
  time-dependent, 225, 228, 230
Cox, D.R., 28, 33, 57, 62, 125, 143, 147, 149, 210, 233, 234, 315, 325, 345, 350, 369, 371, 372
Cox, G.M., 210, 220
CRAN, 237
Cummings, P., 115, 143, 284, 285
Cunningham, W.H., 185, 194, 220
Curtis, D., 12, 18

D'Agostino, R.B., 90
Dalgaard, P., 248, 252
Daniel, S., 220, 321, 325
Darnell, R., 248, 252
Darwin, C., 95, 110
data sets, 250
Dawid's notation for conditional independence, 71
Dawid, A.P., 71, 90
Day, N.E., 369
de Leuuw, J., 115, 141
Dehejia, R.H., 22, 58, 62, 90
Dell'Amico, M., 173, 185
DeMets, D.L., 57, 62
Derigs, U., 220, 230, 234
Des Jarlais, D.C., 20, 145, 372
design sensitivity, **269**, 294, 299, 354, 358
  and dose-response, 303–308
  and split samples, 315–317
  and unit heterogeneity, 280
  for coherence, 301
  optimal, 302
  with uncommon effects, 296
devices, 116, 299, 354
difference-in-differences, 11, 219
differential effect of two treatments, 129, 358
Diggle, P.J., 369
Diprete, T.A., 90
disambiguation, 116, 345
  and elaborate theories, 329
  planning analysis, 329
discontinuity design, 115
disparities, 220, 321
distance matrix, 168
  in R, 242
  time-dependent, 225, 230
distribution free statistic, 39
Dokouhaki, P., 351
Dorn, H.F., 4

dose-weighted signed rank statistic, 127, 303, 308
doses
  and design sensitivity, 303
  Are they any good?, 358
  dose-response, 124, 125, 127, 128, 299, 303–308
  fixed, 126
  just two, as different as possible, 125
  measured with error, 125, 307
  potential, 133
Drake, C.M., 94, 100, 112, 122, 145
Dretske, F., 105, 110
drive for system, 108
  diet and prostate cancer, 108
  epidemiology of dementia, 109
  health insurance, 109
Du, J.T., 141
Duan, N., 20, 145, 372
Duflo, E., 69, 90
Dynarski, S.M., 11, 19, 69, 90, 237, 252

$e(\mathbf{x})$, 72
Ebrahim, S., 3, 19
Edmonds, J., 220
efficiency, 364
Eguchi, S., 90
Eichengreen, B., 123, 142
Eickhoff, I., 94, 274
elaborate theories, 329, 345
  advocated by R.A. Fisher, 329
  ambition not penalized, 346
  and disambiguation, 329
  and sensitivity analysis, 346
  planning analyses to evaluate, 329, 330, 346
elaborate theory
  coherence, 329, 339
  equivalence and difference, 343
  with two control groups, 329, 339
encouragement design, 131–133
equivalence test, 341, 358
Erasmus
  Nota res mala optima, 318
Escobar, G.E., 186, 221, 228, 235, 318, 325, 343–345, 351
Estroff, S.E., 324, 325
Evans, A.S., 369
Evans, L., 10, 19, 129, 142, 369
Even-Shoshan, O., 93, 153, 154, 161, 186, 195, 197, 205, 221, 228, 235, 298, 318, 325, 343–345, 351
Ewens, W.J., 12, 20
exact matching, 156
  compared to fine balance, 203

examples
  AAA randomized trial, 132, 134, 135
  advertising and prices, 104
  Alzheimer disease, 12
  Alzheimer disease and NSAIDs, 130
  ambient light and auto accidents, 114
  apolipoprotein E, 12
  data sets, 250
  diet and prostate cancer, 108
  disability insurance, 124
  DNA adducts from aluminum production, 290
  DNA strand breaks and welding, 79
  DNA-protein cross-links, 163, 198
  Dutch famine, 101
  Furman v. Georgia, 114, 115
  Galileo's experiments, 96
  gangs and violence, 231
  gold standard, 123
  growing up in a poor neighborhood, 114, 115
  gun buybacks, 105, 106
  handguns, 100, 122, 124
  heart transplant, 223
  heroin or cocaine addiction, 103
  interstitial cystitis, 224, 341
  joint taxation, 114, 115
  lead mining, 116, 330, 331
  Maimonides' rule, 7, 8, 67, 118–121, 125, 133–137, 302, 308, 309
  maturity at discharge from the NICU, 228, 318–321, 343–345
  minimum wage, 102, 213, 283
  money for college, 10, 11, 237
  motorcycle helmets, 283
  National Supported Work Demonstration, 22
  National Supported Work Experiment, 287
  nature's natural experiment, 11
  ovarian cancer, 153, 187, 197, 287
  paint as a cause of genetic damage, 125, 303
  permanent income hypothesis, 98
  public versus private high schools, 317
  road hazards, 283
  seat belts, 9, 10, 129
  sleep and auto accidents, 115
  twins with differing education, 282
  vitamins, 3
exclusion criteria, 6, 224
exclusion restriction, 132, 135, 139, 365
exclusive hypotheses, 334, 345, **348**
  for equivalence tests, 335
  in equivalence testing, 342
exiting the treatment group, 7

experimental economics, 107
external validity, *see* validity

$\mathscr{F}$, 27
factorial design, 128
fast Fourier transform, 88
Fatal Accident Reporting System (FARS), 10
favorable situation, 295, 365
  defined, **266**
  for coherence, 301
  for unit heterogeneity, 277
Feldman, H.I., 91
Feller, W., 205
Fenech, M., 90, 142
Fermi, L., 98, 110
Feyerabend, P., 97, 109, 110
fine balance, 157, 197
  compared to exact matching, 190, 203
  compared to propensity scores, 197
  exact, 198
  how to do it, 199
  not exact, but best possible, 201
  with multiple controls, 200
Fishbein, M., 20, 145, 372
Fisher's exact test, 164, 189
Fisher, R.A., 18, 19, 35, 57, 62, 329, 345, 346, 369, 371
  and Mill's method of difference, 275, 281
  randomization as the reasoned basis for inference, 33
  sharp null hypothesis, 26
Flannagan, M.J., 114, 145
Fleiss, J.L., 185
Fleming, T.R., 157, 161
focused hypothesis, 120
Fodor, J., 95, 110
Foster, E.M., 90
Francis, B., 248, 252
Frangakis, C.E., 62, 90, 141, 143, 371
Franklin, A., 103, 110
Freedman, D.A., 57, 62, 369, 371
Freeman, H.E., 370
Friedman, L.M., 57, 62
Friedman, M., 98, 110, 123, 142, 369, 371
full matching, 179, 365
fullmatch, 178, 183, 238, 243
Furberg, C.D., 57, 62

Γ, 77
$\widetilde{\Gamma}$, **269**
Gabriel, K.R., 332, 333, 346, 347, 351
Gail, M., 107, 111, 223, 224, 233, 234, 345, 350
Gaist, P., 20, 145, 372

Galileo, 96–98, 110
Gangl, M., 90
Gansky, S.A., 332, 346, 347, 351
Garcia, G., 125, 143
Gardeazabal, J., 141
Garfinkel, H., 325
Gastwirth, J.L., 16, 19, 90, 94, 358
Gayer, T., 18, 19
Geertz, C., 323, 325
Gelman, A., 369, 371
generalized linear models, 240
generating function, 88
generic bias, 131
Ghosh, M., 293, 298
Gibbons, J., 52
Gibbons, R.D., 194, 195, 234
Gilbert, P.B., 93
Giovannucci, E., 108–111
Glazer, A., 104, 110
Goetghebeur, E., 142
Goldberger, A.S., 309, 310, 317, 325
Gonsebatt, M.E., 143
Greehouse, S.W., 371
Green, K.M., 184, 186
Greenland, S., 91, 283, 284, 370
Greenstone, M., 18, 19
Greevy, R., 7, 19, 62, 91, 133–135, 139, 142, 221
Grodstein, F., 7, 18, 19
Gross, D.B., 91, 98, 99, 110
Gruber, J., 109, 110
Gu, X.S., 185

Haenszel, W., 18, 90, 371
Hahn, J., 90, 91, 115, 142
Halloran, M.E., 29, 62
Hämäläinen, K., 58, 62
Hamermesh, D.S., 18, 19, 91, 142
Hammond, E., 18, 90, 284, 371
Hampel, F., 35, 61
Han, W.J., 91
Hansen, B.B., 91, 173, 178, 183–185, 194, 200, 205, 237, 243, 248, 252, 371
Hargarten, S.W., 105, 111
Harre, R., 96, 110
Haviland, A.M., 86, 91, 183, 185, 231, 234, 358
Heagerty, P., 369
Heckman, J., 19, 58, 62, 142
Heidary, A., 351
Heitjan, D.F., 92
Heller, R., 184, 272, 274, 284, 300–303, 309, 310, 315–317, 324, 325, 340, 350, 358
Hernán, M.A., 7, 18, 19

heterogeneity
 and design sensitivity, 280
 and Mill's method of difference, 275
 and power of a sensitivity analysis, 280
 and sensitivity of point estimates, 281
 and sensitivity to unmeasured biases, 278
 controlled versus uncontrolled, 276
 different kinds, 276
 examples of efforts to reduce heterogeneity, 282
 in a randomized experiment, 276
Hewer, A., 290, 298
Hill, A.B., 18, 19, 91, 116, 118, 119, 121, 122, 124, 125, 142, 371
Hill, J.L., 91, 141, 369
Hinde, J., 248, 252
Hirano, K., 91
Ho, D.E., 91, 110, 142, 185, 248, 252
Hochberg, Y., 346, 350
Hodges, J.L., 43, 44, 46, 47, 49, 62, 274, 281, 284
Hodges-Lehmann point estimate, **43**
 of a multiplicative effect, 46
 of a Tobit effect, 47
 of an additive constant effect, 43
Hoeffding, W., 52, 62
Hoek, H.W., 101, 112
Hoffer, T., 317, 325
Holland, P.W., 19, 62, 131, 142, 144, 276, 285
Holley, R., 16, 19
Hollister, R., 62
Holm, S., 340, 351, 363
Holmes, A.P., 29, 62
Holtgrave, D., 20, 145, 372
Hommel, G., 332, 346, 347, 351
Hornik, R., 94, 209, 213, 220, 221
Hotz, V.J., 58, 62
Howard, S., 125, 143, 372
Howson, C., 103, 110
Hsu, J.C., 332, 335, 342, 346, 347, 350, 351
Hubbard, R.L., 103, 111
Huber, P., 35, 58, 61, 62
Huber-White correction, 371
Hudgens, M.G., 29, 62
Hughes, J.N., 234, 235
Hungarian method
 for the assignment problem, 173
Hursting, S.D., 107, 112
Hwang, J.T.G., 335, 342, 347, 351

i, **23**
I, **23**
Imai, K., 91, 110, 142, 185, 194, 213, 220, 248, 252

Imbens, G.W., 19, 69, 90, 91, 115, 131, 136, 139, 141–143, 213, 220
in 't Veld, B.A., 143
instrument, 7, 131, 365
  strong, 134
  weak, 134, 136
  When are they useful?, 139
instrumental variable, 365
intention-to-treat analysis, 134
interference between units, 28
  inference in the presence of, 29
internal validity, see validity
intersection-union test, 335, 343
IV, 365

j, **23**
Jacobsen, S.L., 185
Jacobsohn, L., 220
Jaeger, D.A., 134, 141
Jakopin, C.B., 116, 117, 141, 330, 331, 333, 334, 336, 337, 350
Joffe, M.M., 91, 143, 148, 149, 213, 221
Johnson, B.A., 91
Joreskog, K.G., 309, 310
Jurečková, J., 35, 62

Kahn, H.S., 101, 111
Kalton, G., 220
Karmanov, V.G., 185, 195
Katan, M.B., 19
Keele, L.J., 87, 91
Kemper, P., 62
Kempthorne, O., 26, 57, 62
Khoury, M.J., 369
Kieser, M., 335, 342, 343, 347, 350
Kilgore, S., 317, 325
Kimmel, S.E., 91
King, G., 91, 185, 194, 248, 252
Klopfer, S.O., 184, 185, 252
known effects, 69, 121, 357
Koch, G., 332, 346, 347, 351
Kochin, L.A., 123, 142
Koepsell, T.D., 369
Kosanke, J.L., 185
Krieger, A.M., 16, 19, 90, 93, 94, 358
Kronmal, R.A., 91
Kropf, S., 332, 346, 347, 351
Krueger, A.B., 18, 90, 102, 110, 141, 213, 220, 250, 252, 283, 284, 371
Kuhn, E.M., 105, 111
Kuhn, H.W., 173, 185
Kuhn, T., 97, 111
Kundo, D., 3, 19

Laird, N.M., 13, 19
LaLonde, R.J., 21, 22, 62, 250, 252
LaLumia, S., 115, 143
Lambe, M., 115, 143
Landis, J.R., 235, 351
Landrum, M.B., 92
Langefeld, C.D., 12, 18
Langen, V., 94, 274
Laska, E.M., 332, 347, 351
latent trajectory groups, 231
Lauritzen, S.L., 369
Lavy, V., 7, 18, 67, 90, 118, 121, 125, 133, 135, 137, 139, 141, 250, 252, 308–310, 339
Lawlor, D.A., 3, 19
Lee, M.J., 91
Lee, S.J., 91
Lehmacher, W., 340, 351
Lehmann alternative, 293
Lehmann, E.L., 43, 44, 46, 47, 49, 52, 57, 62, 185, 274, 281, 284, 285, 289, 293, 297, 298, 332, 343, 347, 351
Lemieux, T., 115, 143
Levin, B., 185
Li, F., 143
Li, G., 110
Li, H.G., 90
Li, Y.F.P., 221, 224, 232, 234, 341, 351
Liang, K-Y., 369
Lilienfeld, A., 18, 90, 369, 371
Lilienfeld, D.E., 369
Lin, D.Y., 91
Little, J., 369
Liu, H-K., 335, 342, 347, 351
Loeys, T., 142
Logan, B., 343, 351
Logan, R., 7, 18, 19
logistic, 366
logit, 167, 240, 366
  ordinal, 212
Lorch, S.L., 93, 186, 221, 228, 235, 318, 325, 343–345, 351
lowess, 158, 160, 366
Lu, B., 62, 91, 94, 140, 143, 209, 210, 213, 220, 221, 230, 233, 234
Ludwig, J., 143
Lumey, L.H., 101, 111

m-estimate, 35, 58
m-test, 58
  sensitivity analysis, 357
Maclure, M., 283, 285
MacMahon, B., 370
Magnanti, T.L., 185

Index 379

Mahalanobis distance, 170, 171
  limitations of, 171
  rank-based, 171
Mahalanobis, P.C., 185
Maindonald, J., 185, 248, 252
Make your theories elaborate, *see* elaborate theories
Manduchi, E., 184
Manski, C.F., 91, 92, 103, 111, 143, 370
Manson, J.E., 7, 18, 19
Mantel, N., 125, 143, 372
Mantel-Haenszel test, 246
Marcus, R., 332, 333, 346, 347, 351
Marcus, S.M., 92, 194, 195, 234
marginal distribution, 366
Marguart, J.W., 115, 143
Maritz, J.S., 35, 58, 62
Mark, S.D., 92
Marolla, F., 101, 112
Masjedi, M.R., 351
matching
  almost exact, 190
  and seemingly innocuous confounding, 321
  and thick description, 322
  before randomization in experiments, 220
  bias due to incomplete matching, 85
  caliper, 168
  efficiency, 184
  exact, 156, 192, 355
  fine balance, 197
  fixed ratio, 175, 178
  fixed versus variable ratio, 178
  for the propensity score, 157
  full, 179, 184
  ideal, 66
  in R, 237
  nonbipartite, 207
  on many covariates, 72
  optimal, 172, 184
  pair, 164
  population before, 65
  risk-set, 223, 233
  tapered, 220, 321
  to control an interaction, 156
  variable controls, 175–178
    boxplot for, 356
  with doses of treatment, 209
  with multiple controls, 175
  with several groups, 210, 321
  with time-dependent covariates, 225, 228, 230
  with two control groups, 210
  without groups, 207
matching before randomization, 220

Maynard, R., 62
McCaffrey, D.F., 92
McCullagh, P., 252
McCullough, M.L., 109, 111
McGinnis, R.E., 12, 20
McKillip, J., 92, 122, 143
McNeil, B.J., 92
McPherson, K., 125, 143, 341, 351, 372
Medoff-Cooper, B., 186, 221, 228, 235, 318, 325, 343–345, 351
Mehrotra, D.V., 93
Meisner, M.J., 332, 347, 351
method of difference, *see* Mill, J.S.
Meyer, B.D., 19, 92, 143, 372
Mi, L., 186, 221, 228, 235, 318, 325, 343–345, 351
Michels, K.B., 7, 18, 19
micronuclei, 117
Mill, J.S., 97, 111, 147, 149, 285
  method of difference, 275, 281
Miller, D.L., 143
Millman, A., 186, 221, 228, 235, 318, 325, 343–345, 351
Milyo, J., 104, 111, 115, 143
Ming, K., 178, 183–185, 252, 358
missing data, 193
Mitra, N., 92
Mitrou, P.N., 111
modularity
  as an aid to transparency, 148
Mohammadi, F., 351
Monte Carlo, 366
Morgan, S.L., 370
Morral, A.R., 92
Mosher, R.E., 93, 186, 325
Mosteller, F., 35, 341, 350
Mullainathan, S., 69, 90
Mullen, P.D., 20, 145, 372
multiple controls, 175
multiple operationalism, *see* coherence

naïve model
  if naïve, why is it important?, 75
naïve model for treatment assignment, *see* treatment assignment
Nagin, D.S., 86, 91, 92, 183, 185, 231, 234, 235, 358, 370
natural experiment, 5, **67**, 69, 75, 139
nature's natural experiment, 11
Navarro-Lozano, S., 142
Nelder, J.A., 252
Neumark, D., 102, 111
Newey, W.K., 92
Neyman, J., 26, 57, 62

Nichols, T.E., 29, 62
Nie, C.L., 105, 111
Nieuwbeerta, P., 234, 235
nonbipartite matching, 207
    discarding some subjects, 211
    in risk-set matching, 230
    pairing an odd number of subjects, 211, 215
noncompliance, 7, 132, 134
Normand, S-L., 92, 194, 195, 234
Norvell, D.C., 284, 285
Nota res mala optima, 318
Nozick, R., 97, 111
null distribution, 31

O'Brien, M.E., 105, 111
O'Brien, P.C., 157, 161
opportunity, 113, 353
optimal matching, 172
optmatch, 173, 178, 183, 200, 238, 243
    how to get it, 249
ordinal logit model, 212
Oreopoulos, P., 115, 143
Orlin, J.B., 185
Orwin, R., 220
ostensibly comparable, 328, 366
outcome, 5, 367
outcomes
    distinguished from covariates, 26

P_{max}, 81
Pagano, M., 63, 88, 92
Paik, M.C., 185
Papadimitriou, C.H., 186, 221
Park, S.Y., 111
pattern matching, *see* coherence
Pearl, J., 370
Peirce, C.S., 104, 111
penalties
    in R, 252
    to implement almost exact matching, 191
    to impose a design constraint, 214
    to improve covariate balance, 192
penalty function, 214
    in R, 251
    to implement a caliper, 174
Pepper, J.V., 103, 106, 111, 112
Pequegnat, W., 20, 145, 372
Pericak-Vance, M., 12
Peritz, E., 332, 333, 346, 347, 351
Peto, J., 125, 143, 372
Peto, R., 125, 143, 148, 149, 372
Petrie, C.V., 106, 112
Phillips, D.H., 290, 298
π_ℓ, 65

Piantadosi, S., 57, 63
Piesse, A., 220
Pike, M., 125, 143, 372
Pinto, D., 125, 143
planned analysis, 327, 354
Platt, J., 97, 109, 111
plausible alternatives, 6, 354
Plott, C.R., 111
Polsky, D., 153, 154, 161, 186, 195, 197, 205, 298
Polya, G., 103, 109, 111
Popper, K.R., 98, 109, 111, 147, 149
power, 367
    in a randomized experiment, 258
    of a sensitivity analysis, **265**
    of a test, **257**
    of Wilcoxon's signed rank test, 261
    suppositional collapse, 259–261
Pratt, J., 52
Prentice-Wilcoxon test for censored pairs, 157
primary analysis, 327, 354
primary endpoint, 327
probability generating function, 88
propensity score, 14, 157, 165–167, 367
    and balancing observed covariates, 72, 166
    and unobserved covariates, 73
    definition, **72**
    estimates work better, 74
    estimation in R, 240
    estimation of, 167
    evaluating a model for, 356
    logit model for, 167
    poor predictions of treatment, 356
    versus exact matching, 190
    why randomization is better, 73
    with missing covariate values, 193
    with ordered doses of treatment, 212
    with several groups, 212
    with time-dependent covariates, 230, 233
propensity scores
    compared to fine balance, 197
Propert, K.J., 221, 224, 232, 234, 235, 341, 351
proportional hazards model, 230, 233
protocol, 7, 327
Psaty, B.M., 91
Pulleyblank, W.R., 185, 194, 220

quantile-quantile plot, 291, 367
quasi-experiment, 6, 69, 75
Quine, W.V.O., 108, 111

R, 237
    books about, 248

brief introduction, 248
convolve, 88
data.frame, 249
finding keywords, 248
help, 249
how to get optmatch, 249
how to get R, 248
Mahalanobis distance matrix, 250
NA, 249
R is free, really, 248
reading data into R, 249
using penalty functions, 251
R_ℓ, 65
R_{ij}, **23**, 26
(r_{Tij}, r_{Cij}), **26**
R Development Core Team, 186, 221, 252
R, **25**
$(r_{T\ell}, r_{C\ell})$, **65**
r$_C$, **26**
r$_T$, **26**
Rachlin, H., 105, 111
Randall, T.C., 153, 154, 161, 186, 195, 197, 205, 298
randomization
    and internal validity, 56
    defined, 27
    reasoned basis for inference in experiments, 33
randomization inference
    using Huber's m-estimates, 58
    using the mean, 31
    using Wilcoxon's signed rank statistic, 36
randomized experiment
    example, 21
Rao, C.R., 301, 310
Rapaport-Kelz, R., 93, 186, 325
Rapkin, B., 20, 145, 372
Rauma, D., 115, 141
reasons for effects, 104, 358
    example, 105, 107
    importance of, 107
    object for empirical investigation, 105
    theoretical, 107, 213
    versus direct evidence of effect, 107, 213
regression discontinuity design, *see* discontinuity design
Reid, N., 57, 62, 210, 369
Reitmeir, P., 340, 351
replication, 104, 139, 358
    example, 102, 103
    goal of, 104
    of biases, 101
restriction
    as an aid to transparency, 148

Rettore, E., 141
Reynolds, K.D., 92, 119, 143
Ridder, G., 91
Ridgeway, G., 92
Ritov, Y., 92
Ritter, G., 235
Rivara, F.P., 124, 145
Robertson, L.S., 283, 285, 370
Robins, J.M., 7, 18, 19, 92, 93, 234, 235, 370
Robinson, J., 26, 57, 63
Rogers, W.H., 35, 61
Rohe, A., 220
Romano, P.S., 94
Romer, D., 111
Rosen, A.K., 69, 94
Rosenthal, R., 142, 370
Rosenzweig, M.R., 18, 20, 93, 144, 372
Rosnow, R.L., 142, 370
Ross, R.N., 153, 154, 161, 186, 195, 197, 205, 298
Rossi, P.H., 370
Rothman, K.J., 125, 144, 370
Rotnitzky, A., 92, 93
Rouse, C., 282, 284
Roychoudhuri, R., 144
Ruberg, S.J., 335, 342, 347, 351
Rubin, D.B., 18, 20, 26, 62, 63, 71, 72, 90, 92, 93, 131, 141–143, 145, 161, 165, 166, 184, 186, 193–195, 205, 252, 253, 370–372
Rutter, M., 18, 20, 93, 141, 144, 370
Ryle, G., 323, 325

s_i, 37
Sacerdote, B.I., 143
Sachs, J., 123, 142
Saenger, G., 101, 112
Salsburg, D.S., 55, 61, 63, 285, 288, 292, 297, 298
Samuelson, P.A., 95, 111
SAS, 173, 200
Saunders, A.M., 12, 20
Savage, I.R., 16, 20
Scharfstein, D., 92
Schatzkin, A., 107, 111
Schmechel, D., 12, 20
Schoket, B., 290, 298
Schrijver, A., 185, 194, 220
Schuirmann, D.L., 335, 342, 347, 351
Schwartz, A.J., 123, 142
Schwartz, S., 153, 154, 161, 186, 195, 197, 205, 298
seemingly innocuous confounding, 317–321
Sekhon, J.S., 20, 186

Sen, P.K., 35, 62
Sennett, R., 114, 144
sensitivity analysis, 14, 70, 75, **76**, 354, 367
   and unit heterogeneity, 278
   better than: association does not imply causation, 77
   example, 79
   for a confidence interval, 84
   for a point estimate, 84
   for an equivalence test, 341
   for an instrument, 137, 139
   for coherence, 340
   for differential effects, 131
   for superiority/equivalence test, 343
   for testing in order, 349, 350
   how to do it, 82
   interpretation of Γ, 357
   multiple controls, 357
   power of, 265
   with a binary outcome, 357
   with a censored outcome, 357
   with an m-test, 357
   with covariance adjustment, 357
   with time-dependent covariates, 233
sequentially exclusive partition, 347, **348**
   exclusive hypotheses, 334
   for coherence of two outcomes, 340
   for equivalence tests, 335
   for superiority/equivalence tests, 344
   with two control groups, 333, 336, 338
setting up the comparisons, 14
sgn
   with ties, 54
   without ties, 36
Shadish, W.R., 18, 20, 56, 57, 62, 63, 93, 144, 253, 370, 372
Shaffer, J.P., 33, 63, 333, 346, 347, 351
sharp null hypothesis of no effect
   defined, 26
   test of, 29
Shepherd, B.E., 93
Shestov, D.B., 101, 111
Shimkin, M., 18, 90, 371
sib-TDT, 12
Siddique, J., 194, 195, 234
signed rank statistic, 36
Silber, J.H., 7, 19, 62, 69, 91, 93, 94, 133–135, 139, 142, 144, 153, 154, 156–161, 186, 195, 197, 205, 220, 221, 228, 235, 287–289, 298, 318–325, 342–345, 351, 358
simplifying the conditions of observation, 148

Small, D., 58, 63, 93, 118, 137, 144, 145, 184, 272, 274, 284, 285, 300–303, 308–310, 315–317, 324–326, 340, 350, 358
Smith, G.D., 3, 19
Smith, H.L., 183, 186
Smith, J.A., 58, 63
Smith, P., 125, 143, 372
Smith, V.L., 111
Sobel, M.E., 29, 63, 145, 372
Sommer, A., 145
Sorensen, J.R., 115, 143
Souleles, N.S., 91, 98, 99, 110, 112
Sparen, P., 101, 111
specificity, see unaffected outcome
Spielman, R.S., 12, 20
split samples, 303, 315–317, 354
stability in the absence of treatment, 11
Staley, K.W., 97, 112
Steiglitz, K., 186, 221
Stein, Z., 101, 112
Stephenson, W.R., 49, 55, 63, 289, 293, 294, 297, 298
Stigler, G.J., 213, 221
stochastically larger, 82, 292
Stolley, P.D., 93
Stone, M., 315, 326
Stone, R., 93
Strittmatter, W.J., 12, 20
strong instrument, 134
strongly ignorable treatment assignment, see treatment assignment
structure assumption, 347
Stuart, E.A., 63, 91, 145, 184–186, 194, 195, 234, 248, 252
study protocol, 7
Suissa, S., 224, 235
Sullivan, J.M., 114, 145
Summers, L.H., 141, 145
surrogate outcomes, see reasons for effects
surrogate units (e.g., laboratory animals), see reasons for effects
Susser, E., 101, 111, 112, 370
Susser, M., 18, 20, 101, 103, 112, 148, 149, 317, 370
SUTVA, see interference between units
symbols, 359
systematic variation, 118, 276
Szapocznik, J., 20, 145, 372
Szulanski, G., 283, 285

T, **36**
Tamhane, A., 343, 351
Tan, Z., 145
tapered matching, 220, 321

TDT, 12
Ten Have, T.R., 58, 63, 91, 194, 195, 234
test
   equivalence, 334, 341, 358
   intersection-union, 335, 343
   level of, 365
   power of, **257**, 367
   size of, 368
   superiority/equivalence, 343
testing hypotheses in order, 349
thick description, 322, **323**
Thistlethwaite, D., 115, 145
Thomas, N., 93
Thomas, Y.F., 103, 111
ties, 38, 39, 54
time-dependent covariates, 230
   distance matrix, 225, 230
   example, 225, 228
   propensity score, 230
   unobserved, 233
Tobin, J., 47, 63
Tobit effect, *see* treatment effect
Todd, P., 58, 63, 115, 141, 142
Tollenaar, N., 235
Toniolo, P., 163, 164, 185, 198, 204
TOST, 335, 342
Toth, P., 173, 185
trajectory groups, 231
transmission disequilibrium test, 12
transparency, **147**, 353
treatment assignment
   and the propensity score, 72
   naïve model, 70, **70**, 71
   strongly ignorable, 71
treatment effect
   additive constant, **40**, 41
   differential, 129, 358
   dilated, 297
   direct evidence of, 106
   general, 44
   heterogeneous, 55, 308
   hypothesis of no effect, 26
   interference between units, 28
   multiplicative, 46
   proportional to dose, 135, 305
   reason for, 106
   superiority/equivalence, 343
   Tobit, 47
   uncommon but dramatic, 55
treatment effects
   attributable, 49
Tremblay, R.E., 86, 91, 185, 231, 234, 235
Trichopoulos, D., 370
Tritchler, D., 63, 88, 92
Trochim, W.M.K., 94, 119, 145
Trudeau, M.E., 93, 186, 325
Tsiatis, A.A., 91
Tufte, E., 370
Tukey, J.W., 35, 61, 319, 326, 328, 351
two stage least squares, 134
two-one-sided test procedure, 335, 342

u_ℓ, 65
u_{ij}, **25**
U-statistic, 52
u, **25**
unaffected outcome, 121
unbiased estimate, 368
unbiased test, 368
uncensored, 368
Uusitalo, R., 58, 62

Vagero, D., 101, 111
validity
   internal versus external, 56
Van der Klaauw, W., 115, 142
van der Laan, M., 234, 235, 370
van der Pal-de Bruin, K.M., 101, 111
van Dyk, D.A., 213, 220
van Eeden, C., 127, 145, 310
Vandenbroucke, J., 3, 18, 20, 94, 141, 145, 153, 161, 322, 326, 372
VanderWeele, T., 94
Velayati, A.A., 351
Vincze, I., 290, 298
Volpp, K.G., 69, 94
Vuori, J., 58, 62

Wahba, S., 22, 58, 62, 90
waiting list, 115
Waldfogel, J., 91, 104, 111, 115, 143
Wang, L.S., 94
Wassmer, G., 340, 351
weak instrument, 134, 136
Weed, D.L., 107, 112
Weiss, L., 63
Weiss, N., 94, 121, 122, 125, 145, 369
Welch, B.L., 26, 35, 56, 57, 63
Wellford, C.F., 106, 112
Werfel, U., 79, 94, 274
Wermink, H., 235
Wermuth, N., 369
West, S.G., 18, 20, 92, 119, 121, 143, 145, 234, 235, 372
West, W., 234, 235
Westlake, W.J., 335, 342, 347, 351
Wilcox, M., 13, 19
Wilcoxon's signed rank statistic, **36**, 303

Wilcoxon's signed rank test, 158
Wilcoxon, F., 35, 63, 298
Wild, C.P., 309, 311
Wilk, M.B., 26, 57, 63
Willett, W., 370
Willett, W.C., 7, 18, 19
Williams, B., 108, 112
Winship, C., 370
Wintemute, G.J., 69, 94, 100, 105, 111, 112, 122, 124, 145
Winter, S.G., 283, 285
Withers, R.L., 105, 111
Wolpin, K.I., 18, 20, 93, 144, 372
Wooldridge, J.M., 19, 91, 250, 253, 370
Wright, M.A., 94, 100, 112, 122, 124, 145
Wright, P.G., 283, 285
Wulfsohn, M., 235
Wynder, E., 18, 90, 371

$\mathbf{x}_\ell$, **65**
$\mathbf{x}_{ij}$, **24**

Xu, X., 221

Y_i, **24**
Y, 25
Yoon, F., 332, 351
Yu, B.B., 94

Z_ℓ, 65
Z_{ij}, **23**
Z, **25**
$\mathscr{Z}$
   as a set, 27
   as an event, 27
Zanutto, E., 63, 94, 209, 213, 221
Zeger, S., 145, 369
Zellner, A., 112
Zhang, X., 93, 186, 325
Zhitkovich, A., 163, 164, 185, 198, 204
Zybert, P.A., 101, 111
Zylberberg, A., 102, 110, 213, 220

springer.com

Observational Studies

Paul R. Rosenbaum

Content: Observational Studies.- Randomized Experiments.- Overt Bias in Observational Studies.- Sensitivity to Hidden Bias.- Known Effects.- Multiple Reference Groups in Case-Referent Studies.- Multiple Control Groups.- Coherence and Focused Hypotheses.- Constructing Matched Sets and Strata. - Some Strategic Issues.

2002. 2nd ed. XIV, 375 p. 13 illus. Hardcover, Springer Series in Statistics
ISBN: 978-0-387-98967-9

Statistical Analysis of Designed Experiments, Third Edition

Helge Toutenburg and Shalabh

Content: Introduction.- Comparison of two samples.- The linear regression model.- Single-factor experiments with fixed and random effects.- More restrictive designs.- Incomplete block designs.- Multifactor experiments.- Models for categorical response variables.- Repeated measures model.- Cross-over design. - Statistical analysis of incomplete data.

2009. 3rd ed. XVI, 612 p. Hardcover, Springer Texts in Statistics
ISBN: 978-1-4419-1147-6

A Comprehensive Guide to Factorial Two-Level Experimentation

Robert Mee

Content: Introduction to full factorial designs with two-level factors.- Analysis of full factorial experiments.- Common randomization restrictions.- More full factorial design examples.- fractional factorial designs: the basics.- Fractional factorial designs for estimating main effects.- Designs for estimating main effects and some two-factor interactions.- Resolution V fractional factorial designs.- Augmenting fractional factorial designs.- Fractional designs with randomization restrictions.- More fractional factorial design examples.- Response surface methods and second-order designs.- Special topics regarding the design.- Special topics regarding the analysis.

2009., XVI, 544 p. Hardcover
ISBN: 978-0-387-89102-6

Easy Ways to Order ▶ Call: Toll-Free 1-800-SPRINGER • E-mail: orders-ny@springer.com • Write: Springer, Dept. S8113, PO Box 2485, Secaucus, NJ 07096-2485 • Visit: Your local scientific bookstore or urge your librarian to order.